Management of Prader-Willi Syndrome

Third Edition

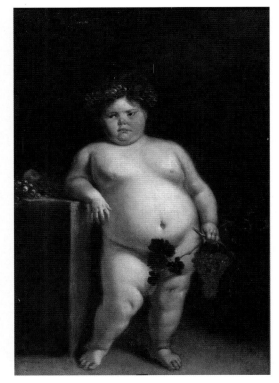

La Monstrua vestida

La Monstrua desnuda

These paintings by the Spanish baroque court painter Juan Carreño de Miranda (1616-1685) depict Doña Eugenia Martinez Vallejo at approximately 6 years of age (c. 1680), after she was brought to the court of Charles II of Spain and given the moniker "La Monstrua." It is said that upon viewing these paintings in the Museo del Prado, Madrid, Dr. Andrea Prader immediately recognized features consistent with Prader-Willi syndrome. Although we may never know if this child did in fact have PWS, the paintings and contemporaneous 17th century descriptions of her condition strongly support this supposition. Indeed, these paintings are often regarded as the earliest illustrations of PWS. La Monstrua has continued to be an intriguing figure, as evidenced by an outdoor sculpture in Avilés created in 1997 by the Spanish artist Amado González Hevia "Favila" (b. 1954; www.amadofavila.com) based on these baroque portraits.

Management of Prader-Willi Syndrome

Third Edition

Merlin G. Butler, MD, PhD
Chief, Section of Medical Genetics and Molecular Medicine, William R. Brown/Missouri Chair of Medical Genetics and Molecular Medicine; Professor of Pediatrics, University of Missouri-Kansas City School of Medicine, Children's Mercy Hospitals and Clinics, Kansas City, Missouri

Phillip D.K. Lee, MD
Chief Scientific Officer, Immunodiagnostic Systems Ltd., Boldon, Tyne and Wear, United Kingdom

Barbara Y. Whitman, PhD
Professor of Pediatrics, Saint Louis University School of Medicine, Cardinal Glennon Children's Hospital, St. Louis, Missouri

Editors

With 125 Illustrations, 11 in Full Color

 Springer

Merlin G. Butler, MD, PhD
Chief, Section of Medical Genetics
   and Molecular Medicine
William R. Brown/Missouri
   Chair of Medical Genetics and
   Molecular Medicine
and
Professor of Pediatrics
University of Missouri-Kansas
   City School of Medicine
Children's Mercy Hospitals
   and Clinics
Kansas City, MO 64108
USA

Phillip D.K. Lee, MD
Chief Scientific Officer
Immunodiagnostic Systems Ltd.
Boldon, Tyne and Wear
NE 35 9PD
UK

Barbara Y. Whitman, PhD
Professor of Pediatrics
Saint Louis University School
   of Medicine
Cardinal Glennon Children's Hospital
St. Louis, MO 63104
USA

Supported by an unrestricted grant from Pharmacia Corporation, a division of Pfizer, Inc.

Library of Congress Control Number: 2005930806

ISBN 10: 0-387-25397-1
ISBN 13: 978-0387-25397-8

Printed on acid-free paper.

Printed in the United States of America.   (BS/MVY)

9 8 7 6 5 4 3 2 1

springeronline.com

Foreword I

In 1988, when the first edition of *Management of Prader-Willi Syndrome* was published, my daughter turned eight years old. That year, due to advances in chromosomal analysis, Gabriella was diagnosed with Prader-Willi syndrome (PWS). After searching for answers so long, we were relieved to finally have a name attached to the challenges we faced. At that time, we were especially grateful for this handbook and for the tools it provided to assist us in confronting the daily management of this disorder. However, reading the prognosis for Gabriella's (and our family's) future was disheartening. Although we remained determined to provide her with the best possible opportunities for living her life to the fullest, it was difficult to see even a glimmer of sunshine.

During the past two decades, I have watched this small spark of light grow into an expanding ray of hope. There have been countless advances in the diagnosis and management of PWS, and many of our children have reaped the benefits of these new possibilities. With the advent of the Internet, the world of PWS expanded, families connected, and awareness of the syndrome was suddenly available to anyone who had access to a computer. Early diagnosis, early education, and timely treatment provide the key to a better quality of life for people with PWS. In many countries, several options for treatment are available; in other nations, it is still impossible to receive a diagnosis or even get necessary medical services. The International Prader-Willi Syndrome Organization (IPWSO) provides education, and it encourages nations to share their knowledge and resources in order to narrow this disparity of services. The third edition of *Management of Prader-Willi Syndrome* is a fine example of collaboration, which will benefit patients throughout the world.

Now we are at the cusp of innovative and extraordinary scientific discoveries. As this book becomes a reference around the world, it is our hope that this comprehensive resource of expertise will provide the impetus and inspiration for even more exploration and answers to the complexity of this disorder.

On behalf of our global family, the IPWSO, I wish to express our heartfelt thanks to the editors, authors, and others who graciously put so many hours into making this book possible. This edition is exceptional because it represents direct and indirect contributions from a network of professionals and parents from across continents. Their combined efforts will be heard around the world and will make a difference to so many families in so many significant and extraordinary ways. Families no longer need to feel isolated, and with knowledge in hand, they can better face the challenges of PWS. This management book provides tangible assistance and represents hope for the future, turning the dream of a better life for people with PWS into a reality.

Pamela F. Eisen
President, International Prader-Willi Syndrome Organization
(IPWSO)

Foreword II

It seems yesterday to me, but it was 27 years ago when I was told that my son Daniele had Prader-Willi syndrome. I felt lost and confused then, because the little information I found available in Italy was not encouraging at all; I could not believe my lovely son was going to die soon, as the literature said. Inside Daniele's eyes I saw his joy of being part of this world. He was ready and willing to face and share, because he had, as he proved, a lot to give. My wife and I decided to do everything we could not to accept the tragic fate it seemed we had to face. Today I must thank my son's eyes for giving me such an inspiration to struggle and keep struggling as President of the International Prader-Willi Syndrome Organization (IPWSO). I am proud to have represented an organization whose now 60 member countries tirelessly work through their qualified and enthusiastic professional and parent delegates. This close cooperation between parents and professionals, with active participation and sharing information together, makes our organization a unique example in the world. It is clear that the results we have achieved are extremely important, giving all involved families hope and positive references for the future of their children.

The rapid expanding of our contacts worldwide through the Internet puts us in contact every day with affected families from different countries, cultures, and realities. It is incredible how desperate mothers become inspired, how discouraged presidents or delegates of local associations become both inspired and efficient, and how doctors become important sources of information for their colleagues previously unknown to them, share scientific information, and cooperate in research worldwide after an active e-mail exchange. There is still a lot to do. We estimate that many, perhaps most, of the affected individuals worldwide are unknown; too many affected people sadly die before we are able to locate them and provide support. Member countries of IPWSO are designated on a world map with a flag. Large areas such as in Eastern Europe, Africa, and Asia are still without a flag; we keep working hard to spread our voice. In some countries, as recently happened with China, the scientific contacts with doctors participating in international scientific meetings or studying abroad are the right channel to follow for opening heavy or hidden gates.

This third edition of *Management of Prader-Willi Syndrome* is yet another excellent tool for soundly informing professionals about the syndrome, especially in countries that lack information or where no parent organization yet exists. It is an example of the steady and hard work of people from multiple disciplines dedicated to study and research on PWS throughout the world. I am proud and honored to write a few words as introduction to this book, hoping its distribution will cover all continents and reach as many professionals as possible. On behalf of the IPWSO Board, I wish to thank all authors for their precious contribution to making this third edition an important milestone in our mission to know more about the syndrome that affects special children of special parents, and who are followed by special doctors.

Giorgio Fornasier
Past President, Director of Program Development, International
Prader-Willi Syndrome Organization (IPWSO)

Preface

In the first edition of *Management of Prader-Willi Syndrome*, published in 1988, Dr. Hans Zellweger wrote in the foreword, "I . . . have reached the conclusion that PWS is one of the two most grave ailments I have encountered. . . . "Fortunately, if Hans were still alive, I think he would find that this is no longer the case. Although still complex to manage and challenging at times, the world of Prader-Willi syndrome today is much more hopeful. Our son, Matt, was a teenager at the time Dr. Zellweger wrote that "Family systems deteriorate and life becomes hell for all concerned." Although we did cringe at those words at the time, I can now smile and breathe a sigh of relief to know that we, and many other parents, have delightful young adult children who are a joy to know.

Rereading the 1988 foreword made me appreciate just how far we have come. Many of our children are now diagnosed as newborns thanks to increased awareness and the availability of accurate, confirmatory genetic testing. Parents are now educated about the syndrome much earlier and in far greater depth; as a result they are now far more sophisticated advocates for their children with physicians, schools, and other care providers. Parents insist on, and frequently provide, more comprehensive education of professionals. Early identification and strong advocacy, combined with in-depth research has resulted in many of our children now being tall (thanks to growth hormone), slim (thanks to early intervention and good environmental controls), and happy (thanks to good behavior management and psychotropic medication). This is a new generation of children with many options for treatment and management.

This third edition of *Management of Prader-Willi Syndrome* is crucial because the world is bursting with new alternatives for diagnosis and treatment, but that information typically gets to professionals in bits and pieces. In this book, the editors present the current understanding of cause and diagnosis and pull together a comprehensive approach to treatment from the wisdom of many renowned experts in the field of PWS.

With the Internet, we also now have an entire world of expert knowledge to pull from—and an entire world of PWS for which we are responsible. At the PWSA (USA) office, pleas for help do not just come from states like Kentucky, California, and Texas anymore. They also come from countries such as India, Romania, Tanzania, and Brazil. As we reach across continents with the stroke of a key, our goal is that this book also is far-reaching, through the bookshelves of libraries across the world.

Our praise and gratitude go to the many authors who donated their time and expertise to this book—and a very special thank you goes to our section editors, Merlin G. Butler, MD, PhD, Phillip D.K. Lee, MD, and Barbara Y. Whitman, PhD, and to our managing editor, Linda Keder. Hopefully, I will be around long enough to be able to say someday, "My goodness, that third edition was certainly dismal!" May each generation step on the shoulders of the last, to peer over the wall of reality of their day in order to view a brighter future for their children with PWS.

Janalee Heinemann, MSW
Executive Director, Prader-Willi Syndrome Association (USA)

Acknowledgments

The editors wish to express their appreciation to the many colleagues who contributed their time and expertise in authoring chapters for this edition of *Management of Prader-Willi Syndrome*. We have benefited from our interactions with each of the authors before and during the production of the book and look forward to continued collaborations. The editors also thank Dr. Barbara Lippe for her personal and professional efforts in improving the recognition and treatment of individuals with PWS and the leadership of both the Prader-Willi Syndrome Association (USA) and the International Prader-Willi Syndrome Organization (IPWSO) for their advocacy and their ongoing support of this project. We wish to express our special thanks to all of the patients and their families with whom we have interacted over the years. These interactions have motivated us to strive harder in gaining a better understanding of the causes of Prader-Willi syndrome and in developing treatment plans to assist those affected. We are grateful to them for teaching us about the condition and how it impacts on their daily lives.

The editors would also like to thank the many individuals who have provided outstanding assistance and motivation in completing this book, including assistants and colleagues in our individual workplaces. We are indebted to Linda Keder, managing editor, for her guidance and enormous hard work. Her untiring nature and meticulous skills for perfection have led us along the journey from the planning stage to revising and editing of this edition. Her skills as an established editor and her dedication and commitment to excellence are unsurpassed. We also would like to thank Janalee Heinemann, executive director of PWSA (USA), for her patience with us as editors and for her dedication to completing tasks in a timely fashion. We would be remiss if we did not acknowledge Paula Callaghan, senior editor of the clinical medicine section at Springer SBM publishers. Her experience and wealth of information about the publishing process have been very helpful.

Lastly, our deepest gratitude goes to our respective families for their support and encouragement during the many hours spent in the production of this book. Much of the time required for writing and editing occurred at the end of the day when our other job responsibilities were finished. In addition, we appreciate their patience, support, and understanding during all the years of our professional lives and in allowing us to enjoy the fruits of our labor.

Merlin G. Butler, MD, PhD
Phillip D.K. Lee, MD
Barbara Y. Whitman, PhD

Contents

Part I   Diagnosis and Genetics

Part II   Medical Physiology and Treatment

Appendices

Contributors

Randell C. Alexander, MD, PhD
Professor of Clinical Pediatrics, Morehouse School of Medicine, Department of Pediatrics, Atlanta, GA 30310, USA

Karin Buiting, PhD
Institut für Humangenetik, Universitätsklinikum Essen, Essen, Germany

Merlin G. Butler, MD, PhD
Chief, Section of Medical Genetics and Molecular Medicine, William R. Brown/Missouri Chair of Medical Genetics and Molecular Medicine; Professor of Pediatrics, University of Missouri-Kansas City School of Medicine, Children's Mercy Hospitals and Clinics, Kansas City, MO 64108, USA

Aaron L. Carrel, MD
Associate Professor of Pediatrics, University of Wisconsin Medical School, Madison, WI 53792, USA

Suzanne B. Cassidy, MD
Clinical Professor of Pediatrics, University of California, Irvine, Orange, CA 92868, USA

Naomi Chedd, MS, LMHC
Education Consultant and Mental Health Counselor in private practice, Brookline, MA 02446, USA

Steve Drago, BS, CABA
Associate Director, ARC of Alachua County, Gainesville, FL 32606, USA

Urs Eiholzer, MD
Director, Institute Growth Puberty Adolescence, Zürich, Switzerland

Pamela F. Eisen
President, International Prader-Willi Syndrome Organization (IPWSO), Wormleysburg, PA 17043, USA

Kenneth J. Ellis, PhD
Professor of Pediatrics and Director, Body Composition Laboratory, Children's Nutrition Research Center, Baylor College of Medicine, Houston, TX 77030, USA

Giorgio Fornasier
Director of Program Development, International Prader-Willi Syndrome Organization (IPWSO), 32020 Limana, Italy

Janice L. Forster, MD
Former Consulting Psychiatrist, The Children's Institute Prader-Willi Syndrome and Behavior Disorders Program; Physician in private practice, Pittsburgh, PA 15228, USA

Toni Goelz, BS
Staff Physical Therapist, Cardinal Glennon Children's Hospital, St. Louis, MO 63104, USA

Barbara J. Goff, EdD
Consultant for Prader-Willi Syndrome; Assistant Professor of Education, Westfield State College, Westfield, MA 01086, USA

Linda M. Gourash, MD
Formerly Chief Attending Physician, The Children's Institute Prader-Willi Syndrome and Behavior Disorders Program; Physician in private practice, Pittsburgh, PA 15228, USA

Louise R. Greenswag, RN, PhD
Consultant for Prader-Willi Syndrome, Division of Developmental and Behavioral Medicine, Center for Disabilities and Development, University of Iowa Hospitals and Clinics, Iowa City, IA 52242, USA

James E. Hanchett, MD[†]
Formerly Consultant to The Children's Institute Prader-Willi Syndrome and Behavior Disorders Program, Pittsburgh, PA 15217, USA

Jeanne M. Hanchett, MD
Assistant Professor, Department of Pediatrics, University of Pittsburgh School of Medicine; The Children's Institute of Pittsburgh, Pittsburgh, PA 15217, USA

[†] Deceased.

Janalee Heinemann, MSW
Executive Director, Prader-Willi Syndrome Association (USA), Sarasota, FL 34242, USA

Robert M. Hodapp, PhD
Co-Director, Family Research Program, Vanderbilt Kennedy Center; Professor, Special Education, Peabody College, Nashville, TN 37203, USA

Bernhard Horsthemke, PhD
Professor of Human Genetics, Institut für Humangenetik, Universitäts-klinikum Essen, Essen, Germany

Kevin Jackson, PhD, BCBA
Senior Behavior Analyst, Agency for Persons with Disabilities, Gainesville, FL 32602, USA

Ellie Kazemi
Graduate Student, Psychological Studies in Education, University of California, Los Angeles, Graduate School of Education and Information Studies, Los Angeles, CA 90095, USA

Phillip D.K. Lee, MD
Formerly Professor of Pediatrics, David Geffen School of Medicine at University of California, Los Angeles (UCLA), Los Angeles, CA 90095, USA; currently Chief Scientific Officer, Immunodiagnostic Systems Ltd., Boldon, Tyne and Wear, NE 35 9PD, UK

Karen Levine, PhD
Clinical Director, Autism Services, North Shore ARC, Danvers, MA 01923, USA

Barbara A. Lewis, PhD, CCC–SLP
Associate Professor, Department of Pediatrics, School of Medicine, Case Western Reserve University, Rainbow Babies and Children's Hospital, Cleveland, OH 44106, USA

Shawn E. McCandless, MD
Assistant Professor, Genetics and Pediatrics, Center for Human Genetics, University Hospitals of Cleveland and Case Western Reserve University, Cleveland, OH 44106, USA

Harriette R. Mogul, MD, MPH
Professor of Medicine, Chief, Section of Obesity and Osteoporosis, Department of Medicine, New York Medical College, Valhalla, NY 10595, USA

Kristin G. Monaghan, PhD
Director, DNA Diagnostic Laboratory, Department of Medical Genetics, Henry Ford Hospital and Medical Center, Detroit, MI 48202, USA

Ann O. Scheimann, MD
Assistant Professor of Pediatrics, Johns Hopkins School of Medicine, Baltimore, MD 21287; Adjunct Assistant Professor of Pediatrics, Baylor College of Medicine, Houston, TX 77030, USA

Barbara R. Silverstone, JD
Staff Attorney, National Organization of Social Security Claimants' Representatives, Englewood Cliffs, NJ 07632, USA

Travis Thompson, PhD
Executive Program Director, Minnesota Autism Center, Minneapolis, MN 55413, USA

Daniel L. Van Dyke, PhD
Co-Director, Cytogenetics Laboratory, Mayo Clinic, Rochester, MN 55905, USA

Robert H. Wharton, MD[†]
Spalding Rehabilitation Hospital, Boston, MA 02114, USA

Barbara Y. Whitman, PhD
Professor of Pediatrics, Saint Louis University School of Medicine, Cardinal Glennon Children's Hospital, St. Louis, MO 63104, USA

David A. Wyatt
Crisis Intervention Counselor, Prader-Willi Syndrome Association (USA), Sarasota, FL 34242, USA

Mary K. Ziccardi, BS
Administrator, The MENTOR Network, REM Ohio, Inc., Cleveland, OH 44125, USA

[†] Deceased.

Introduction

Why a third edition of *Management of Prader-Willi Syndrome*? The purpose remains the same as that of previous editions, to reach professionals, parents, and extended families—virtually anyone interested in learning about Prader-Willi syndrome (PWS). And, since the last edition was published, there have been major advances in clinical management and research into the complex issues associated with this syndrome. Genetic tests ensure identification with 99% accuracy, endocrine studies have shown the benefits of growth hormone therapy, and innovative strategies are being designed to enhance psychosocial adaptation. Professionals and parents of affected children now have more in-depth knowledge, recognize the long-term consequences of this condition, and share common goals.

A historical backdrop provides a perspective on what has transpired since PWS was first described in 1956 and how far knowledge has progressed. For more than 30 years after being identified, only the major features were recognized. No cause was known. Few physicians had heard of it. The obesity was considered uncontrollable and life expectancy short. Children presented then—as now—as limp, sleepy, non-responsive babies, who are poor eaters and slow to achieve developmental milestones. Typically, doctors might have found a paragraph or two about PWS in a few medical references but little about management. Diagnosis was made by clinical evaluation. Most individuals were identified late in childhood if at all. Management was often mediocre.

During the 1980s several important events occurred. Holm, Sulzbacher, and Pipes published the book *Prader-Willi Syndrome*, a compendium of comprehensive findings from a conference on the syndrome in 1981. That same year, Dr. David Ledbetter discovered the deletion of information on the 15th chromosome in some affected individuals. In 1983, Dr. Merlin Butler recognized that the deletion causing PWS occurred in the chromosome 15 inherited from the father. Six years later, Dr. Robert Nicholls identified uniparental disomy 15 (both 15s from the mother) as the genetic explanation for most of the phenotypically similar subjects without the chromosome 15 deletion and proposed the role of genomic imprinting.

Even though scientific research advanced and descriptive articles about PWS appeared in the literature, bewildered parents continued to struggle to find help. In 1987, when we established a PWS clinic at the University of Iowa to serve families, it became clear that case management was limited and fragmented in scope. Professionals as well as families needed an additional resource. The challenge was to determine what information would be most valuable. What we originally conceived as a small booklet of facts to explain PWS to the parents attending the clinic evolved into the first edition of *Management of Prader-Willi Syndrome*. Welcomed and fully supported by the Prader-Willi Syndrome Association (USA), the success of this first volume received worldwide distribution due, in large part, to the leading authorities that contributed chapters. Seven years later, burgeoning research in genetics, endocrinology, the use of psychotropic medications, and innovative approaches to management led to the publication of the second volume. This third edition is an indication that the boundaries of what is known about PWS have expanded even further.

The multifaceted characteristics of PWS remain the same. The syndrome is an eating rather than a weight disorder; the overpowering hunger is a physiological fact of life. And, unlike those with exogenous obesity, most persons with PWS never feel full. Difficulties arise when this lack of satiety is compounded by the fact that most affected individuals do not have the cognitive capacity to comply with unsupervised food intake regimens. Furthermore, while food issues are a critical component of management, equally important are the emotional and cognitive features that affect how individuals think, feel, and react. Children, usually affable and compliant until late toddlerhood, tend to become emotionally labile as they grow older, demonstrating aberrant behaviors, regardless of the presence or absence of obesity. With appropriate nutritional management, those who previously would have died young due to obesity now live well into adulthood and, with their families, confront issues as developmentally disabled adults that require comprehensive, sensitive management.

As with any condition, diagnosis is merely the beginning. Ongoing case management tailored to the needs of children with PWS and their families is essential. For example, recognition early in life can lead to specific strategies to augment developmental milestones. Recognizing this struggle, the U.S. national parent organization, the Prader-Willi Syndrome Association (USA) is helping bridge the gap between parents and professionals by initiating the learning process. With knowledge of what typically happens and possible variations on this theme, frequently noted concerns may be anticipated and ameliorated.

The text of this new edition has been divided into three sections:
Part I: Diagnosis and Genetics, edited by Merlin G. Butler, MD, PhD, explains and discusses the complex genetics of PWS and procedures for clinical and genetic diagnostic evaluations, as well as providing an overview of the clinical findings and natural history of PWS.

Part II: Medical Physiology and Treatment, edited by Phillip D.K. Lee, MD, provides an overview of medical issues and management across the life span, with separate chapters devoted to growth hormone issues and treatment and to gastrointestinal issues, obesity, and body composition.

Part III: Multidisciplinary Management, edited by Barbara Y. Whitman, PhD, focuses on the neurodevelopmental, neurocognitive, and developmental aspects of PWS and the accompanying life span management issues, including behavior and crisis management, educational concerns, and vocational and residential options.

We believe that this book, edited and written by the very best in the field and updated with the latest research, is likely to be the single best resource of information and support for the next several years. There is no yardstick to measure the challenges that these children and their families face, but professionals and parents alike will be able to achieve a depth of understanding and grasp of details about PWS that were an unachievable dream only two decades ago. Although we have come far, there is still no magic cure. We are hopeful that a genetic intervention will eventually be found.

Louise R. Greenswag, RN, PhD
Randell C. Alexander, MD, PhD
Editors, *Management of Prader-Willi Syndrome*, First and Second
Editions

Part I

Diagnosis and Genetics

1

Clinical Findings and Natural History of Prader-Willi Syndrome

Merlin G. Butler, Jeanne M. Hanchett, and Travis Thompson

Historical and Genetic Overview

Prader-Willi syndrome (PWS) is a complex genetic condition characterized by a range of mental and physical findings, with obesity being the most significant health problem. PWS is considered to be the most common genetically identified cause of life-threatening obesity in humans and affects an estimated 350,000–400,000 people worldwide. The Prader-Willi Syndrome Association (USA) is aware of more than 3,500 persons with this syndrome in the United States from an estimated pool of 17,000–22,000.[37] Prader-Willi syndrome has an estimated prevalence in the range of 1 in 8,000 to 1 in 20,000 individuals.[17,42] Although PWS is thought to be one of the more common disorders seen for genetic services, pediatricians may only encounter a few patients during their lifelong clinical practice. It is present in all races and ethnic groups but reported disproportionately more often in Caucasians.[38] Most cases are sporadic; however, at least 20 families have been reported in the literature with more than one affected member, including twins. The chance for recurrence is estimated to be less than 1%.[43] However, in some families where defective control of differentially expressed genes through errors in the process of genomic imprinting (described in detail in Chapter 3) is present in the PWS child and in the father, then the recurrence may be as high as 50% for having additional children with PWS.[12,13]

Prader-Willi syndrome was apparently first documented in an adolescent female by J. Langdon Down in 1887,[54] but it was not described in the medical literature until about 70 years later. This adolescent female had mental deficiency, obesity, short stature, and hypogonadism, and the condition was referred to as polysarcia. In 1956, Prader, Labhart, and Willi reported nine individuals (five males and four females between the ages of 5 and 23 years) with similar clinical findings.[110] Since the 1970s, the disorder has been referred to as the Prader-Willi syndrome.

In 1981, Ledbetter and others[93] first reported the cause as an interstitial deletion of the proximal long arm of chromosome 15 at region q11–q13 in the majority of subjects studied using high resolution chromosome analysis (Figure 1.1). The syndrome became one of the first genetic syndromes to be attributed to a chromosome microdeletion detectable with new high resolution chromosome or cytogenetic methods allowing for more precision than previously available. Butler and Palmer in 1983[35] were the first to report that the origin of the chromosome 15 deletion was *de novo*, or due to a new event, and not present in either parent. The chromosome 15 leading to the deletion was donated from the father in all cases, as determined by studying minor normal variations (chromosome polymorphisms) in chromosome structure. This puzzling observation of finding only a paternal deletion was later clarified by molecular genetic techniques and the reporting of a maternal deletion in a separate clinical condition recognized as Angelman syndrome (AS). The molecular genetic findings will be discussed later.

The cytogenetic deletion of chromosome 15 was seen in approximately 60% of persons with PWS while the remaining subjects showed normal chromosome 15s, translocations, or other abnormalities of chromosome 15.[17,32,93] Butler and others in 1986[32] also reported clinical differences in those PWS subjects with and without the chromosome 15 deletion. Those individuals with the deletion were more homogeneous in their clinical presentation and were hypopigmented. Because of the

Chromosome 15

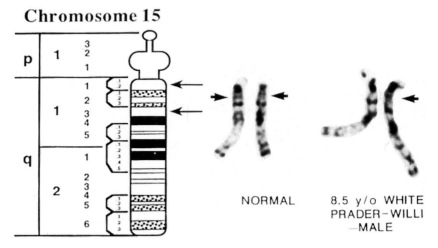

NORMAL   8.5 y/o WHITE
PRADER–WILLI
—MALE

Figure 1.1. A prometaphase or high resolution chromosome 15 ideogram (left) showing the pattern for chromosome arms (p for short arm and q for long arm) and designated bands with two representative chromosome 15 pairs (right). The arrows on the ideogram indicate the deletion breakpoints at bands 15q11 and 15q13. The 15q12 band is indicated by the arrow on each member of the chromosome 15 pair that is normal. The left member of the chromosome 15 pair from the 8.5-year-old male with Prader-Willi syndrome shows the deletion. Modified from Butler, Meaney, and Palmer, *American Journal of Medical Genetics* 1986;23:793–809.[32]

hypopigmentation and other eye findings (e.g., strabismus), visual evoked potential studies of specific brain wave patterns stimulated by visual images seen by the patient were performed by Creel et al.[48] They reported abnormal visual evoked potentials in PWS subjects similar to what was seen in individuals with classic albinism. They further reported misrouting of the optic nerve as it crossed over at the chiasma and described that optic nerve fibers crossed over 20 degrees or more from the temporal side of the retina at the chiasma instead of projecting to the same or ipsilateral side of the brain hemisphere. This occurred more often in PWS subjects with hypopigmentation. Misrouting of the optic nerve can lead to strabismus and/or nystagmus, which are frequently seen in PWS.

Molecular genetic methods using a technique referred to as Southern hybridization of newly identified polymorphic DNA markers isolated from the 15q11–q13 chromosome region were studied by Nicholls, Butler, and others in 1989.[107] PWS individuals with normal-appearing chromosomes were surprisingly found to have both chromosome 15s from the mother. This finding was referred to as maternal uniparental disomy 15 (i.e., both 15s from the mother) or UPD (Figure 1.2). Studies have shown that both members of the chromosome 15 pair were inherited from the mother as a result of an error in egg production due to nondisjunction in meiosis; the chromosome 15 pair fails to separate normally when the chromosome number is reduced by one half in production of the gamete. When a single chromosome 15 is received from the father in the normal sperm then trisomy 15 results, or three chromosome 15s are present in the fertilized egg. Subsequently, loss of

Figure 1.2. Polymerase chain reaction amplification of genomic DNA using D15S822 locus from the 15q11-q13 region from a Prader-Willi syndrome family with normal chromosome studies in the Prader-Willi syndrome individual. The mother (on the left), the Prader-Willi syndrome individual (in the middle), and the father (on the right) each show two DNA bands representing the presence of the D15S822 locus in each chromosome 15 (nondeleted status). The DNA pattern from the mother and Prader-Willi syndrome individual are identical but no DNA signal from chromosome 15 was inherited from the father. The Prader-Willi syndrome individual has two chromosome 15s from the mother and no chromosome 15 from the father, demonstrating maternal disomy 15 or both 15s from the mother.

the father's chromosome 15 occurs in early pregnancy. This loss of a single chromosome 15 in the cells rescues the fetus. If the trisomic 15 finding had continued in each cell of the fetus, a spontaneous miscarriage would have occurred. Chromosome abnormalities, specifically trisomy 15 and other trisomy and monosomy events, are common causes of spontaneous miscarriages.

PWS and Angelman syndrome, an entirely different clinical disorder generally due to a maternal deletion of the 15q11–q13 region, were the first examples of genomic imprinting in humans, or the differential expression of genetic information depending on the parent of origin. Thus, gene activity differences exist in maternal and paternal chromosome 15s that impact on the clinical phenotype or outcome.

There are two forms of maternal disomy 15 in PWS, maternal isodisomy and maternal heterodisomy. Maternal isodisomy refers to the presence of identical genetic material (e.g., came from the same chromosome 15 in the mother during production of the egg), and maternal heterodisomy refers to the presence of genetic material from two different chromosome 15s received from the mother. Children with PWS due to maternal isodisomy may be at increased risk of having a second genetic condition if the mother is a carrier of a recessive disorder on chromosome 15 such as Bloom syndrome. Bloom syndrome is an autosomal recessive condition where the gene is located outside of the 15q11–q13 region but both members of the gene pair (alleles) are abnormal. If the mother carries an abnormal gene allele for Bloom syndrome on one of her chromosome 15s, she is unaffected because the gene allele on her other chromosome 15 is normal. However, if the chromosome 15 region or segment containing the abnormal recessive gene allele for Bloom syndrome is donated to the child by the mother in an isodisomy fashion (i.e., both chromosome 15s contain the same identical genetic information or alleles), then PWS results as well as Bloom syndrome.

PWS is thought to be due to a contiguous gene condition involving several genes, with nearly one dozen genes or transcripts mapped to the 15q11–q13 region known to be imprinted; most are paternally expressed (active) or maternally silent. Several of these genes are candidates for causing features recognized in PWS including *SNURF/ SNRPN, NDN, snoRNAs, MKRN3,* and *MAGEL.*[2,8,9,106] The best characterized paternally expressed gene studied to date is *SNRPN* (small nuclear ribonucleoprotein N). A second DNA sequence of *SNRPN* is termed *SNURF* (*SNRPN* upstream reading frame), consisting of gene exons 1–3, which encompasses the imprinting center or the genetic locus that controls the regulation of imprinting, or turning on and off the activity of specific genes throughout the chromosome 15q11–q13 region. The proteins coded for by *SNRPN* and *SNURF* are not present in subjects with PWS. Further, chromosome translocations involving the paternal chromosome 15 that disrupt *SNURF* have been reported in subjects with typical features of PWS.[37] Thus, a disruption of this gene locus will cause loss of function of paternally expressed genes in this region (e.g., *MAGEL2*) and loss of proteins coded by the genes. Hence, many of these paternally expressed genes in this chromosome area play a role in brain development and function, key for producing the clinical phenotype recognized in PWS. Detailed descriptions of the 15q11–q13 genes, their respective roles in PWS (and AS), and imprinting will be described in more detail elsewhere (see Chapter 3).

Genomic imprinting appears important for growth in humans and other mammals. Many conditions now thought to be due to genomic imprinting, such as Beckwith-Weidemann and Russell-Silver syndromes, present with either overgrowth or growth retardation.[19] Many imprinted genes identified to date are involved with regulation of cell proliferation or differentiation. They may also play a role in development of tumors. Some imprinted genes are known as fetal growth factors; thus, a close relationship exists between genomic imprinting and early fetal development. Generally, maternal genes have a tendency to suppress growth of the developing fetus while paternal genes enhance growth. The competition for expression of maternal and paternal genes has evolved to control growth of the fetus, and fine-tuning of expression of genes depends on the parent of origin with a complex balance of maternal and paternal genes. A disruption of the genetic balance results in growth anomalies in the fetus.

Clinical Presentation and Diagnosis

Many clinical features in PWS may be subtle or non-specific while other features are more characteristic for the disorder. The primary features of PWS include infantile hypotonia, feeding difficulties, mental deficiency, hypogonadism, behavior problems (temper tantrums, stubbornness, obsessive-compulsive disorder), hyperphagia and early childhood onset of obesity, small hands and feet, endocrine disturbances including recently identified growth hormone deficiency, and a characteristic facial appearance (small upturned nose, narrow bifrontal diameter, dolichocephaly, down-turned corners of the mouth, sticky saliva, almond-shaped eyes, and strabismus) (Figure 1.3, see color insert). Table 1.1 lists the clinical manifestations, frequency, and time period when they occur. Some of the characteristics are subtle but well recognized by clinical geneticists trained in dysmorphology; however, a greater awareness by pediatricians, other physicians, and health care providers now exists. Because of the better recognition and awareness of PWS by the medical community during the past 10 years and more accurate and reliable genetic testing, the diagnosis is made earlier than in the past and extensive diagnostic procedures avoided. Many children with PWS were not diagnosed in the past until rapid weight gain leading to obesity was evident and the presence of specific learning/behavioral problems was observed. For example, in the mid-1980s Butler reported that the average age at diagnosis for PWS was greater than 6 years of age.

In 1993, Holm, Cassidy, Butler, and others[85] developed consensual diagnostic criteria to assist in the diagnosis of PWS using major and

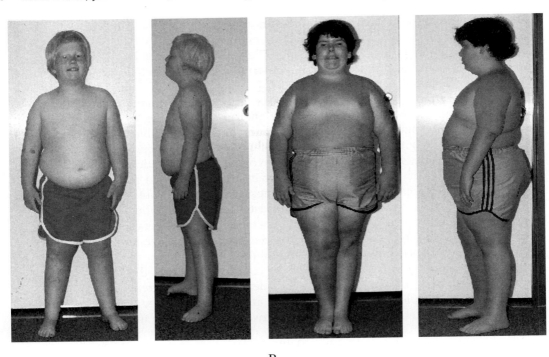

A   B

Figure 1.3. Frontal and profile views of two males (patient A is 8.5 years of age, with the chromosome 15q11–q13 deletion [seen in Figure 1.1]; patient **B** is 11 years of age, with maternal disomy 15) with Prader-Willi syndrome. Note the typical facial appearance (e.g., narrow bifrontal diameter, almond-shaped eyes, triangular mouth), small hands and feet, characteristic obesity, and hypopigmentation (seen in patient **A** with the 15q11–q13 deletion). Modified from Butler, *American Journal of Medical Genetics* 1990;35:319–332.[17] (See color insert.)

minor features and established a scoring system for patients presenting with features seen in this syndrome. The scoring system consisted of three categories (major, minor, and supportive criteria) and scoring was based on a point system. The major criteria were weighed at one point each while minor criteria were weighed at one half point each. Supportive criteria received no points but may be helpful to confirm the diagnosis. The diagnostic criteria for PWS and recommendations for their current use as an indicator for genetic testing are discussed in Chapter 2.

Cytogenetic studies using fluorescence *in situ* hybridization (FISH) of chromosome 15q11–q13 DNA probes became available in the 1990s and have been helpful in detecting deletions (Figure 1.4, see color insert). If the DNA probes isolated from the 15q11–q13 region do not hybridize, or attach, to this chromosome region, then the deletion status is confirmed. These studies have shown that 70% of subjects

Table 1.1. Summary of Clinical Findings in Individuals Reported with Prader-Willi Syndrome*

Time period when clinical manifestations first appear	Clinical manifestation	Affected/total patients	Overall (%)
Pregnancy and delivery	Reduced fetal activity	137/181	76
	Nonterm delivery	83/203	41
	Breech presentation	56/212	26
Neonatal and infancy	Delayed milestones	405/412	98
	Hypogenitalism/hypogonadism	270/285	95
	Hypotonia	504/538	94
	Feeding problems	445/479	93
	Cryptorchidism	240/273	88
	Narrow bifrontal diameter	138/184	75
	Low birth weight (<2.27 kg)	68/226	30
Childhood	Mental deficiency	504/517	97
	Obesity	287/306	94
	Small hands and feet	237/286	83
	Skin picking	261/330	79
	Short stature (<–1 SD)	232/306	76
	Almond-shaped eyes	151/202	75
	Strabismus	259/494	52
	Delayed bone age	74/148	50
	Scoliosis	159/360	44
	Personality problems	161/397	41
	Early dental caries/enamel hypoplasia	56/141	40
Adolescence and adulthood	Menstruation	38/98	39
	Reduced glucose tolerance/diabetes mellitus	74/371	20
	Seizures	40/199	20

* Number of males 286; number of females 211.

Adapted from M. G. Butler, "Prader-Willi syndrome: current understanding of cause and diagnosis," *American Journal of Medical Genetics* 1990; 35:319–332.[17]

with PWS have the typical chromosome 15q11–q13 deletion (Figure 1.1) while about 25% have normal-appearing chromosomes. Currently, DNA tests of children with PWS and normal-appearing chromosomes with normal FISH studies, in comparison with parental DNA, have shown maternal disomy 15, or UPD, in the vast majority (Figure 1.2). A small percentage (1%–3%) of those PWS subjects with normal-appearing chromosomes and normal biparental chromosome 15 inheritance have either very small atypical deletions (undetectable with FISH) or mutations (defects) such as microdeletions of the imprinting center. All three genetic causes (deletion, maternal disomy, and imprinting defects) can be identified by abnormal methylation testing. Other rare genetic causes include translocations or inversions involving the chromosome 15q11–q13 region. Table 1.2 lists the genetic testing available to identify the genetic cause in PWS subjects. Molecular and cytogenetic testing for this syndrome will be described in more detail elsewhere (see Chapter 4).

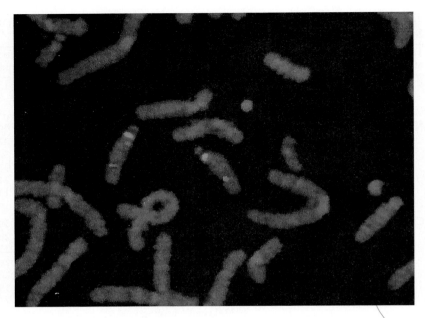

Figure 1.4. Representative fluorescence *in situ* hybridization (FISH) using a SNRPN probe from the chromosome 15q11-q13 region (red color), a centromeric probe from chromosome 15 (green color), and a distal control probe from chromosome 15q (red color) showing the absence of the SNPRN signal close to the centromere on the deleted chromosome 15 from a subject with Prader-Willi syndrome. (See color insert.)

Table 1.2. Genetic Laboratory Testing for Prader-Willi Syndrome

1. **Chromosome analysis to rule out translocations**
2. **Fluorescence *in situ* hybridization (FISH) analysis using DNA probes from chromosome 15q11–q13**
3. **DNA methylation testing of SNRPN gene with polymerase chain reaction (PCR)**
4. **DNA microsatellite analysis using PCR to confirm a paternal 15q11–q13 deletion and size (type I or type II) in PWS or to identify maternal disomy 15; DNA from the patient and each parent will be required.**
5. **Gene expression studies of imprinted genes (e.g., paternally expressed [maternally silent] SNRPN gene) from the 15q11–q13 region using reverse transcriptase-PCR. The lack of paternally expressed genes will confirm the diagnosis of PWS but will not assign the genetic subtype (e.g., deletion or maternal disomy).**
6. **DNA replication pattern analysis of genes from the 15q11–q13 region varies depending on the parental source of chromosome 15.**

Tests 1, 2, and 3 are usually performed in subjects presenting to the genetics clinic with features of PWS, while tests 4, 5, and 6 are research-based tests and available in research laboratories only.

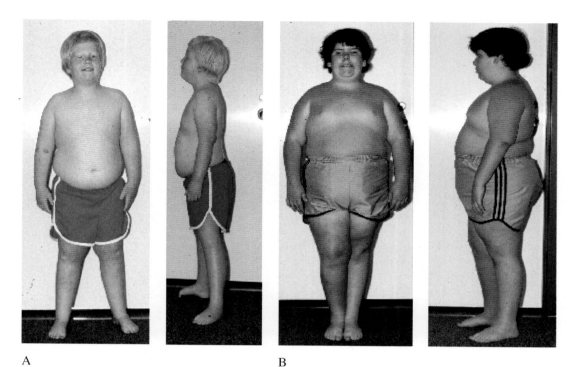

A   B

Figure 1.3. Frontal and profile views of two males (patient A is 8.5 years of age, with the chromosome 15q11–q13 deletion [seen in Figure 1.1]; patient **B** is 11 years of age, with maternal disomy 15) with Prader-Willi syndrome. Note the typical facial appearance (e.g., narrow bifrontal diameter, almond-shaped eyes, triangular mouth), small hands and feet, characteristic obesity, and hypopigmentation (seen in patient **A** with the 15q11–q13 deletion). Modified from Butler, *American Journal of Medical Genetics* 1990;35:319–332.[17]

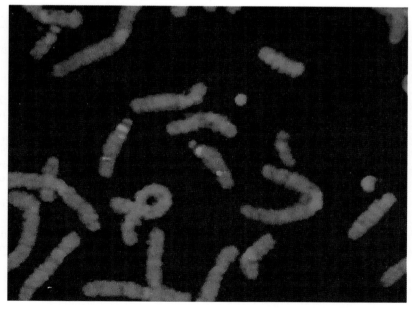

Figure 1.4. Representative fluorescence *in situ* hybridization (FISH) using a SNRPN probe from the chromosome 15q11-q13 region (red color), a centromeric probe from chromosome 15 (green color), and a distal control probe from chromosome 15q (red color) showing the absence of the SNPRN signal close to the centromere on the deleted chromosome 15 from a subject with Prader-Willi syndrome.

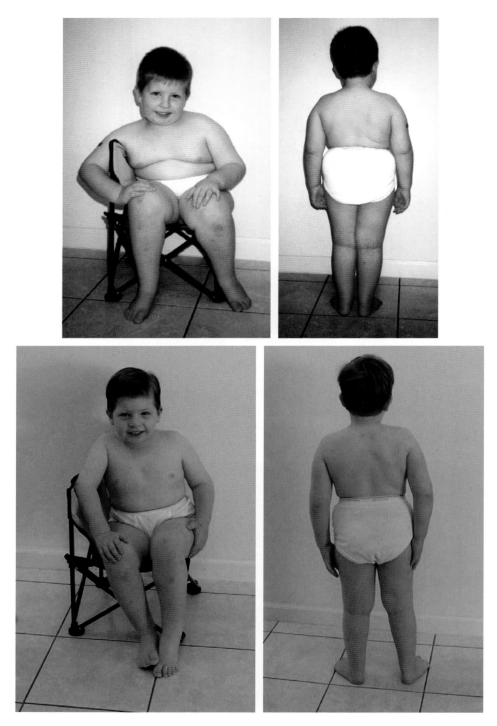

Figure 1.5. Five-year-old boy with Prader-Willi syndrome before (top) and after 3 months of growth hormone (GH) treatment (bottom). Note the improved body habitus, muscle bulk, and reduced fat. The patient was able to increase total caloric intake, and he had increased activity and wakefulness. Reprinted with permission from S.B. Cassidy, "Prader-Willi syndrome in the new millennium: introduction," *The Endocrinologist* 2000;10(4) Suppl 1: 1S–2S. Copyright©2000 by Lippincott Williams & Wilkins.

Clinical Stages and Natural History

The course and early natural history of PWS can be divided into two distinct clinical stages. The first stage occurs during the neonatal and early infancy periods and is characterized by varying degrees of hypotonia, a weak cry, a narrow forehead, developmental delay, temperature instability, a poor suck reflex, sticky saliva, feeding difficulties sometimes requiring gastrostomy or stomach tube placement, hypogonadism, and underdevelopment of the sex organs. Failure-to-thrive is noted during this first stage. The hypotonia is thought to be central in origin, non-progressive, and on the average begins to improve between 8 and 11 months of age.[32] Infants with PWS sit independently at 11–12 months, crawl at 15–16 months, walk at 24–27 months, and talk (10 words) at 38–39 months.[32,72] The delay in achieving motor milestones appears to relate more to psychomotor development than to excessive obesity. Language appears to be the most delayed of the developmental milestones.[32,72]

The second stage usually begins around 2 years of age and is characterized by continued developmental delay or psychomotor retardation and onset of hyperphagia leading to obesity. Other features noted during the second stage may include speech articulation problems, foraging for food, rumination, unmotivated sleepiness (found in greater than 50% of subjects), physical inactivity, decreased pain sensitivity, skin picking and other forms of self-injurious behavior, prolonged periods of hypothermia, strabismus, hypopigmentation, scoliosis, obstructive sleep apnea, and abnormal oral pathology (enamel hypoplasia, dental caries, malocclusion, and decreased saliva).[37]

Early in the second stage, infants and toddlers are usually easy-going and affectionate, but in about one half of PWS subjects personality problems develop between 3 and 5 years of age. Temper tantrums, depression, stubbornness, obsessive compulsivity, and sudden acts of violence of varying degrees may be observed during this stage. These behavioral changes may be initiated by withholding of food but may occur with little provocation during adolescence or young adulthood. Poor peer interactions, immaturity, and inappropriate social behavior may also occur during this time.[17,57,58,61,127]

Pregnancy and Delivery

Decreased fetal movement is noted by the mothers in nearly all PWS pregnancies. About one fourth of babies with PWS are delivered in breech presentation. Approximately one half of babies with PWS are born pre- or post-term (2 weeks earlier or later than the anticipated delivery date). Mild prenatal growth retardation is noted, with an average birth weight of 2.8 kg. Low birth weight is seen in about 30% of deliveries.[17]

Infancy

Because of generalized hypotonia, neonates with PWS are profoundly floppy and their presentation can be mistaken for other disorders such as Werdnig-Hoffman or trisomy 18 syndromes, a brain anomaly or bleeding, a metabolic disturbance, an infectious agent, or other environmental insult. The hypotonia may lead to asphyxia and, in the past, has led to extensive medical evaluations including muscle biopsies and repeated brain imaging studies. Most of the structural brain imaging studies and muscle biopsies performed on infants are thought to be normal or not diagnostic for a specific syndrome.

Most infants with PWS have a weak or absent cry, little spontaneous movement, hyporeflexia, excessive sleepiness, and poor feeding due to diminished swallowing and sucking reflexes that may necessitate gavage feedings lasting for several months. However, feeding difficulties generally improve by 6 months of age. Failure-to-thrive and poor weight gain are common features seen in infants with PWS. Due to failure-to-thrive, feeding difficulties, and hormone deficiencies, the growth of the PWS infant may fall below the 3rd centile for weight. Temperature instability may be present during early infancy with high or low body temperatures. The bouts of hyperthermia or fever may trigger medical investigations to pursue an infectious agent, which is not often found. Infants with PWS have significant respiratory abnormalities including sleep-related central and obstructive apneas and reduced response to changes in oxygen and carbon dioxide levels.[4,108,136] PWS subjects tend to be relatively insensitive to pain including that caused by obtaining blood samples. Cryptorchidism (undescended testicles) and a hypoplastic penis and scrotum are frequently seen in males and hypogenitalism in females. Cryptorchidism may require treatment (hormonal and/or surgery), and a referral to a pediatric urologist is warranted (see Chapter 5). In addition, the hands and feet may be small at birth.

Mild dysmorphic features are recognized during the infancy period, particularly involving the face. These features include a narrow forehead; mild upward slanting of the palpebral fissures; a long, narrow-appearing head (dolichocephaly); a small upturned nose; a thin upper lip; sticky saliva; and down-turned corners of the mouth. Many of the dysmorphic facial features are reminiscent of those due to midline central nervous system defects, while others such as a triangular mouth may be related to hypotonia and poor prenatal neuromuscular functioning. PWS subjects may have diminished facial mimic activity due to the muscular hypotonia as they grow older. They may also present with hypopigmentation for their family background, including fair skin and light hair color noted during the infancy period. Hypopigmentation relative to family background is a feature in about three fourths of PWS subjects and correlates with those individuals having the 15q11–q13 deletion and loss of the *P* gene (which causes oculocutaneous albinism type II) located in this region.[16]

As noted above, infants with PWS tend to sit at 11–12 months and crawl at 15–16 months on average. Developmental milestones are typically delayed, and the delay continues into childhood.

Early Childhood

Although infants with PWS may be tube-fed during infancy, by 18 months to 2 years of age their feeding behavior changes radically and an insatiable appetite may develop, which causes major somatic and psychological changes in early childhood. Along with global developmental delays, temper tantrums, difficulty in changing routines, stubbornness, controlling or manipulative behavior, and obsessive-compulsive behavior may become more apparent during childhood. Lying, stealing, and aggressive behavior are common during the childhood years and may continue into adolescence and adulthood. Children with PWS may be less agreeable, less open to new ideas and experiences, and more dependent than typically developing children. In addition, they are generally less physically active than other children their age. Frequently, medical and/or behavior management advice is sought to treat the behavioral problems. Medications such as serotonin reuptake inhibitors have been of benefit in controlling behavior and psychiatric disturbances in some children with PWS, particularly as they become older.[42,43]

On average, toddlers with PWS learn to walk at about 27 months of age. By the time they enter kindergarten, they are nearly always overweight and short for age. However, growth hormone treatment significantly improves body composition, stature, and energy level. PWS children should be evaluated and treated for endocrine abnormalities such as hypothyroidism and growth retardation. In addition, enamel hypoplasia and dental caries are frequently seen during childhood in the syndrome. During early childhood PWS children may also develop nystagmus or strabismus, but the most common recognized eye finding in PWS is myopia,[64] followed by decreased visual acuity and impaired stereoscopic vision; the latter finding is more common in PWS subjects with maternal disomy 15.[65] The hypopigmentation becomes more pronounced during childhood, particularly in those PWS subjects with the 15q11–q13 deletion.

The small hands and feet may not be apparent at birth but become more recognized during mid-childhood. Almond-shaped eyes may be more noticeable during childhood due to the periorbital tissue shape. A characteristic body habitus or posture, including sloping shoulders, heavy mid-section, and *genu valgus* with straight lower leg borders and sparing of distal extremities from excess fat deposition, is usually present by toddlerhood.

About one third of children with PWS function in the low-normal intellectual range (70–100 IQ), and the remaining PWS children (and adults) function in the mild-to-moderate range of mental retardation (50–70 IQ).[37,125] The average IQ is 65, with a range of 20 to 100. Academic achievement is poor for cognitive ability. During the first 6 years of life, children with PWS often do not achieve normal levels of cognition, motor, or language development. There are reported differences in behavior, academic, and intelligence testing between the PWS subjects with the typical chromosome 15 deletion versus those with maternal disomy, which will be discussed later in this chapter (see also Chapter 8).

Many children with PWS begin school in mainstream settings. About 5% attend regular school until secondary level, but the intellectual impairment and potential behavioral problems present in the majority of children with the syndrome require special education and support services. By elementary school age, children with PWS may steal or hide food at home or during school to be eaten later. It is common for those with PWS to have relatively strong reading, visual, spatial, and long-term memory skills but relatively weak math, sequential processing, and short-term memory skills. Verbal skills are relative strengths, particularly in those PWS subjects with maternal disomy, although speech articulation is often poor, with nasal or slurred characteristics. An unusual skill with jigsaw puzzles and fine motor skills are particularly common in those with the typical 15q11–q13 deletion. Recent studies have documented genotype/phenotype differences between the two genetic subtypes (deletion vs. maternal disomy), which may impact on these skills. Additional psychobehavioral findings in children with PWS will be discussed in more depth elsewhere (see Chapters 8 and 12).

Adolescence and Adulthood

Normal puberty is absent or delayed in both males and females with PWS. Gonadotropin hormone production by the brain is low, and other endocrine disturbances may be identified in this subject population. As a result, adolescents and young adults with PWS look young for their chronologic age. The occurrence of reduced growth hormone secretion and hypogonadotropic hypogonadism in the majority of children and adolescents with PWS, along with an insatiable appetite and high pain threshold, would suggest a hypothalamic pituitary dysfunction.

Hypogonadism and hypogenitalism occur in the vast majority of males and females with PWS and become more evident during adolescence. The hypogonadism is due to hypothalamic hypogonadotropism, since it is often associated with low levels of gonadotropin released by the brain needed for gonadal development and function. The degree of hypogonadism is variable from patient to patient but more marked in males. Cryptorchidism occurs with a hypoplastic penis and scrotum in males and can be identified in early infancy, but the hypoplastic labia minora and clitoris in females may not be as easily recognized, although they are cardinal features of this syndrome. Sertoli cells and a variable number of Leydig and germinal cells are usually present in the testicles, although there have been no reports of fertility in males. In addition, the tubules are usually small and atrophic. Penile size increases modestly in many males during the third or fourth decade of life, but testicular size remains small. In males with palpable testes, the size is seldom greater than 6ml in volume. Treatment of the small penis with topical or parenteral testosterone has been effective in achieving penile growth in PWS males.[40] However, mature genital development in males is rarely seen. Gonadotropin treatment may also be helpful in treating PWS males with cryptorchidism. Cryptorchid testes may descend spontaneously in some patients during childhood and puberty, but surgical intervention may be indicated. Precocious development of pubic and axillary hair occurs frequently as a consequence of premature adrenarche. Beard and body hair are variable, occurring later than normal, if at all. Beard growth is absent in about 50% of men.[45] Additional endocrine-related findings and more in-depth discussions are presented in later chapters (see Part II).

Menarche is often late or does not occur in females with PWS. In 98 appropriately aged females reported in the literature,[17] 38 developed spontaneous menstruation (see Table 1.1). In a study of mostly adult PWS females, breast development was normal in about one half, with onset between 9 and 13 years of age.[72] Primary amenorrhea (absence of menstruation) was found in about 70% of females and oligomenorrhea (infrequent menstruation) in the remaining. Age of menarche is extremely wide, from 7.5 to 38 years. In a series of 106 females between the ages of 15 and 63, 13 had been given hormones to induce menses. Very few of the women who had spontaneous menarche had regular menses; most had scant and infrequent menses.[76] Pubic hair was normal in 40% of females studied.[72]

Two women with presumed PWS were reported in the 1970s with established pregnancies (one woman was pregnant on two separate occasions).[92] However, the diagnosis of PWS was not confirmed with genetic testing in the two reported females. More recently, at least two women with documented PWS, confirmed by cytogenetic and molecular genetic methods, have been reported with established pregnancies. One woman with PWS was 33 years of age and had suspected maternal disomy (i.e., abnormal methylation testing and normal chromosome study with FISH). She gave birth to a healthy girl by C-section delivery after an estimated 41 weeks gestation.[2] The other adult female had the 15q deletion and gave birth, not surprisingly, to an infant with Angelman syndrome due to the child's receiving the 15q deletion from its mother.[116] The mother with PWS and suspected maternal disomy had low cerebrospinal fluid concentration of 5–hydroxyindoleacetic acid, a serotonin metabolite.[2] Due to behavioral problems before pregnancy, she had been medically treated with serotonergic drugs, which may have influenced gonadotropin release to induce hormonal conditions required for pregnancy and possibly reflecting low serotonin synaptic transmission in the brain. Warnock et al.[133] also reported onset of menses in two adult females with PWS when treated with fluoxetine, a serotonin reuptake inhibitor that increases the available serotonin for brain function. Therefore, fertility in women with PWS is exceedingly rare, but reproductive issues should be addressed in reproductive-age females with PWS.

Approximately 90% of subjects with PWS without growth hormone treatment will have short stature by adulthood. The average adult male without growth hormone therapy is 155 cm (61 inches) tall, and the adult female averages 147 cm (58 inches).[31] Growth standards for Caucasian PWS males and females aged 2 to 22 years, without a history of growth hormone therapy, were reported by Butler and Meaney in 1991[31] and are used in the clinical setting in the United States to monitor growth parameters (see Appendix C). These growth standards illustrate the growth pattern anomalies in this syndrome compared with healthy subjects. For example, the 50th centile for height in the PWS group fell below the normal 5th centile by the age of 12 to 14 years, while the 50th centile for weight in the PWS group approximated the 95th centile in the healthy control group. Height may follow the 10th centile or below until the age of 10 years for females and 12 years for males, at which time the height velocity often declines relative to normal due to the lack of a growth spurt. Growth standards for PWS from other countries have also been reported (see Appendix C). Growth hormone therapy may increase the growth spurt and impact positively on the ultimate height and body composition in the PWS subject (see Chapter 7).

Inverse correlations have been reported with linear measurements (e.g., height, hand and foot lengths) and age, indicating a deceleration of linear growth with increasing age relative to normal individuals.[30] A relative deceleration in growth of certain craniofacial dimensions (e.g., head circumference, head length) was also suggested.[98] Therefore, dolichocephaly or a long, narrow head shape may be considered an early diagnostic sign for PWS.

Small hands with thin, tapering fingers and small feet (acromicria) are seen during infancy and childhood and become more pronounced during adolescence and adulthood. There is a straight ulnar border of the hands. Foot length tends to be more affected than hand length, compared with normative standards. The average adult shoe size for males not treated with growth hormone is size 5 and for females size 3. Therefore, short stature and small hands and feet are present in the majority of PWS subjects, although the frequency in blacks is lower.[86] There is relative sparing of the hands and feet from obesity, with fat distribution often appearing to end abruptly at the ankle and wrist. Scoliosis may also become more pronounced during the adolescent period, and kyphosis may be present by early adulthood.

Without intervention, adolescents with PWS may weigh 250 to 300 pounds by their late teens, which can lead to a shortened life from the complications of obesity. Overeating can lead to immediately life-threatening events such as stomach rupturing. By late adolescence, some with PWS begin stealing food from stores and rummaging through discarded lunch bags or trash cans to find partially eaten left-over food or inedible food items (e.g., bags of sugar, frozen food). Some parents find it necessary to lock the refrigerator and cabinets containing food to prevent excessive eating. Despite these precautions, youth with PWS may pry open locked cabinets to gain access to food. In the past, many patients with PWS died before the age of 30 years. The eating behavior, complications of obesity that can reduce the life expectancy of a person with PWS, and cognitive impairments will preclude normal adult independent living. Behavioral and psychiatric problems interfere with quality of life in adulthood and may require medical treatment and behavioral management. However, if weight is adequately controlled, life expectancy may be normal. Hence, caloric diet restriction is lifelong and important to control the obesity and its complications. Continued consultation with a dietitian with experience in Prader-Willi syndrome is recommended.

Adults with PWS have generalized mild hypotonia and decreased muscle bulk and tone, poor coordination, and often decreased muscle strength. However, muscle electrophysiological and biopsy studies are generally normal or non-specific. The decreased muscle tone and muscle mass contribute to the lower metabolic rate, leading to physical inactivity and obesity, although thyroid function tests have generally been within normal range but not adequately studied to date.[17]

Sleep disorders and respiratory dysfunction such as hypoventilation and oxygen desaturation are common from childhood to adulthood for the PWS subject. Adolescents with PWS have a tendency to fall asleep during the day, particularly when they are inactive. They do not sleep soundly and may awaken often during the night; some may forage for food.

Behavioral and learning problems may become more prominent during the teenage years, particularly temper tantrums and obsessions. Typical adolescent rebelliousness is often exaggerated in those with PWS, particularly over access to food. Psychotropic agents can be helpful in controlling abnormal behavior, but no specific medication has been universally effective in controlling abnormal behavior or food-seeking behavior. The use of medication to control behavior problems will be discussed in more detail elsewhere (see Chapter 12). It should be noted that children and adults with PWS are typically affectionate and outgoing, like to please others, and seek positive attention.

Typical behavior problems include rigidity of personality, perseveration in conversation, tantrums, obsessive-compulsive symptoms, and noncompliance. Occasionally these worsen during adulthood. In some adults, symptoms of psychosis can be seen. Acute psychosis can be seen in young adulthood in about 10% of PWS patients. Recently, individuals with maternal disomy were reported to have a higher risk of developing psychoses during adulthood.[132] Detailed genotype/phenotype correlations addressing these and other issues will be discussed elsewhere.

When an adult with PWS is surrounded by caregivers who have been trained in the management of PWS, behaviors can be managed. Many persons can become happy members of their community. When the environment is one in which caregivers are not aware of the specific needs of adults with PWS, behavioral deterioration can be expected.

The adult with PWS may have goals for himself or herself similar to those of any other person entering adulthood: establishing vocational goals, deciding where to live, and desiring to become independent in decision-making. This sometimes leads to conflict and frequently causes an exacerbation of health and behavior problems. For most persons with PWS, formal education ends between ages 18 and 21. If vocational training has been successfully introduced before then, a smooth transition to the world can occur. Unfortunately, many individuals do not have this available to them and a gap exists between completion of school and entrance into a job-training program. This loss of daytime structured activities usually results in behavioral deterioration, and health problems may occur or increase. The challenge to family members and caregivers is establishing an appropriate environment for the adult, which includes a supervised living arrangement (with food restrictions), a vocational setting appropriate for the skills and behavior of the young adult, and professional sources knowledgeable about PWS. As in childhood, limiting access to food is essential, as is skill in management of typical behavior problems. If this does not happen, deterioration in health occurs.

Assignment of legal guardianship to another adult is frequently helpful and necessary to assure a safe environment for adults with PWS. Decisions regarding living arrangements and availability of food are usually made unwisely by persons with the syndrome, and therefore it is in the best interest of the person to have guardianship legally assigned to a parent or other adult.

Some persons with PWS live into their seventh decade of life. A person who died at age 71 years was described in 1994,[39] and a second individual at 68 years of age was described in 2000.[18] Undoubtedly, there are other older persons with this syndrome who have not been reported in the literature. A review of causes of death in persons with PWS in 1996 indicated that obesity-hypoventilation syndrome caused the majority of deaths. This review included the experience of seven physicians who cared for 665 persons with the syndrome.[77] Twenty-five deaths had occurred, 14 of them related to obesity complications and the remainder due to several other causes. The average age at death was 23 years. The syndrome does not include any specific internal organ abnormality; therefore, if obesity is controlled and overeating avoided, life expectancy should be similar to that of other individuals with mild mental deficiencies.

Obesity and Related Problems

Children with PWS become overweight by 2 to 4 years of age, without appropriate intervention, as a consequence of overeating due to an insatiable appetite and compulsive behavior related to food. Weight gain worsens over time due to the fact that individuals with PWS require fewer calories (40% to 70% reduction compared with non-obese controls), and it can become life-threatening if not controlled. About one third of subjects with PWS weigh more than 200% of their ideal body weight,[99–101,115] and without intervention significant morbidity and mortality may occur from the complications of obesity such as cardio-pulmonary compromise.

Specific obesity-related findings may be seen in PWS subjects including heart failure, hypertension, thrombophlebitis and chronic leg edema, orthopedic problems, abnormal lipid profiles, and Type 2 diabetes mellitus. Premature development of atherosclerosis with severe coronary artery disease has been reported in PWS subjects.[91] Complications of obesity include skin rashes (particularly in fat folds), ulcers of the skin (most commonly in the lower legs), and cellulitis (particularly in lower extremities). Other obesity-related problems may include obstructive sleep apnea and narrowing of the airway, pathologic fractures, reduced physical activity, impaired respiratory function and hypoventilation, high carbon dioxide levels, specific endocrine disturbances, risks from general anesthesia, and hypometabolism.

The primary health issues in PWS are exacerbated by obesity and related findings. When weight is kept under control, there are few serious health issues. If the weight increases, diseases associated with obesity may appear. For example, Type 2 diabetes mellitus is seen in 25% to 30% of PWS adults who become morbidly obese.[15] Diabetes is difficult to control with medication if food restriction is inadequate. When food intake is reduced to the appropriate number of calories, diabetes usually comes under control quickly. Symptoms of the disease disappear in most cases and blood sugar values become more normal. If this cannot be achieved, complications of diabetes may occur in a few years after the diabetes is detected. These complications may include retinopathy, neuropathy, kidney failure, and amputations. However, insulin resistance is lower in PWS subjects and insulin sensitivity is higher compared with obese individuals without PWS.[123]

Morbid obesity often causes obesity-related hypoventilation, which can be a serious problem and demands attention. Loud snoring, inability to sleep flat in bed, and shortness of breath with minimal exertion become noticeable. Hypoventilation and sleep apnea can reduce the oxygen level in the blood, necessitating pulmonary devices to increase oxygenation during sleep. If this is not reversed, right-sided heart failure (cor pulmonale) may ensue. Rapid weight gain, swelling of the ankles, marked shortness of breath, cyanosis, and decreased activity are signs of cardiac decompensation. Although rapid weight loss and a program of physical exercise can reverse this cascade of events, hospitalization usually becomes necessary (see Chapter 17). A reevaluation of the living environment is essential when this occurs.

Appetite Disorder

As previously stated, obesity is due to several factors such as hyperphagia, a lower metabolic rate, persistent hunger, decreased perception of satiety, and an uncontrollable appetite with impaired emesis.[17,40,84] The intense preoccupation with food, food craving, lack of satiation, and incessant food seeking in PWS are the most striking features of the syndrome. There are clinical reports of individuals with PWS engaging in pica and eating unpalatable food items (e.g., frozen food), although people with PWS generally prefer the same foods as most other people, primarily sweet carbohydrates. It is also clear that they are willing to consume unpalatable items and even engage in pica when food access is sufficiently restricted.

We have previously demonstrated that people with PWS have elevated plasma gamma aminobutyric acid (GABA) levels,[60] possibly upregulated following the loss of GABA receptor subunit genes from the chromosome 15q11–q13 region, with resulting GABA-A receptor dysfunction. GABA acts as a major brain neurotransmitter involved with inhibition. GABA-ergic compounds generally cause disinhibition, leading to even less impulse control.[6] People with PWS should have little impulse control when faced with availability of food, which is commonly claimed by parents and clinicians.

Only recently have studies investigated neuropeptide or endocrine factors contributing to eating behavior.[50] For example, leptin produced by adipose tissue and cholecystokinin produced by the gut are mediators of satiety and eating behavior. However, no significant differences were found in fasting plasma levels for either leptin or cholecystokinin in PWS subjects compared with obese controls.[23,33,109] In addition, no significant differences in plasma levels were found in the two PWS genetic subtypes (typical deletion and maternal disomy). It appears that leptin levels in PWS follow the pattern observed in the normal population and are positively correlated with fat mass. Recently, ghrelin, an endogenous ligand or protein that attaches to the growth hormone secretagogue receptor (a hypothalamic G-protein-coupled receptor) produced by enteroendocrine cells of the stomach, has been found to be involved in energy balance and appetite stimulation. Adults and children with PWS were recently reported to have fasting plasma ghrelin levels three to five times higher than in non-PWS obese or nonobese subjects.[50,78] Infants with PWS as early as 3 months of age also have elevated ghrelin levels.[21] Genes from ghrelin and its receptors are also expressed in different regions of the brain in PWS and control subjects.[124] These observations need confirmation and additional testing before we know their role in hyperphagia and the insatiable appetite seen as cardinal features in PWS.

Metabolism and Energy Expenditure

The morbid obesity associated with PWS is the result of chronic imbalance between energy intake and energy expenditure. It is challenging to objectively measure these two components of energy balance in PWS due to the mental and physical status of these patients, but this is essential to determine the energy expenditure per subject and specific treatment options. Preliminary studies in PWS have attempted to measure energy expenditure. For example, daily energy expenditure was reported to be 47% lower in PWS subjects compared with controls, but this difference was reduced to 14% when allowances were made for differences in fat-free mass. Basal or resting energy expenditure, corrected for fat-free mass, was similar in PWS patients compared with obese controls, suggesting that the low daily energy expenditure in PWS was mainly a result of a reduced fat-free mass (muscle) and possibly a lower level of physical activity. However, electromyograms, motor nerve conduction velocity tests, serum creatine phosphokinase, and results of light microscopy studies of muscles are usually normal in PWS,[84] while specialized histochemistry studies of muscles demonstrate type II muscle fiber atrophy consistent with disuse.[1]

Few metabolic studies are known to explain the obesity in PWS, particularly relating to adipose tissue metabolism. Thyroid hormone, lipid profiles, glucocorticoid, and amino acid levels in PWS are comparable to those in obese individuals without PWS, although reduced glucose tolerance is reported in one fifth of patients with PWS.[11,34,36,75]

In six PWS patients no abnormalities in fat metabolism or transport were identified in response to norepinephrine, insulin, and glucose.[7] While fat cells were found to be larger than in control individuals, uptake of fat and fatty acid composition in adipose tissue was normal and the fat cell number was not increased.[7,69,74] Levels of adipose tissue lipoprotein lipase, an enzyme that regulates uptake and storage of triglycerides, were increased 10–fold in fat biopsy specimens from seven PWS subjects compared with those of control individuals, when adjusted for percent ideal body weight and fat cell size.[117] This enzyme may be elevated in PWS, but additional research is needed.

Serum cholesterol and triglyceride levels are generally thought to be normal in PWS, but fat biopsy specimens reported from nine subjects with PWS showed that triglycerides make up more than 98% of the fat.[105] There was a demonstrated threefold increase in long-chain poly-unsaturated fatty acids, which suggested a resistance to lipolysis or breakdown of fat. In addition, body composition and substrate utiliza-tion studies in 11 PWS subjects showed a higher percentage of adipose tissue, but an apparent normal substrate utilization was found with a normal percentage of fat used for basal metabolism when compared with other obese control individuals.[105] The roles of the elevated long-chain fatty acids and increased levels of adipose tissue lipase in the development of obesity in PWS suggested from fat biopsy specimens are not understood. In addition, the relationship, if any, to the genetic alterations of chromosome 15 is unclear.

Muscle is metabolically active tissue, while adipose or fat stores energy. Therefore, the low muscle mass in PWS significantly contrib-utes to hypotonia, reduced physical activity, and low energy expendi-ture. These factors lead to storage of energy in the form of excess adipose tissue. In one study, PWS subjects expended approximately 50% less energy than healthy obese controls. Daily energy expenditure can be measured using a whole-room calorimeter to record oxygen utilization and carbon dioxide production to determine the metabolic rate and energy expenditure. Lower levels were found in PWS subjects compared with obese controls. However, no information was recorded for differences in fat-free mass or muscle.[37]

Hill, Butler, and others[83] reported resting metabolic rates in lean, obese, and PWS subjects using indirect calorimetry following an over-night fast. The PWS subjects had the lowest resting metabolic rate, fol-lowed by lean then obese subjects. The usual strong relationship observed between resting energy expenditure and fat-free mass was not found in the children studied, compared with non-PWS lean and obese controls. The study further suggested that the initial low rates of energy expenditure in subjects with PWS was independent of fat-free mass, but once patients gained a large amount of weight, the relation-ship appeared to normalize. Recently, Goldstone et al.[70] reported on 13 women with PWS, aged 20 to 38 years, and 45 control women, aged 18 to 56 years, and found differences in resting metabolic rate. Such dif-ferences could be explained by the abnormal body composition in PWS, suggesting that energy expenditure is normal at the tissue level in this syndrome.

Reduced physical activity described in PWS subjects was suggested by Schoeller et al.[115] as a major cause of the decreased energy expenditure requirements. Although subjects with PWS have hypotonia during infancy, which lessens during the first 1 to 2 years of life, a residual amount remains throughout life. This low muscle tone and lack of coordination may favor a sedentary lifestyle. However, few studies have attempted to quantify the levels of physical activity in PWS either in a free setting or in a controlled institutional care setting. It is thought that children with PWS are less physically active during play compared with normal children. To assess physical activity, Nardella et al.[104] studied children with PWS and controls attending a summer camp over a 2–week period by using portable activity meters and pedometers. They reported a wide range of activity levels in subjects with PWS compared with normal children. These results were inconclusive and furthermore did not address the situation in a free setting.

The role of altered energy expenditure and decreased metabolic rate as the causation of obesity, along with excessive caloric intake, decreased energy expenditure, and reduced physical activity leading to morbid obesity in PWS subjects, will require additional research. Growth hormone deficiency and its therapeutic use to stimulate growth, particularly to increase stature and muscle mass while decreasing fat mass and improving pulmonary function in PWS subjects, will require further long-term studies, but significant benefits have been reported. Growth hormone therapy and related endocrine and metabolic issues will be described elsewhere (see Chapter 7).

Onset and Measurement of Obesity

The onset of obesity occurs between ages 1 and 6 years, depending on the definition of obesity, with an average age of onset by 2 years in PWS subjects without growth hormone treatment. This onset follows a period of failure-to-thrive, feeding difficulties, and hypotonia during the neonatal and early infancy period, which is designated as the first stage of PWS. The abnormal feeding noted during infancy may be due to central nervous system or brain dysfunction, particularly an abnormality of the hypothalamus, although gross neuropathologic studies have not identified a consistent brain lesion in this area. However, limited brain imaging studies with positron emission tomography for evaluating brain metabolism have indicated decreased glucose metabolism in the parietal lobes of the brain and hypothalamus in at least one adult with PWS.[29] In addition, recent studies have shown abnormal oxytocin levels in brains of PWS subjects[97] and high plasma levels of GABA,[60] which will require further testing.

Skinfold measurements of PWS infants do show excessive fat at an early age (6 months) and before the infant is judged to be obese by weight/height parameters.[22] The subcutaneous fat pattern is characteristically centrally located over the buttocks, trunk, and thighs but spares the distal regions of the extremities with the use of skinfold measurements.[40,99,100,115] A two- to threefold increased fat mass compared with the general population has been reported, using skinfold measurements to determine percent body fat.[100] This patterning has been substantiated using more current and accurate measures of obesity such as dual-energy X-ray absorptiometry (DXA); for example, 40% to 60% body fat was found using DXA in a recent study, which is about two- to threefold higher than in the general population.[26] There are several methods in use to determine body composition, particularly fat, such as bioelectrical impedance, which measures fat-free mass and fat mass; skinfold thickness, which measures subcutaneous fat mass at various sites; magnetic resonance imaging (MRI) of lean and fat tissues; and DXA (or DEXA), which provides a measure of fat tissue, lean tissue, and bone density. Of the methods listed, DXA is considered the most accurate in determining body composition in PWS or in healthy subjects. A thorough discussion of body composition measurement options is included later in this volume (see Chapter 6).

The fatness pattern appears to be sex-reversed, with males having more fat than females.[99,100] Historically, individuals with PWS were described as morbidly obese with body mass index (BMI, defined as weight in kilograms divided by height in meters squared in adults) reported[11] as high as 47 ± 4.2, compared with a normal BMI range between 21 and 27. There is a tendency for internal or visceral fat areas, as determined by MRI studies, to be lower in PWS subjects compared with obese subjects, as reported by Goldstone et al.[71] Goldstone et al. studied 13 adult females with PWS and 14 obese control women and found significantly reduced visceral fatness or adiposity in the PWS females, which is independent of their total adiposity or use of exogenous sex steroid treatment. This was in contrast to that expected by their physical inactivity, hypogonadism, growth hormone deficiency, and psychiatric problems. However, Talebizadeh and Butler[123] observed an overall trend for decreased visceral fat area but not significantly different in the PWS subjects compared with controls. They observed that PWS subjects with higher visceral fat area may be at an increased risk for obesity-related complications, compared with PWS subjects without an elevated visceral fat area.

The cause of the reduced visceral adiposity in PWS may reflect hormone imbalance, hypothalamic dysfunction, or genetic influence on body fat distribution. The visceral fat is positively correlated with glucose and triglyceride levels in both PWS and obese subjects and contributes to insulin resistance. A different peripheral-visceral fat storage pattern seen in PWS subjects compared with obese controls may account for the abnormal fat storage and lipolysis in PWS. In addition, insulin levels are generally lower in PWS subjects compared with obese individuals.

Weight Management

Weight control through diet restriction, exercise programs, and/or hormone replacement is a constant key management issue throughout the life span of an individual with PWS. Diet is historically the cornerstone for controlling the obesity seen in PWS. Caloric restriction of 6–8 calories/cm of height will be required to allow for weight loss, while 10–12 calories/cm of height usually maintains weight in children with PWS, particularly in those not undergoing growth hormone therapy.

Historically, the calorie requirement to maintain weight is about 60% of normal. A low-calorie, well-balanced diet of 1,000–1,200 kcal/day combined with regular exercise would be advised, but this is difficult to implement in children and adults with PWS due to the insatiable appetite and food-seeking behavior. Supplemental vitamins and calcium are recommended. Unfortunately, no medications or surgical procedures have had long-term effectiveness in controlling appetite and obesity in PWS. An exercise regime should also contribute to weight loss and may improve hypotonia, respiratory problems, and excessive daytime sleepiness and apnea common in PWS subjects. In addition, locking refrigerators and food cupboards may be required to prevent the person with PWS from obtaining additional food and from eating inappropriately. With the increasing use of growth hormone therapy in the pediatric PWS population, weight, stature, and caloric intake require close supervision by a multidisciplinary team.

Surgical procedures such as gastroplasty (reducing the size of the stomach) and intestinal bypass to correct or prevent obesity have met with only limited success in PWS (see Chapter 6). Stomach rupture has been reported as a cause of death in PWS due to overeating,[77] and surgically decreasing the size of the stomach may further increase the risk for the stomach to rupture. Therefore, these surgical procedures are not recommended at this time in subjects with PWS.

As concluded, obesity is the most significant health problem in PWS and is an increasingly common trait affecting one half of the general adult population in the United States.[63,81] It is on the rise in children with dramatic increases during the past decade. Obesity is a risk factor in five of the top 10 causes of death (heart disease, stroke, diabetes, atherosclerosis, and malignancies) in this country. There are several genetic syndromes besides PWS with obesity as a major component. These include: Cohen, Bardet-Biedl, Albright hereditary osteodystrophy, Borjeson-Forssman-Lehmann, Alstrom, Carpenter, and fragile X syndromes; cytogenetic or chromosome syndromes (e.g., 1p36 deletion, Smith-Magenis syndrome, Down syndrome, sex chromosome aneuploidy); and mutations of several obesity-related genes (e.g., leptin, leptin receptor, melanocortin 4–receptor, pro-opiomelanocortin [POMC], prohormone convertase-I, peroxisome proliferator-activated receptor-gamma, and beta-3 adrenergic receptor).[10,37] However, because of the substantial body of research on PWS and a prevalence rate higher than most, PWS is the best example of an obesity syndrome to study in order to gain a better understanding of the genetics of obesity, as well as behavioral dysfunction, and genotype/phenotype correlations.

Other Medical Findings

Bone Density

Other health concerns in individuals with PWS include reduced bone density and osteoporosis, which may lead to fractures.[26,41,114] Bone mineral density is a function of a dynamic process of bone resorption and production of mineralized bone matrix. The degradation of bone matrix or collagen can be determined by measuring urinary N-telopeptides of type I collagen, while total bone mineral content and body composition (e.g., percent fat) can be determined by dual energy X-ray absorptiometry (DXA). Butler et al.[26] reported data on bone density, anthropometric measurements, and biochemical markers of bone turnover in subjects with PWS or simple obesity. Significantly decreased total bone and spine mineral density and total bone mineral content in PWS subjects (ages 10–44 years) were found compared with normative data and similarly aged controls. However, no significant difference in urinary N-telopeptide levels was found between the PWS subjects and obese controls. This suggested a possible lack of depositing bone mineral during growth, when bones should become more dense (e.g., in adolescence), more so than bone loss in the subjects with PWS. This may be due to decreased production of sex or growth hormones and/or long-standing hypotonia with decreased physical activity.

Oral and Dental Issues

Dental or oral findings have received limited study in PWS. Thick, sticky saliva is a consistent finding detectable during the neonatal period in PWS subjects regardless of the genetic subtype. Therefore, the gene(s) causing the salivary problem may be influenced by genetic imprinting. Salivary flow in PWS subjects is approximately 20% of that reported in controls.[79] Therefore, xerostomia or a dry mouth is frequently seen in PWS subjects. The salivary ions and protein are present in increased amounts, which may reflect a concentration effect relative to decreased water in the saliva.

Normal salivation produces a buffering capacity, induces clearance of substances in the mouth by swallowing, protects the teeth from extrinsic and intrinsic acids that initiate dental erosion, and allows for a capacity to remineralize partially demineralized enamel.[139] Therefore, syndromes with salivary dysfunction such as PWS predispose to tooth wear and underline the importance of normal salivation in the protection of teeth against wear by erosion, attrition, and abrasion. Dental malocclusion of varying degrees frequently occurs in PWS and may require orthodontic alignment. An altered craniofacial development and maxillary hypoplasia appears to predispose to dental malocclusion and possibly a narrow upper airway. The role of the dentist is considered important to monitor and treat the dental problems frequently seen in PWS. Additional oral findings are discussed in Chapter 6.

Growth and Growth Hormone

Growth in people with PWS is generally characterized by initial failure-to-thrive, which may require gavage or tube-feeding during stage one of development, followed by a period of normal growth rate with heights below the 50th centile and often below the 10th centile. The lack of a growth spurt during adolescence results ultimately in mild short stature in both males and females. Without growth hormone treatment, adult males with PWS are generally no taller than typical females in relationship to normative data. Bone age is generally delayed but may be normal or advanced. However, observed delays or accelerations of bone age generally return toward normal during adolescence or adulthood.

Intrafamilial and midparental PWS correlations and heritability estimates of anthropometric or physical variables from PWS subjects and their parents were undertaken by Butler et al.[27] in order to determine the effects of the genetic background on several growth parameters in this syndrome. The data suggested that taller parents with longer foot length have taller-than-average PWS children with longer feet. These physical characteristics are apparently influenced by the genetic background, while soft tissue parameters such as arm and calf circumferences and skinfolds showed lower heritability estimates.

Growth hormone deficiency or insufficiency is now recognized as a key feature of PWS and will be discussed in major detail elsewhere. However, one study of children with PWS showed that 98% had subnormal growth hormone levels over a 24–hour period, and 91% displayed subnormal response to growth hormone stimulation.[3] This insufficiency affects both stature and body composition in PWS subjects. Individuals with simple obesity without PWS also show reduced growth hormone secretion during provocative testing when compared with non-obese subjects, but the growth hormone insufficiency in PWS is not considered to be secondary to obesity. Individuals with simple obesity tend to be taller than normal due to elevated levels of another growth-promoting protein produced by the liver called insulin-like growth factor I (IGF-I).[87] IGF-I is low in children with PWS.[96,112]

Early studies in the 1980s of growth hormone replacement in children with PWS provided strong evidence that growth velocity was substantially improved. Treatment with growth hormone doses of 0.125 mg/kg/week showed an increased growth rate of 130%, 300%, and 600% of baseline in 3 of 4 treated children with PWS.[95] This study allowed for larger and more rigorous investigations of PWS subjects, which further delineated the presence of growth hormone insufficiency in children with PWS and provided convincing evidence that growth hormone replacement not only increases stature but also improves lean muscle mass, decreases body fat, increases exercise capacity and physical activity, and improves respiratory function. For example, a 25% reduction in total body fat was achieved with a corresponding 30% increase in fat-free mass, measurable by DXA and bioelectrical impedance, in children with PWS.[96] In addition, some parents reported improved behavior in their child with PWS undergoing growth hormone therapy, although additional long-term studies are needed. The positive changes in body composition (Figure 1.5, see color insert) appear to be due to growth hormone treatment and not due to changes in caloric intake or exercise regime. Potential side effects have been recognized, including the risk for Type 2 diabetes mellitus and worsening of scoliosis in PWS subjects. Growth hormone appears to have a

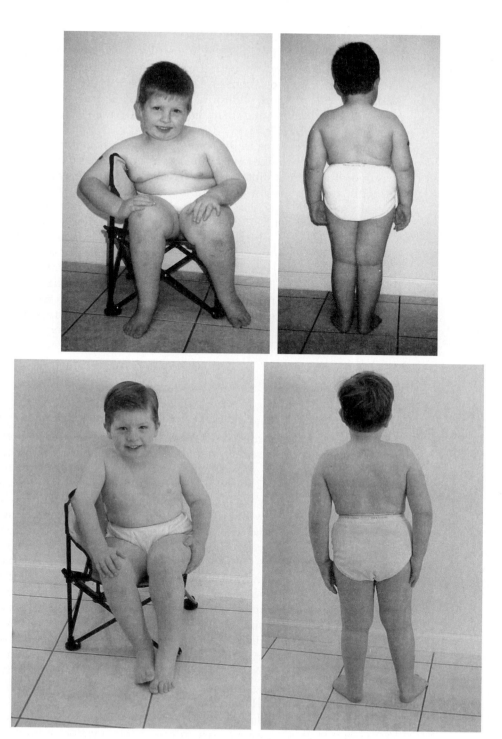

Figure 1.5. Five-year-old boy with Prader-Willi syndrome before (top) and after 3 months of growth hormone (GH) treatment (bottom). Note the improved body habitus, muscle bulk, and reduced fat. The patient was able to increase total caloric intake, and he had increased activity and wakefulness. Reprinted with permission from S.B. Cassidy, "Prader-Willi syndrome in the new millennium: introduction," *The Endocrinologist* 2000;10(4) Suppl 1: 1S–2S. Copyright©2000 by Lippincott Williams & Wilkins. (See color insert.)

high level of safety in the child with PWS and holds promise for potential benefits in adults with the syndrome not previously treated with growth hormone, although there is a paucity of data in adults. Additional information regarding use, benefits, and potential side effects of growth hormone treatment will be described in more detail elsewhere (see Chapter 7).

Many clinical features in PWS support growth hormone deficiency, which has been documented by a low peak growth hormone response to provocative stimulation tests, decreased random growth hormone secretion, and low serum insulin-like growth factor-I (IGF-I) levels in several studies in over 300 affected children.[14] Additionally, 40% to 100% of children with PWS have fulfilled the criteria for growth hormone deficiency, which is generally defined as a peak growth hormone level of less than 10 micrograms/liter in response to one or two stimulation tests. Along with insufficient growth hormone secretion, PWS subjects have a dysfunctional hypothalamic-pituitary-gonadal axis, which may contribute to thermoinstability and a high pain threshold and appears to play a role in the abnormal appetite and incomplete sexual development. This axis and disturbances seen in PWS will be discussed in more detail elsewhere (see Chapter 5).

Blood Studies

Both normal and elevated levels of cholesterol and lipids have been reported in subjects with PWS.[7,11,36,94] Amino acid levels have also been reported to be normal in PWS subjects.[34] For example, Butler et al.[36] reported a comparison study of plasma lipid, cholesterol, glucose, and insulin levels in 26 subjects with Prader-Willi syndrome (deletion and nondeletion) and 32 individuals with simple obesity. The average percentage of ideal body weight (IBW) for the reported PWS group was 175.6 ± 68.0 compared with 150.3 ± 43.8 for the obese subjects. Fasting plasma lipid, glucose, and insulin levels were not significantly different between the two subject groups, but insulin levels were higher than the normative non-obese ranges. However, no significant correlations in lipid levels were found in subjects with PWS (deletion or nondeletion) compared with obese individuals for either age or percentage of IBW. No consistent abnormalities were reported in fasting plasma amino acid or urine organic acid levels in PWS subjects compared with obese controls, and similarly for thyroid hormone, spontaneous cortisol, and ACTH levels. Basal prolactin levels were reported within the normal range in PWS subjects. However, plasma gamma amino-butyric acid (GABA), dehydroepiandrosterone (DHEA), and DHEA-sulfate levels were reportedly elevated in PWS subjects.[14,47,60]

Butler et al.[25] reported a 6–year-old PWS male with maternal disomy 15 and ethylmalonic aciduria, which raised the question of an inborn error of metabolism, particularly a mitochondrial abnormality. Fasting plasma coenzyme Q10, an oxygen scavenger utilized in the mitochondria or the power plant of the cell for the electron transport chain involved in cellular metabolism and energy expenditure, was measured in PWS subjects and compared with obese and non-obese subjects.[24] No significant differences were found in the coenzyme Q10 levels in PWS subjects compared with similarly aged obese subjects; however, lower coenzyme Q10 levels were found in the PWS subjects compared with non-obese controls. The role of the mitochondria in decreased energy expenditure and the lower metabolic rate in PWS will require further testing.

Brain Studies

Although most autopsy studies in PWS subjects have been unremarkable, recent reports have suggested that the paraventricular nucleus (a region of the brain involved in the control of appetite and sexual behavior) may be reduced in size, with fewer oxytocin-expressing neurons. Further studies by Swaab[121] showed a 30% reduction in the growth hormone releasing hormone (GHRH) neurons in the arcuate nucleus, a key brain region for the release of neuropeptides involved in eating behavior. These changes in the hypothalamic region may be sufficient to impair the regulation of food intake and may be the result of a defective protein (or lack of a protein) due to the chromosome 15 abnormality and interference in gene regulation. This protein derangement may interfere with several systems through gene transcription errors, thereby affecting a regulatory protein impacting on neurotransmitter or neurohormone levels. An example is the *SNRPN* gene, which is expressed in the brain and involved with splicing messenger RNA. This paternally expressed gene is found in the 15q11–q13 region and deleted in the majority of PWS subjects or with lack of expression in those with maternal disomy. This loss may impact on the function of the hypothalamus and lead to altered hypothalamic neuron and neuroendocrine function. The differential expression of a second gene called *7B2* in the hypothalamus of subjects with PWS may also be affected by this genetic derangement.[67] However, the gene expression for ghrelin, peptide YY, and their receptors was studied and found to be present in brain tissue from both PWS and control subjects.[124]

The product of gene *7B2* is a neuroendocrine chaperone protein that interacts with prohormone convertase PC2, involved in the regulated secretory pathway in the brain impacting on function. The *7B2* gene is located in the 15q13–q14 region, close to the 15q11–q13 region involved in PWS. Therefore, alteration of the 15q11–q13 region (e.g., deletion) may impact on regulation or activity of the *7B2* gene in PWS subjects, as recently reported.[8,9] Hence, Gabreels et al.[67] studied the presence of the neuroendocrine 7B2 protein in the supraoptic and paraventricular nucleus of the hypothalamus of five subjects with PWS using antibodies against various segments of the 7B2 precursor polypeptide. Three of the five PWS subjects showed no reaction to the 7B2 antibody, MON-12, while 30 control subjects showed a positive reaction. Thus, there was a clear modification of 7B2 expression in some PWS subjects, indicating altered neuroendocrine function. In a similar report, these authors showed no antibody reaction in these brain nuclei against processed vasopressin, another brain-released hormone, but did show reactivity against the vasopressin precursor.[66] Preliminary molecular genetic data on the *7B2* gene in eight individuals with PWS showed abnormalities (M.G. Butler, unpublished data), but additional studies are warranted. In addition, preliminary genome-wide expression using gene microarray technology showed over- and underexpression of dozens of lipid metabolism and neurodevelopmental genes, when comparing brain and somatic tissue of PWS subjects matched with controls (M.G. Butler, unpublished data).[8,9,124]

Positron emission tomography (PET) scans and magnetic resonance imaging (MRI) studies in PWS have revealed disturbances in the hypothalamic region of the brain, suggesting possible dysfunction.[29] Additionally, abnormal cortical development was reported using 3-D MRI in PWS subjects.[138] Specialized proton magnetic resonance spectroscopy (MRS) of the brain in subjects with PWS and MRI images revealed mild abnormalities including slight ventriculomegaly, cortical atrophy, and a small brain stem. Other methods to examine brain chemistry and metabolites include N-acetylaspartate choline (NAA/Cho) and N-acetylaspartate/creatine (NAA/Cr) ratios, which were decreased in subjects with PWS, although the Cho/Cr ratio did not differ from control subjects. Thus, neuron loss or dysfunction was suggested in PWS.[80] Parietal lobe pathology detected on 1H-magnetic resonance spectroscopy may also be associated with more global brain damage and loss of cognitive function.

Cognitive and Behavioral Findings

Decreased intellectual functioning was among the four original defining characteristics of PWS.[55,110,140] IQs have ranged from 12 to 100 in previous studies in PWS,[32,49,59,75,88,140] but the average IQ is typically in the mild range of intellectual disability (55–70 IQ). The distribution includes a few reports within the 85–100 IQ range and other reports within the profound-to-severe range of mental retardation (IQ < 40). Greenswag[72] reported the IQ in a large survey of PWS subjects and found that IQ was greater than 85 in about 5% of subjects, about 25% had borderline mental retardation, 35% had mild retardation, 25% had moderate mental retardation, and about 5% had severe mental retardation.

Although approximately 30% of PWS individuals have IQs in the normal or borderline range, cognitive dysfunction is nearly always present. However, body weight may correlate with IQ in PWS individuals. For example, Crnic and colleagues[49] reported that individuals with PWS who were never obese had significantly higher IQ scores (mean = 80.2) than PWS subjects who were currently obese (mean = 57.3) or had been obese and lost weight (mean = 59.9).

Interestingly, individuals with PWS may have greater-than-expected abilities to recognize and evaluate figures and shapes and strengths in tasks requiring integration of stimuli in a spatial relationship. The reports of superior puzzle-solving ability in PWS individuals would be consistent with this observation.[56,84] Dykens[56] studied children with PWS in order to characterize their presumed ability at jigsaw puzzles. She reported relative strengths on standardized visual-spatial tasks such as object assembly, and the scores were significantly higher in PWS subjects compared with age- and IQ-matched control subjects with mixed mental retardation, but below those of age-matched normal children with average IQs. In contrast, children with Prader-Willi syndrome scored similarly to normal peers on word searches and outperformed them on jigsaw puzzles, placing more than twice as many pieces as the typically developing group.

Warren and Hunt[135] found that children with PWS performed less well on a picture recognition task than children with mental retardation of unknown etiology matched for chronological age and IQ. The two groups of children performed similarly on a task meant to measure access to long-term memory. There are few studies on learning and memory in PWS, although it is thought that children with PWS have a deficit in short-term visual memory but not in long-term visual storage.[135] As mentioned earlier, visual perception, organization, and puzzle-solving skills are reported as relative strengths in some people with PWS, including stronger visual memory in the PWS subjects with maternal disomy compared with those with typical deletions[89] (see section on Genotype-Phenotype Relationships, below).

Early studies have reported that reading abilities may be better than arithmetic abilities in most PWS subjects, although both are deficient.[120] Greenswag[72] reported that 75% of 232 PWS individuals analyzed had received special education services and typically performed at the sixth grade or lower level in reading and at the third grade or lower in mathematics. Reading problems in PWS may be exacerbated by visual perceptual deficits, particularly in the maternal disomy subjects. While PWS individuals with typical deletions usually have a lower verbal IQ, those with maternal disomy exhibit a deficit in visual processing on some tasks, which may offset their greater verbal ability. However, Dykens et al.[59] reported that adolescents and adults with PWS have overall standard academic achievement scores higher than their ability measures. In addition, a study by Kleppe et al.[90] revealed multiple articulation errors (dysarthria), reduced intelligibility, and delayed language skills (vocabulary, syntax, and morphologic abilities) in children with PWS.

Dykens et al.[59] reported that daily living skills may change with increasing age in PWS. Thompson and Butler (unpublished data) found that PWS subjects and IQ-matched controls differed substantially in their degree of independent community living skills. The PWS subjects appeared significantly less competent than controls. While this may indicate biologically-based differences in cognitive ability, it may also reflect the far more restricted lives that most individuals with PWS lead (and therefore more limited opportunities to develop skills) due to the concerns that caregivers have about access to food in uncontrolled settings.

Butler, Thompson, and colleagues[126] collected clinical, genetic, cognitive, academic, and behavioral data from 49 individuals with PWS (22 males, 27 females; 27 with the typical 15q deletion, 21 with maternal disomy 15, and 1 with an imprinting defect), ranging in age from 10 to 50 years, and 27 control participants. We found that PWS participants obtained significantly lower scores than control subjects in performance IQ (p < 0.05), visual-motor skills (p < 0.01), and adaptive functioning as assessed by the broad independence (p < 0.01), community independence (p < 0.01), and motor skill (p < 0.001) dimensions of the Scales of Independent Behavior. No significant differences were found for measures of academic achievement. With respect to maladaptive behaviors, PWS participants demonstrated significantly higher levels of self-injury on the Reiss (p < 0.05) and higher scores on the general maladaptive index (p < 0.05) of the Scales of Independent Behavior, indicating worse behavior.

Behavioral Issues

Several behavioral and psychological findings can be present in individuals with PWS including temper tantrums, stubbornness, hoarding, manipulative behavior, depression, emotional lability, arguing, worrying, compulsive behavior, skin picking, difficulty adapting to new situations, difficulty relating to peers, poor social relationships, low self-esteem, and difficulty in detecting social cues from other people.[40,72,75,84,120,137] Individuals with PWS are generally more verbally aggressive and self-assaultive, but less sexually inappropriate, than study controls. Children with PWS may be more immature and clumsy and more likely to be disliked and teased by their peers.[131] However, they also can be caring and affectionate. Not all individuals with PWS will have the behavior problems that have been reported. A further description and discussion of the behavioral characteristics will be presented elsewhere (see Part III).

Self-Injurious Behavior

Skin picking and other forms of self-abusive behavior are found at some time in most persons with this syndrome and also can be seen in individuals with autism and other related developmental disorders. Infections in the involved skin areas can occur. Occasionally, an area may be picked for several years. Serious health problems from persistent self-injury may occur and include eye poking, subdural hemorrhage from forceful head banging, infections from self-inflicted skin picking, and anorectal disease from rectal picking and digging.[5] Studies may reveal one or more ulcers in the rectum, which can be confused with inflammatory bowel disease.

Self-injurious behaviors can be among the most clinically problematic behaviors associated with Prader-Willi syndrome. Previous studies have found self-injurious behavior (most notably skin picking) to be a prevalent behavioral problem in 69% of adolescents with Prader-Willi syndrome[137] and in 81% of adults.[129] Symons et al.[122] surveyed families of 62 persons (24 males, 38 females; mean age of 18 years; age range 3 to 44 years) with Prader-Willi syndrome and determined the prevalence, topographies, and specific body locations of self-injurious behavior. Self-injury was reported for 81% of the participants, with over 800 self-injury body sites recorded. Skin picking was the most prevalent form of self-injury (82%), followed by nose picking (28%), hand biting (17%), head banging (14%), hair pulling (9%), and rectal picking (6%).

The front of the legs and head were disproportionately targeted as preferred self-injury body sites.

Among school-age students with mental retardation, Symons et al.[122] found that the most common forms of self-injury were biting, head banging, and hitting, whereas picking and pinching were more common among the PWS subjects. Non-PWS individuals with mental retardation tend to direct self-injury toward the head, while PWS subjects tend to distribute self-injury widely across the body including specific finger, arm, and leg areas. Other forms of self-abusive behavior have been reported, including trichotillomania, pushing pins and tacks into skin, and pulling out nails.[5] Onset of these behaviors is variable, usually during childhood but sometimes not until adolescence or early adulthood.

The etiology and pathophysiological mechanisms of self-injury are poorly understood. The primary pharmacological approaches for treating self-injury in persons with mental retardation and related developmental disabilities are aimed at the dopaminergic system (i.e., neuroleptics), the serotonergic systems (i.e., serotonin reuptake inhibitors), or the endogenous opioid peptide systems (i.e., opiate antagonists).[6,128] Some pharmacological approaches to treating self-injury among persons with Prader-Willi syndrome have been useful.[119,134]

Compulsive Behavior

Dykens et al.[58] reported that compulsive symptoms are found in up to 60% of persons studied with PWS. Obsessive-compulsive disorder (OCD) [32,72] symptoms usually begin in adolescence or adulthood in the non-PWS population, and the compulsive behaviors are believed to reduce or prevent anxiety. A large body of evidence has implicated serotonergic mechanisms as underlying the clinical manifestations of OCD in other neuropsychiatric populations.[103] Pharmacologic agents that inhibit serotonin reuptake (e.g., clomipramine) have produced clinical improvements as well as changes in peripheral measures of serotonin function,[62] and use of these agents has improved compulsivity and reduced OCD symptoms in some individuals with PWS.[51,82] Similarities in the number, type, and severity of compulsive behaviors in adults with PWS and those with OCD have been reported.[59] However, compulsive behavior in typically developing children decreases from 2 to 6 years, while compulsivity increases in children with PWS during this age range.[52,53] There appears to be a temporal correlation between appetite onset and compulsivity and tantrums in children with PWS. While people with PWS have compulsive disorder that overlaps with that seen in other conditions (e.g., OCD, autism), it may also involve different mechanisms.

On the Yale-Brown Obsessive Compulsive Scale (Y-BOCS), Butler, Thompson, and colleagues[126] found group differences between PWS and control participants for the total ($p < 0.05$) and compulsions ($p < 0.01$) scores, as well as for specific aspects of severity of compulsions: time spent performing compulsions ($p < 0.01$), length of compulsion-free intervals ($p < 0.01$), interference with daily activities ($p < 0.01$), and degree of control over compulsions ($p < 0.01$). In contrast, significant group differences were not obtained for any Yale-Brown Obsessive Compulsive Scale obsession scores, or for internal aspects of compulsive behaviors, such as the individual's effort to resist against compulsions.

It appears that people with PWS have a compulsive disorder related, but probably not identical, to compulsive disorder seen in other conditions such as autism, Tourette's syndrome, and obsessive-compulsive disorder alone. We further speculate that lack of GABA-ergic inhibition of dopaminergic and serotonergic neurons in the orbitofrontal and prefrontal cortex and head of the caudate nucleus may be implicated in the compulsive symptoms seen due to the following: (1) the presence of three GABA-A receptor genes (alpha 5, beta 3, gamma 3) in the chromosome 15q11-q13 region; (2) our findings of elevated levels of GABA in plasma of individuals with PWS and Angelman syndrome (as previously indicated, possibly reflecting up-regulation of GABA due to improper binding at GABA receptors); and (3) recent gene expression studies with paternal bias of the GABA receptor genes.[8,9] There are reports that psychotropic medications reduce obsessive-compulsive disorder symptoms in other people and tend to reduce tantrums and compulsive behavior in people with PWS,[51, 119, 134] but this area has been poorly studied in PWS.

Genotype-Phenotype Relationships

Clinical, Behavior, and Cognition

Phenotypic clinical differences among individuals with Prader-Willi syndrome having separate genetic subtypes (typical deletion, maternal disomy, or imprinting defect) are important to study to learn about the role of specific genes and their clinical outcomes. The correlation of specific clinical manifestations and genetic findings will enhance our understanding of genomic imprinting and genotype/phenotype correlations in PWS. The rarity of individuals with imprinting defects has limited their comparison to those individuals with typical deletions or maternal disomy for genetic and clinical correlation studies.

The phenotypic spectrum of PWS is quite variable and may be dependent on the genetic subtype. In addition, genotype/phenotype differences may be helpful in guiding the clinician in the evaluation of patients with suspected PWS and in providing prognostic counseling for families once the diagnosis (and specific genetic subtype) of PWS is established.

Hypopigmentation (i.e., lighter hair, eye, and skin colors compared with other family members at a similar age) has been noted to occur at a higher frequency in PWS subjects with the typical chromosome 15 deletion[16,17,32] and is found to be associated with a deletion of the *P* gene, which is involved in pigment production, localized at the distal end of the chromosome 15q11–q13 region. Molecular genetic studies have shown that hypopigmentation does not correlate with the DNA haplotype pattern in the region of the *P* gene.[118] In addition, the *P* gene is not considered to be imprinted and is expressed in both chromosome 15s but deleted from one member of the pair in PWS (and AS) subjects with the typical 15q11–q13 chromosome deletion, leading to a decreased amount of pigment.[8,9]

Previous studies have revealed that individuals with PWS with the chromosome 15 deletion were more homogeneous than other PWS subjects in their clinical presentation and anthropometric or physical measurements (e.g., radiographic measurements of the bones of the hand), as reflected in the metacarpophalangeal pattern profile.[28] Those with the 15q deletion (males and females) also were more homogeneous than nondeletion subjects with respect to dermatoglyphic plantar findings of the foot, with a lack of plantar interdigital II-IV patterns with almost exclusively hallucal distal loops.[111] Early studies were conducted before the recognition of maternal disomy 15 as the most common cause of the nondeletion status among people with PWS.

With the advent of molecular testing for all subjects with PWS, several studies have been reported to compare the major genetic subtypes (deletion and maternal disomy) with clinical findings. For example, Gillessen-Kaesbach et al.[68] noted lower birth weights in individuals with PWS and the typical deletion subtype, while an increased maternal age was found in the case of those with maternal disomy. This latter observation was consistent with the source of the extra chromosome 15 from the mother during egg production causing trisomy 15, or three chromosome 15s in the fetus, followed by loss of the chromosome 15 donated from the father during early pregnancy, leading to maternal disomy 15 in the fetus at birth. Mitchell et al.[102] also reported a shorter birth length in PWS males with maternal disomy, compared with males having the 15q deletion, and a shorter course of gavage feeding with a later onset of hyperphagia in PWS females with maternal disomy. In addition, Cassidy et al.[44] observed that people with PWS and maternal disomy were less likely to have the typical facial appearance and less likely to show certain behavioral features of PWS, including skin picking, skill with jigsaw puzzles, a high pain threshold, and articulation problems. No significant differences were found between the groups (deletion and maternal disomy) in most other clinical findings including neonatal hypotonia, need for gavage feeding, cryptorchidism, genital hypoplasia, small hands and feet, scoliosis, dental anomalies, sticky saliva, behavioral disturbances, hyperphagia, decreased vomiting, or sleep disorder. Gunay-Aygun et al.[73] reported that the diagnosis of PWS among individuals with maternal disomy was typically reported later than among those with a deletion, possibly due to a milder phenotype in maternal disomy subjects.

People with PWS may hoard and arrange items excessively. Dykens et al.[57] described differences between the genetic subtypes in PWS subjects such that the deletion group had higher scores using the Child Behavior Checklist, indicating more compulsive symptoms and more symptom-related distress. In addition, Symons et al. characterized the self-injurious behavior in 62 PWS subjects via a questionnaire survey.[122] PWS individuals with the typical 15q11-q13 deletion injured at significantly more body sites than did individuals with maternal disomy 15. Skin picking was the most common form of self-injury. Thompson and

Butler[126] found that PWS subjects with typical deletions exhibited significantly greater self-injury than both a control group and a maternal disomy subgroup. The typical deletion subgroup also displayed higher compulsivity scores than the control group and spent more time engaging in compulsive behavior. Their compulsive rituals interfered with their daily living, and they were less able to control their compulsive behavior. Both PWS genetic subgroups showed significantly greater global severity of compulsive behaviors than the control group, while the typical deletion group consistently showed the most severe symptoms of OCD. Subjects with maternal disomy showed an intermediate level of OCD symptoms. Additional studies will be needed to confirm this observation. In addition, Vogels et al.[132] recently reported that psychoses in Prader-Willi syndrome subjects occur more often in those adults with maternal disomy 15 compared with those with the deletion.

Subgroup comparisons in a study by Roof, Butler, Thompson and others[113] revealed additional differences between PWS deletion and maternal disomy participants in measures of intelligence and academic achievement administered to 38 individuals with PWS (16 males and 22 females; 24 with deletion and 14 with maternal disomy). PWS subjects with maternal disomy 15 had significantly higher verbal IQ scores than those with the deletion (p < 0.01). The magnitude of difference in verbal IQ was 9.1 points (69.9 versus 60.8 for maternal disomy and deletion PWS subjects, respectively). Only 17% of the subjects with the 15q11-q13 deletion had a verbal IQ ≥ 70, whereas 50% of those with maternal disomy had a verbal IQ ≥ 70. However, performance IQ scores did not differ between the two PWS genetic subtype groups (62.2 versus 64.7 for maternal disomy and deletion PWS subjects, respectively). The full scale IQ did not differ significantly between the two groups (64.1 versus 61.0 for maternal disomy and deletion, respectively). Specific subtest differences were noted in numeric calculation skill, attention, word meanings, factual knowledge, and social reasoning, with the maternal disomy PWS subgroup scoring higher than the typical deletion subgroup. Deletion PWS subgroup subjects scored higher than the maternal disomy subgroup on the object assembly subtest, which further supports specific visual-perceptual skills being a relative strength for the deletion subgroup. This may explain anecdotal accounts of subjects with PWS having an uncanny ability to assemble jigsaw puzzles and the study reported by Dykens.[56] Dykens found that, within Prader-Willi syndrome, puzzle proficiency was not predicted by age, IQ, gender, degree of obesity, or obsessive-compulsive symptoms but by the genetic status (specifically higher in the deletion subgroup).

The mechanisms whereby certain skills appear to be preserved in the maternal disomy subgroup have yet to be identified. Whether this phenomenon is caused by genomic imprinting versus the nonimprinting status of genes in the 15q11-q13 region is not known. The presence of more active or expressed genes in maternal disomy individuals may be due to possible dosage mechanisms. If only one allele, or member of a gene pair, is normally expressed (i.e., one allele is active on the mother's chromosome 15 but inactive on the father's chromosome 15), and there are two active copies instead of one copy (e.g., maternal disomy 15), then more gene product is produced, which may be advantageous. Evidence to date further documents the difference between verbal and performance IQ score patterns among subjects with PWS and the deletion versus the maternal disomy subtype.

Visual Perception and Visual Memory

Discrimination of shape of motion testing was performed by Fox et al.[65] wherein forms were generated by random dot elements that varied in element density and temporal correlation. This testing was done in four participant groups (PWS deletion, PWS maternal disomy, comparison subjects, and normal controls). The procedure uses white dots presenting on a computer monitor that blink on and off randomly. Imbedded within the randomly presented dots is a fixed array of dots with a defined form (e.g., the letter "E"), moving slowly from side to side. The array may vary in element density and degree of correlation of the blinking dots. Performances of normal controls exceeded that of all other groups (78% correct, $p < 0.009$). The typical PWS deletion (66%) and equivalent controls (59%) did not differ significantly. However, performance of the maternal disomy group was significantly worse (38%) than any of the other groups ($p < 0.04$). The inferior performance of the maternal disomy group may be attributed to receiving two active alleles of maternally expressed genes influencing development of the visual system. Other possibilities include the requirements of paternally expressed genes, residual mosaic trisomy in brain tissue, or complex interactions including specific ratios of differentially spliced gene products.[65] Alternatively, since we know people with PWS have elevated plasma GABA, and it has been shown from other studies that excessive GABA levels have deleterious effects on retinal functioning, it is possible that visual signal strengths could be compromised at the level of initial input, which would manifest itself as a perceptual deficit.

Joseph et al.[89] reported substantial differences in visual-spatial memory among PWS individuals with maternal disomy compared with those with deletions. The rate of short-term memory decay among individuals with maternal disomy was considerably slower than in either PWS individuals with typical deletions or matched controls. The study involved 17 individuals with PWS—7 with deletions, 10 with maternal disomy—and 9 matched controls. Each participant performed a visual recognition task. A series of color digital photographs was presented; most were presented twice, and the remainder appeared only once. Photographs presented twice were separated by 0, 10, 30, 50, or 100 intervening photographs. After viewing each photograph, participants indicated whether or not the photograph had been presented previously. This procedure was conducted twice, once using photographs of foods and the second time using non-food items. As the number of intervening photographs increased between the first and second presentation, participants were less likely to remember having seen the photograph previously. Performance by the maternal disomy participants was less affected by increasing the number of intervening photographs relative to the other two groups (deletion and non-PWS controls), suggesting superior visual recognition memory.

Phenotypes Associated with Longer Versus Shorter Typical Deletions

The majority of people with PWS have a paternally derived interstitial deletion of the 15q11–q13 chromosome region including 3.5 million to 4 million base pairs of DNA. Two proximal breakpoints (BP1 and BP2) where the typical deletion occurs have been reported in this region. Type I deletions (involving BP1) are larger than type II deletions (involving BP2) by about 500 kilobases of DNA.[130] Clinical, anthropometric, and behavioral data were analyzed in 12 PWS subjects (5 males, 7 females; mean age 25.9 ± 8.8 years) with type I deletion and 14 PWS subjects (6 males, 8 females; mean age 19.6 ± 6.5 years) with type II deletion, determined by presence or absence of DNA markers between BP1 and BP2.[20] PWS subjects with the longer typical deletions scored significantly higher in tests measuring self-injurious and maladaptive behaviors compared with PWS subjects with shorter typical deletions. In addition, obsessive-compulsive behavior was more evident in PWS subjects with longer deletions. It appears that loss of genetic material between breakpoints BP1 and BP2 significantly increases the severity of behavioral and psychological problems in this syndrome. Four genes have been identified and located between BP1 and BP2.[46] They may play a role in brain development or function accounting for our observed clinical differences. One of those genes, *NIPA-1*, is expressed in brain tissue and will require further investigation to determine its role in causing specific clinical findings in PWS.

Conclusions

Prader-Willi syndrome is a prototypic contiguous gene disorder with several genes in the 15q11–q13 region contributing to the phenotype. PWS and Angelman syndrome were the first examples in humans of genomic imprinting, or due to the difference of genetic expression depending on the parent of origin. Significant behavioral differences distinguish the two major genotypes (typical deletion versus maternal disomy). Generally, those with typical deletions of the entire region have had the most severe behavioral phenotype with more skin picking, lower verbal IQ, and hypopigmentation. Those with maternal disomy are more impaired on visual perception but have superior visual recognition memory. We have recently found that among those with typical deletions, there are two subtypes, a longer (type I) and shorter (type II) deletion. The phenotypes of individuals with the longer deletion are more severe in general than those with shorter deletions. Within the region between the longer and shorter deletion breakpoints is the *NIPA-1* gene, which is expressed in brain tissue and may be implicated in these differences.[20]

Many features in persons with PWS suggest a hypothalamic dysfunction: hyperphagia, sleep disorders, deficient growth hormone secretion, and hypogonadism. In addition, children with damage to the hypothalamus such as following craniopharyngioma brain tumor surgery can show features similar to PWS including an increased appetite, obesity, learning and behavior problems, a decreased growth rate, and endocrine disturbances. This relationship should be further pursued in brain pathology and neuroanatomical studies including brain imaging (e.g., functional MRI and optical topography).

A multidisciplinary approach is needed to treat individuals with PWS, regardless of the age of the patient. Primary care physicians such as pediatricians, family physicians, or internists should be able to treat most patients with PWS in consultation with a clinical geneticist, endocrinologist, dietitian, and other experts as needed. Additional information about PWS can be obtained by contacting a local genetics center and through the national and local Prader-Willi syndrome associations.

Most individuals with this syndrome can be healthy if diagnosed early and a treatment plan is in place to avoid the complications of uncontrolled obesity. Additional needs depend on the overall health of the child and the age at diagnosis. Exercises to increase coordination, balance, and strength are important but should be kept simple and low in number at the beginning and gradually increase over time. Consultation with the patient's doctor and physical therapist is recommended before undertaking an exercise program at home or in the school setting. A better understanding of energy expenditure and energy balance in PWS is clearly needed.

There are currently no consistent, well-established behavioral or psychological management methods for overcoming all behavioral problems associated with PWS. The greatest success appears to be obtained by capitalizing upon the inherent compulsivity of people with PWS in devising highly predictable daily routines, the use of behavior management reward systems to promote exercise and diet control, and providing choice whenever feasible within the structured daily schedule. Tasks to improve social skills such as taking turns or working together with peers may be useful. These tasks should be incorporated into the classroom setting by working closely with the school educators and administrators.

Selective serotonin reuptake inhibitors may be helpful in reducing tantrums, while atypical neuroleptics that produce minimal weight gain may be of some value in treating aggressive outbursts that may occur in some PWS individuals. Skin picking appears to be regulated by a different neurochemical mechanism than other compulsive symptoms, which may implicate the lack of GABA-ergic inhibition at the brain level. No systematic trials of GABA agonists have been reported to date. Pharmacological treatments for weight control and behavior problems have met with only modest results. A better understanding of genome-wide microarray expression in brain and peripheral tissues and recognition of disturbed interconnected gene pathways in PWS may lead to additional treatment modalities.

It is often difficult but vital to find appropriate services to meet the needs of the immediate and extended family as well as teachers. Since nearly one third of the people with PWS have a low normal IQ, they often do not qualify for mental retardation services yet are poorly served within most psychiatric programs. Many parents and advocacy organizations have developed specialized residential programs for young adults with PWS to meet this need.

With recent approval of growth hormone for treating individuals with PWS, many children are now on therapy and positive results have been reported, specifically with an increased muscle mass and strength, improved respiratory function, decreased fat mass, increased physical activity and energy expenditure, and taller stature. Potential side effects relating to growth hormone treatment are under long-term study in the PWS population. PWS subjects may have scoliosis, which could be exacerbated by rapid growth in children with or without PWS regardless of growth hormone treatment. Therefore, checking and monitoring for scoliosis should be performed by their physician on a regular basis before and while receiving growth hormone. Growth hormone treatment in non-PWS populations may be associated with an increased risk for slipped capital femoral epiphysis, a condition that is associated with obesity. Growth hormone treatment also may be associated with an increased incidence of pseudotumor cerebri. Thus, careful monitoring for these findings should be routine in following children with PWS.

Obesity is a major health problem in PWS with an increased risk for Type 2 diabetes mellitus in individuals with PWS and other complications related to obesity. Moreover, growth hormone treatment may decrease insulin sensitivity and further increase the risk for non-insulin-dependent diabetes mellitus. Therefore, all children with PWS and obesity should be carefully monitored for glucose intolerance and diabetes mellitus regardless of growth hormone treatment.

It is reasonable to believe that the improvement in body composition noted in children with PWS on growth hormone therapy should lower the risk for co-morbid diseases (e.g., diabetes, high blood pressure, cardiovascular disease). A long life span with a better quality of life would be anticipated. There is a high probability that the growth hormone/insulin growth factor axis deficiency seen in children with PWS is also present in adults. Therefore, adults may also benefit from growth hormone therapy, but there is a paucity of data in treating adults with PWS at this time. Understanding the genetic cause and pathophysiology of Prader-Willi syndrome should allow for better treatment options for the future and a better quality of life for those affected with this condition.

References

1. Afifi AK, Zellweger H. Pathology of muscular hypotonia in the Prader-Willi syndrome. *Journal of Neurological Sciences.* 1969;9:49–61.
2. Akefeldt A, Tornhage CJ, Gillberg C. A woman with Prader-Willi syndrome gives birth to a healthy baby girl. *Developmental Medicine and Child Neurology.* 1999;41(11):789–790.

3. Angulo M, Castro-Magana M, Mazur B, Canas JA, Vitollo PM, Sarrantonio M. Growth hormone secretion and effects of growth hormone therapy on growth velocity and weight gain in children with Prader-Willi syndrome. *Journal of Pediatric Endocrinology and Metabolism.* 1996;9(3):393–400.

4. Arens R, Omlin KJ, Livingston FR, Liu J, Keens TG, Ward SL. Hypoxic and hypercapnic ventilatory responses in Prader-Willi syndrome. *Journal of Applied Physiology.* 1994;77:2224–2230.

5. Barghava SA, Putnam PE, Kocoshis SA, Rowe M, Hanchett JM. Rectal bleeding in Prader-Willi syndrome. *Pediatrics.* 1996;97(2):265–267.

6. Barron J, Sandman CA. Paradoxical excitement to sedative-hypnotics in mentally retarded clients. *American Journal of Mental Deficiency.* 1985;90(2):124–129.

7. Bier DM, Kaplan SL, Havel RJ. The Prader-Willi syndrome: regulation of fat transport. *Diabetes.* 1977;26:874–881.

8. Bittel DC, Kibiryeva N, Talebizadeh Z, Butler MG. Microarray analysis of gene/transcript expression in Prader-Willi syndrome: deletion versus UPD. *Journal of Medical Genetics.* 2003;40:568–574.

9. Bittel DC, Kibiryeva N, Talebizadeh Z, Driscoll DJ, Butler MG. Microarray analysis of gene/transcript expression in Angelman syndrome: deletion versus UPD. *Genomics.* 2005;85:85–91.

10. Bray GA. Obesity illustrated. In: Bray GA, ed. *An Atlas of Obesity and Weight Control.* London: Parthenon Publishing Group; 2003.

11. Bray GA, Dahms WT, Swerdloff RS, Fiser RH, Atkinson RL, Carrel RE. The Prader-Willi syndrome. *Medicine.* 1983;62:59–80.

12. Buiting K, Färber C, Kroisel P, et al. Imprinting centre deletions in two PWS families: implications for diagnostic testing and genetic counseling. *Clinical Genetics.* 2000;58:284–290.

13. Buiting K, Groß S, Lich C, Gillessen-Kaesbach G, El-Maarri O, Horsthemke B. Epimutations in Prader-Willi and Angelman syndrome: a molecular study of 136 patients with an imprinting defect. *American Journal of Human Genetics.* 2003;72:571–577.

14. Burman P, Ritzén EM, Lindgren AC. Endocrine dysfunction in Prader-Willi syndrome: a review with special reference to GH. *Endocrine Reviews.* 2001;22(6):787–799.

15. Butler JV, Whittington JE, Holland AJ, Boer H, Clarke D, Webb T. Prevalance of, and risk factors for, physical ill-health in people with Prader-Willi syndrome: a population-based study. *Developmental Medicine and Child Neurology.* 2002;44(4):248–255.

16. Butler MG. Hypopigmentation: a common feature of Prader-Labhart-Willi syndrome. *American Journal of Human Genetics.* 1989;45:140–146.

17. Butler MG. Prader-Willi syndrome: current understanding of cause and diagnosis. *American Journal of Medical Genetics.* 1990;35(3):319–332.

18. Butler MG. A 68-year-old white female with Prader-Willi syndrome. *Clinical Dysmorphology.* 2000;9(1):65–67.

19. Butler MG. Imprinting disorders: non-Mendelian mechanisms affecting growth. *Journal of Pediatric Endocrinology and Metabolism.* 2002;5:1279–1288.

20. Butler MG, Bittel DC, Kibiryeva N, Talebizadeh Z, Thompson T. Behavioral differences among subjects with Prader-Willi syndrome and type I or type II deletion and maternal disomy. *Pediatrics.* 2004;113(3 Pt1): 565–573.

21. Butler MG, Bittel DC, Talebizadeh Z. Plasma peptide YY and ghrelin levels in infants and children with Prader-Willi syndrome. *Journal of Pediatric Endocrinology and Metabolism.* 2004;17(9):1177–1184.

22. Butler MG, Butler RI, Meaney FJ. The use of skinfold measurements to judge obesity during the early phase of Prader-Labhart-Willi syndrome. *International Journal of Obesity.* 1988;12:417–422.

23. Butler MG, Carlson MG, Schmidt DE, Feurer ID, Thompson T. Plasma cholecystokinin levels in Prader-Willi syndrome and obese subjects. *American Journal of Medical Genetics.* 2000;95(1):67–70.

24. Butler MG, Dasouki M, Bittel D, Hunter S, Naini A, DiMauro S. Coenzyme Q10 levels in Prader-Willi syndrome: comparison with obese and non-obese controls. *American Journal of Medical Genetics.* 2003;119A(2):168–171.

25. Butler MG, Dasouki M, Hunter S, Gregersen N, Naini A, DiMauro S. A child with Prader-Willi syndrome and ethylmalonic aciduria. 4th International Prader-Willi Syndrome Scientific Conference, St. Paul, MN; June 27, 2001.

26. Butler MG, Haber L, Mernaugh R, Carlson MG, Price R, Feurer ID. Decreased bone mineral density in Prader-Willi syndrome: comparison with obese subjects. *American Journal of Medical Genetics.* 2001;103:216–222.

27. Butler MG, Haynes JL, Meaney FJ. Intra-familial and mid-parental child correlations and heritability estimates of anthropometric measurements in Prader-Willi syndrome families. *Dysmorphology and Clinical Genetics.* 1990;4:2–6.

28. Butler MG, Kaler SG, Meaney FJ. Metacarpophalangeal pattern profile analysis in Prader-Willi syndrome. *Clinical Genetics.* 1982;22:315–320.

29. Butler MG, Kessler RM. Positron emission tomography of three adult patients with Prader-Willi syndrome. *Dysmorphology and Clinical Genetics.* 1992;6:30–31.

30. Butler MG, Meaney FJ. An anthropometric study of 38 individuals with Prader-Labhart-Willi syndrome. *American Journal of Medical Genetics.* 1987;26:445–455.

31. Butler MG, Meaney FJ. Standards for selected anthropometric measurements in Prader-Willi syndrome. *Pediatrics.* 1991;88:853–860.

32. Butler MG, Meaney FJ, Palmer CG. Clinical and cytogenetic survey of 39 individuals with Prader-Labhart-Willi syndrome. *American Journal of Medical Genetics.* 1986;23(3):793–809.

33. Butler MG, Moore J, Morawiecki A, Nicolson M. Comparison of leptin levels in Prader-Willi syndrome and control individuals. *American Journal of Medical Genetics.* 1998;75(1):7–12.

34. Butler MG, Murrell JE, Greene HL. Amino acid levels in Prader-Willi syndrome and obese individuals. *Dysmorphology and Clinical Genetics.* 1990;4:18–22.

35. Butler MG, Palmer CG. Parental origin of chromosome 15 deletion in Prader-Willi syndrome. *Lancet.* 1983;1:1285–1286.

36. Butler MG, Swift LL, Hill JO. Fasting plasma lipid, glucose and insulin levels in Prader-Willi syndrome and obese individuals. *Dysmorphology and Clinical Genetics.* 1990;4:23–26.

37. Butler MG, Thompson T. Prader-Willi syndrome: clinical and genetic findings. *The Endocrinologist.* 2000;10:3S-16S.

38. Butler MG, Weaver DD, Meaney FJ. Prader-Willi syndrome: are there population differences? *Clinical Genetics.* 1982;22(5):292–294.

39. Carpenter PK. Prader-Willi syndrome in old age. *Journal of Intellectual Disabilities Research.* 1994;38:529–531.

40. Cassidy SB. Prader-Willi syndrome. *Current Problems in Pediatrics.* 1984;14:1–55.

41. Cassidy SB. Prader-Willi syndrome: characteristics, management, and etiology. *The Alabama Journal of Medical Sciences.* 1987;24(2):169–175.

42. Cassidy SB. Prader-Willi syndrome. *Journal of Medical Genetics.* 1997;34(11): 917–923.

43. Cassidy SB, Dykens E, Williams CA. Prader-Willi and Angelman syndromes: sister imprinted disorders. *American Journal of Medical Genetics.* 2000;97(2):136–146.

44. Cassidy SB, Forsythe M, Heeger S, et al. Comparison of phenotype between patients with Prader-Willi syndrome due to deletion 15q and uniparental disomy 15. *American Journal of Medical Genetics.* 1997;68: 433–440.

45. Cassidy SB, Rubin KG, Mukaida CS. Genital abnormalities and hypogonadism in 105 patients with Prader-Willi syndrome. *American Journal of Medical Genetics.* 1987;28:922–923.

46. Chai JH, Locke DP, Eichler EE, Nicholls RD. Evolutionary transposition of 4 unique genes mediated by flanking duplicons in the Prader-Willi/ Angelman syndrome deletion region. *American Journal of Human Genetics.* 2002;71:A395.

47. Chasalow FI, Blethen SL, Tobash JG, Myles D, Butler MG. Steroid metabolic disturbances in Prader-Willi syndrome. *American Journal of Medical Genetics.* 1987;28:857–864.

48. Creel DJ, Bendel CM, Wiesner GL, Wirtschafter JD, Arthur DC, King RA. Abnormalities of the central visual pathways in Prader-Willi syndrome associated with hypopigmentation. *New England Journal of Medicine.* 1986;314:1606–1609.

49. Crnic KA, Sulzbacher S, Snow J, Holm VA. Preventing mental retardation associated with gross obesity in the Prader-Willi syndrome. *Pediatrics.* 1980;66(5):787–789.

50. Cummings DE, Clement K, Purnell JQ, et al. Elevated plasma ghrelin levels in Prader-Willi syndrome. *Nature Medicine.* 2002;8(7):643–644.

51. Dech B, Budow L. The use of fluoxetine in an adolescent with Prader-Willi syndrome. *Journal of the American Academy of Child and Adolescent Psychiatry.* 1991;30(2):298–302.

52. Dimitropoulos A, Feurer ID, Butler MG, Thompson T. Emergence of compulsive behavior and tantrums in children with Prader-Willi syndrome. *American Journal of Mental Retardation.* 2000;106(1):39–51.

53. Dimitropoulos A, Feurer ID, Roof E, et al. Appetitive behavior, compulsivity, and neurochemistry in Prader-Willi syndrome. *Mental Retardation and Developmental Disabilities Research Review.* 2001;6(2):125–130.

54. Down JL. In: *Mental Affections of Childhood and Youth.* London: Churchill Publisher; 1887:172.

55. Dunn HG. The Prader-Labhart-Willi syndrome: review of the literature and report of nine cases. *Acta Paediatrica Scandinavica Supplement.* 1968;186:1–9.

56. Dykens EM. Are jigsaw puzzle skills "spared" in persons with Prader-Willi syndrome? *Journal of Child Psychology and Psychiatry.* 2002; 43(3):343–352.

57. Dykens EM, Cassidy SB, King BH. Maladaptive behavior differences in Prader-Willi syndrome; due to paternal deletion versus maternal uniparental disomy. *American Journal of Mental Retardation.* 1999;104: 67–77.

58. Dykens EM, Leckman JF, Cassidy SB. Obsessions and compulsions in Prader-Willi syndrome. *Journal of Child Psychology and Psychiatry.* 1996;37: 995–1002.

59. Dykens EM, Hodapp RM, Walsh K, Nash LJ. Profiles, correlates, and trajectories of intelligence in Prader-Willi syndrome. *Journal of the American Academy of Child and Adolescent Psychiatry.* 1992;31(6):1125–1130.

60. Ebert MH, Schmidt DE, Thompson T, Butler MG. Elevated plasma gamma-aminobutyric acid (GABA) levels in individuals with either Prader-Willi syndrome or Angelman syndrome. *The Journal of Neuropsychiatry and Clinical Neurosciences.* 1997;9(1):75–80.

61. Feurer ID, Dimitropoulos A, Stone WL, Butler MG, Thompson T. The latent variable structure of the Compulsive Behavior Checklist in people with Prader-Willi syndrome. *Journal of Intellectual Disability Research.* 1998;42:472–480.

62. Flament MF, Rapoport JL, Murphy DL, Beng CJ, Lake CR. Biochemical changes during clomipramine treatment of childhood obsessive-compulsive disorder. *Archives of General Psychiatry.* 1987;44(3):219–225.

63. Flegal KM, Carroll MD, Kucamarski RJ. Overweight and obesity in the United States: prevalence and trends. *International Journal of Obesity and Related Metabolic Disorders.* 1998;22:39–47.

64. Fox R, Butler MG, Sinatra RB. Visual capacity and Prader-Willi syndrome. *Journal of Pediatric Ophthalmology and Strabismus.* 1999;36:1–7.

65. Fox R, Yang GS, Feurer ID, Butler MG, Thompson T. Kinetic form discrimination in Prader-Willi syndrome. *Journal of Intellectual Disability Research.* 2001;45:317–325.

66. Gabreels BA, Swaab DF, de Kleijn DP, et al. Attenuation of the polypeptide 7B2, prohormone convertase PC2, and vasopressin in the hypothalamus of some Prader-Willi patients: indications for a processing defect. *Journal of Clinical Endocrinology and Metabolism.* 1998;83(2):591–599.

67. Gabreels BA, Swaab DF, Seidah NG, van Duijnhoven HL, Martens GJ, van Leeuwen FW. Differential expression of the neuroendocrine polypeptide 7B2 in hypothalamus of Prader-(Labhart)-Willi syndrome patients. *Brain Research.* 1994;657(1–2):281–293.

68. Gillessen-Kaesbach G, Robinson W, Lohmann D, Kaya-Westerloh S, Passarge E, Horsthemke B. Genotype-phenotype correlation in a series of 167 deletion and non-deletion patients with Prader-Willi syndrome. *Human Genetics.* 1995;96(6):638–643.

69. Ginsberg-Fellner F. Growth of adipose tissue in infants, children and adolescents: variations in growth disorders. *International Journal of Obesity.* 1981;5(6):605–611.

70. Goldstone AP, Brynes AE, Thomas EL, et al. Resting metabolic rate, plasma leptin concentrations, leptin receptor expression, and adipose tissue measured by whole-body magnetic resonance imaging in women with Prader-Willi syndrome. *American Journal of Clinical Nutrition.* 2002;75(3): 468–475.

71. Goldstone AP, Thomas EL, Brynes AE, et al. Visceral adipose tissue and metabolic complications of obesity are reduced in Prader-Willi syndrome female adults: evidence for novel influences on body fat distribution. *Journal of Clinical Endocrinology and Metabolism.* 2001;86(9):4430–4338.

72. Greenswag LR. Adults with Prader-Willi syndrome: a survey of 232 cases. *Developmental Medicine and Child Neurology.* 1987;29(2):145–152.

73. Gunay-Aygun M, Heeger S, Schwartz S, Cassidy SB. Delayed diagnosis in patients with Prader-Willi syndrome due to maternal uniparental disomy 15. *American Journal of Medical Genetics.* 1997;71(1):106–110.

74. Gurr MI, Jung RT, Robinson MP, James WP. Adipose tissue cellularity in man: the relationship between fat cell size and number, the mass and distribution of body fat and the history of weight gain and loss. *International Journal of Obesity.* 1982;6(5):419–436.

75. Hall BD, Smith DW. Prader-Willi syndrome. A resume of 32 cases including an instance of affected first cousins, one of whom is of normal stature and intelligence. *Journal of Pediatrics.* 1972;81(2):286–293.

76. Hanchett JM. Menstrual periods in Prader-Willi syndrome women. *American Journal of Medical Genetics.* 1996;64:577.

77. Hanchett JM, Butler M, Cassidy SB, et al. Age and causes of death in Prader-Willi syndrome patients. *American Journal of Medical Genetics.* 1996;2:211.

78. Haqq AM, Farooqi IS, O'Rahilly S, et al. Serum ghrelin levels are inversely correlated with body mass index, age, and insulin concentrations in normal children and are markedly increased in Prader-Willi syndrome. *Journal of Clinical Endocrinology and Metabolism.* 2003;88(1):174–178.

79. Hart PS. Salivary abnormalities in Prader-Willi syndrome. *Annals of the New York Academy of Sciences.* 1998;842:125–131.

80. Hashimoto T, Mori K, Yoneda Y, et al. Proton magnetic resonance spectroscopy of the brain in patients with Prader-Willi syndrome. *Pediatric Neurology.* 1998;18(1):30–35.

81. Hedley AA, Ogden CL, Johnson CL, Carroll MD, Curtin LR, Flegal KM. Prevalence of overweight and obesity among U.S. children, adolescents, and adults, 1999–2002. *Journal of the American Medical Association.* 2004; 291(23):2847–2850.

82. Hellings JA, Warnock JK. Self-injurious behavior and serotonin in Prader-Willi syndrome. *Psychopharmacology Bulletin.* 1994;30:245–250.

83. Hill JO, Kaler M, Spetalnick B, Reed G, Butler MG. Resting metabolic rate in Prader-Willi syndrome. *Dysmorphology and Clinical Genetics.* 1990;4: 27–32.

84. Holm VA. The diagnosis of Prader-Willi syndrome. In: Holm VA, Sulzbacher SJ, Pipes PL, eds. *Prader-Willi Syndrome.* Baltimore, MD: University Park Press; 1981:27–44.

85. Holm VA, Cassidy SB, Butler MG, Hanchett JM, Greenberg F. Prader-Willi syndrome: consensus diagnostic criteria. *Pediatrics.* 1993;91:398–402.

86. Hudgins L, Geer JS, Cassidy SB. Phenotypic differences in African Americans with Prader-Willi syndrome. *Genetics in Medicine.* 1998;1(3):49–51.

87. Iranmenesh A, Lizarralde G, Veldhuis JD. Age and relative adiposity are specific determinants of the frequency and amplitude of growth hormone (GH) secretion bursts and the half-life of endogenous GH in healthy men. *Journal of Clinical Endocrinology.* 1991;73:1081–1088.

88. Jancar J. Prader-Willi syndrome (hypotonia, obesity, hypogonadism, growth and mental retardation). *Journal of Mental Deficiency Research.* 1971;15(1):20–29.

89. Joseph B, Egli M, Sutcliffe JS, Thompson T. Possible dosage effect of maternally expressed genes on visual recognition memory in Prader-Willi syndrome. *American Journal of Medical Genetics.* 2001;105(1):71–75.

90. Kleppe SA, Katsyama KM, Shipley KG, Foushee DR. The speech and language characteristics of children with Prader-Willi syndrome. *Journal of Speech and Hearing Disorders.* 1990;55(2):300–309.

91. Lamb AS, Johnson WM. Premature coronary artery atherosclerosis in a patient with Prader-Willi syndrome. *American Journal of Medical Genetics.* 1987;28(4):873–880.

92. Laxova R, Gilderdale S, Ridler MA. An aetiological study of 53 female patients from subnormality hospital and of their offspring. *Journal of Mental Deficiency Research.* 1973;17:193–225.

93. Ledbetter DH, Riccardi VM, Airhart SD. Deletions of chromosome 15 as a cause of the Prader-Willi syndrome. *New England Journal of Medicine.* 1981;304(6):325–329.

94. Lee PDK. Endocrine and metabolic aspects of Prader-Willi syndrome. In: Greenswag LR, Alexander RC, eds. *Management of Prader-Willi Syndrome.* 2nd ed. New York, NY: Springer-Verlag; 1995.

95. Lee PDK, Wilson DM, Hintz RL, Rosenfield RG. Growth hormone treatment of short stature in Prader-Willi syndrome. *Journal of Pediatric Endocrinology.* 1987;2:31–34.

96. Lindgren AC, Hagenas L, Muller J, et al. Growth hormone treatment of children with Prader-Willi syndrome affects linear growth and body composition favourably. *Acta Paediatrica.* 1998;87:28–31.

97. Martin A, State M, Anderson GM, et al. Cerebrospinal fluid levels of oxytocin in Prader-Willi syndrome: a preliminary report. *Biology and Psychiatry.* 1998;44(12):1349–1352.

98. Meaney FJ, Butler MG. Craniofacial variation and growth in the Prader-Labhart-Willi syndrome. *American Journal of Physical Anthropology.* 1987;74(4):459–464.

99. Meaney FJ, Butler MG. Assessment of body composition in Prader-Labhart-Willi syndrome. *Clinical Genetics.* 1989;35(4):300.

100. Meaney FJ, Butler MG. Characterization of obesity in the Prader-Labhardt-Willi syndrome: fatness patterning. *Medical Anthropology Quarterly.* 1989;3:294–305.

101. Meaney FJ, Butler MG. The developing role of anthropologists in medical genetics: anthropometric assessment of the Prader-Labhardt-Willi syndrome as an illustration. *Medical Anthropology.* 1989;10:247–253.

102. Mitchell J, Schinzel A, Langlois S, et al. Comparison of phenotype in uniparental disomy and deletion Prader-Willi syndrome: sex specific differences. *American Journal of Medical Genetics.* 1996;65(2):133–136.

103. Murphy DL, Zohar J, Benkelfat C, Pato MT, Pigott TA, Insel TR. Obsessive-compulsive disorder as a 5-HT subsystem-related behavioral disorder. *British Journal of Psychiatry Supplement.* 1989;8:15–24.

104. Nardella MT, Sulzbacher S, Worthington-Roberts BS. Activity levels of persons with Prader-Willi syndrome. *American Journal of Mental Deficiency.* 1983;87:498–505.

105. Nelson RA, Huse DM, Holman RT, et al. Nutrition, metabolism, body composition, and response to the ketogenic diet in Prader-Willi syndrome. In: Holm VA, Sulzbacher S, Pipes PL, eds. *Prader-Willi Syndrome.* Baltimore, MD: University Park Press; 1981:105–120.

106. Nicholls RD, Knepper, JL. Genome organization, function, and imprinting in Prader-Willi and Angelman syndromes. *Annual Review of Genomics and Human Genetics.* 2001;2:153–175.

107. Nicholls RD, Knoll JHM, Butler MG, Karum S, Lalande M. Genetic imprinting suggested by maternal heterodisomy in nondeletion Prader-Willi syndrome. *Nature.* 1989;342:281–285.

108. Nixon GM, Brouillette RT. Sleep and breathing in Prader-Willi syndrome. *Pediatrics and Pulmonology.* 2002;34:209–217.

109. Pietrobelli A, Allison DB, Faith MS, et al. Prader-Willi syndrome: relationship of adiposity to plasma leptin levels. *Obesity Research.* 1998;6(3):196–201.

110. Prader A, Labhart A, Willi H. Ein syndrom von adipositas, kleinwuchs, kryptorchismus und oligophrenie nach myatonieartigem zustand im neugeborenenalter. *Schweizerische Medizinische Wochenschrift.* 1956;86:1260–1261.

111. Reed T, Butler MG. Dermatologic features in Prader-Willi syndrome with respect to chromosomal findings. *Clinical Genetics.* 1984;25:341–346.

112. Ritzen EM, Bolme P, Hall K. Endocrine physiology and therapy in Prader-Willi syndrome. In: Cassidy SB, ed. *Prader-Willi Syndrome and Other*

Chromosome 15q Deletion Disorders. New York, NY: Springer-Verlag; 1992: 153–169.

113. Roof E, Stone W, MacLean W, Feurer ID, Thompson T, Butler MG. Intellectual characteristics of Prader-Willi syndrome: comparison of genetic subtypes. *Journal of Intellectual Disabilities Research.* 2000;44:1–6.

114. Rubin K, Cassidy SB. Hypogonadism and osteoporosis. In: Greenswag LR, Alexander RC, eds. *Management of Prader-Willi Syndrome.* New York, NY: Springer; 1988.

115. Schoeller DA, Levitsky LL, Bandini LG, Dietz WW, Walczak A. Energy expenditure and body composition in Prader-Willi syndrome. *Metabolism.* 1988;37(2):115–120.

116. Schulze A, Mogensen, H, Hamborg-Petersen B, Graem N, Ostergaard JR, Brodum-Nielsen K. Fertility in Prader-Willi syndrome: a case report with Angelman syndrome in the offspring. *Acta Paediatrica.* 2001; 90(4):455–459.

117. Schwartz RS, Brunzell JD, Bierman EL. Elevated adipose tissue lipoprotein lipase in the pathogenesis of obesity in Prader-Willi syndrome. In: Holm VA, Sulzbacher S, Pipes PL, eds. *Prader-Willi Syndrome.* Baltimore, MD: University Park Press; 1981:137–143.

118. Spritz RA, Bailin T, Nicholls RD, et al. Hypopigmentation in Prader-Willi syndrome correlates with P gene deletion but not with haplotype of the hemizygous P allelle. *American Journal of Medical Genetics.* 1997;71: 57–62.

119. Stein DJ, Keating J, Zar HJ. A survey of the phenomenon and pharmacotherapy of compulsive and impulsive-aggressive symptoms in Prader-Willi syndrome. *Journal of Neuropsychiatry.* 1994;6:23–29.

120. Sulzbacher S, Crnic KA, Snow J. Behavior and cognitive disabilities in Prader-Willi syndrome. In: Holm VA, Sulzbacher SJ, Pipes PL, eds. *Prader-Willi Syndrome.* Baltimore, MD: University Park Press; 1981:147–159.

121. Swaab DF. Prader-Willi syndrome and the hypothalamus. *Acta Paediatrica Supplement.* 1997;423:50–54.

122. Symons FJ, Butler MG, Sanders MD, Feurer ID, Thompson T. Self-injurious behavior and Prader-Willi syndrome: behavioral forms and body locations. *American Journal of Mental Retardation.* 1999;104:260–269.

123. Talebizadeh Z, Butler MG. Insulin resistance and obesity-related factors in Prader-Willi syndrome: comparison with obese subjects. *Clinical Genetics.* 2004;67:230–239.

124. Talebizadeh Z, Kibiryeva N, Bittel DC, Butler MG. Ghrelin, peptide YY and their receptors: gene expression in brain subjects with and without Prader-Willi syndrome. *International Journal of Molecular Medicine.* 2005; 15:707–711.

125. Taylor RL. Cognitive and behavioral characteristics. In: Caldwell ML, Taylor RL, eds. *Prader-Willi Syndrome: Selected Research and Management Issues.* 2nd ed. New York, NY: Springer-Verlag; 1988.

126. Thompson T, Butler MG. Prader-Willi syndrome: clinical, behavioral and genetic findings. In: Wolraich ML, ed. *Disorders of Development and Learning.* 3rd ed. Hamilton, Ontario: B.C. Decker, Inc.; 2003.

127. Thompson T, Butler MG, MacLean WE, Joseph B. Prader-Willi syndrome: genetics and behavior. *Peabody Journal of Education.* 1996;71:187–212.

128. Thompson T, Gray DB. Destructive behavior in developmental disabilities: diagnosis and treatment. In: Thompson T, Gray DB, eds. *Behavior in Developmental Disabilities.* Thousand Oaks, CA: Sage Publishers; 1994.

129. Thornton L, Dawson KP. Prader-Willi syndrome in New Zealand: a survey of 36 affected people. *New Zealand Medical Journal.* 1990;103 (885):97–98.

130. Ungaro P, Christian SL, Fantes JA, et al. Molecular characterization of four cases of intrachromosomal triplication of chromosome 15q11–q14. *Journal of Medical Genetics*. 2001;38(1):26–34.

131. van Lieshout CF, De Meyer RE, Curfs LM, Fryns JP. Family contexts, parental behavior, and personality profiles of children and adolescents with Prader-Willi, fragile X, or Williams syndrome. *Journal of Child Psychology and Psychiatry*. 1998;39(5):699–710.

132. Vogels A, Matthijs G, Legius E, Devriendt K, Fryns JP. Chromosome 15 maternal uniparental disomy and psychosis in Prader-Willi syndrome. *Journal of Medical Genetics*. 2003;40(1):72–73.

133. Warnock JK, Clayton AH, Shaw HA, O'Donnell T. Onset of menses in two adult patients with Prader-Willi syndrome treated with fluoxetine. *Psychopharmacology Bulletin*. 1995;11:239–242.

134. Warnock JK, Kestenbaum T. Pharmacologic treatment of severe skin-picking behaviors in Prader-Willi syndrome. Two case reports. *Archives of Dermatology*. 1992;128(12):1623–1625.

135. Warren J, Hunt E. Cognitive processing in children with Prader-Willi syndrome. In: Holm VA, Sulzbacher SJ, Pipes PL, eds. *Prader-Willi Syndrome*. Baltimore, MD: University Park Press; 1981:161–178.

136. Wharton RH, Loechner KJ. Genetic and clinical advances in Prader-Willi syndrome. *Current Opinions in Pediatrics*. 1996;8:618–624.

137. Whitman BY, Accardo P. Emotional symptoms in Prader-Willi syndrome adolescents. *American Journal of Medical Genetics*. 1987;28(4):897–905.

138. Yoshii A, Krishnamoorthy KS, Grant PE. Abnormal cortical development shown by 3D MRI in Prader-Willi syndrome. *Neurology*. 2002;59(4): 644–645.

139. Young W, Khan F, Brandt R, Savage N, Razek AA, Huang Q. Syndromes with salivary dysfunction predispose to tooth wear: case reports of congenital dysfunction of major salivary glands, Prader-Willi, congenital rubella, and Sjogren's syndromes. *Oral Surgery, Oral Medicine, Oral Pathology, Oral Radiology, and Endodontics*. 2001;92(1):38–48.

140. Zellweger H, Schneider HJ. Syndrome of hypotonia-hypomentia-hypogonadism-obesity (HHHO) or Prader-Willi syndrome. *American Journal of Diseases of Children*. 1968;115(5):588–598.

2

Diagnostic Criteria for Prader-Willi Syndrome

Shawn E. McCandless and Suzanne B. Cassidy

Diagnostic criteria for Prader-Willi syndrome (PWS) were originally established in 1993 by consensus of a group of highly experienced physicians and psychologists.[7] These criteria were developed to carefully establish the basis for diagnosis to allow for accurate management and genetic counseling, and to ensure uniform clinical diagnosis for future molecular investigations in defining the cause of the syndrome and development of diagnostic tests. As diagnostic testing with FISH (fluorescence *in situ* hybridization) and other molecular techniques became a reality, the validity of the criteria was confirmed.[3,4,5,8,13,18] At the same time, the need for precise clinical diagnostic criteria was superseded by the need for guidelines to identify appropriate indications for molecular testing.[6] Today, this is the most important use of diagnostic criteria. This chapter will review the diagnostic criteria, evaluate the current status of these criteria, and review approaches to diagnosis in light of the natural history of the disorder. The goal of diagnostic criteria now should be to maximize early identification while limiting costs for unnecessary testing.

Diagnosing Prader-Willi Syndrome Using Diagnostic Criteria

The diagnostic criteria developed in 1993 through a consensus process are shown in Table 2.1.[7] Clinical diagnosis is based on a scoring system that assigns one point each for major criteria and one half point for minor criteria. The diagnosis is confirmed in an individual of 3 years or older when at least five major criteria are present and there is a total of at least eight points. Because many of the diagnostic criteria are not present in the first several years of life, diagnosis in a child under 3 years of age requires the presence of only four major criteria and at least five total points.

Experience demonstrated that the diagnostic criteria allowed for confidence that the diagnosis was correct but that they did not confirm the diagnosis in some individuals who truly did have the disorder.[5] This

Table 2.1. Consensus Diagnostic Criteria for Prader-Willi Syndrome*

Major criteria

1. Neonatal and infantile central hypotonia with poor suck, gradually improving with age
2. Feeding problems in infancy with need for special feeding techniques and poor weight gain/ failure to thrive
3. Excessive or rapid weight gain on weight-for-length chart (excessive is defined as crossing two centile channels) after 12 months but before 6 years of age; central obesity in the absence of intervention
4. Characteristic facial features with dolichocephaly in infancy, narrow face or bifrontal diameter, almond-shaped eyes, small-appearing mouth with thin upper lip, down-turned corners of the mouth (3 or more required)
5. Hypogonadism—with any of the following, depending on age:
   a. Genital hypoplasia (male: scrotal hypoplasia, cryptorchidism, small penis and/or testes for age [<5th percentile]; female: absence or severe hypoplasia of labia minora and/or clitoris)
   b. Delayed or incomplete gonadal maturation with delayed pubertal signs in the absence of intervention after 16 years of age (male: small gonads, decreased facial and body hair, lack of voice change; female: amenorrhea/oligomenorrhea after age 16)
6. Global developmental delay in a child younger than 6 years of age; mild to moderate mental retardation or learning problems in older children
7. Hyperphagia/food foraging/obsession with food
8. Deletion 15q11–q13 on high resolution (>650 bands) or other cytogenetic/molecular abnormality of the Prader-Willi chromosome region, including maternal disomy

Minor criteria

1. Decreased fetal movement or infantile lethargy or weak cry in infancy, improving with age
2. Characteristic behavior problems—temper tantrums, violent outbursts and obsessive/ compulsive behavior; tendency to be argumentative, oppositional, rigid, manipulative, possessive, and stubborn; perseverating, stealing, and lying (5 or more of these symptoms required)
3. Sleep disturbance or sleep apnea
4. Short stature for genetic background by age 15 (in the absence of growth hormone intervention)
5. Hypopigmentation—fair skin and hair compared to family
6. Small hands (<25th percentile) and/or feet (<10th percentile) for height age
7. Narrow hands with straight ulnar border
8. Eye abnormalities (esotropia, myopia)
9. Thick, viscous saliva with crusting at the corners of the mouth
10. Speech articulation defects
11. Skin picking

Supportive findings (increase the certainty of diagnosis but are not scored)

1. High pain threshold
2. Decreased vomiting
3. Temperature instability in infancy or altered temperature sensitivity in older children and adults
4. Scoliosis and/or kyphosis
5. Early adrenarche
6. Osteoporosis
7. Unusual skill with jigsaw puzzles
8. Normal neuromuscular studies

* Scoring: Major criteria are weighted at one point each. Minor criteria are weighted at one half point. Children 3 years of age or younger: Five points are required for diagnosis, four of which should come from the major group. Children 3 years of age to adulthood: Total score of eight is necessary for the diagnosis. Major criteria must comprise five or more points of the total score.

Source: V. A. Holm et al., "Prader-Willi syndrome: consensus diagnostic criteria."[7] Reproduced with permission from *Pediatrics*, Vol. 91(2), p. 399, Copyright ©1993 by the AAP.

was further confirmed in a critical review of the diagnostic criteria by Gunay-Aygun et al.,[6] who showed that 15 of 90 patients with a molecularly confirmed diagnosis of PWS did not meet the requirements of the diagnostic criteria. The authors pointed out that the criteria were developed to precisely identify affected individuals for counseling and for involvement in studies to develop molecular tests. Now that highly sensitive and specific molecular testing is available, the purpose of diagnostic criteria has shifted to identifying those individuals for whom diagnostic testing is indicated.[6] Therefore, new criteria were suggested to prompt specific genetic testing for PWS (see Table 2.2). Other authors[18] have identified "core" criteria, the absence of any one of which strongly predicts negative molecular genetic testing. These criteria are: a weak suck in the neonatal period, a weak cry and reduced activity in the neonatal period, absence or rarity of vomiting, and thick saliva. Again, the presence of these criteria is age dependent, limiting their usefulness in the very young patient.

The new criteria shown in Table 2.2 take into account the evolving clinical course of the disorder. Recognition of the natural history of a condition is especially important now that specific management strategies are known to alter the long-term consequences of the disorder. Other chapters in this book will document the importance of early diagnosis in maximizing potential growth and performance in individuals with PWS while minimizing long-term complications.

As noted previously, there are two distinct phases recognized in the life of an individual with PWS. The findings associated with these two phases affect the likelihood that an individual with PWS will satisfy

Table 2.2. Proposed Revised Criteria To Prompt Diagnostic Testing for Prader-Willi Syndrome

Age	Features Suggesting PWS
Birth to 2 years	Hypotonia and weak suck
Early childhood	Appropriate neonatal history of hypotonia and weak suck Global developmental delay
Later childhood	Appropriate neonatal history of hypotonia and weak suck Global developmental delay Excessive appetite and lack of satiety (hyperphagia), obesity (if food not limited)
Adult	Mild or borderline mental retardation Excessive appetite and lack of satiety (hyperphagia), obesity (if food not limited) Hypothalamic hypogonadism Typical behavior (especially obsessive-compulsive features, skin picking, or temper tantrums)

Adapted from M. Gunay-Aygun et al., *Pediatrics*, 2001;108(5):E92.[6]

the diagnostic criteria. Phase 1 begins in the uterine environment with decreased fetal movement and a high rate of malposition (breech), leading to the need for uterine manipulation (e.g., external version to move the breech fetus to a headfirst position for delivery) and surgical delivery (Caesarian section). In the newborn and in early infancy there is profound hypotonia, excessive sleepiness, failure to wake for feeding, and lack of crying. This is generally accompanied by a poor or weak suck, difficulty latching on to the breast leading to high failure rate for breast-feeding, poor intake from a bottle, and poor weight gain. Most infants require tube feedings for weeks to months, rarely persisting beyond 6 to 9 months, in order to avoid or minimize the poor weight gain.

Of the eight major diagnostic criteria put forward by the consensus group,[7] only four relate to clinical findings found in the first 2 years of life (numbers 1, 2, 4 and 5 in Table 2.1). Another major criterion, presence of a cytogenetic abnormality, may be found in some but is unlikely to be found unless the diagnosis of PWS is already being considered. Therefore, the diagnostic criteria are much less useful in the first 2 years of life, at the time when the proper diagnosis is critical for early intervention to prevent complications. This recognition of the limitations of the original diagnostic criteria has prompted the current recommendation that testing be considered for all infants with hypotonia and a poor suck (Table 2.2). PWS appears to be one of the most common identifiable causes of marked neonatal hypotonia. This was shown in a German study that identified PWS, confirmed molecularly, in 29 of 65 hypotonic infants tested.[5] This was subsequently confirmed by Richer et al.[16] in an 11-year retrospective study of newborns admitted to the neonatal intensive care unit of a children's hospital, which showed that PWS was second only to hypoxia as a cause of hypotonia (6 of 33 patients with central hypotonia).

The differential diagnosis in the first phase of neonatal hypotonia includes a wide variety of other chromosomal rearrangements that are mostly detectable by high resolution chromosome analysis or subtelomeric FISH in addition to spinal muscular atrophy type I (Werdnig-Hoffmann disease), congenital myotonic dystrophy, peroxisome biogenesis disorders (e.g., Zellweger syndrome), congenital central nervous system malformations or anoxic injury, and rarely benign congenital hypotonia.[1] Many other disorders may present with congenital hypotonia, most of which are easily distinguished by other clinical findings. The differential diagnosis of infantile hypotonia is beyond the scope of this chapter.

By the later part of the first year and early second year of life, the feeding difficulties are usually resolving, the infant begins to have better weight gain, and there is slow but steady developmental progress. This "honeymoon period" may last 2 or 3 years before the onset of the second phase of the clinical course of PWS, the hyperphagic phase. At this time recognition of significant developmental delay may bring the child with PWS to medical attention, although there is a tendency to delay diagnostic testing at this point because the feeding problems and hypotonia appear to be improving relative to the first year of life.

The second distinct phase in the life of a person with PWS is characterized by the onset of excessive appetite, or more accurately, lack of satisfaction of hunger after eating. This is accompanied by a failure to progress beyond several normal behavioral phases of the 2-year-old, including the need for routines in the schedule, difficulty with transitions between activities, and rapidly escalating loss of emotional control (tantrums) during disagreements or disappointments. At the same time there may be recognition of unusual food-related behaviors such as foraging or hoarding food. Often there is obsessive thinking about food and verbal repetition of questions about food (e.g., "When will it be time for lunch?"). The combination of rapid weight gain and behavioral issues often brings the child to medical attention and suggests the diagnosis of PWS to the medical provider. Unfortunately, this often occurs after the child has become massively obese. It is not uncommon to see the plot of the child's weight on a standard growth curve go from below the 3rd percentile to above the 95th percentile in less than a year.

At this time, the consensus clinical diagnostic criteria are more likely to be positive. The hypogonadism is clearly present, though it may be overlooked by physicians, especially in prepubertal females. It is generally more obvious in males than females because of the high rate of undescended or nonpalpable testes and small penis. The female external genitalia may be underdeveloped, with very small labia minora and clitoris. If not appreciated in childhood, the hypogonadism becomes readily apparent in late adolescence with the lack or arrest of development of secondary sexual characteristics. Interestingly, there is often premature adrenarche, with development of fine pubic hair, underarm hair, odoriferous sweat, and sometimes acne.

The characteristic facial appearance is often more readily distinguished in later childhood as well, particularly in the presence of obesity. Anecdotally, it appears that treatment with growth hormone may alter the facial appearance somewhat, making diagnosis more difficult based on that finding. The presence of the minor criteria of short stature and small hands and feet may also be altered by therapy with growth hormone.

Later in childhood the global developmental delay becomes more obvious as the motor issues begin to resolve. In the school-age child specific learning strengths and weaknesses may be recognized on the background of global cognitive dysfunction, leading the informed educator to raise the possibility of the diagnosis.

How Sensitive and Specific Are the Diagnostic Criteria?

The consensus diagnostic criteria are highly specific for the diagnosis in later childhood. There are few convincing reports in the literature of individuals who clearly meet the diagnostic criteria but have normal molecular studies,[18] but it is possible that injury to the developing hypothalamus or pituitary could cause some of the typical PWS findings (e.g., Cushing syndrome). Experience suggests that these patients can be distinguished from molecularly confirmed patients with PWS on the basis of clinical grounds, although we are aware of no confirmation studies.

The published consensus diagnostic criteria are less sensitive for molecularly confirmed subjects with PWS. Gunay-Aygun et al.[6] found that 15 of 90 (16.7%) patients confirmed by methylation testing or FISH studies did not meet consensus diagnostic criteria. These included 14 subjects with deletions and 1 subject with uniparental maternal disomy, and all were over 3 years of age. Other authors have found better sensitivity.[3] Some investigators have tried to refine the diagnostic criteria retrospectively by identifying those features found in all confirmed subjects.[18] These so-called core criteria have not been evaluated prospectively. What is clear from all of these studies, though, is the importance of the characteristic neonatal findings of hypotonia and poor feeding. Absence of these findings makes the likelihood of molecular confirmation of PWS very low, although validation studies are needed.

Another interesting finding by Gunay-Aygun et al.[6] is the recognition that many of the so-called minor diagnostic features of the consensus criteria (see Table 2.1) are actually more sensitive in identifying PWS than some of the major criteria. Specifically, the characteristic facial appearance is not highly sensitive for the diagnosis of PWS, nor does it appear to be highly specific.

It seems most appropriate, in light of the highly sensitive and specific molecular genetic testing available, to use clinical indicators to help select individuals for laboratory testing. The newer proposed indications, as suggested by Gunay-Aygun et al.,[6] are shown in Table 2.2.

Several conditions other than PWS are associated with obesity, hyperphagia, developmental delay, and hypotonia in older children and adults. Among the more commonly seen are Albright hereditary osteodystrophy (Online Mendelian Inheritance in Man[15] #103580), fragile X syndrome (OMIM #309550), Bardet-Biedl syndrome (OMIM #209900), Alstrom syndrome (OMIM #203800), Cohen syndrome (OMIM #216550), Beckwith-Wiedemann syndrome (OMIM #130650), and other chromosomal rearrangements. Generally, these disorders can be distinguished clinically, although in some cases molecular testing to rule out PWS may be indicated in the process of making another diagnosis of exclusion (e.g., some cases of Cohen syndrome). There are other less common syndromes involving obesity and developmental abnormalities that also should be distinguished, but because they are less readily recognizable they may require confirmation that PWS is not the correct diagnosis. These include MOMO (macrosomia, obesity, macrocephaly, ocular abnormalities) syndrome (OMIM #157980), Urban-Rogers-Meyer syndrome (OMIM #264010), and syndromes reported by Camera et al.[2] and Vasquez et al.[17] In addition, as with neonatal hypotonia, there are a number of chromosomal abnormalities associated with these findings that are detectable by high resolution chromosome analysis and subtelomeric FISH.

Role of Diagnostic Criteria in the Laboratory Evaluation of PWS

Because of the need for early initiation of appropriate management strategies, diagnostic assessment ideally should occur early in life, when the primary clinical abnormality is profound hypotonia. Finding an individual with infantile hypotonia, for example, should prompt obtaining diagnostic laboratory testing for PWS.

The diagnostic evaluation for possible PWS is most cost-effective when performed in stepwise fashion.[12] The first step is confirmation of PWS using a methylation-sensitive DNA method, utilizing either methylation-sensitive PCR-based or Southern blotting.[9,10,13] We note that in practice both approaches occasionally have false negative and false positive results, so it is reasonable to consider repeating diagnostic testing using a different methodology in individuals who meet diagnostic criteria but have a previous negative test result. The recently developed diagnostic approach that measures expression of the *SNRPN* gene by reverse transcription PCR has not yet been shown to be as complete in ascertaining PWS.[14]

Once abnormal methylation is established the pattern will define whether the abnormality is consistent with PWS or Angelman syndrome. The next step is to perform FISH analysis to identify microdeletions of chromosome 15q11–q13 utilizing, at minimum, the SNRPN probe that can identify both the common large deletions as well as some smaller deletions involving the region around the imprinting center. If no deletion is found, the DNA is obtained from the affected individual and both parents, if possible (although it can often be done with DNA from only one parent), for uniparental disomy studies. If this is also normal, then the only remaining explanation for the abnormal methylation pattern is an imprinting defect, which may be inherited or a new event *(de novo)*.

As noted elsewhere in this textbook, rarely PWS can result from chromosome translocations involving the PWS critical region of chromosome 15, again with potentially significant implications for recurrence risk in the parents and extended family. Since full chromosome analysis is also needed if the methylation pattern is abnormal, in order to conduct FISH for the deletion and to rule out a translocation, the authors routinely order chromosome analysis at the same time as PWS methylation analysis. This also serves to identify other chromosomal abnormalities causing a phenotype similar to that of PWS (see, e.g., McCandless et al.[11]).

Changes in the availability of specific molecular testing for PWS make documentation of the specific cause mandatory to confirm the diagnosis and prevent recurrences in those rare families with an inherited form of the disorder. Likewise, the development of effective treatment strategies, such as growth hormone therapy, means that early diagnosis is critical to providing the best possible outcome. Therefore, the role of the clinical diagnostic criteria has changed from that of carefully defining the disorder to one of identifying those individuals for whom further testing is indicated. Judicious interpretation of the medical literature and further critical study will allow the diagnosis of PWS to be made early in the vast majority of patients while limiting the expense of testing in those at low likelihood of being affected. The priority for testing is in infants with marked hypotonia and feeding problems, two highly sensitive markers of PWS.

References

1. Alexander RC, Van Dyke DC, Hanson JW. Overview of Prader-Willi syndrome. In: Greenswag LR, Alexander RC, eds. *Management of Prader-Willi Syndrome.* 2nd ed. New York, NY: Springer-Verlag; 1995:3–17.
2. Camera G, Marugo M, Cohen MM Jr. Another postnatal-onset obesity syndrome. *American Journal of Medical Genetics.* 1993;47(6):820–822.
3. Christianson AL, Viljoen DL, Winship WS, de la Rey M, van Rensburg EJ. Prader-Willi syndrome in South African patients—clinical and molecular diagnosis. *South African Medical Journal.* 1998;88(6):711–714.
4. Chu CE, Cooke A, Stephenson JB, et al. Diagnosis in Prader-Willi syndrome. *Archives of Disease in Childhood.* 1994;71(5):441–442.
5. Gillessen-Kaesbach G, Groß S, Kaya-Westerloh S, Passarge E, Horsthemke B. DNA methylation based testing of 450 patients suspected of having Prader-Willi syndrome. *Journal of Medical Genetics.* 1995;32(2):88–92.
6. Gunay-Aygun M, Schwartz S, Heeger S, O'Riordan MA, Cassidy SB. The changing purpose of Prader-Willi syndrome clinical diagnostic criteria and proposed revised criteria. *Pediatrics.* 2001;108 (5):E92.
7. Holm VA, Cassidy SB, Butler MG, et al. Prader-Willi syndrome: consensus diagnostic criteria. *Pediatrics.* 1993;91(2):398–402.
8. Hou JW, Wang TR. Prader-Willi syndrome: clinical and molecular cytogenetic investigations. *Journal of the Formosan Medical Association.* 1996;95(6):474–479.
9. Kubota T, Das S, Christian SL, Baylin SB, Herman JG, Ledbetter DH. Methylation-specific PCR simplifies imprinting analysis. *Nature Genetics.* 1997;16(1):16–17.
10. Kubota T, Sutcliffe JS, Aradhya S, et al. Validation studies of SNRPN methylation as a diagnostic test for Prader-Willi syndrome. *American Journal of Medical Genetics.* 1996;66(1):77–80.
11. McCandless SE, Cassidy SB, Driscoll DJ, et al. Cytogenetic and molecular abnormalities in 387 patients referred for evaluation of Prader-Willi and Angelman syndromes. *American Journal of Human Genetics.* 1997;61(4): A31.
12. Monaghan KG, Wiktor A, Van Dyke DL. Diagnostic testing for Prader-Willi syndrome and Angelman syndrome: a cost comparison. *Genetics in Medicine.* 2002;4(6):448–450.
13. Muralidhar B, Butler MG, Methylation PCR analysis of Prader-Willi syndrome, Angelman syndrome, and control subjects. *American Journal of Medical Genetics.* 1998;80(3):263–265.
14. Muralidhar B, Marney A, Butler MG. Analysis of imprinted genes in subjects with Prader-Willi syndrome and chromosome 15 abnormalities. *Genetics in Medicine.* 1999;1(4):141–145.
15. Online Mendelian Inheritance in Man, OMIM™. McKusick-Nathans Institute for Genetic Medicine, Johns Hopkins University (Baltimore, MD) and National Center for Biotechnology Information, National Library of Medicine (Bethesda, MD), 2000. Available at: http://www.ncbi.nlm.nih. gov/omim/.

16. Richer LP, Shevell MI, Miller SP. Diagnostic profile of neonatal hypotonia: an 11-year study. *Pediatric Neurology.* 2001;25(1):32–37.

17. Vasquez SB, Hurst DL, Sotos JF. X-linked hypogonadism, gynecomastia, mental retardation, short stature, and obesity—a new syndrome. *Journal of Pediatrics.* 1979;94(1):56–60.

18. Whittington J, Holland A, Webb T, Butler J, Clarke D, Boer H. Relationship between clinical and genetic diagnosis of Prader-Willi syndrome. *Journal of Medical Genetics.* 2002;39(12):926–932.

3

Molecular Genetic Findings in Prader-Willi Syndrome

Karin Buiting and Bernhard Horsthemke

Prader-Willi syndrome (PWS), the most common genetic cause of marked obesity in humans,[17] is due to loss of expression of paternal genes from the 15q11–q13 region under the control of an imprinting center. PWS and Angelman syndrome, an entirely different clinical condition due to lack of maternally expressed genes, were the first examples in humans of genomic imprinting. There are three recognized genetic subtypes in PWS, including paternally derived interstitial deletions of the 15q11–q13 region, maternal uniparental disomy 15 (both 15s from the mother), and imprinting defects.

Genomic Imprinting

The chromosomal region 15q11–q13 contains a cluster of genes that are expressed from the paternal or maternal chromosome only (Figure 3.1). This peculiar expression pattern is a consequence of genomic imprinting, which is an epigenetic process by which the paternal and the maternal germ lines mark specific chromosome regions. In each generation, the parental imprints are erased and reset according to the sex of the individual (Figure 3.2A). The mechanisms underlying genomic imprinting are not completely understood, but parent-of-origin-specific DNA methylation plays an important role in this process. DNA methylation refers to the addition of a methyl group (CH_3) to carbon atom 5 of cytosine (Figure 3.2B and C). Only cytosines followed by guanine are methylated. The sequence 5'-CG-3'/3'-GC-5' is palindromic, and the dinucleotides are either methylated or unmethylated on both strands. *De novo* DNA methylation is catalyzed by the DNA methyltransferases 3A and 3B (DNMT3A and DNMT3B). After DNA replication, the newly synthesized DNA strand is methylated by the maintenance DNA methyltransferase 1 (DNMT1). In general, methylation of the promoter region of a gene means gene silencing.

Within the imprinted region in 15q11–q13, most genes are methylated and silenced on the maternal chromosome and expressed from the unmethylated paternal allele only. A paternally derived deletion of this

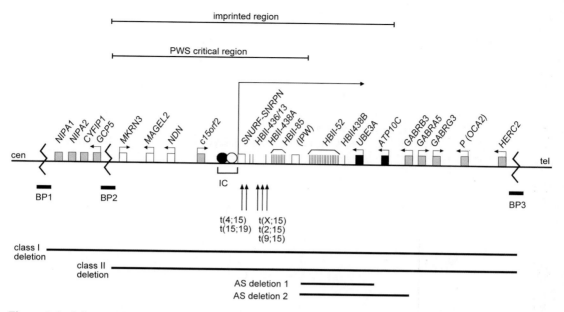

Figure 3.1. Schematic overview of human chromosomal region 15q11–q13. White boxes represent genes expressed from the paternal chromosome only; black boxes represent genes expressed from the maternal chromosome only; and gray boxes represent genes expressed from both chromosomes. Orientation of transcription, or gene expression, is indicated by horizontal arrows. The snoRNA genes are indicated as grouped or ungrouped vertical lines and the two critical imprinting center (IC) elements for AS (black circle) and PWS (white circle) are shown. Vertical arrows show the positions of translocation breakpoints on chromosome 15 that have been reported for 5 patients with PWS. The breakpoint cluster regions are indicated as black bars. The extension of the class (type) I and class (type) II deletions and the atypical deletions in patients AS deletion 1[35] and AS deletion 2[16] are drawn as horizontal lines. The proximal deletion breakpoints of AS deletion 1 and AS deletion 2 define the distal boundary of the PWS critical region.

region, the absence of a paternal chromosome 15 in maternal uniparental disomy 15, or the silencing of the paternal alleles by an imprinting defect lead to a complete loss of function of the paternally expressed 15q11–q13 genes and Prader-Willi syndrome (Figure 3.3). It is still a matter of debate whether PWS is caused by the loss of function of a single gene or of several genes, with growing evidence supporting the involvement of more than one gene.

Deletions and Translocations

A 4 Mb *de novo* interstitial deletion of the paternal chromosome [del(15)(q11q13)], which includes the entire imprinted domain plus several nonimprinted genes, is found in about 70% of subjects with PWS. The deletion is visible by high resolution chromosome banding analysis and was first described by Ledbetter and colleagues in 1981.[45] The deletion occurs at a frequency of about 1 in 10,000 newborns and is probably one of the most common deletions observed in humans. In a few patients, the region is deleted as the result of an unbalanced translocation. By studying the inheritance of chromosomal polymorphisms, Butler and Palmer in 1983[19] demonstrated that the deletion

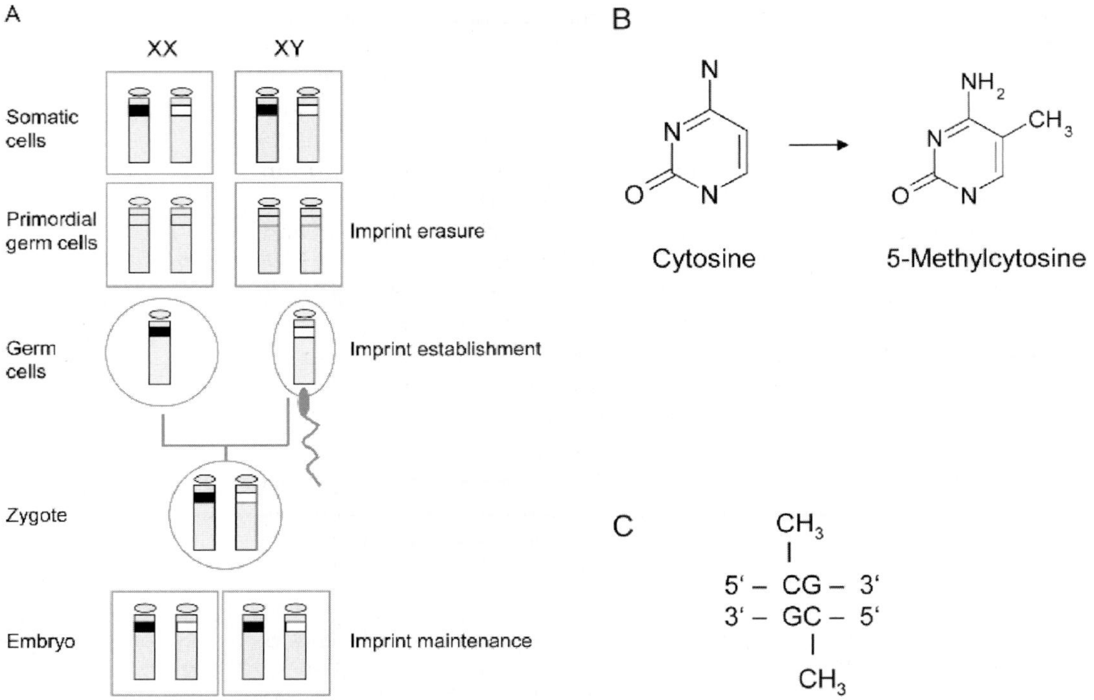

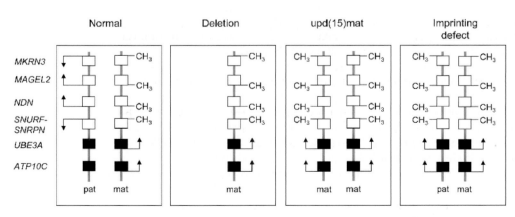

Figure 3.2. A. Genomic imprinting. During gametogenesis, parental imprints are erased and reset according to the sex of the individual. Imprints are maintained during mitotic cell divisions. White boxes represent paternal imprint; black boxes represent maternal imprint; gray boxes represent no imprint. **B.** Methylated cytosine. **C.** Methylation of CG dinucleotides in both DNA strands.

Figure 3.3. Genetic lesions in PWS. White boxes represent genes expressed from the paternal chromosome only, and black boxes represent genes expressed from the maternal chromosome only. Orientation of transcription, or gene expression, is indicated by vertical arrows.

always occurs on the chromosome 15 inherited from the father. This finding was confirmed in 1989 at the molecular level by Knoll and co-workers,[43] who also demonstrated that chromosome 15 deletions in Angelman syndrome (AS), which affect the same chromosome region, always occur on the maternal chromosome. It is now known that AS results from the loss of function of the *UBE3A* gene, which in the brain is expressed from the maternal chromosome only (Figure 3.1).

At the molecular level, two classes or types of deletions (types I and II) can be distinguished. In both types, the distal breakpoints are close to, but telomeric to the *P* gene (breakpoint region 3, BP3, Figure 3.1).[25,43,44,51] In type I deletions (30%–40% of patients), the proximal breakpoint is centromeric to the marker *D15S541* (breakpoint region 1, BP1). In type II deletions (60%–70% of patients), the proximal break-point is between *D15S541* and *D15S543* (breakpoint region 2, BP2).[26,43] In order to analyze the chromosomal mechanisms underlying deletions in PWS, Carrozzo et al.[20] and Robinson et al.[57] genotyped families of PWS and AS patients with the help of microsatellite markers flanking the common deletion region. Both groups obtained evidence that the deletions can occur by crossover events between the two homologous chromosome 15s (interchromosomal) or between different regions of one chromosome 15 (intrachromosomal). The recurrence risk is very low if the parents have normal chromosomes.

The clustering of the deletion breakpoints in most of the patients with del(15)(q11q13) suggested that there are duplicated sequences that are susceptible to nonhomologous crossovers. Interestingly, proximal 15q is involved in other cytogenetic rearrangements also. For example, inv dup(15) with breakpoints in q11.2 or q13 account for approximately 50% of all supernumerary marker chromosomes.[27,57,70] Less frequent are duplications, triplications, and inversions of this chromosomal region.[62]

The first evidence for duplicated sequences came from the identification of a gene family *(D15F37/HERC2)*, which was found to have multiple expressed copies within the breakpoint cluster regions in proximal 15q.[11,12,39] Using combinations of molecular and cytogenetic methods,[1,25] it could be demonstrated that all duplicated copies of this gene family and additional expressed and nonexpressed sequences are part of large duplicated sequence stretches of 200–400 kb in size.

Balanced translocations involving chromosome 15 are extremely rare in PWS. So far, five paternally derived *de novo* translocations have been reported, all of which disrupt the *SNURF-SNRPN* gene in either of two locations (Figure 3.1). In two patients with typical PWS, the translocation breakpoint disrupts *SNURF.*[42,66] In three patients with some features of PWS, the translocation breakpoints clustered in a region distal to *SNRPN.*[28,63,73]

Maternal Uniparental Disomy

To find out more about the molecular defect in PWS patients without del(15)(q11q13), Nicholls, Butler and colleagues in 1989 studied DNA samples from such families with the help of chromosome 15 markers.[52] They identified two probands that lacked a paternal chromosome 15 and found that they had two different maternal chromosome 15s [uniparental heterodisomy 15, upd(15)mat]. This finding established that the PWS chromosomal region is subject to genomic imprinting and that PWS is caused by the absence of a paternal contribution of genes on chromosome 15. As the paternally expressed genes are silent on the maternal chromosome, the loss of a paternal chromosome cannot be compensated by a second maternal chromosome 15. Subsequent studies revealed that upd(15)mat is the second most frequent finding in PWS and accounts for approximately 25% of all cases.

Maternal uniparental disomy arises in most cases from the postzygotic correction of a meiotic error. During meiosis, the diploid set of chromosomes (n = 46) is reduced to a haploid set (n = 23). Nondisjunction of the homologous chromosome 15s during female meiosis I or nondisjunction of the two sister chromatids during meiosis II results in an oocyte with two chromosome 15s or no chromosome 15. Fertilization of an oocyte with two chromosome 15s by a normal sperm with one chromosome 15 leads to a zygote trisomic for chromosome 15. This condition is not compatible with normal development but can be rescued by loss of one chromosome 15. In two thirds of cases, one of the two maternal chromosome 15s will be lost from the trisomic cell. This results in a normal set of chromosomes. If, however, the paternal chromosome is lost, the cell is left with two maternal chromosomes [upd(15)mat]. As judged by the state of DNA markers close to the centromere, most cases result from a meiosis I error. In these cases, the marker closest to the centromere is heterozygous. If the patient is homozygous at this marker, although the mother is heterozygous, a meiosis II error is likely. As a consequence of recombination events prior to the segregation of the chromosomes, heterodisomic and isodisomic segments alternate along the chromosome. As in other nondisjunction cases, the risk of uniparental disomy (UPD) increases with maternal age.[58] The recurrence risk is very low if the parents have normal chromosome 15s.

Imprinting Defects

A few patients with PWS (about 1%) have apparently normal chromosome 15s of biparental inheritance, but the paternal chromosome carries a maternal imprint. This leads to a complete loss of the paternally expressed genes in 15q11–q13. Thus, the functional consequence of the incorrect imprint is identical to that of maternal uniparental disomy.

In approximately 15% of these patients with PWS and an imprinting defect, the incorrect imprint is the result of a microdeletion affecting the imprinting center (IC).[10,15,54,60,65] The IC overlaps with the *SNURF-SNRPN* gene (Figure 3.1) and regulates in *cis* DNA methylation, gene expression, and chromatin structure of the whole imprinted domain. The IC appears to consist of two elements. One element is defined by

Table 3.1. Estimated Frequency and Recurrence Risk for Specific Genetic Subtypes in Prader-Willi Syndrome

	Deletion 15q11–q13 (*de novo*)	Uniparental disomy 15	Imprinting defect	Balanced translocations
Frequency	70%	25%	~1%	<0.1%
Recurrence risk	<1%	<1%	0–50%	Unknown

IC deletions in patients with AS and an imprinting defect. The shortest region of overlap (AS-SRO) is 880 bp and maps 35 kb centromeric to the *SNURF-SNRPN*.[14] The IC element affected by these deletions is necessary for establishing the maternal imprint. It probably interacts with the second element, which is defined by IC deletions in patients with PWS and an imprinting defect. The shortest region of overlap (PWS-SRO) is 4.3 kb in size and spans exon 1 of *SNURF-SNRPN*.[54] The IC element affected by these deletions is required for the maintenance of the paternal imprint during early embryogenesis.[2,30]

Most of the IC deletions are familial mutations. A deletion of the PWS-SRO can be transmitted silently through the female germ line, but leads to an incorrect maternal imprint on the paternal chromosome when inherited from a male. Familial IC deletions are associated with a 50% recurrence risk (see Table 3.1). In the case of a *de novo* deletion, the recurrence risk is not increased when it occurred after fertilization, but it can be up to 50% when the father has a germ line mosaicism.

The majority of patients with PWS and an imprinting defect (85%) have no IC deletion or point mutation of the PWS-SRO.[9,13] This indicates that the imprinting defect occurred spontaneously in the absence of a DNA sequence change. Interestingly, in all informative cases the chromosome carrying the incorrect imprint was inherited from the paternal grandmother.[9,13] These data suggest that the imprinting defect results from a failure to erase the maternal imprint in the father's germ line. In contrast to patients with an IC deletion, some of these patients share the same paternal chromosome with a healthy sibling. This indicates that the recurrence risk for another child with the disorder is very low.

Genes in the 15q11–q13 Region

The proof for genomic imprinting of 15q11–q13 has come from the identification of genes that are expressed from the paternal or maternal chromosome only (Figure 3.1). Within the 2 Mb imprinted domain, four paternally expressed genes encoding a protein have been identified, *MKRN3*, *MAGEL2*, *NDN*, and the *SNURF-SNRPN*. Paternal-only expression of these genes is regulated by parent-of-origin-specific DNA methylation of the promoter regions of each gene. Whereas the active paternal allele is unmethylated, the inactive maternal allele is methylated. Parent-of-origin-specific DNA methylation can be used to confirm the clinical diagnosis of PWS patients with a deletion of 15q11–q13, uniparental disomy, or an imprinting defect. Recently, several paternally expressed small nucleolar (sno) RNA genes have been identified. Telomeric to the paternally expressed domain are two maternally expressed genes, UBE3A[41,48] and ATP10C.[36,49]

MAKORIN3 (MKRN3, formerly ZNF127) is a ubiquitously expressed intronless gene that defines the most telomeric border of the imprinted domain of human 15q11–q13.[23,40] It encodes a putative RING zinc finger transcription factor, which belongs to the MAKORIN gene family and may function as a ribonucleoprotein. Just telomeric to MKRN3 are two other intronless genes, MAGEL2 and NECDIN (NDN). Both genes encode proteins that are part of the melanoma-associated antigen (MAGE) protein family.[3,38,46,47] MAGEL2 is expressed only in the brain and placenta.[5]

In the mouse, Necdin (Ndn) is expressed predominantly in postmitotic neurons with the highest expression in the hypothalamus and other brain regions at late embryonic and early postnatal stages. However, the human NDN gene was found to be expressed in all tissues studied, with highest expression in the brain and placenta. It is upregulated during neuronal differentiation, and in vitro experiments have shown that over-expression of this gene leads to suppression of cell proliferation. Data obtained from different mouse models suggest that this gene may contribute to respiratory problems observed in patients with PWS (see below).

The most complex gene in 15q11–q13 is SNURF-SNRPN. The original gene was found to consist of 10 exons, which encode two different proteins.[34,53] Exons 1 to 3 encode SNURF, a small polypeptide of unknown function, while exons 4 to 10 encode SmN, a spliceosomal protein involved in mRNA splicing in the brain. Exon 1 and the promoter region overlap with the IC. In the past few years, many more 5′ and 3′ exons of SNURF-SNRPN have been identified. These exons have two peculiar features: they do not have any protein coding potential, and they occur in many different splice forms of the primary transcript. Alternative transcripts containing novel 5′ exons were described by Dittrich et al.[29] and characterized in detail by Färber et al.[31] These transcripts start at two sites that share a high degree of sequence similarity and span the AS-SRO. Additional 3′ exons were described by Buiting et al.[8] and Runte et al.[59] This analysis also showed that the IPW exons, which were previously thought to represent an independent gene,[71] are part of the SNURF-SNRPN transcription unit. Some of these splice variants are found predominately in the brain and span the UBE3A gene in an antisense orientation. In contrast to the paternally expressed genes in 15q11–q13, maternal-only expression of UBE3A in brain is not regulated by DNA methylation. It is tempting to speculate that the antisense transcript silences the paternal allele of UBE3A.

Interestingly, the IC-*SNURF-SNRPN* transcript also serves as a host for several snoRNAs, which are encoded within introns of this complex transcription unit. They are processed from the primary transcript, and as this is made from the paternal chromosome only, the snoRNAs show the same imprinted expression pattern. The genes are present as single copy genes (*HBII-13*, *HBII-437*, *HBII438A* and *HBII-438B*) or as multi-gene clusters (*HBII-85* with 27 gene copies and *HBII-52* with 47 gene copies).[22,59] In contrast to other snoRNAs, which are usually involved in the modification of ribosomal RNAs, these snoRNAs do not have a region complementary to ribosomal RNA and might be involved in the modification of mRNAs. In *HBII-52*, 18 nucleotides are complementary to the serotonin receptor 2C mRNA.[22] It is possible that these snoRNAs are involved in the editing and/or alternative splicing of this mRNA.

In two unrelated families, a small deletion spanning *UBE3A* and the *HBII-52* gene cluster has been identified. Whereas maternal transmission of the deletion leads to AS, paternal transmission is not associated with an obvious clinical phenotype. This excludes the *HBII-52* snoRNAs from a role in PWS. The *HBII-85* gene cluster is distal to three balanced translocation breakpoints in patients with some features of PWS. As *HBII-85* is not expressed in these patients,[73] these snoRNAs may play a role in PWS.

In addition to the imprinted genes, several nonimprinted genes have been identified in the chromosomal region affected by the common large deletion (Figure 3.1). These genes may modify the PWS phenotype, and some are responsible for other genetic disorders. One is the oculocutaneous albinism type II (*OCA2*) gene. Hypopigmentation is a frequent finding in PWS and AS patients with a common large deletion.[64] This nonimprinted phenotype is associated with the deletion of one *OCA2* gene copy and may be caused by a gene dosage effect. However, Bittel et al.[3,4] did not obtain any evidence, by expression profiling, of reduced *OCA2* mRNA levels in lymphoblastoid cell lines from PWS and AS deletion patients. They did observe several transcripts from within the PWS/AS region, e.g., the GABA receptor subunit genes *GABRA5* and *GABRB3*—that had less than half the control level of expression in the deletion cell lines. These genes are probably expressed at higher levels from the paternal allele than the maternal allele. A maternal bias of expression was seen for four genes including *UBE3A* and *ATP10C*. Different expression levels of some of these genes in UPD versus deletion cells may underlie the phenotypic differences seen in PWS patients with different chromosomal defects.

Patients with a class or type I deletion, but not patients with a type II deletion, are hemizygous for four nonimprinted genes located between the deletion breakpoint cluster regions BP1 and BP2. These are the genes *NIPA1*, *NIPA2*, *CYFIP1*, and *GCP5*.[23] Heterozygous mutations in *NIPA1* lead to autosomal dominant spastic paraplegia.[55] Since spastic paraplegia has never been observed in patients with PWS or AS and a type I deletion, it is likely that a mutant protein and not reduced gene dosage leads to the disorder. Recently, clinical differences have been reported by Butler et al.[18] in PWS subjects with type I deletions versus type II deletions. Type I deletion subjects have more behavioral problems than type II deletion PWS subjects.

Downstream Genes from the 15q11–q13 Region

It is likely that the loss of function of the paternally expressed genes in 15q11–q13 directly or indirectly affects the activity of other genes. The identification of downstream genes may help to understand the pathogenesis of PWS. To date, only one study has addressed this question. Using high density microarrays, Horsthemke et al.[37] compared the gene expression profiles of normal and upd(15) fibroblasts isolated from a mosaic patient. By comparing cell lines from the same individual, Horsthemke et al.[37] circumvented the problem of interindividual variation in gene expression. However, there are also certain drawbacks associated with the study: cloned fibroblasts can show considerable interstrain variation, and the major symptoms in PWS are related to brain and not fibroblast dysfunction. The overall gene expression profiles were highly similar, indicating that the chromosome 15 status did not have a major effect on the global gene expression pattern in fibroblasts. Among the genes showing reduced expression in upd(15)mat cells was *SCG2*, which encodes for secretogranin II. This protein is mainly found in the core of catecholamine-storage vesicles within cells of the neuroendocrine system. It is a precursor of secretoneurin, which induces dopamine release.[61] *SCG2* is an interesting candidate gene, because dopamine is known to regulate food intake by modulating food reward via the meso-limbic circuitry of the brain.[44] Recent imaging studies in humans[69,74,75] have linked dopamine with eating behavior and obesity, and Wang et al.[69] have suggested that obese individuals may perpetuate pathological eating as a means to compensate for a decreased reward. To substantiate the notion that a defect in dopamine-modulated food reward circuits contributes to the development of hyperphagia in PWS, it will be necessary to replicate these findings in other patients and in mouse models. In addition, Bittel et al.[3,4] reported altered expression of genes/transcripts distal to the 15q11–q13 region using microarray technology in PWS and AS deletion subjects, which may reflect the different chromatin patterns in these subjects compared with normal.

Mouse Models of Prader-Willi Syndrome

Human chromosome region 15q11–q13 is evolutionarily related to mouse chromosome region 7C. The conserved synteny of the "PWS genes" makes it possible to study the effects of uniparental disomy and chromosomal deletions, as well as knock-outs for individual genes and regulatory elements in mice. These studies may help to unravel the pathogenesis of PWS.

The first mouse model for maternal uniparental disomy was described by Cattanach and colleagues in 1992.[21] Mice with a maternal duplication of 7C are smaller compared with their wild type littermates and die 2 to 8 days after birth. The early postnatal lethality is possibly associated with a reduced suckling activity. This effect is therefore consistent with the feeding problems and failure-to-thrive in newborns with PWS.

Nicholls and colleagues in 1999 reported a transgene mouse model with a chromosome deletion of 7C.[32] The size of the deletion is similar to the deletions in PWS patients. Mice inheriting the deletion from their father show failure-to-thrive and growth retardation and die within the first week of life.

Several knock-outs of individual genes have been described also. Jong et al.[40] demonstrated that a disruption of *Mkrn3* had no phenotypic effect in the mouse. The resulting mouse was viable and fertile, suggesting that *Mkrn3* does not have a significant role in PWS.

The *Ndn* gene was studied in three different mouse models. In the first study,[67] no obvious phenotype was seen. However, in two other studies, Gerard et al.[33] and Muscatelli et al.[50] found that in certain mouse strains the paternal transmission of a *null* allele strain leads to postnatal lethality with variable penetrance. The lethality of the mutant mice arose from a respiratory defect, suggesting that absence of *NDN* expression may contribute to the observed respiratory abnormalities in individuals with PWS through a suppression of central respiratory drive.[56] Surviving mice in all three studies had no overt phenotype and were non-obese and fertile.

Furthermore, Muscatelli et al.[50] found behavioral and hypothalamic alterations in *Ndn*-deficient mice: *Ndn* mutant mice displayed increased skin scraping in an open field test and improved spatial learning and memory in the Morris water maze, possibly mimicking the skin picking and improved spatial memory typical of PWS. These findings suggest that the NECDIN protein may be responsible for at least a subset of the multiple clinical features present in PWS.

Brannan and her colleagues showed that a disruption of the *Snrpn* gene is without any obvious phenotypic effect.[6,76] The same is true for a *Snurf* deletion, as demonstrated by Tsai et al.,[68] but White et al.[72] suggested that the *Snrpn*-deficient mice may have a deficiency of dopamine in the striatum.

To study the IC, which overlaps with *Snrpn* exon 1, three different deletions were engineered. Bressler et al.[7] found that a 0.9 kb deletion of the exon 1 region of *Snurf-Snrpn* had no effect, whereas a paternally transmitted 4.8 kb deletion led to postnatal lethality in about 50% of mutant mice. Surviving mutant mice were viable and healthy.

The paternal transmission of a 42 kb deletion (exons 1–6 of *Snurf-Snrpn* plus 23 kb upstream sequence) resulted in 100% postnatal lethality. The mutant mice were smaller than their wild type littermates, exhibited hypotonia and died within the first days of life. Again, strain-specific differences were observed.[6,24] By breeding chimeric males with females of another strain, some viable offspring were obtained. Surviving mice were smaller than normal, but fertile and not obese. Expression studies of the paternally expressed genes at birth, 1 week, 2 weeks, and 3 weeks of age revealed a leaky expression of all these genes. Even more surprising was the finding that, based on studies of the *Snrpn* locus, the leaky expression was derived from the maternal allele. From this it can be concluded that certain mouse strains do not silence imprinted genes as completely as other strains. Furthermore, it appears that only low levels of expression of PWS candidate genes are required to overcome most of the typical features of PWS.

Two mouse models with a deletion distal to the *Snurf-Snrpn* gene have been reported. One deletion spans from *Snurf-Snrpn* exon 2 to *Ube3a*.[68] Mice with a paternally inherited deletion showed hypotonia, growth retardation, and 80% postnatal lethality. All surviving animals were fertile and non-obese. No obvious phenotype was found in mice harboring a paternally derived deletion spanning from the *Ipw* exons to *Ube3a*.[68] This suggests that the critical region for many PWS features maps between *Snrpn* and the *Ipw* exons. This interval does not contain any protein-coding genes but only the snoRNA genes *MBII-85, MBII-13* and *MBII-437*. Therefore, these snoRNAs are good candidates for hypotonia, failure-to-thrive, and growth retardation.

A common feature in the mouse models that showed a phenotypic effect is postnatal lethality, and apart from the *Ndn*-deficient mice, all exhibit growth retardation. Only the *Ndn*-mutant mice showed hypothalamic alterations and features mimicking respiratory problems in PWS babies. Hypotonia was reported for both an IC-deletion mouse and a deletion including snoRNA genes. In none of the mouse models was obesity and/or hypogonadism found. However, the UPD and large deletion mice do not survive the first week of life, and therefore these animals cannot be studied at later developmental stages.

References

1. Amos-Landgraf JM, Ji Y, Gottlieb W, et al. Chromosome breakage in the Prader-Willi and Angelman syndromes involves recombination between large, transcribed repeats at proximal and distal breakpoints. *American Journal of Human Genetics.* 1999;65:370–386.
2. Bielinska B, Blaydes SM, Buiting K, et al. De novo deletions of the *SNRPN* exon 1 region in early human and mouse embryos result in a paternal to maternal imprint switch. *Nature Genetics.* 2000;25:74–78.
3. Bittel DC, Kibiryeva N, Talebizadeh Z, Butler MG. Microarray analysis of gene/transcript expression in Prader-Willi syndrome: deletion versus UPD. *Journal of Medical Genetics.* 2003;40:568–574.
4. Bittel DC, Kibiryeva N, Talebizadeh Z, Driscoll DJ, Butler MG. Microarray analysis of gene/transcript expression in Angelman syndrome: deletion versus UPD. *Genomics.* 2005;85:85–91.
5. Boccaccio I, Glatt-Deeley H, Watrin F, Roeckel N, Lalande M, Muscatelli F. The human *MAGEL2* gene and its mouse homologue are paternally expressed and mapped to the Prader-Willi region. *Human Molecular Genetics.* 1999;8:2497–2505.
6. Brannan CI, Chamberlain S. Mouse models of Prader-Willi syndrome. 2001 International Prader-Willi Syndrome Scientific Conference, St. Paul, MN, USA.
7. Bressler J, Tsai TF, Wu MY, et al. The *SNRPN* promoter is not required for genomic imprinting of the Prader-Willi/Angelman domain in mice. *Nature Genetics.* 2001;28:232–240.

8. Buiting K, Dittrich B, Endele S, Horsthemke B. Identification of novel exons 3′ of the human *SNRPN* gene. *Genomics.* 1997;40:132–137.

9. Buiting K, Dittrich B, Groß S, et al. Sporadic imprinting defects in Prader-Willi syndrome and Angelman syndrome: implications for imprint-switch models, genetic counseling, and prenatal diagnosis. *American Journal of Human Genetics.* 1998;63:170–180.

10. Buiting K, Färber C, Kroisel P, et al. Imprinting centre deletions in two PWS families: implications for diagnostic testing and genetic counseling. *Clinical Genetics.* 2000;58:284–290.

11. Buiting K, Greger V, Brownstein BH, et al. A putative gene family in 15q11–13 and 16p11.2: possible implications for Prader-Willi and Angelman syndromes. *Proceedings of the National Academy of Science USA.* 1992;89: 5457–5461.

12. Buiting K, Groß S, Ji Y, Senger G, Nicholls RD, Horsthemke B. Expressed copies of the MN7 (D15S37) gene family map close to the common deletion breakpoints in the Prader-Willi/Angelman syndromes. *Cytogenetics and Cell Genetics.* 1998;81:247–253.

13. Buiting K, Groß S, Lich C, Gillessen-Kaesbach G, El-Maarri O, Horsthemke B. Epimutations in Prader-Willi and Angelman syndrome: a molecular study of 136 patients with an imprinting defect. *American Journal of Human Genetics.* 2003;72:571–577.

14. Buiting K, Lich C, Cottrell S, Barnicoat A, Horsthemke B. A 5-kb imprinting center deletion in a family with Angelman syndrome reduces the shortest region of deletion overlap to 880 bp. *Human Genetic.* 1999;105: 665–666.

15. Buiting K, Saitoh S, Groß S, et al. Inherited microdeletions in the Angelman and Prader-Willi syndromes define an imprinting center on human chromosome 15. *Nature Genetics.* 1995;9:395–400.

16. Bürger J, Horn D, Tonnies H, Neitzel H, Reis A. Familial interstitial 570 kbp deletion of the *UBE3A* gene region causing Angelman syndrome but not Prader-Willi syndrome. *American Journal of Medical Genetics.* 2002;111:233–237.

17. Butler MG. Prader-Willi syndrome: current understanding of cause and diagnosis. *American Journal of Medical Genetics.* 1990;35:319–332.

18. Butler MG, Bittel DC, Kibiryeva N, Talebizadeh Z, Thompson T. Behavioral differences among subjects with Prader-Willi syndrome and type I or type II deletion and maternal disomy. *Pediatrics.* 2004;113:565–573.

19. Butler MG, Palmer CG. Parental origin of chromosome 15 deletion in Prader-Willi syndrome. *Lancet.* 1983;1:1285–1286.

20. Carrozzo R, Rossi E, Christian SL, et al. Inter- and intrachromosomal rearrangements are both involved in the origin of 15q11–q13 deletions in Prader-Willi syndrome. *American Journal of Human Genetics.* 1997;61:228–230.

21. Cattanach BM, Barr JA, Evans EP, et al. A candidate mouse model for Prader-Willi syndrome which shows an absence of *SNRPN* expression. *Nature Genetics.* 1992;2:270–274.

22. Cavaillé J, Buiting K, Kiefmann M, et al. Identification of brain-specific and imprinted small nucleolar RNA genes exhibiting an unusual genomic organization. *Proceedings of the National Academy of Science USA.* 2000; 7:14311–14316.

23. Chai JH, Locke DP, Greally JM, et al. Identification of four highly conserved genes between breakpoint hotspots BP1 and BP2 of the Prader-Willi/Angelman syndromes deletion region that have undergone evolutionary transposition mediated by flanking duplicons. *American Journal of Human Genetics.* 2003;73:898–925.

24. Chamberlain SJ, Johnstone KA, DuBose AJ, et al. Evidence for genetic modifiers of postnatal lethality in PWS-IC deletion mice. *Human Molecular Genetics.* 2004;13(23):2971–2977.

25. Christian SL, Fantes JA, Mewborn SK, Huang B, Ledbetter DH. Large genomic duplicons map to sites of instability in the Prader-Willi/ Angelman syndrome chromosome region (15q11–q13). *Human Molecular Genetics.* 1999;8:1025–1037.

26. Christian SL, Robinson WP, Huang B. et al. Molecular characterization of two proximal deletion breakpoint regions in both Prader-Willi and Angelman syndrome patients. *American Journal of Human Genetics.* 1995;57: 40–48.

27. Crolla JA, Harvey JF, Slitch LF, Dennis NR. Supernumerary marker 15 chromosomes: a clinical, molecular and FISH approach to diagnosis and prognosis. *Human Genetics.* 1995;95:161–170.

28. Conroy JM, Grebe TA, Becker LA, et al. Balanced translocation 46,XY,t(2;15)(q37.2;q11.2) associated with atypical Prader-Willi syndrome. *American Journal of Human Genetics.* 1997;61:388–394.

29. Dittrich B, Buiting K, Korn B, et al. Imprint switching on human chromosome 15 may involve alternative transcripts of the *SNRPN* gene. *Nature Genetics.* 1996;14:163–170.

30. El-Maarri O, Buiting K, Peery EG. Maternal methylation imprints on human chromosome 15 are established during or after fertilization. *Nature Genetics.* 2001;27:341–344.

31. Färber C, Dittrich B, Buiting K, Horsthemke B. The chromosome 15 imprinting centre (IC) region has undergone multiple duplication events and contains an upstream exon of *SNRPN* that is deleted in all Angelman syndrome patients with an IC microdeletion. *Human Molecular Genetics.* 1999;8:337–343.

32. Gabriel J, Merchant M, Ohta T, et al. A transgene insertion creating a heritable chromosome deletion mouse model of Prader-Willi and Angelman syndromes. *Proceedings of the National Academy of Sciences USA.* 1999;96:9258–9263.

33. Gerard M, Hernandez L, Wevrick R, Stewart CL. Disruption of the mouse *Necdin* gene results in early post-natal lethality. *Nature Genetics.* 1999;23:199–202.

34. Gray TA, Saitoh S, Nicholls RD. An imprinted, mammalian bicistronic transcript encodes two independent proteins. *Proceedings of the National Academy of Sciences USA.* 1999;96:5616–5621.

35. Hamabe J, Kuroki Y, Imaizumi K, et al. DNA deletion and its parental origin in Angelman syndrome patients. *American Journal of Medical Genetics.* 1991;41(1):64–68.

36. Herzing LB, Kim SJ, Cook EH, Ledbetter DH. The human aminophospholipid-transporting ATPase gene *ATP10C* maps adjacent to *UBE3A* and exhibits similar imprinted expression. *American Journal of Human Genetics.* 2001;68:1501–1505.

37. Horsthemke B, Nazlican H, Hüsing J, et al. Somatic mosaicism for maternal uniparental disomy 15 in a girl with Prader-Willi syndrome: confirmation by cell cloning and identification of candidate downstream genes. *Human Molecular Genetics.* 2003;12:2723–2732.

38. Jay P, Rougeulle C, Massacrier A, et al. The human necdin gene, *NDN*, is maternally imprinted and located in the Prader-Willi syndrome chromosomal region. *Nature Genetics.* 1997;17:357–361.

39. Ji Y, Walkowicz M, Buiting K, et al. The ancestral gene for transcribed, low-copy repeats in the Prader-Willi/Angelman region encodes a large protein

implicated in protein trafficking that is deficient in mice with neuromuscular and spermiogenic abnormalities. *Human Molecular Genetics*. 1999;8: 533–542.

40. Jong MTC, Gray TA, Ji Y, et al. A novel imprinted gene, encoding a RING zinc-finger protein, and overlapping antisense transcript in the Prader-Willi syndrome critical region. *Human Molecular Genetics*.1999;8:783–793.

41. Kishino T, Lalande M, Wagstaff J. *UBE3A/E6-AP* mutations cause Angelman syndrome. *Nature Genetics*. 1997;15:70–73.

42. Knoll JH, Nicholls RD, Magenis RE, Glatt K. Angelman syndrome: Three molecular classes identified with chromosome 15q11q13-specific DNA markers. *American Journal of Human Genetics*. 1990;47:149–155.

43. Knoll JHM, Nicholls RD, Magenis RE, Graham JM Jr, Lalande M, Latt SA. Angelman and Prader-Willi syndromes share a common chromosome 15 deletion but differ in parental origin of the deletion. *American Journal of Medical Genetics*. 1989;32:285–290.

44. Kuwano A, Mutirangura A, Dittrich B, et al. Molecular dissection of the Prader-Willi/Angelman syndrome region (15q11–13) by YAC cloning and FISH analysis. *Human Molecular Genetics*. 1992;1:417–425.

45. Ledbetter DH, Riccardi VM, Airhart SD, Strobel RJ, Keenan SB, Crawford JD. Deletions of chromosome 15 as a cause of the Prader-Willi syndrome. *New England Journal of Medicine*. 1981;304:325–329.

46. Lee S, Kozlov S, Hernandez L, et al. Expression and imprinting of *MAGEL2* suggest a role in Prader-Willi syndrome and the homologous murine imprinting phenotype. *Human Molecular Genetics*. 2000;9:1813–1819.

47. MacDonald H, Wevrick, R. The *necdin* gene is deleted in Prader-Willi syndrome and is imprinted in human and mouse. *Human Molecular Genetics*. 1997;11:1873–1878.

48. Matsuura T, Sutcliffe JS, Fang P, et al. De novo truncating mutations in E6-AP ubiquitin-protein ligase gene (*UBE3A*) in Angelman syndrome. *Nature Genetics*. 1997;5:74–77.

49. Meguro M, Kashiwagi A, Mitsuya K, et al. A novel maternally expressed gene, *ATP10C* encodes a putative aminophospholipid translocase associated with Angelman syndrome. *Nature Genetics*. 2001;28:19–20.

50. Muscatelli F, Abrous DN, Massacrier A, et al. Disruption of the mouse *Necdin* gene results in hypothalamic and behavioral alteration reminiscent of the human Prader-Willi syndrome. *Human Molecular Genetics*. 2000;9: 3101–3110.

51. Mutirangura A, Jayakumar A, Sutcliffe JS, et al. A complete YAC contig of the Prader-Willi/Angelman chromosome region (15q11–q13) and refined localization of the *SNRPN* gene. *Genomics*. 1993;18:546–552.

52. Nicholls RD, Knoll JH, Butler MG, Karam S, Lalande M. Genetic imprinting suggested by maternal uniparental heterodisomy in nondeletion Prader-Willi syndrome. *Nature*. 1989;342:281–285.

53. Özcelik T, Leff S, Robinson W, et al. Small nuclear ribonucleoprotein polypeptide N (*SNRPN*), an expressed gene in the Prader-Willi syndrome critical region. *Nature Genetics*. 1992;2:265–269.

54. Ohta T, Gray TA, Rogan PK, et al. Imprinting-mutation mechanisms in Prader-Willi syndrome. *American Journal of Human Genetics*. 1999;64:397–413.

55. Rainier S, Chai JH, Tokarz D, Nicholls RD, Fink JK. NIPA1 Gene mutations cause autosomal dominant hereditary spastic paraplegia (SPG6). *American Journal of Human Genetics*. 2003;73:967–971.

56. Ren J, Lee S, Pagliardini S, et al. Absence of Ndn, encoding the Prader-Willi syndrome-deleted gene necdin, results in congenital deficiency of central

respiratory drive in neonatal mice. *Journal of Neuroscience.* 2003;23:1569–1573.

57. Robinson WP, Dutly F, Nicholls RD, et al. The mechanisms involved in formation of deletions and duplications of 15q11–q13. *Journal of Medical Genetics.* 1998;35:130–136.

58. Robinson WP, Spiegel R, Schinzel AA. Deletion breakpoints associated with the Prader-Willi and Angelman syndromes (15q11–q13) are not sites of high homologous recombination. *Human Genetics.* 1993;91:181–184.

59. Runte M, Hüttenhofer A, Groß S, Kiefmann M, Horsthemke B, Buiting K. The IC-*SNURF-SNRPN* transcript serves as a host for multiple small nucleolar RNA species and as an antisense RNA for *UBE3A*. *Human Molecular Genetics.* 2001;10:2687–2700.

60. Saitoh S, Buiting K, Rogan PK, et al. Minimal definition of the imprinting center and fixation of a chromosome 15q11–q13 epigenotype by imprinting mutations. *Proceedings of the National Academy of Science USA.* 1996;93:7811–7815.

61. Saria A, Troger J, Kirchmair R, Fischer-Colbrie R, Hogue-Angeletti R, Winkler H. Secretoneurin releases dopamine from rat striatal slices: a biological effect of a peptide derived from secretograninII (chromograninC). *Neuroscience.* 1993;54:1–4.

62. Schinzel AA, Brecevic L, Bernasconi F, et al. Intrachromosomal triplication of 15q11–q13. *Journal of Medical Genetics.* 1994;31:798–803.

63. Schulze A, Hansen C, Skakkebaek NE, Brondum-Nielsen K, Ledbetter DH, Tommerup N. Exclusion of SNRPN as a major determinant of Prader-Willi syndrome by a translocation breakpoint. *Nature Genetics.* 1996;12(4):452–454.

64. Spritz RA, Bailin T, Nicholls RD, et al. Hypopigmentation in the Prader-Willi syndrome correlates with *P* gene deletion but not with haplotype of the hemizygous *P* allele. *American Journal of Medical Genetics.* 1997;71:57–62.

65. Sutcliffe JS, Nakao M, Christian S, et al. Deletions of a differentially methylated CpG island at the *SNRPN* gene define a putative imprinting control region. *Nature Genetics.* 1994;8:52–58.

66. Sun Y, Nicholls RD, Butler MG, Saitoh S, Hainline BE, Palmer CG. Breakage in the *SNRPN* locus in a balanced 46,XY,t(15;19) Prader-Willi syndrome patient. *Human Molecular Genetics.* 1996;5:517–524.

67. Tsai TF, Armstrong D, Beaudet AL. Necdin-deficient mice do not show lethality or the obesity and infertility of Prader-Willi syndrome. *Nature Genetics.* 1999;22:15–16.

68. Tsai TF, Jiang Y, Bressler J, Armstrong D, Beaudet AL. Paternal deletion from *Snrpn* to *Ube3a* in the mouse causes hypotonia, growth retardation and partial lethality and provides evidence for a gene contributing to Prader-Willi syndrome. *Human Molecular Genetics.* 1999;8:1357–1364.

69. Wang GJ, Volkow ND, Logan J, et al. Brain dopamine and obesity. *Lancet.* 2001;357:354–357.

70. Webb T. Inv dup(15) supernumerary marker chromosomes. *Journal of Medical Genetics.* 1994;31:585–594.

71. Wevrick R, Kerns JA, Francke U. Identification of a novel paternally expressed gene in the Prader-Willi syndrome region. *Human Molecular Genetics.* 1994;3:1877–1882.

72. White RA, Vorontosova E, Chen R, et al. A behavioral and correlated neurochemical mouse phenotype related to the absence of Snrpn, a locus associated with Prader-Willi syndrome. *American Journal of Human Genetics Supplement.* 2002;71:467.

73. Wirth J, Back E, Hüttenhofer A, et al. A translocation breakpoint cluster disrupts the newly defined 3' end of the *SNURF-SNRPN* transcription unit on chromosome 15. *Human Molecular Genetics.* 2001;10:201–210.

74. Volkow ND, Wang GJ, Fowler JS, et al. "Nonhedonic" food motivation in humans involves dopamine in the dorsal striatum and methylphenidate amplifies this effect. *Synapse.* 2002;44:175–180.

75. Volkow ND, Wang GJ, Maynard L, et al. Brain dopamine is associated with eating behaviors in humans. *International Journal for Eating Disorders.* 2003; 33:136–142.

76. Yang T, Adamson TE, Resnick JL, et al. A mouse model for Prader-Willi syndrome imprinting-centre mutations. *Nature Genetics.* 1998;19:25–31.

4

Laboratory Testing for Prader-Willi Syndrome

Kristin G. Monaghan and Daniel L. Van Dyke

In 1981, Ledbetter and co-workers[45] described a cytogenetically visible deletion in proximal 15q in four patients with Prader-Willi syndrome (PWS). Proximal 15q was targeted for analysis because several previously published Prader-Willi cases exhibited an isochromosome 15q or a translocation with proximal 15q breakpoints.[12,34] Since the first report of the cytogenetic deletion of proximal 15q in PWS, several cytogenetic and molecular genetic techniques have been used to further characterize the chromosome 15q11–q13 region and for laboratory diagnostic purposes in a range of PWS subjects having a variety of abnormal chromosome findings.

Using several chromosome staining methods to evaluate chromosome 15 short arm variants, Butler and Palmer[16] showed that the PWS deletion preferentially involves the paternally inherited chromosome 15. Butler[13] subsequently reported that short arm or C-band variants appeared to be more common in chromosome 15s with a deletion than in normal 15s.

Nicholls, Butler, and co-workers[56] studied the parental origin of the chromosome 15 pair in two PWS patients who did not exhibit a 15q deletion. They were the first to describe uniparental disomy (UPD) in PWS and concluded that genetic imprinting in this region must play a causal role in both PWS and Angelman syndrome (AS). One of their PWS patients carried a familial robertsonian 13;15 translocation. What they could not recognize at the time was the causal relationship between robertsonian translocations and UPD.[3]

Aside from these important early studies, a large variety of balanced and unbalanced chromosome 15 rearrangements have been described in PWS patients.[12,17] Most of the chromosome 15 rearrangements associated with PWS have resulted in deletion of proximal 15q, and in many of these cases deletion of the paternally inherited chromosome 15 was demonstrated. Some unusual chromosomal causes of PWS that have been described include intrachromosomal triplication[71] and an unbalanced reciprocal translocation resulting in maternal disomy for proximal 15q.[58] In contrast to the deletion cases, PWS and AS patients who carry an additional dicentric chromosome [+dic(15)(q11-q13)] also have UPD, which accounts for their phenotype.[61] Patients with a +dic(15)(q11–q13) without UPD have a different phenotype.[51,59]

Hulten et al.[37] described a family segregating a translocation involving chromosome 15 and 22 [t(15;22)(q13;q11)] in affected PWS and AS children. The affected children exhibited the same 15q deletion and 22q duplication, except that the PWS children inherited the translocation from their carrier father, and the AS child inherited it from his carrier mother.

A landmark report by Dittrich et al.[26] described the abnormal DNA methylation pattern in PWS and AS. They digested patient DNA with the methylation-sensitive HindIII/HpaII restriction enzymes and performed Southern blot analysis with the PW71 (D15S63) probe. Normal subjects exhibited a 6.0 kb and 4.4 kb band, whereas PWS subjects exhibited only the maternal 6.0 kb band and AS subjects exhibited only the paternal 4.4 kb band.

In 1994, Lerer et al.[47] reported their use of probe PW71 to detect parent-of-origin-specific differences in DNA methylation at the *D15S63* locus in PWS patients with 15q deletions or UPD. Using similar methods, Butler[14] evaluated a group of 27 suspected PWS patients. Thirteen had a deletion by G-banding and FISH. One patient had a balanced 15;19 translocation, four had UPD, and two had no parental studies but exhibited abnormal methylation, so probably had UPD. Two patients with normal karyotypes exhibited a normal methylation pattern and so remained unexplained. Testing ruled out PWS in the remaining five patients. The patient with the t(15;19) exhibited a normal methylation pattern, but the FISH results were consistent with a chromosome 15 break within the PWS critical region.[67] This important patient appears to represent the first with a definite PWS diagnosis and a normal methylation pattern. A few similar PWS patients have been described with a balanced translocation and a normal DNA methylation pattern.[17] In the cases that have been evaluated, the translocation breakpoint was within the *SNRPN* gene or between the *SNRPN* and *IPW* genes.[22,43,63,67] Each of these translocations represented a new mutation in the patient.

Several reports have described multiplex PWS families.[50,52] In each case, the methylation pattern of affected PWS individuals and carrier family members, such as fathers, was abnormal and consistent with a diagnosis of PWS. The affected family members inherited an atypically small deletion—as small as 7.5 kb—from their father, and the unaffected carriers inherited the deletion from their mother. Such deletions are important to identify because there is a 50% risk of PWS in children of carrier males. For example, the family described by McEntagart et al.[50] exhibited the usual PWS methylation pattern. Microsatellite analysis was employed to distinguish between UPD and a deletion. Four microsatellite markers within the PWS critical region demonstrated normal biparental inheritance, but two other markers exhibited only a maternal allele. Additional family studies confirmed an inherited deletion of the markers *D15S128* and *D15S63*. The unaffected father and paternal grandmother of the proband exhibited a typical AS methylation pattern, or having no maternal contribution, and this inheritance pattern has been observed in other families (cf. Buiting et al.[8]). In these families, detailed genetic services and counseling are recommended.

Ohta et al.[57] described sporadic PWS patients with abnormal methylation but without evidence of UPD, a deletion involving the PWS critical region, or a deletion of the imprinting center. In contrast to the imprinting center deletion families, these cases appear to have no significant recurrence risk. Buiting et al.[9] summarized their observations of 51 PWS and 85 AS patients who exhibited an imprinting defect (abnormal methylation pattern) but who did not have evidence of the usual deletion or UPD. Seven of the 51 PWS patients were shown to have a deletion involving the imprinting center, and in five of these subjects the father carried the same deletion on the chromosome 15 that he inherited from his mother. Three of the PWS patients with an imprinting center deletion had a positive family history. In contrast, none of the remaining 44 patients had an affected relative. Sequence analysis of the PWS-SRO (PWS-shortest region of deletion overlap) for 32 of the 44 nondeletion patients revealed no clinically significant mutations—only benign, single nucleotide polymorphisms. Parental and grandparental origin studies were most consistent with PWS having been caused by a rare and sporadic failure during spermatogenesis to erase the maternal imprint from the PWS/AS imprinting center.

FISH Probes for Prader-Willi Syndrome Diagnosis

Several FISH probe kits are commercially available that employ a probe encompassing the SNRPN and IC region, and a control probe localized to distal 15q. Metaphase FISH analysis is expected to reveal a normal pattern with a SNRPN/IC signal and a control signal on each chromosome 15, or a deletion pattern with a control signal on each chromosome 15 but only one SNRPN/IC signal. A normal FISH pattern does not exclude UPD or a microdeletion involving the IC only. Some but not all chromosome 15 rearrangements are detectable using FISH (e.g., isochromosome 15q or extra dicentric 15).

Cytocell (www.cytocell.com) manufactures the Aquarius Prader-Willi/Angelman Region Probe with Control Probe (15qter). This probe kit targets SNRPN and the imprinting center. Vysis (www.vysis.com) manufactures four PWS/AS region probe kits. Among them, the LSI Prader-Willi/Angelman Region Probe (SNRPN) targets SNRPN and includes control probes for the centromere region and 15q22 (PML).

Molecular Analysis for Prader-Willi Syndrome

Molecular genetic testing for Prader-Willi syndrome can be divided into four categories: methylation analysis by Southern blot or polymerase chain reaction (PCR), reverse transcription PCR (RT-PCR) to detect SNRPN expression, microsatellite analysis to detect uniparental disomy, and specialized studies using various molecular techniques for the identification of imprinting defects. Each method has advantages and disadvantages compared with other methods (Table 4.1). For some patients, a combination of several methods may be needed to establish a diagnosis and determine the etiology (Figure 4.1). Genetic testing and genetic counseling for Prader-Willi syndrome are important not only to confirm the diagnosis but also to determine and discuss the recurrence risk of PWS with other family members. The recurrence risk is less than 1% if the proband has a large deletion, uniparental disomy, or a nondeletion imprinting defect. Imprinting defects that are due to a microdeletion in the imprinting center (IC) are associated with a 50% risk of PWS in the siblings of a proband. Familial PWS deletions are found on the paternal chromosome of the proband and on the maternal chromosome 15 of the phenotypically normal father.

Molecular genetic tests currently used for the diagnosis of PWS are based on the difference in the methylation status of the maternal and paternal chromosome 15s. The difference occurs within genomic areas consisting of a high proportion of the dinucleotide CpG, referred to as CpG islands. CpG islands are located throughout the genome and are usually associated with the promoters of genes. Methylation of the CpG island occurs on cytosines. Unmethylated promoters are generally found in active genes, whereas inactive genes usually have a methylated promoter region. This mechanism is involved in normal human development, X-chromosome inactivation and imprinting.[77] With respect to imprinting of the PWS region on chromosome 15, the maternal chromosome is normally methylated and thus inactive, whereas the paternal chromosome is unmethylated and active. The inheritance of one methylated, inactive PWS locus and one unmethylated, active PWS locus is crucial for normal human development.

Southern Blot Analysis

Initially, genetic testing for PWS involved chromosome analysis and DNA analysis of several restriction fragment length polymorphisms (RFLPs) and microsatellites to identify deletions and UPD. This was time consuming and not useful for some families due to a lack of informative polymorphic markers. Furthermore, high-resolution karyotypes were unreliable for the detection of a PWS deletion.[66]

The first clinically available DNA test for Prader-Willi syndrome involved Southern blot analysis, a method still used in many molecular diagnostic laboratories. Southern blot analysis involves several steps, the first of which is digesting DNA with restriction enzymes. The restriction fragments are then separated on the basis of size by agarose gel electrophoresis, with the smaller fragments migrating through the gel faster than larger ones. The DNA is then denatured to produce single stranded fragments that are transferred to a nitrocellulose or nylon membrane (Southern blot) on which they become immobilized. The single stranded DNA on the Southern blot is then hybridized to a single stranded DNA probe (usually several hundred base pairs to several kilobases in length) that has been radioactively labeled. The single stranded DNA probe binds to its complementary single stranded

Table 4.1. Comparison of Methods Used for the Molecular Genetic Diagnosis of Prader-Willi Syndrome

Method	Advantages	Disadvantages
Southern blot	>99% detection rate* Simple to perform Reliable, extensively used	Requires 5 µg of DNA Limited by enzyme restriction sites analyzed Does not determine etiology of PWS* Radioactivity required 2–3 week turnaround time False positives due to incomplete restriction enzyme digestion
mPCR	Nonisotopic Avoids pitfall of restriction site polymorphisms 3–5 day turnaround time ~200 ng of DNA required	Bisulfite treatment of DNA is time consuming Need to adjust relative primer concentration to achieve amplicons of similar intensity
Fluorescence melting curve analysis	Nonisotopic 2–3 day turnaround time No transfer of PCR products or electrophoresis involved ~200 ng of DNA required	Bisulfite treatment of DNA required Expensive equipment Appropriate primer design is necessary
BRA	≤100 ng of DNA No concerns regarding preferential amplification of PCR products	More labor intensive than mPCR Risk of false negative results in PWS due to incomplete digestion
Methylation-sensitive restriction digest followed by PCR	Rapid turnaround time Nonisotopic No bisulfite DNA treatment required 500 ng DNA required	Limited by enzyme restriction sites analyzed Risk of false negative results in PWS due to incomplete digestion Labor intensive (2 digests and 3 multiplex PCRs per sample)
SNRPN expression analysis (RT-PCR)	Nonisotopic 1–2 day turnaround time Should detect 100% of PWS, regardless of etiology (compared with a >99% detection rate for the above methods)	Not diagnostic for AS RNA is more labile than DNA Special laboratory handling required for RNA

* None of the methods included in this table will determine the PWS etiology, and all will detect >99% of subjects with PWS.

Note: Fluorescence *in situ* hybridization (FISH) is not included as a molecular genetics technique.

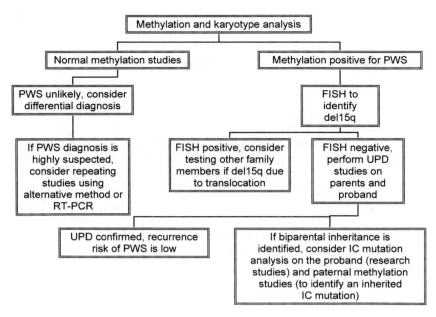

Figure 4.1. Laboratory testing for diagnosis and characterization of Prader-Willi syndrome.

DNA sequence on the nylon membrane. Excess probe is then washed off of the blot, and the membrane is exposed to X-ray film, a process called autoradiography. The film is then developed, with the autoradiogram revealing the location on the original agarose gel where the probe was bound, or hybridized, corresponding to the allele sizes of the restriction fragments.

Southern blot analysis for the diagnosis of PWS requires the use of two restriction enzymes, one that will cleave DNA regardless of the methylation status of the DNA (maternal and paternal chromosome 15), and another enzyme that will only cleave its recognition sequence if the DNA is unmethylated (paternal chromosome 15). Many different probes have been used for the diagnosis of PWS including the zinc finger gene, *ZNF127 (D15S9), PW71B (D15S63),* and the gene for the small nuclear ribonucleoprotein N *(SNRPN)* exon 1 (summarized by Buchholz et al.[5]). Southern blots hybridized with the ZNF127 probe are difficult to interpret and thus are not used in most clinical laboratories. PW71B and SNRPN have been used extensively in clinical laboratories because they give a clear difference between the maternal and paternal alleles.

Methylation analysis by Southern blotting using either the PW71B or SNRPN probes will detect over 99% of subjects with PWS.[4] The SNRPN and PW71B probes have been validated as diagnostic tests for PWS.[42] Few subjects have been reported in the literature with typical PWS features and normal methylation studies. One example involved an individual with a classic PWS phenotype, normal SNRPN and PW71B methylation studies, and normal UPD studies. Chromosome analysis detected a *de novo* balanced translocation involving 15q11–q13 disrupting the *SNRPN* locus, which was confirmed by SNRPN expression studies.[67]

Disadvantages of Southern blot analysis include: a large amount of high quality, high molecular weight DNA must be used (usually ~5 μg); and information is only provided about those methylated regions within the PWS sequence recognized by methylation sensitive restriction enzymes. In addition, radioactivity is usually involved, although nonradioactive Southern blotting using chemiluminescent detection has been described.[69] Furthermore, because the methylation test does not distinguish among deletions, uniparental disomy, and IC defects, reflex testing is required to define the etiology of the PWS. The PW71B and SNRPN probes are reliable as the primary approach to PWS molecular diagnosis. However, PWS patients have been reported with discordant results with the two probes.[5,6] Thus, for PWS patients with a classic phenotype who have a negative methylation result using one probe, studies using either the opposite probe or the same probe with different restriction enzymes should be performed.

PW71B

PW71, located at *D15S63*, is an imprinted locus of unknown function that resides within the PWS/AS critical region. The probe, PW71B, used for methylation studies[24] is a 365bp HaeIII fragment cloned into the SmaI site of the plasmic vector, pUC19. The probe is excised by digestion with EcoRI and HaeIII and is available through the American Type Culture Collection (ATCC/NIH Repository #99412/99413). PW71B includes the original PW71 probe[26] and gives better hybridization signals than the original probe.[24] At least one HpaII and one CfoI restriction enzyme recognition sequence are methylated on the maternal, but not paternal, chromosome 15. Combinations of restriction enzymes that can be used with PW71B include BglII/CfoI, HindIII/CfoI, and HindIII/HpaII. The maternal and paternal fragment sizes vary based on the combination of restriction endonucleases used (Figure 4.2).

The PW71B methylation test accurately detects most cases of PWS[29]; however, false positive and false negative results can occur as well as technical difficulties in the laboratory.[6,8,48,72] Technical difficulties reported with PW71B are due to the probe being so small. This can be resolved by using a high probe count, low stringency washes, and dextran sulfate, which increases the hybridization signal by preventing nonspecific binding of the probe to the membrane.[24,25] False positives may occur due to incomplete digestion with the methylation sensitive restriction enzyme.[25] To control for the completeness of enzyme digestion, the laboratory should run a normal sample in addition to samples positive for AS and PWS on all Southern blots.

Unusual results have been reported due to a benign polymorphism of North-African origin in the restriction enzyme recognition sequence for BglII.[23] False positive results reported by Buiting et al.[6] were due to a 28kb deletion spanning the *D15S63* locus with a frequency of 1 in 75

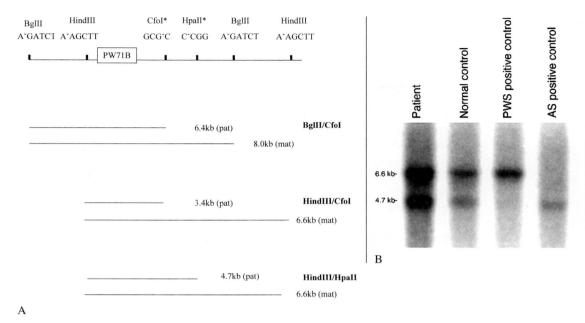

Figure 4.2 A. Restriction map and RFLP sizes surrounding PW71B probe[23,24,25,29] (*MT-sensitive site, methylated site will not cut with enzyme). **B**. Autoradiogram of Southern blot using the probe PW71B with HpaII and HindIII digested DNA. The patient analyzed is negative for Prader-Willi syndrome (PWS); both DNA bands are present. AS = Angelman syndrome.

among the Ashkenazi Jewish population.[64] This deletion is a benign variant with no effect on imprinting in the PWS region; it arose due to a recombination event between repetitive *Alu* sequences. False negative results for PW71B and SNRPN have been reported due to mosaicism for a 15q11–q13 deletion, detected by FISH studies but masked in methylation studies by a normal cell population.[48]

SNRPN

Exons 1–3 within the SNRPN upstream reading frame encode a protein product of unknown function termed "SNURF" (SNRPN upstream reading frame). Exons 4–10 encode SNRPN (SmN), which functions in mRNA splicing. The *SNRPN* gene is expressed only from the paternal allele[32]; there is extensive methylation of the maternal SNRPN CpG island, which represses transcription of SNRPN from this allele. This differential methylation is the basis of the SNRPN molecular diagnostic test for PWS.[33,68] The CpG island of the *SNRPN* gene contains several NotI sites that are methylated on the maternal chromosome. The probe used for analysis is a 0.9 kb NotI fragment cut from a 4.2 kb XbaI fragment containing SNRPN CpG island and is available from the American Type Culture Collection (ATCC Repository #95678, #95679). Following digestion of genomic DNA with NotI and XbaI, Southern blot analysis of a normal maternal chromosome results in a 4.2 kb band, consistent with the methylation of the NotI sites (which will not cut with the enzyme). A normal paternal chromosome results in a 0.9 kb band, because the unmethylated NotI sites are cut by the enzyme (Figure 4.3).

For SNRPN Southern blot analysis, blood should be collected in either EDTA or ACD tubes. DNA extracted from blood collected in sodium heparin tubes may not be completely digested with NotI, resulting in a weak paternal band and a possible false positive result for PWS.[40] In addition to a weak paternal band, additional bands larger than the expected 4.2 kb band are also seen in DNA extracted from sodium heparinized blood samples. Studies to determine the completeness of the digest can be performed if the laboratory suspects incomplete NotI digestion.[25,40]

Polymerase Chain Reaction (PCR)

Several methods involving PCR analysis for the molecular diagnosis of PWS have been developed, including methylation-specific PCR, fluorescence melting curve analysis, bisulfite restriction analysis, and reverse transcription PCR. These techniques have several advantages

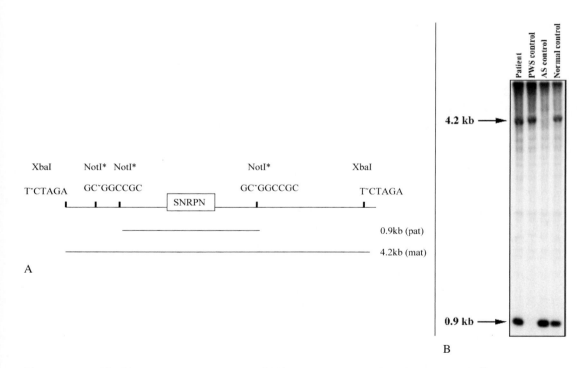

Figure 4.3 A. XbaI/NotI restriction map and RFLP sizes surrounding SNRPN probe[68] (*MT-sensitive site, methylated site will not cut with enzyme). **B.** Autoradiogram of Southern blot using the probe SNRPN with NotI and XbaI digested DNA. The patient analyzed is negative for Prader-Willi syndrome (PWS); both the 4.2 kb and the 0.9 kb DNA bands are present. AS = Angelman syndrome.

over Southern blot analysis for the diagnosis of PWS (Table 4.1), including a rapid turnaround time and smaller amount of DNA template required.[39] As with Southern blot analysis, each method will detect virtually all cases of PWS.

Methylation-Specific PCR (mPCR)

In 1997, Zeschnigk and co-workers demonstrated that >95% of the CpG dinucleotides around SNRPN exon 1 are methylated on the maternal chromosome and not methylated on the paternal chromosome. The template for methylation-PCR (mPCR) is bisulfite treated DNA. Bisulfite converts unmethylated cytosines to uracil.[35] PCR primers specific to the modified and unmodified DNA sequence were designed to distinguish between the maternal and paternal alleles.[39,41,78,79]

Following an overnight incubation of DNA treatment with bisulfite, multiplex PCR using maternal- and paternal-specific primers analyzed by gel electrophoresis produces two different amplicons, one corresponding to the maternal allele and the other corresponding to the paternal allele. The amplicon sizes vary depending on the primers used. Using the primers described by Kubota et al.,[41] normal individuals show a paternal product of 100 bp and a maternal product of 174 bp, whereas PWS patients show only the maternal 174 bp product and AS patients show only the 100 bp paternal product (Figure 4.4). Untreated

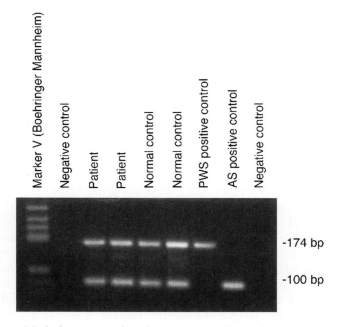

Figure 4.4. Methylation-specific PCR (mPCR) analysis of the SNRPN promoter using method described by Kubota et al.[41] Lane 1 (far left): size standard (marker V, Boehringer Mannheim). Lane 2: no DNA or negative control. Lanes 3–4: patients analyzed for Prader-Willi syndrome (PWS); both are normal with the maternal (174 bp) and paternal (100 bp) bands or amplicons present. Lanes 5–6: normal controls. Lane 7: PWS positive control. Lane 8: AS positive control. Lane 9 (far right): no DNA or negative control.

DNA does not result in a PCR product since the primers are not complementary to the unmodified DNA sequence.[41] Denaturing high performance liquid chromatography (DHPLC) and primer extension ion pair reverse phase HPLC are alternative methods to gel electrophoresis for the analysis of mPCR products.[2,49]

PCR using bisulfite treated DNA produces accurate results in PWS patients, in complete agreement with Southern blot analysis.[11,54,79] This method requires only a small amount of DNA template, which can be extracted from whole blood, lymphoblasts, dried blood cards, cultured amniocytes, cultured chorionic villi, heparinized blood, and archived fixed-cell suspensions remaining from FISH studies.[41,75] This method analyzes the methylation status of the *SNRPN* gene along many CpG dinucleotides, avoiding the potential pitfall of restriction site polymorphisms[23] and incomplete digests. One drawback of mPCR is that the smaller, paternal allele is preferentially amplified in this assay, which may result in a false negative result for PWS or a false positive result for AS. To obtain amplicons of similar intensity, the relative primer concentrations need to be adjusted with three- to fourfold less paternal primers than maternal primers.[41,79] Skewed signal intensity may also reflect mosaicism for UPD, which can be confirmed by microsatellite analysis of individual cells.[36]

Fluorescence Melting Curve Analysis

Bisulfite treated genomic DNA amplified with primers specific for SNRPN can also be analyzed by melting curve analysis.[77] This method is based on the difference in melting temperatures between the maternal and paternal PCR amplicons. Double-stranded DNA denatures at a temperature dependent on its sequence, with G-C base pairs melting at a higher temperature than A-T base pairs. Using this technology, amplification is performed in a thermal cycler connected to a fluorometer or LightCycler, eliminating the need for gel electrophoresis. This method has all of the benefits of mPCR with an even shorter turnaround time. However, careful primer design is essential to a successful assay, and this method utilizes expensive equipment that may not be available in all clinical molecular laboratories.

PCR Using DNA Digested with Methylation-Sensitive Restriction Enzymes

A variant of mPCR involves the digestion of genomic DNA with methylation-sensitive restriction enzymes followed by PCR amplification of the SNRPN promoter.[21] Two separate digests are performed: one with NotI, which will only digest unmethylated DNA; and another with McrBC, which will only digest methylated DNA. Three separate multiplex PCRs are then performed using uncut DNA, NotI treated DNA, and McrBC treated DNA as templates. Each PCR contains two sets of primers, one specific for SNRPN and another control primer set specific for a nonimprinted gene on another chromosome. SNRPN sequences are not amplified from McrBC treated DNA from PWS patients or from NotI treated DNA from AS patients. Though reported to be 100% sensitive and specific for PWS and AS caused by deletions and UPD, this assay will only detect CpG methylation that occurs in methylation-sensitive restriction sites and is prone to false negative results secondary to incomplete NotI digestion. This method is nonisotopic, requires a minimal amount of DNA, and does not require bisulfite treatment of the DNA.

Bisulfite Restriction Analysis (BRA)

Bisulfite restriction analysis (BRA) was developed as an alternative to mPCR.[73,74] This method involves two rounds of methylation-independent nested PCR amplification of bisulfite treated genomic DNA, followed by CfoI digestion. Bisulfite treatment destroys the CfoI recognition sequence on the paternal allele, so that the paternal amplicon will remain undigested whereas the maternal allele will be cleaved. There is the risk of a false negative result in a PWS patient due to incomplete digestion, and BRA is more labor intensive than mPCR, requiring the bisulfite treatment of DNA, two separate PCRs, a restriction enzyme digest, and gel electrophoresis. However, BRA eliminates the concerns of preferential amplification of the paternal allele in mPCR and can also be used on fixed-cell suspensions remaining from cytogenetic studies.[74]

Reverse Transcription PCR (RT-PCR)

Although Southern blot, mPCR and its variants will detect >99% of PWS cases, a few cases have been reported in the literature with a classic PWS phenotype, normal methylation studies, and balanced 15q11q13 translocations.[22,43,63,67] In these cases, disruption of the *SNRPN* locus by the translocation was revealed by abnormal SNRPN expression. Expression studies of the *SNRPN* gene are used as another means for the clinical molecular diagnosis of PWS.[32] The SNRPN expression test is rapid and reliable in the molecular diagnosis of PWS.[18,76] This assay involves the reverse transcription of RNA extracted from blood collected in either sodium EDTA or sodium heparin tubes. Multiplex-PCR is performed on the cDNA product with two sets of primers, one specific for the *SNRPN* mRNA and one specific for a constitutively expressed X-linked gene (*WASP*). The products are analyzed by agarose gel electrophoresis and visualized by staining with ethidium bromide. Absence of the SNRPN product in the presence of the control amplicon indicates lack of *SNRPN* expression, hence a diagnosis of PWS.

Similar to mPCR, RT-PCR is nonisotopic, requires only a small amount of blood, and has a short turnaround time (1–2 days). RT-PCR is predicted to detect all cases of PWS, regardless of etiology, but is not informative in diagnosing AS. This method has been shown to correctly identify cases of PWS due to deletion, UPD, imprinting defects, and translocations (including those that are not detectable by standard methylation analysis), although relatively few cases of PWS due to IC microdeletions and translocations have been studied. RT-PCR involves the isolation and handling of RNA, which requires dedicated "RNAse free" laboratory areas. Another drawback of working with RNA is that it degrades more rapidly than DNA. False positive or negative results have not been reported using RT-PCR; however, theoretically a false negative result could occur in a patient with normal levels of nonfunctional SNRPN protein.[18]

Uniparental Disomy

Once the diagnosis of PWS has been established using SNRPN methylation or expression analysis, additional studies are needed to determine the molecular mechanism of PWS, provide accurate genetic counseling, and establish recurrence risks for other family members (Figure 4.1). Over 70% of subjects with PWS have deletions; therefore, FISH studies should be the initial reflex test performed. If FISH studies are negative (normal) then the possibility of UPD, accounting for about 20% to 25% of PWS subjects, should be considered.[44] Uniparental disomy studies involve the amplification of multiple microsatellites within the PWS critical region as well as microsatellites outside of the critical region (to distinguish between an IC deletion and UPD). DNA from the proband and both parents is required for UPD analysis. If only one maternal allele is present within the critical region as well as proximal and distal loci, isodisomy is implied. If both maternal alleles are present, heterodisomy is implied. Maternal inheritance within the critical region, with maternal and paternal alleles present in the outside regions, will occur if a deletion is present. A number of microsatellites within and adjacent to the critical region can be used for this analysis.[4,7,15,19,28,30,47,55] Information regarding many of the microsatellites is available on the Genome Database Web site (www.gdb.org[27]), including primer sequences and references. When possible, microsatellites can be amplified by multiplex PCR to save time and reagents. The PCR products can be radioactively labeled and analyzed by polyacrylamide gel electrophoresis. Alternatively, microsatellites can be amplified by multiplex PCR using fluorescence labeled primers and analyzed on an automated DNA sequencer.[46]

Imprinting Defects

PWS patients with positive methylation DNA studies, a normal karyotype and FISH studies, and biparental inheritance of chromosome 15 are assumed to have an imprinting defect. This occurs in less than 5% of PWS cases. Approximately 15% of the imprinting defects are due to microdeletions within the IC, the majority of which are inherited, with a 50% recurrence risk.[8,9,10,57] An abnormal methylation pattern in the unaffected father of a PWS proband is indicative of a familial IC deletion. Methylation studies of a carrier father will reveal an AS methylation pattern. A normal methylation pattern in the father of a PWS proband is suggestive of a nondeletion case, which accounts for the other 85% of PWS imprinting defects.[8,9]

Nondeletion imprinting defects occur during spermatogenesis when the imprint on the father's maternally inherited chromosome 15 is not erased.[7,9] Although the proband inherits one chromosome 15 from each parent, the paternally inherited chromosome carries the grandmaternal imprint, thus the proband has two maternally imprinted chromosomes.[50] The mechanism for this type of defect is not known. No IC point mutations have been identified in patients with a nondeletion imprinting defect.[7,9,57]

High-Resolution FISH Probes

Tharapel et al.[70] reported their use of primed *in situ* labeling (PRINS) to identify deletions of *SNRPN*, *GABRB3*, and *DGCR2/TUPLE1*. This method has potential use to identify small deletions within the PWS imprinting center.

High-resolution FISH probes to test for constitutional and acquired chromosome imbalances have been developed.[38,62] The probes can identify small deletions involving only the imprinting center, including those that have been reported in multiplex PWS families. Their PWS-SRO probe performed as expected in typical PWS deletion cases and in a case of familial PWS in which a deletion was confined to the PWS imprinting center.[38]

Prenatal Diagnosis

Prenatal diagnosis of PWS and genetic counseling should be offered when the father is a microdeletion carrier, as the recurrence risk for PWS in this situation is 50%. A carrier female is not at risk to have children with PWS; however, she is at risk to have affected grandchildren through her carrier sons. Prenatal testing is also indicated for PWS due to a paternal translocation, as there is an increased recurrence risk. For PWS due to large deletions, nondeletion IC defects, and UPD, the recurrence risk is small (<1%); however, prenatal testing may be offered to reassure the parents.[1]

PWS prenatal testing should be considered in women undergoing routine prenatal testing for fetal chromosome abnormalities by CVS or amniocentesis when any of the following are detected: 15q deletions, mosaic trisomy 15, familial or *de novo* translocations or marker chromosomes involving chromosome 15 (including iso-chromosome 15).[1,44,60,65] All of these situations are associated with an increased risk of 15q deletions or UPD.

Microsatellite analysis, which requires fetal as well as parental DNA, can be used to test for UPD or small deletions. Microsatellite analysis may also be used for preimplantation genetic diagnosis to detect PWS IC deletions. Methylation analysis can be performed on fetal tissue to detect >99% of PWS cases. PW71B is known to be hypomethylated in amniocytes and chorionic villi, which could result in a false positive diagnosis of AS. ZNF127 and PW71B are not diagnostic for prenatal specimens.[24,31,42,65,72] SNRPN analysis by mPCR or Southern blot analysis is the method of choice for prenatal PWS diagnosis.[4,42]

Selecting a Reference Laboratory

The American College of Medical Genetics (ACMG) and American Society of Human Genetics (ASHG) have recommended that PWS/AS genetic testing include either: (1) Southern hybridization with SNRPN or PW71B probes followed by FISH, UPD, or imprinting studies in positive cases, or (2) FISH and methylation analysis performed simultaneously, followed by UPD studies if needed.[1] In either scenario, chromosome analysis is recommended to identify a translocation or other abnormality. It is recommended that the methods used take into account a number of factors, including the availability of testing, known test results for a given patient, and physician experience. When the likelihood that a patient will test positive for PWS is less than 50%, it is less expensive to perform DNA methylation studies, with FISH and UPD studies done only if PWS or AS is diagnosed.[53] However, whenever an expedient diagnosis is crucial—for instance, in a hypotonic newborn—simultaneous FISH and DNA testing should be considered.

We surveyed U.S. clinical molecular genetics laboratories listed on the GeneTests web site (www.geneclinics.org[20]) and found that about half perform PWS testing by PCR methods and half utilize Southern blot analysis, with SNRPN being the most common probe used (K.G. Monaghan, unpublished data). About one third of the labs offer microsatellite analysis for UPD studies, and two thirds offer FISH for PWS. Approximately half of the labs offer prenatal diagnosis for PWS. This is consistent with a recent American College of Medical Genetics/College of American Pathologists (ACMG/CAP) molecular genetics survey (MGL-A, 2003) in which 46% of labs reported using Southern blot analysis, 46% used PCR, 4% performed SNRPN expression studies by RT-PCR, and a minority used a combination of methods including Southern blot and PCR or Southern blot and SNRPN expression studies.

Laboratories in the U.S. can be inspected and accredited by the CAP laboratory accreditation program and Clinical Laboratory Improvement Act (CLIA). In addition, proficiency testing is available for many genetic conditions, including cytogenetics, FISH, and PWS/AS DNA testing. Of the labs participating in a recent MGL-CAP survey, 93% (43/46) correctly genotyped the PWS/AS specimens included in the proficiency test. In the 2002 CY-C CAP survey, all participating laboratories correctly scored the PWS unknown case. Although participation in inspections and proficiency testing do not exclude the possibility of technical or clerical errors, participation in these voluntary inspections and lab surveys indicates the laboratory's commitment to quality.

Other considerations when selecting a reference laboratory include the cost and reporting time. In addition to the many technical advantages of mPCR (Table 4.1), our survey found that PCR offers a short turnaround time (3–5 days) and a lower price (~US$250–$400) compared with Southern blot analysis (2–3 weeks; ~US$300–$500) and FISH (3–5 days; ~US$300–$500).

Finally, when a laboratory report is inconsistent with the clinical presentation of a patient, it is important to keep in mind the pitfalls of genetic testing. An abnormal result can be masked by polymorphisms within restriction enzyme recognition sequences (refer to section on Southern blotting), mosaicism for 15q11–q13 deletions,[48] or mosaicism for UPD.[36] Incomplete restriction enzyme digests can result in a false positive for PWS, while degraded DNA can result in a false positive for AS when analyzed by Southern blotting. A normal polymorphism within a primer sequence can interfere with PCR amplification. Sample mix-ups can occur in the clinic or in the laboratory. Testing for the wrong disease can result in a noninformative negative result, in which case the differential diagnosis should be considered. Mistaken paternity will yield uninformative results for microsatellite analysis. Contamination of samples is especially problematic for PCR-based assays and contamination of fetal specimens with maternal DNA can result in inaccurate prenatal results. Laboratories should perform additional studies to exclude maternal cell contamination of any prenatal specimen. If there is any doubt in the patient's result it is reasonable to repeat testing using an alternative methodology or to request a second specimen.

References

1. ASHG/ACMG (American Society of Human Genetics/American College of Medical Genetics) Test and Technology Transfer Committee. Diagnostic testing for Prader-Willi and Angelman syndromes: report of the ASHG/ACMG Test and Technology Transfer Committee. *American Journal of Human Genetics.* 1996;58:1085–1088.
2. Baumer A, Wiedemann U, Hergersberg M, Schinzel A. A novel MSP/DHPLC method for the investigation of the methylation status of imprinting genes enables the molecular detection of low cell mosaicisms. *Human Mutation.* 2001;17:423–430.
3. Berend SA, Horwitz J, McCaskill C, Shaffer LG. Identification of uniparental disomy following prenatal detection of Robertsonian translocations and isochromosomes. *American Journal of Human Genetics.* 2000;66:1787–1793.
4. Buchholz T, Jackson J, Robson L, Smith A. Evaluation of methylation analysis for diagnostic testing in 258 referrals suspected of Prader-Willi or Angelman syndromes. *Human Genetics.* 1998;103:535–539.
5. Buchholz T, Jackson J, Smith A. Methylation analysis at three different loci within the imprinted region of chromosome 15q11–13. *American Journal of Medical Genetics.* 1997;72:117–119.
6. Buiting K, Dittrich B, Dworniczak B, et al. A 28-kb deletion spanning D15S63 (PW71) in five families: a rare neutral variant? *American Journal of Human Genetics.* 1999;65:1588–1594.
7. Buiting K, Dittrich B, Groß S, et al. Sporadic imprinting defects in Prader-Willi syndrome and Angelman syndrome: implications for imprint-switch models, genetic counseling, and prenatal diagnosis. *American Journal of Human Genetics.* 1998;63:170–180.
8. Buiting K, Farber C, Kroisel P, et al. Imprinting center deletions in two PWS families: implications for diagnostic testing and genetic counseling. *Clinical Genetics.* 2000;58:284–290.
9. Buiting K, Groß S, Lich C, Gillessen-Kaesbach G, El-Maarri O, Horsthemke B. Epimutations in Prader-Willi and Angelman syndromes: a molecular study of 136 patients with an imprinting defect. *American Journal of Human Genetics.* 2003;72:571–577.
10. Buiting K, Saitoh S, Groß S, Dittrich B, Schwartz S, Nicholls RD, Horsthemke B. Inherited microdeletions in the Angelman and Prader-Willi syndromes define an imprinting center on human chromosome 15. *Nature Genetics.* 1995;9:395–400.
11. Buller A, Pandya A, Jackson-Cook C, et al. Validation of a multiplex methylation-sensitive PCR assay for the diagnosis of Prader-Willi and Angelman syndromes. *Molecular Diagnosis.* 2000;5:239–243.

12. Butler MG. Prader-Willi syndrome: current understanding of cause and diagnosis. *American Journal of Medical Genetics.* 1990;35:319–332.

13. Butler MG. Are specific short arm variants or heteromorphisms over-represented in the chromosome 15 deletion in Angelman or Prader-Willi syndrome patients? *American Journal of Medical Genetics.* 1994;50:42–44.

14. Butler MG. Molecular diagnosis of Prader-Willi syndrome: comparison of cytogenetic and molecular genetic data including parent of origin dependent methylation DNA patterns. *American Journal of Medical Genetics.* 1996;61:188–190.

15. Butler MG, Christian SL, Kubota T, Ledbetter DH. A five year-old white girl with Prader-Willi syndrome and a submicroscopic deletion of chromosome 15q11-q13. *American Journal of Medical Genetics.* 1996;65:137–141.

16. Butler MG, Palmer CG. Parental origin of chromosome 15 deletion in Prader-Willi syndrome. *Lancet.* 1983;I:1285–1286.

17. Butler MG, Thompson T. Prader-Willi syndrome: clinical and genetic findings. *The Endocrinologist.* 2000;10:3S-16S.

18. Carrel AL, Huber S, Allen DB, Voelkerding KV. Assessment of SNRPN expression as a molecular tool in the diagnosis of Prader-Willi syndrome. *Molecular Diagnosis.* 1999;4:5–10.

19. Cassidy SB, Forsythe M, Heeger S, et al. Comparison of phenotype between patients with Prader-Willi syndrome due to deletion 15q and uniparental disomy 15. *American Journal of Medical Genetics.* 1997;68:433–440.

20. Children's Health System and University of Washington, Seattle. The Gene Tests page. Available at: http://www.geneclinics.org. Accessed August 25, 2003.

21. Chotai KA, Payne SJ. A rapid, PCR-based test for the differential molecular diagnosis of Prader-Willi and Angelman syndromes. *Journal of Medical Genetics.* 1998;35:472–475.

22. Conroy JM, Grebe TA, Becker LA, et al. Balanced translocation 46, XY, t(2;15)(q37.2;q11.2) associated with atypical Prader-Willi syndrome. *American Journal of Human Genetics.* 1997;61:388–394.

23. Cuisset L, Vasseur C, Jeanpierre M, Delpech M, Noseda G, Ponsot, G. Potential pitfall in Prader-Willi syndrome and Angelman syndrome molecular diagnosis. *American Journal of Medical Genetics.* 1998;80:543–545.

24. Dittrich B, Buiting K, Groß S, Horsthemke B. Characterization of the methylation imprint in the Prader-Willi syndrome chromosome region. *Human Molecular Genetics.* 1993;2:1995–1999.

25. Dittrich B, Buiting K, Horsthemke B. PW71 methylation tests for Prader-Willi and Angelman syndromes. *American Journal of Medical Genetics.* 1996;61:196–197.

26. Dittrich B, Robinson WP, Knoblauch H, et al. Molecular diagnosis of the Prader-Willi and Angelman syndromes by detection of parent-of-origin specific DNA methylation in 15q11-q13. *Human Genetics.* 1992;90:313–315.

27. DKFZ Heidelburg, Germany. The Genome DataBase mirror site. Available at: http://www.gdb.org. Accessed October 30, 2003.

28. Fridman C, Varela MC, Kok F, Setian N, Koiffmann, CP. Prader-Willi syndrome: genetic tests and clinical findings. *Genetic Testing.* 2000;4(4):387–392.

29. Gillessen-Kaesbach G, Groß S, Kaya-Westerloh S, Passarge E, Horsthemke B. DNA methylation based testing of 450 patients suspected of having Prader-Willi syndrome. *Journal of Medical Genetics.* 1995;32:88–92.

30. Glatt K, Sinnett D, Lalande M. The human γ-aminobutyric acid receptor subunit β3 and α5 gene cluster in chromosome 15q11-q13 is rich in highly polymorphic (CA)n repeats. *Genomics.* 1994;19:157–160.

31. Glenn CC, Deng G, Michaelis RC, et al. DNA methylation analysis with respect to prenatal diagnosis of the Angelman and Prader-Willi syndromes and imprinting. *Prenatal Diagnosis.* 2000;20:300–306.

32. Glenn CC, Porter KA, Jong MTC, Nicholls RD, Driscoll DJ. Functional imprinting and epigenetic modification of the human SNRPN gene. *Human Molecular Genetics.* 1993;2:2001–2005.

33. Glenn C, Saitoh S, Jong MTC, et al. Gene structure, DNA methylation, and imprinted expression of the human SNRPN gene. *American Journal of Human Genetics.* 1996;58:335–346.

34. Hawkey CJ, Smithies A. The Prader-Willi syndrome with a 15/15 translocation. *Journal of Medical Genetics.* 1976;13:152–156.

35. Herman E, Graff JR, Myohanen S, Nelkin BD, Baylin SB. Methylation-specific PCR: a novel PCR assay for methylation status of CpG islands. *Proceedings of the National Academy of Science USA.* 1996;93:9821–9826.

36. Horsthemke B, Nazlican H, Husing J, et al. Somatic mosaicism for maternal uniparental disomy 15 in a girl with Prader-Willi syndrome: confirmation by cell cloning and identification of candidate downstream genes. *Human Molecular Genetics.* 2003;12:2723–2732.

37. Hulten M, Armstrong S, Challinor P, et al. Genomic imprinting in an Angelman and Prader-Willi translocation family. *Lancet.* 1991;338:638–639.

38. Knoll JHM, Rogan PK. Sequence-based, in situ detection of chromosomal abnormalities at high resolution. *American Journal of Medical Genetics.* 2003;121A(3):245–257.

39. Kosaki K, McGinniss MJ, Veraksa AN, McGinnis WJ, Jones KL. Prader-Willi and Angelman syndromes: diagnosis with a bisulfite-treated methylation-specific PCR method. *American Journal of Medical Genetics.* 1997;73:308–313.

40. Kubota T, Aradhya S, Macha M, et al. Analysis of parent of origin specific DNA methylation at SNRPN and PW71 in tissues: implication for prenatal diagnosis. *Journal of Medical Genetics.* 1996;33:1011–1014.

41. Kubota T, Das S, Christian SL, Baylin SB, Herman JG, Ledbetter DH. Methylation-specific PCR simplifies imprinting analysis. *Nature Genetics.* 1997;16:16–17.

42. Kubota T, Sutcliffe JS, Aradhya S, et al. Validation studies of SNRPN methylation as a diagnostic test for Prader-Willi syndrome. *American Journal of Medical Genetics.* 1996;66:77–80.

43. Kuslich CD, Kobori JA, Mohapatra G, Gregorio-King C, Donlon TA. Prader-Willi syndrome is caused by disruption of the SNRPN gene. *American Journal of Human Genetics.* 1999;64:70–76.

44. Ledbetter DH, Engel E. Uniparental disomy in humans: development of an imprinting map and its implications for prenatal diagnosis. *Human Molecular Genetics.* 1995;4:1757–1764.

45. Ledbetter DH, Riccardi VM, Airhart SD, Strobel RJ, Keenan BS, Crawford JD. Deletions of chromosome 15 as a cause of the Prader-Willi syndrome. *New England Journal of Medicine.* 1981;304:325–329.

46. Lee MJ, Nishio H, Nagai T, Okamoto N, Yuki T, Sumino K. Molecular genetic analysis of the Prader-Willi syndrome using fluorescent multiplex PCR of dinucleotide repeats on chromosome 15q11-q13. *Clinica Chimica Acta.* 1998;271:89–96.

47. Lerer I, Meiner V, Pashut-Lavon I, Abeliovich D. Molecular diagnosis of Prader-Willi syndrome: parent of origin dependent methylation sites and non-isotopic detection of (CA)n dinucleotide repeat polymorphisms. *American Journal of Medical Genetics.* 1994;52:79–84.

48. Malzac P, Moncla A, Pedeillier K, Vo Van C, Girardot L, Voelckel MA. Atypical molecular findings identify limits of technical screening tests for

Prader-Willi and Angelman syndrome diagnoses. *American Journal of Medical Genetics.* 1998;78:242–244.

49. Matin MM, Baumer A, Hornby DP. An analytical method for the detection of methylation differences at specific chromosomal loci using primer extension and ion pair reverse phase HPLC. *Human Mutation.* 2002;20:305–311.

50. McEntagart ME, Webb T, Hardy C, King MD. Familial Prader-Willi syndrome: case report and a literature review. *Clinical Genetics.* 2000;58:216–233.

51. Mignon C, Malzac P, Moncla A, et al. Clinical heterogeneity in 16 patients with inv dup 15 chromosome: cytogenetic and molecular studies, search for an imprinting effect. *European Journal of Human Genetics.* 1996;4:88–100.

52. Ming JE, Blagowidow N, Knoll JHM, et al. Submicroscopic deletion in cousins with Prader-Willi syndrome causes a grandmatrilineal inheritance pattern: effects of imprinting. *American Journal of Medical Genetics.* 2000;92:19–24.

53. Monaghan KG, Wiktor A, Van Dyke DL. Diagnostic testing for Prader-Willi syndrome and Angelman syndrome: a cost comparison. *Genetics in Medicine.* 2002;44:448–450.

54. Muralidhar B, Butler MG. Methylation PCR analysis of Prader-Willi syndrome, Angelman syndrome, and control subjects. *American Journal of Medical Genetics.* 1998;80:263–265.

55. Mutirangura A, Greenberg F, Butler MG, et al. Multiplex PCR of three dinucleotide repeats in the Prader-Willi/Angelman critical region (15q11–q13): molecular diagnosis and mechanism of uniparental disomy. *Human Molecular Genetics.* 1993;2:143–151.

56. Nicholls RD, Knoll JHM, Butler MG, Karam S, Lalande M. Genetic imprinting suggested by maternal heterodisomy in non-deletion Prader-Willi syndrome. *Nature.* 1989;342:281–284.

57. Ohta T, Gray TA, Rogan PK, et al. Imprinting-mutation mechanisms in Prader-Willi syndrome. *American Journal of Human Genetics.* 1999;64:397–413.

58. Park JP, Moeschler JB, Hani VH, et al. Maternal disomy and Prader-Willi syndrome consistent with gamete complementation in a case of familial translocation (3;15)(p25;q11.2). *American Journal of Medical Genetics.* 1998;78:134–139.

59. Rineer S, Finucane B, Simon EW. Autistic symptoms among children and young adults with isodicentric chromosome 15. *American Journal of Medical Genetics.* 1998;81:428–433.

60. Robinson WP, Langlois S, Schuffenhauer S, et al. Cytogenetic and age-dependent risk factors associated with uniparental disomy 15. *Prenatal Diagnosis.* 1996;16:837–844.

61. Robinson WP, Wagstaff J, Bernasconi F, et al. Uniparental disomy explains the occurrence of the Angelman or Prader-Willi syndrome in patients with an additional small inv dup(15) chromosome. *Journal of Medical Genetics.* 1993;30:756–760.

62. Rogan PK, Cazcarro P, Knoll JHM. Sequence-based design of single-copy genomic DNA probes for fluorescence in situ hybridization. *Genome Research.* 2001;11:1088–1094.

63. Schulze A, Hansen C, Skakkebaek NE, Brondum-Nielsen K, Ledbetter DH, Tommerup N. Exclusion of SNRPN as a major determinant of Prader-Willi syndrome by a translocation breakpoint. *Nature Genetics.* 1996;12:452–454.

64. Silverstein S, Lerer I, Buiting K, Abeliovich D. The 28-kb deletion spanning D15S63 is a polymorphic variant in the Ashkenazi Jewish population. *American Journal of Human Genetics.* 2001;68:261–263.

65. Slater HR, Vaux C, Pertile M, Burgess T, Petrovic V. Prenatal diagnosis of Prader-Willi syndrome using PW71 methylation analysis-uniparental disomy and the significance of residual trisomy 15. *Prenatal Diagnosis.* 1997;17:109–113.

66. Smith A, Prasad M, Deng ZM, Robson L, Woodage T, Trent RJ. Comparison of high-resolution cytogenetics, fluorescence in situ hybridization, and DNA studies to validate the diagnosis of Prader-Willi and Angelman syndromes. *Archives of Disease in Childhood.* 1995;72:397–402.

67. Sun Y, Nicholls RD, Butler MG, Saitoh S, Hainline BE, Palmer CG. Breakage in the SNRPN locus in a balanced 46, XY, t(15;19) Prader-Willi syndrome patient. *Human Molecular Genetics.* 1996;5:517–524.

68. Sutcliffe JS, Nakao M, Christian S, et al. Deletions of a differentially methylated CpG island at the SNRPN gene define a putative imprinting control region. *Nature Genetics.* 1994;8:52–58.

69. Tenner KS, O'Kane DJ. Clinical application of Southern blot hybridization with chemiluminescence detection. *Methods in Enzymology.* 2000;305: 450–466.

70. Tharapel AT, Kadandale JS, Martens PR, Wachtel SS, Wilroy RS, Jr. Prader Willi/Angelman and DiGeorge/Velocardiofacial syndrome deletions: diagnosis by primed in situ labeling (PRINS). *American Journal of Medical Genetics.* 2002;107:119–122.

71. Ungaro P, Christian SL, Fantes JA, et al. A molecular characterization of four cases of intrachromosomal triplication of chromosome 15q11–q14. *Journal of Medical Genetics.* 2001;38:26–34.

72. van den Ouweland AMW, van der Est MN, Wesby-van Swaay E, Tijmenson TSLN, Los FJ, Van Hemel JO. DNA diagnosis of Prader-Willi and Angelman syndromes with the probe PW71 (D15S63). *Human Genetics.* 1995;95:562–567.

73. Velinov M, Jenkins EC. PCR-based strategies for the diagnosis of Prader-Willi/Angelman syndromes. *Methods and Molecular Biology.* 2003;217: 209–216.

74. Velinov M, Gu H, Genovese M, Duncan C, Brown WT, Jenkins E. The feasibility of PCR-based diagnosis of Prader-Willi and Angelman syndromes using restriction analysis after bisulfite modification of genomic DNA. *Molecular Genetics and Metabolism.* 2000;69:81–83.

75. Velinov M, Gu H, Shah K, et al. PCR-based methylation testing for Prader-Willi or Angelman syndromes using archived fixed-cell suspensions. *Genetic Testing.* 2001;5:153–155.

76. Wevrick R, Francke U. Diagnostic test for the Prader-Willi syndrome by SNRPN expression in blood. *Lancet.* 1996;348:1068–1069.

77. Worm J, Aggerholm A, Guldberg P. In-tube DNA methylation profiling by fluorescence melting curve analysis. *Clinical Chemistry.* 2001;47:1183–1189.

78. Zeschnigk M, Lich C, Buiting K, Doerfler W, Horsthemke B. A single-tube PCR test for the diagnosis of Angelman and Prader-Willi syndrome based on allelic methylation differences at the SNRPN locus. *European Journal of Human Genetics.* 1997;5:94–98.

79. Zeschnigk M, Schmitz B, Dittrich B, Buiting K, Horsthemke B, Doerfler W. Imprinted segments in the human genome: different DNA methylation patterns in the Prader-Willi/Angelman syndrome region as determined by the genomic sequencing method. *Human Molecular Genetics.* 1997;6:387–395.

Part II
Medical Physiology and Treatment

5

Medical Considerations in Prader-Willi Syndrome

Urs Eiholzer and Phillip D.K. Lee

In the decade since the previous edition of this text,[2] a wealth of information has accumulated regarding the medical pathophysiology and treatment of PWS. Clinical diagnostic criteria and acceptance of genetic testing in the mid-1990s helped to focus clinical and research attention on individuals with this condition. The identification of a defect in the growth hormone (GH) system and, moreover, demonstration of GH treatment efficacy, provided the first and, to date, only treatment for PWS to attain regulatory approval.

Despite these advances, we still lack systematic information regarding many of the morbidities associated with PWS. Aside from GH-related studies, few rigorously-collected data on sufficient numbers of individuals have been published. Medical care for most individuals with PWS worldwide remains uncoordinated and clinical protocols are largely based on anecdotal experience. These circumstances are undoubtedly due to a combination of the uncommon occurrence of the condition, a relative paucity of specialized treatment facilities with consequent lack of centralized care and data collection, and historical prejudices and beliefs regarding natural history and treatment outcomes. Nevertheless, our understanding of the medical aspects of PWS has reached a new level in the past decade, and there are continuing efforts amongst experts in the field to establish logical, evidence-based treatment modalities.

Whereas the 1995 edition of this text contained only a single chapter concerning medical pathophysiology, entitled "Endocrine and Metabolic Aspects of Prader-Willi Syndrome,"[127] the current edition includes three chapters covering a wider range of medical concerns. This first chapter in Part II provides a general overview of medical concerns in PWS and discussion of conditions that are not addressed in successive chapters. The next chapter (Chapter 6) addresses issues related to obesity, gastrointestinal function, and body composition. Finally, Chapter 7 is dedicated to a review of GH therapy in PWS.

Natural History and Age-Related Morbidity

The medical life history of a typical individual with PWS can be instructive in providing a logical framework for the succeeding discussions. Various "life stage" classifications have been proposed, all generally similar in terms of associated morbidities. The life-staging outlined here is primarily for the sake of illustration. In reality, most PWS-associated morbidities are part of a lifelong continuum.

The first stage is *in utero*. The most notable abnormality at this stage is fetal hypomotility, usually noted by women with previous non-PWS pregnancies. This hypomotility is consistent with hypotonia, a lifelong feature of the syndrome. In male fetuses with PWS, there is an increased risk for testicular maldescent but not other genital abnormalities. Scoliosis may also occur *in utero*, although this is uncommon in our experience. Hypogonadism and scoliosis, like hypotonia, are lifetime features of the syndrome. No other abnormalities have been noted during fetal life. No abnormalities of the placenta, umbilical cord, or amniotic fluid volume or composition have been reported. Increased fetal wastage has not been reported.

The neonate with PWS is usually born at or near term and no specific abnormalities of labor and delivery have been noted. A slightly low, but not usually abnormal, birth weight is characteristic. Cryptorchidism and scoliosis may be noted at birth, along with the characteristic facies and other features as described in preceding chapters.

The most striking medical abnormality in the neonatal period is severe whole-body hypotonia, a condition that is virtually universal for infants with PWS. Decreased limb movements, marked truncal hypotonia, weak cry, and decreased neuromotor reflexes are related characteristics. Dysconjugate eye movements (strabismus, but not nystagmus) may be noted. A weak suck reflex and consequent poor feeding occur in all infants with PWS.

For the remainder of the neonatal period hypotonia and associated morbidities continue to be the number one major medical concern. Failure-to-thrive is a common diagnosis, often leading to valiant efforts to increase body weight gain. However, it should be noted that obesity, defined as increased body fat relative to lean mass, is present even in these underweight infants with PWS[68]; feeding maneuvers may improve weight gain, while doing little to improve mass and function. However, severe undernutrition may have detrimental effects on bone growth and brain development.

Neuromotor and speech delay, the latter probably due to a combination of oromotor hypotonia and central cognitive defects, are also first noted in the neonatal period. Apnea and hypoventilation may be observed in the neonatal period and are nearly universal on formal testing.

The toddler/childhood stage has a markedly late beginning in children with PWS, with most children unable to ambulate until after 2 years of age. Although markedly decreased muscle mass and hypotonia continue during this period, the available muscle fibers gain sufficient function to enable food gathering and oromotor function, greatly increasing the risk for overweight. Problems with apnea and hypoventilation may appear to abate during this period. Scoliosis may progress with ambulation and linear growth. Physical therapy can markedly improve physical function during childhood, although hypotonia is still evident.

One of the paradoxes of childhood growth in PWS is that linear growth will often decelerate as weight gain accelerates, typically beginning at 1 to 3 years of age. The major exceptions are those children who develop premature adrenarche. As discussed in a later section, these children can appear to have normal or accelerated linear growth; however, their final adult height is often severely compromised due to premature epiphyseal closure.

Adolescence is marked by incomplete sexual maturation. Adrenarche is usually normal or early, but male individuals with PWS rarely progress past mid-puberty and girls often fail to have menarche. Because of the hypogonadism and deficiencies in the growth hormone system, the usual pubertal growth spurt is blunted. Scoliosis, if present, can become particularly problematic starting in late childhood. Overweight also tends to progress rapidly, accompanied by increased risks for associated morbidities, such as diabetes mellitus. Unlike with usual exogenous obesity, bone mineral density is often low in adolescents with PWS, a problem that tends to progress through adult life.

The natural medical history of adults with PWS is relatively unstudied. Although most young adults with PWS are capable of living a semi-independent life, significant physical disability is evident. The combination of overweight and hypotonia eventually leads to wheelchair-dependence in many cases and respiratory insufficiency in most cases. In addition, osteoporosis, often caused or accentuated by untreated hormonal deficiencies, and scoliosis can also lead to significant morbidity.

Mortality

The natural life span of individuals with PWS is currently undefined. Experience suggests that survival past the 5th or 6th decade is unusual. For instance, in a survey of 232 adults with PWS, the oldest was 62 years.[78] Perhaps because of the obvious feature of overweight and obesity, an anecdotal assumption has been made that many, if not most, individuals with PWS succumb to cardiovascular complications related to obesity. However, available data suggest that this may not be the case. As reviewed in other sections, insulin resistance does not appear to occur in the majority of adults with PWS. Atherogenic lipid profiles have also not been observed with increased frequency, and atherosclerotic heart disease has not been reported. Personal experience and recent data indicate that the major cause of overall mortality in PWS is respiratory insufficiency or cardiorespiratory failure. In many cases, demise appears to be triggered by acute or chronic pulmonary infection. It appears likely that underlying poor respiratory effort due to hypotonia may be a major contributing factor. This anecdotal impression is supported by available published reviews of cases.

In an early review of cases, Laurance et al.[125] found 24 individuals with PWS alive after age 15 years (15–41 years of age) and 9 deaths; 4 deaths occurred before age 15, and 5 between 17 and 23 years of age. Cardiorespiratory failure was the proximate cause of death in these cases.

A population-based survey in England included 66 individuals with PWS, aged birth to 46 years; 25% were found to have Type 2 diabetes mellitus, and 50% had recurrent respiratory infections.[31,242] Other morbidities included ulcers of the lower extremities (22% of adults), scoliosis (15% in children), and sleep disorders (20%). Limited data from this study indicated an extrapolated lifetime mortality rate of >3% per year.

An international review of mortality in PWS found 13 deaths in individuals under the age of 5 years and 14 deaths in individuals older than 9 years.[185] Of the 13 deaths in younger individuals, 9 were due to respiratory failure and 2 were judged to be "sudden" following onset of gastrointestinal symptoms and fever. In the 14 older patients, pneumonia was contributory in 3 cases and cardiorespiratory failure in 1. Of the other deaths, two were thought to be due to gastric dilatation, two had uncertain cause, and there were single cases of myocardial infarction, stroke, familial cardiomyopathy, femoral thrombosis, spinal myelitis, and malignancy. Nine of the younger and only three of the older patients were autopsied.

In a study of 36 adults with PWS in Australia, there were 10 deaths at a mean age of 33 years (range 20–49 years); 3 were due to pneumonia or cardiorespiratory arrest, 1 was due to respiratory illness complicated by congestive heart failure, 2 were undetermined, and there were single cases of hypoglycemia, myocardial infarction, stroke, and pulmonary embolus.[193]

A population-based study in Flanders of all patients diagnosed by DNA methylation analysis found no individuals with PWS over age 56 years and a marked drop in numbers of cases with age, particularly after age 32 years.[229] Seven cases of death were reviewed: three children succumbed to pneumonia and respiratory failure and four adults died from various causes—respiratory infection, cardiorespiratory failure, stroke, and car accident. Respiratory problems were thought to be possibly contributory in five of eight deaths (5 months to 43 years of age).[200]

Small adrenal size was identified in three of four autopsied patients in this latter series but was not reported in another series of seven autopsies.[185] The possibility of adrenal hypofunction at the time of death is unresolved, although other studies have shown normal adrenal function in PWS.

Medical Concerns

Numerous medical conditions have been associated with PWS, with varying degrees of published documentation. For the purposes of discussion, PWS-associated physiology, pathophysiology, and medical morbidities will be grouped as follows:

This Chapter: Medical Considerations

I. Disorders of Sexual Development and Maturation
    A. Genital hypoplasia and cryptorchidism
    B. Hypogonadism

II. Musculoskeletal Disorders
    A. Hypotonia
    B. Scoliosis
    C. Osteoporosis

III. Respiratory and Circulatory Disorders
    A. Respiratory disorders
    B. Cardiovascular and cerebrovascular systems

IV. Miscellaneous Medical Concerns
    A. Thermoregulatory disorders, autonomic dysfunction, and anesthesia risk
    B. Ophthalmologic disorders
    C. Sensory function
    D. Mitochondrial DNA
    E. Epilepsy
    F. Cancer
    G. Infection

Chapter 6: Gastrointestinal System, Obesity, and Body Composition

I. Gastrointestinal System and Disorders
    A. Oropharynx
        1. Salivation
        2. Feeding and swallowing
    B. Stomach
        1. Mechanical function
        2. Digestion
    C. Intestines
    D. Pancreas and liver

II. Obesity and Nutrition
    A. Overweight and obesity
        1. Diagnosis
        2. Pathogenesis:
            a. Overweight and body composition
            b. Energy expenditure
        3. Associated morbidities
    B. Treatment
        1. Nutritional strategies
        2. Special nutritional considerations
        3. Associated morbidities

III. Measurement of Body Composition
    A. Review of methodologies
    B. Summary comments

Chapter 7: Growth Hormone

I. Background
II. Growth hormone treatment
    A. Efficacy of treatment in children with PWS
    B. GH treatment of adults with PWS
    C. Adverse effects

III. Summary and Conclusions

Based on clinical experience and published studies, conditions that do not occur with increased frequency in PWS include the following:

1. Thyroid dysfunction[1,29,36,51,126,177,206,240]
2. Hyper- or hypoprolactinemia[36,126,160,177,195,206]
3. Adrenocortical dysfunction[29,35,126,195]
4. Abnormal melatonin secretion[177]
5. Parathyroid hormone disorders
6. Intrinsic defects in calcium or vitamin D metabolism (although nutritional deficits can occur)
7. Dyslipidemia[22,170,204]
8. Type 1 diabetes mellitus
9. Autoimmune or immunodeficiency disorders
10. Disorder of the renin/aldosterone/angiotensin[29] and antidiuretic hormone systems
11. Renal and hepatic disorders
12. Primary lung diseases, including asthma
13. Disorders of olfaction and taste[72,171] (not including food preference)
14. Hearing disorders (Auditory and visual processing disorders have been reported.[199])
15. Slipped capital femoral epiphysis[241]
16. Seizure disorders

Disorders of Sexual Development and Maturation

Genital Hypoplasia and Cryptorchidism

Pathophysiology

Genital hypoplasia is a frequent finding in neonates with PWS, occurring in 13 of 15 cases (86.5%) reviewed for derivation of infant diagnostic criteria.[44,105]

In male infants with PWS, cryptorchidism is reported in >90% of cases; most cases are bilateral.[17,54,57] In comparison, cryptorchidism occurs in 5% or less of non-PWS infant males.[19] It is clinically important to distinguish true cryptorchidism from retractile and gliding testes,[19] although the evaluation and treatment of these conditions is similar. Small testes and scrotal hypoplasia, although usually not a true bifid scrotum, are also reported in most cases of PWS,[53] whereas micropenis is reported in less than 50%. Hypospadias and retention of female structures have not been reported. Prostate and ductal systems have not been fully characterized, although anecdotal reports indicate that the testicular accessory structures are often noted to be hypoplastic at surgery. Inguinal hernia can be expected to occur in >90% of cases of cryptorchidism due to persistence of a patent processus vaginalis[128]; however, data in PWS regarding inguinal hernia risk is lacking.

Genital abnormalities in female infants with PWS have been more difficult to characterize, although neonatal labial hypoplasia has been reported.[40] Hypoplasia of the clitoris and/or labia minora was noted in 32 of 42 females with PWS (76%) evaluated at a mean age of 17.5 years. Internal female genital anatomy has not been characterized, although no specific abnormalities of müllerian duct derivatives have been described. In addition, no cases of inappropriate virilization or appearance of male structures has been reported.

Clues to possible mechanisms for the fetal genital malformations in PWS can be gleaned from the nature of the observed abnormalities and knowledge of normal fetal genital maturation, which has been extensively reviewed in the literature[7,144,191] and is only briefly summarized here.

In the male fetus, the first evidence for sexual differentiation, the appearance of the seminiferous cords and Sertoli cells, occurs at ~6–7 weeks gestation. Leydig cell production of testosterone is first detectable before 9 weeks gestation. Müllerian (female) duct regression occurs between 7 and 10 weeks, overlapping with Wolffian (male) duct differentiation between 8 and 12 weeks.

For the male external genitalia, elongation of the genital tubercle is noted by 8 weeks. By 16 weeks gestation, urethral closure with penile formation and fusion of the labioscrotal folds are complete. All of these events, and the formation of the prostate, are dependent upon the production of testosterone by the fetal Leydig cells and conversion to dihydrotestosterone, via the action of 5α-reductase Type II in the target tissues. The external genitalia experience further growth and less significant differentiation after approximately 13 to 14 weeks gestation.

The testes originate at 7 to 8 weeks gestation in proximity to the inguinum and kidneys and are apparently anchored in place by the gubernaculum, a gelatinous structure that attaches the lower pole of each testis to the inguinal region. Testicular descent has been divided into two stages.[110] The first stage, occurring at 8–15 weeks gestation and sometimes termed transabdominal descent, involves cranial migration of the kidneys and other structures while the testes remain in place. The gubernaculum enlarges in an androgen-independent process that may be facilitated by testicular production of insulin-3 (also known as descendin), creating a path in the inguinal canal for future testicular descent. The inguinal canal is formed by differentiation of musculature surrounding the gubernaculum. During the second stage, at 28 to 35 weeks, the patent processus vaginalis protrudes through the internal inguinal ring, thereby transmitting intra-abdominal pressure to the gubernaculum.[109] This is thought to cause movement of the gubernaculum through the inguinal canal into the scrotum, thereby guiding initial testicular descent. As summarized by Hussman and Levy,[109] descent of

the testes is dependent on both gubernaculum-dependent development of the inguinal canal and on the sporadic increases in intra-
abdominal pressure induced by fetal respirations and hiccupping. After
descent, the gubernaculum regresses through an androgen-dependent
mechanism. Androgens may also influence development of musculature involved in formation of the inguinal canal and optimization of
intra-abdominal pressure.

Although fetal gonadotropin production is evident by 9 to 10 weeks,
the early stages of testicular differentiation, androgen production, and
genital development are likely maintained by placental chorionic
gonadotropin, which is the predominant fetal gonadotropin through
gestation. However, during late gestation, production of LH and possibly other fetal pituitary hormones appears to be necessary for completion of external genital development, genital growth and testicular
descent, as evidenced by the common findings of micropenis and
cryptorchidism in cases of congenital hypopituitarism (non-PWS).

It is important to realize that each stage of male fetal sexual differentiation and development is controlled by time-limited processes that
are determined not only by hormonal exposure, but also by specific
genes with time-limited activity.[7] Therefore, each step described above
must occur within a specific "window of opportunity." In addition, the
testicular hormones (insulin-3, testosterone) involved in testicular
descent are largely active on a local or paracrine, rather than systemic,
level. Therefore, unilateral defects in hormone action may lead to unilateral cryptorchidism.

In consideration of this knowledge of male genital embryology, a
tentative theory for the occurrence of male genital abnormalities in
PWS can be generated. Steps that are not abnormal in PWS include
urethral closure and labioscrotal fusion, implying normal testicular
differentiation, normal *in utero* testosterone production, and normal
activity of 5α-reductase through 16 weeks gestation. In addition, the
usual location of undescended testes near the inguinal ring or in the
inguinal canal in PWS-associated cryptorchidism implies normal
androgen-independent transabdominal descent at 8–15 weeks. The
preponderance of bilateral cryptorchidism indicates that unilateral testicular malfunction or structural abnormality is not involved. Finally,
the not uncommon occurrence of bilaterally descended testes and the
relatively low occurrence of micropenis suggest that the genital abnormalities are not intrinsic to the genetic defect per se, but may be secondary to other factors that occur with variable severity from one affected
individual to another. Two major possibilities along these lines are (1)
fetal gonadotropin deficiency, leading to decreased late-gestational
testosterone deficiency; and (2) fetal hypotonia, leading to inadequate
intra-abdominal pressure. Since both hypotonia and a variable, non-
absolute gonadotropin deficiency occur in postnatal life, it appears
likely that both factors are involved in the etiology of the male genital
abnormalities in PWS.

Normal female external genital development has been characterized
by Ammini et al.[9] The labial folds, clitoris and vestibule with single
perineal opening are evident by 11 weeks gestation. Between 11 and 20

weeks, the labia majora, derived from the genital swellings (scrotal anlage in the male), continue to enlarge. However, the labia minora, derived from the genital folds (~penile shaft in the male), and the clitoris, derived from the genital tubercles (~glans penis in the male), show little change during this period. At 23–25 weeks, the labia minora enlarge and protrude past the labia majora, and the dual openings (urethra and vagina) move to the perineum. Thereafter, the labia majora become more prominent. Fetal pituitary gonadotropin levels peak at 20–24 weeks (FSH > LH) in the female, and a maximal number of ovarian primordial follicles are observed at 22–24 weeks, possibly stimulated by the FSH peak. In females with PWS, the major external genital abnormality has been noted to be hypoplastic labia minora and a small clitoris, structures that show maximal growth immediately following the fetal gonadotropin surge. Therefore, it appears likely that, as with males with PWS, the external genital abnormalities in some female infants with PWS may be related to a deficiency of fetal gonadotropin secretion.

Evaluation

Any male neonate with unexplained hypotonia and cryptorchidism with or without micropenis should be suspected of having PWS, especially if there are no other anogenital abnormalities. Unexplained neonatal hypotonia is, of course, a stand-alone criterion for consideration of PWS testing[80]; however, cases of late-diagnosed (older children and adults) PWS presenting with hypotonia and cryptorchidism have been anecdotally observed by the authors. Similarly, in females with hypotonia, the presence of hypoplasia of the labia minora and/or small clitoris should further encourage consideration of testing for PWS.

In cases of cryptorchidism, evaluation for presence of testes and testicular function may be advisable before and after surgical treatment. Consultation with a pediatric urologist and pediatric endocrinologist is highly advisable. If the testicle(s) is completely nonpalpable, an initial inguinal/pelvic ultrasound should be performed. In some cases, a magnetic resonance imaging (MRI) may provide additional information,[58] although this procedure has not usually been necessary in our experience. Brain imaging studies are not particularly useful unless there is other evidence for pituitary or CNS dysfunction; intrinsic structural abnormalities of the central nervous system visible on routine imaging have not been identified in PWS.[35,201]

Single measurements of testosterone, LH, and FSH in the first few days of life or during the expected secondary gonadotropin surge at 30–60 days of age may, if positive, confirm the presence of functional testicular tissue. A testosterone level >20–50 ng/dL (700–1750 pmol/L) is indicative of testosterone production. However, although this random sampling is often useful in non-PWS infants with cryptorchidism, utility may be limited in congenital hypogonadotropic hypogonadism since gonadotropin priming may be necessary for optimal testicular response. Therefore, a negligible random neonatal testosterone level in an infant with PWS does not preclude the existence of functional testicular tissue.

Gonadal stimulation testing to document functional testicular tissue is not usually required, but may be indicated in those patients with true bilateral cryptorchidism and a low random neonatal testosterone level or in older infants/children where the hypothalamic-pituitary-gonadal axis may be expected to be naturally suppressed. Since orchiopexy is usually not performed in the immediate neonatal period, such testing can often be delayed until immediately prior to surgery. A standard "short" testosterone stimulation test involves daily intramuscular injections of chorionic gonadotropin ($1500\,U/m^2/dose$) for 3 to 5 days, with serum testosterone measured at baseline and 24 hours after the final injection. However, since gonadotropin priming may be necessary to achieve testicular secretion, a "long" test may be preferable. A typical regimen is chorionic gonadotropin 500–1000 U by intramuscular injection twice a week (e.g., Monday/Thursday) for 5 weeks (10 doses total) with baseline and 24-hour post-final-dose serum testosterone levels. The baseline level is theoretically not necessary for either the short or long test, although it may provide information about basal secretion and tissue responsiveness. A post-test testosterone level of >100–200 ng/dL is clearly indicative of testicular activity; levels between 20 ng/dL and 100 ng/dL are considered equivocal, although there is presumably some amount of functional tissue present. The long test has the additional theoretical advantages of more prolonged testosterone production, possibly resulting in some penile growth, testicular enlargement, and possible testicular descent. However, these latter effects are rarely, if ever, observed in PWS. Intranasal and subcutaneous gonadotropin administration have also been used for diagnosis and treatment of cryptorchidism; however, reports specific to PWS are limited.[91,112]

Treatment

Micropenis: In the male infant with PWS and micropenis, a short course of low-dose testosterone may be used to improve the appearance and function of the penis.[84] It is often stated that a stretched penile length of <1.5–2.0 cm could lead to difficulty with toilet training and upright urination, as well as complicate peer relationships[40]; however, scientific data relating to these effects is lacking. Treatment of micropenis in infancy is often based on the judgment of the physicians and parents. In non-PWS individuals, there is no evidence that such treatment adversely affects later penile growth potential.[23]

If treatment is elected, a typical treatment regimen is depot testosterone (enanthate or cypionate), 25 mg by intramuscular injection every 3 to 4 weeks for 3 to 6 months. A longer course or higher dose is not recommended since either of these could lead to inappropriate virilization and acceleration of skeletal maturation. Experience suggests that this treatment should be initiated in infancy, preferably before 6 months of age, for maximal effectiveness. There is a theoretical risk for triggering central puberty in older children, although this risk may be minimal in children with PWS. The child should be examined prior to each dose to gauge clinical response and the necessity for additional doses; a stretched penile length ≥2.0 cm is adequate. Particularly in the obese child, the suprapubic fat pad should be fully compressed when obtaining a penile length measurement. On occasion, a few strands of pubic hair and acceleration of linear growth may be observed during the treatment course, but these effects invariably abate within a few weeks after treatment is completed.

Cryptorchidism: Spontaneous descent of a cryptorchid testicle(s) may occur in a prepubertal or pubertal boy with PWS,[88] as it does in a significant proportion of cases of non-PWS cryptorchidism.[19] The converse condition, spontaneous ascent of a descended testicle(s) has not been reported in PWS. In other cases, medical intervention with gonadotropin administration may be successful in achieving testicular descent,[214] although the ultimate success rate of hormonal therapy is low in non-PWS cases of true cryptorchidism[207] and data in PWS is lacking.

In both PWS and non-PWS cases of cryptorchidism, the justification for and timing of surgical exploration and orchiopexy have been controversial. Studies in non-PWS populations indicate that cryptorchidism is associated with a 35- to 48-fold increased risk for testicular cancer.[128] However, it is unclear whether surgical correction of cryptorchidism (orchiopexy) reduces the risk for cancer. Several studies indicate that the risk for testicular cancer is negligible if surgical correction is performed before puberty, whereas the risk for cancer is increased >30-fold thereafter.[99] The majority of such tumors in non-PWS cryptorchid males are testicular carcinoma, although germ cell and Sertoli cell tumors are also reported.[50] Given these statistics, it may be somewhat surprising that only two cases of testicular cancer in PWS have been published, a seminoma in a 40-year-old man[178] and a germ cell neoplasm in a 9-year-old boy with cryptorchidism.[111] Since the risks for maldescent-associated cancer in non-PWS cases may also involve an association with puberty, it is possible that the numbers of observed cases in PWS are low due to early correction and lack of later spontaneous pubertal development.

Facilitation of physical surveillance for torsion and testicular tumors has also been proposed as a justification for orchiopexy.[128] In one study of non-PWS individuals, more than half the cases of torsion of an intra-abdominal testis, usually presenting with pain and discomfort, were associated with germ cell tumor.[176] In PWS, a single case of germ cell neoplasia discovered in a cryptorchid testicle has been reported.[111]

Preservation of testicular function and fertility has been given as another reason for treatment of cryptorchidism. In a study of 1,335 non-PWS boys with cryptorchidism, an increased risk for germ cell absence was noted after 15–18 months of age, and this was associated with adult infertility in a follow-up study, while surgical correction prior to this time appeared to preserve fertility.[50] This may have relevance to PWS, in which lack of germ cells has been found on testicular biopsy.[29,116,235] Preservation of male fertility has not been an emphasis in the treatment of PWS, but the evolving treatment of this condition may soon justify consideration of this possibility.

Among other factors to be considered, an undescended testis located in the inguinum may be at increased risk for traumatic injury. In non-PWS cases, cryptorchidism is associated with a >90% chance for a patent processus vaginalis and consequent increased risk for hernia and hydrocele,[128] thus providing another justification for surgical evaluation and repair. Cosmetic and psychosocial considerations may also be important in some cases.

All cases of cryptorchidism should be evaluated by an experienced pediatric urologist before 6 months of age. Since infants with PWS may present unique surgical and anesthesia risks, surgical procedures may be delayed and medical therapy initially attempted. Orchiopexy may be particularly difficult in PWS due to the degree of maldescent and hypoplastic accessory structures. Either before or after surgery, the testicles may be atretic and nonfunctional, and experience suggests that the risk for post-orchiopexy testicular retraction and degeneration is high. In some cases, removal of the testicular tissue with or without placement of prostheses may be a preferred or necessary option.

Hypogonadism

Pathophysiology

Hypogonadism, evidenced by incomplete pubertal (sexual) development occurs in more than 80% of adolescents and adults with PWS.[105] Puberty normally occurs as a two-system process, beginning after 8 years of age: adrenarche and gonadal maturation, or gonadarche. Adrenarche involves production of adrenal androgens, resulting in secondary sexual hair growth (pubic and axillary hair); this effect is superseded in males by testicular production of testosterone. The control of adrenarche is not completely defined. In boys, gonadarche involves testicular production of testosterone, resulting in masculinization. In girls, gonadarche includes ovarian maturation and estrogenization. Gonadarche is controlled by pituitary production of the gonadotropins, luteinizing hormone (LH), and follicle-stimulating hormone (FSH), which are in turn controlled by hypothalamic production of gonadotropin-releasing hormone (GnRH).

In PWS, adrenarche usually occurs normally in both sexes, although a relative decrease in secondary sexual hair has been anecdotally reported. A significant proportion of children with PWS have premature adrenarche, perhaps related to insulin resistance or decreased insulin sensitivity[122,182,184] (discussed further in Chapter 6).

On the other hand, gonadarche is usually abnormal in PWS. Many female patients with PWS progress through phenotypic puberty (breast development) but fail to have normal menstrual cycles and/or have early menopause. Breast development can progress to full maturation, although 49 of 81 individuals (64%) in a survey study were reported to have small or flat breasts.[78] Breast development is presumably due to estrogens originating from the ovaries and/or peripheral conversion of adrenal precursors; estrogen receptor variants have not been identified in PWS. Virtually all males with PWS have a failure to progress past mid-puberty.

It is likely that the postnatal hypogonadism in PWS is a continuation of hypogonadotropic hypogonadism that begins *in utero*. However, there are virtually no data regarding gonadotropin secretion in infants or pre-teen children with PWS. It is not known, for instance, whether the normal postnatal rise in FSH and LH secretion occurs in infants with PWS. Although gonadotropin response to exogenous GnRH may be absent or low in preteen children with PWS,[251] this may be considered a typical pre-pubertal pattern in non-PWS children and not necessarily indicative of pathology.

In adolescents and adults with PWS, low baseline and GnRH-stimulated gonadotropin levels are a frequently reported finding, arguing for an intrinsic hypogonadotropic hypogonadism, i.e., at the hypothalamic or pituitary level. Primary gonadal failure has not been reported, except in cases of acquired testicular atrophy,[186] and there is a single report of abnormal ovarian histology in PWS.[29] In addition, treatment with gonadotropin or gonadotropin-releasing hormone can result in improvement or normalization of gonadal function,[29,91] implying that gonadal steroidogenesis is not a primary disorder.

On the other hand, multiple observations indicate that irreversible, complete intrinsic gonadotropin deficiency at the pituitary or hypothalamic level may not be involved. Normal or precocious spontaneous menses have been reported,[29,40,117,232] implying cyclic production of gonadotropins. In addition, elevated gonadotropin secretion has been observed in some cases, especially following primary gonadal failure (i.e., with castration or testicular atrophy),[150,186,188,245] and some individuals with PWS have apparently normal gonadotropin responses on stimulation testing.[177]

Clomiphene citrate, a triphenylethylene estrogen antagonist, has been observed to stimulate gonadotropin secretion and gonadal function (including menses and spermatogenesis) in males and females with PWS.[29,88,130,141] In non-PWS individuals, clomiphene is commonly used for induction of gonadotropin secretion and ovulation during treatment of female infertility. As an estrogen analog, clomiphene binds to intracellular estrogen receptors in the arcuate nucleus of the hypothalamus, thereby blocking estrogen inhibition of gonadotropin-releasing hormone (GnRH) secretion, resulting in stimulation of pituitary gonadotropin secretion.[62] Clomiphene also appears to sensitize the pituitary gland to GnRH action, but direct effects of clomiphene to stimulate pituitary gonadotropin secretion have not been found. Finally, clomiphene may also sensitize the ovary to gonadotropin action. Therefore, clomiphene stimulation of gonadotropin secretion and action in individuals with PWS argues against an intrinsic or complete hypothalamic or pituitary GnRH or gonadotropin defect.

Induction of menses in PWS has also reported with use of fluoxetine (Prozac),[236] a selective serotonin reuptake inhibitor (SSRI) not uncommonly used for psychopharmacologic treatment of PWS. Induction of menses has been reported in a 25-year-old woman with PWS during treatment with citalopram (Celexa), another SSRI[8]; this patient subsequently became pregnant after withdrawal of citalopram, although her menses had become "sparse." SSRIs are thought to act by inhibiting neuronal reuptake and metabolism of serotonin (5-hydroxytryptamine, 5-HT), thereby increasing the local concentration and activity of serotonin. Serotonin-secreting neurons originate primarily in the midbrain and extend throughout the central nervous system. The role of serotonin in gonadotropin secretion has not been completely defined; however, it appears to modulate pulsatile secretion of FSH (but perhaps not LH) by affecting pulsatile hypothalamic GnRH secretion.[169,215,227] Therefore, these cases provide additional arguments against an intrinsic complete GnRH/gonadotropin deficiency in PWS.

A possible mechanism for partial, variable deficiency in gonadotropin secretion in PWS is suggested by rodent studies in which disruption of *Ndn*, an imprinted gene located within the PWS region that codes for necdin (neurally differentiated embryonal carcinoma-cell derived factor), results in a 25% reduction in GnRH neurons in the hypothalamus.[149] However, corollary studies in humans have not been reported.

Male fertility has not been reported in PWS and testicular biopsy has been reported to show Sertoli cell-only histology[29,116,235]—i.e., presence of Sertoli cells, which provide the milieu for spermatogenesis, and Leydig cells, which produce testosterone, but absence of germinal cells (spermatogonia). It is uncertain whether this is due to lack of LH stimulation of Leydig cell testosterone production (a requirement for spermatogenesis) or lack of FSH stimulation of Sertoli cell mitosis and/or some other factor(s). The reported induction of biopsy-confirmed spermatogenesis with clomiphene in a young adult with PWS[88] suggests that the abnormal histology is not due to an intrinsic testicular defect.

Fertility in PWS may also be limited by social factors. Most adults with PWS are not married and, presumably, have limited sexual contact. Social limitations to fertility may be altered as advances in medical and behavioral treatment improve physical outcomes and lifestyle choices.

Evaluation

Adolescents with PWS will usually have hypogonadotropic hypogonadism, unless a primary gonadal defect is present. For research purposes or in equivocal cases, serum FSH and LH levels can be measured at timed intervals following intravenous administration of GnRH.[250,251] Clomiphene pretreatment may augment gonadotropin secretion.

In cases where testicular atrophy or necrosis is suspected, random simultaneous measurements of testosterone (in boys), FSH, and LH are usually sufficient; elevations of gonadotropins and low levels of testosterone can be expected if there is primary gonadal failure.[150,186,188,245] Primary ovarian failure has not been reported in PWS; however, elevated gonadotropin levels would be expected in such a case.

Measurements of serum testosterone may be useful in guiding hormone replacement therapy in males. Estrogen (estradiol) measurements in females are less useful due to large inter- and intra-individual temporal variability as well as limited assay sensitivity.

Fertility testing is rarely, if ever, clinically warranted in PWS. Despite the obesity, there does not appear to be an increased occurrence of polycystic ovary syndrome in PWS.

Treatment

Virtually all males with PWS fail to progress past mid-puberty and testosterone levels are usually low relative to the adult normal range.

However, androgen replacement therapy has been somewhat contro-versial, primarily due to anecdotal concerns that testosterone treatment might cause inappropriately aggressive behavior or worsen other behavioral problems, although there are no published data showing such a relationship. On the other hand, testosterone therapy, combined with psychotherapy, was reported to improve behavior in an adoles-cent with PWS.[212] In addition, detailed studies in non-PWS eugonadal and hypogonadal men have not shown an effect of physiologic or moderately supraphysiologic levels on measures of aggression.[159] Our experience suggests that testosterone replacement should be started at a biologically appropriate time, i.e., in early adolescence; late initiation of therapy may be associated with adverse psychosocial outcome, perhaps due to an impact on the socialization process.

Normal adult testosterone levels are considered to be necessary for preservation of bone mass and prevention of osteoporosis in males.[218] In addition, testosterone has positive effects on muscle mass and strength in hypogonadal men.[21] Since osteoporosis and hypotonia are major causes of morbidity in PWS, replacement therapy would appear to be medically advisable.

In recent years, new delivery systems for testosterone replacement have become available.[21] The traditional intramuscular injection of 200 mg depot testosterone once or twice monthly may no longer be preferred in most cases since this causes an acute supraphysiologic rise over 24 hours, followed by a steep decline to subnormal levels over 2 to 3 weeks. If supraphysiologic levels of testosterone have even minor effects on mood and behavior, then the depot injection may cause inap-propriate swings in these parameters. Testosterone patches are now available at several dose levels. The patches are changed daily, can be applied to any skin area (the back, thighs and buttocks are particularly convenient), and the 5-mg patch usually achieves low-normal adult testosterone levels. Anecdotal experience indicates that pruritis and skin picking at the patch site have not been a problem, even in patients who have skin picking in other areas. Finally, testosterone gel prepara-tions can be useful in selected cases, although precautions must be taken to avoid cross-contamination of adult females and prepubertal children.

Estrogen replacement therapy, usually combined with progestin, is an established therapy for non-PWS hypogonadal women. This treat-ment has been shown to have multiple benefits, including preservation of bone mass and reduction of fracture risk. However, estrogen replace-ment in women with PWS is not as well established. Most women with PWS have evidence of estrogen effect (breast development) and menses are not uncommonly reported. Osteoporosis is a documented morbid-ity in adult women with PWS; however, there are no published data showing efficacy of estrogen replacement. Nevertheless, estrogen replacement may be advisable in women with PWS, particularly in the presence of amenorrhea or oligomenorrhea and/or osteoporosis.

Hypogonadism in women can be treated with any of a number of approved oral or transdermal forms of estrogen in combination with progestins or micronized progesterone. Commercial birth control preparations are often used for convenience, are anecdotally well tolerated, and are probably the most commonly prescribed modality in our experience. However, in view of the low reported resting energy expenditure and predisposition for adiposity of Prader-Willi syndrome patients and the report of divergent effects of oral vs. transdermal estrogen on lipid oxidation and fat accumulation,[162] it could be argued that non-oral forms of estrogen might be physiologically preferable in this population. In addition, in consideration of the increasing use of growth hormone therapy in PWS, the growth hormone replacement dose required for comparable response in growth hormone deficient non-PWS women taking oral estrogen was more than double that of women on transdermal estrogen.[49] Thus, treatment with transdermal estrogen with oral progestin (to reduce endometrial cancer risk) or a combined estrogen-progestin patch may be preferred, although there is limited experience and no published data using these preparations in PWS.

Testosterone or estrogen replacement therapy should be coupled with careful monitoring of clinical and behavioral status, and bone mineral density. Testosterone levels may be useful in adjusting the patch or gel dose. Monitoring of other biochemical parameters (coagulation profiles, liver enzymes, prostate-specific antigen) may be advisable in some cases. Prostate or gynecologic examinations should also be considered, as appropriate.

In either males or females with PWS who are unable to tolerate gonadal steroid replacement, bisphosphonate therapy can be considered for treatment or prevention of osteoporosis, as has been recommended for similar situations in non-PWS women.[154] Bisphosphonate treatment may also be used as an adjunct to gonadal steroid replacement in cases with severe osteoporosis.

Contraception per se is probably not necessary for most individuals with PWS, although it may be a consideration for women with menstrual cycles. However, sexual activity and risks for sexually-transmitted disease are a reality and concern in the treatment of PWS. Therefore, all adolescents and adults with PWS should receive appropriate education and counseling.

Musculoskeletal Disorders

Hypotonia

Hypotonia is a universal feature of PWS, manifesting in fetal life as hypomotility and continuing through postnatal existence. Neonatal hypotonia is considered to be the key criterion for clinical suspicion of PWS in neonates.[77,80,105,209] The proximate etiology of the hypotonia is decreased muscle mass, but the pathogenesis of this condition is unknown. Neuromuscular studies, including electromyography and nerve conduction velocity studies, are usually normal; this feature of hypotonia with normal neuromuscular studies was included in the clinical diagnostic criteria.[105] Ultrasound analysis showed no abnormality in muscle echogenicity in eight infants with PWS (9 days to 18 months old). A few autopsied cases showed cerebellar changes that could cause motor dysfunction[94,95]; however, these findings do not explain the decreased muscle mass and related neuromuscular features.

Histologic and ultrastructural studies of muscle fibers in PWS are limited. No morphologic changes are identified on routine staining and microscopy and the fibers appear mature.[5,196] In one infant, there was a relative paucity of Type 1 fibers and a preponderance of small Type 2 fibers.[14] In a study of 11 PWS infant muscle biopsies, the Type 1 fibers showed increased size variability and nonspecific abnormalities of Type 2 fibers were noted.[196] However, examination of muscle fibers generally shows minimal, if any, abnormality.[5,6,196] Ultrastructural analysis in one case showed increased subsarcomeric accumulation of normal-appearing mitochondria and sarcomeric Z-line changes. The composite data from these limited studies appear to be more consistent with a peripheral muscular disorder or disuse than to a central nervous system myopathy. A possible contribution of a deficiency in insulin-like growth factor-I (IGF-I), a growth hormone-related growth factor, is suggested by the correlation of IGF-I levels with fat-free mass.[217]

Physical therapy is an essential element of treatment for Prader-Willi syndrome. However, data relating physical therapy and improvement in neuromuscular function in PWS are notably lacking. Due to the lack of muscle mass, both strength and endurance are deficient, and the range of innate activity can be quite variable.[152] The usual functional improvement in muscle function during infancy and childhood may be due to strengthening of existing muscle fibers rather than to significantly increased muscle mass per se. One study indicates that fat mass may be inversely correlated with physical activity levels in PWS[217]; a relationship of activity and fat-free mass was not reported. A 6-month aerobic exercise program (walking) led to significant improvement in aerobic capacity, coupled with decreased body fat and weight, in six adults with PWS as compared with a control group (n = 5).[190] A study using brief, daily calf muscle training (repetitive heel lowering on a household stair) in 17 children with PWS (aged 4.4–18.8 years) found a significant decrease in calf skinfold thickness and significant increases in calf circumference, physical activity, and endurance (physical capacity).[69] These data provide objective evidence that physical therapy can improve muscle mass and function in PWS.

Growth hormone treatment has also been shown to improve muscle strength and endurance in PWS. This topic is discussed in more detail in Chapter 7. Adrenarche and testosterone replacement therapy are subjectively associated with improved muscle function in some patients; however, objective data in PWS are not available.

Scoliosis

Scoliosis, a commonly observed disorder in PWS, is defined as a lateral spinal curvature (Cobb's angle) of >10 degrees on standing radiograph.[172] Cases may be observed *in utero* and in the newborn period, suggesting early initiation of pathogenetic mechanisms. However, experience indicates that most cases are diagnosed clinically in childhood, as with non-PWS scoliosis.[172] Notable progression occurs during late childhood and adolescence and continues into adult life, often exacerbated by osteoporosis. Laurance et al.[125] found scoliosis in 15 of 24 adults with PWS, 2 of whom received surgical treatment. A survey of PWS individuals in Italy showed 42 of 72 (59%) with scoliosis, including 12 of 16 deletion cases (75%).[170] A population survey in the U.K. found a 15% prevalence of "severe" scoliosis across all age groups.[31] In a survey study of 232 adults with PWS, nearly half reported scoliosis that was significant enough to require treatment.[78] Other reports estimate that >80% of individuals with PWS are affected.[106]

The pathogenesis of scoliosis in PWS is not defined but most likely falls into the category of neuromuscular scoliosis. Congenital or acquired primary vertebral abnormalities are not commonly observed, and other causes, such as trauma or degenerative disease, are also not likely. Although many cases of congenital scoliosis in PWS have been reported, there has been no association with cardiac, spinal cord, renal, or other malformations as occurs with idiopathic congenital scoliosis.[15] Scoliosis is often noted in PWS individuals who are not overweight,[197] indicating that weight is not a causative factor.

Neuromuscular scoliosis has been divided into neuropathic and myopathic subcategories.[140] Neuropathic scoliosis is associated with disorders of the upper or lower motor neurons (e.g., cerebral palsy and poliomyelitis, respectively). Myopathic scoliosis is associated with primary muscle disorders, such as congenital hypotonia and muscular dystrophy. Abnormal kyphosis or lordosis (posterior or anterior spinal angulation, respectively, as viewed from the side) is also frequently noted in neuromuscular scoliosis, as it is in PWS. Unlike the more common idiopathic adolescent scoliosis, neuromuscular scoliosis tends to progress through adult life, as observed with PWS scoliosis. Therefore, the described and anecdotal characteristics of scoliosis in PWS most closely resemble myopathic neuromuscular scoliosis, i.e., due to hypotonia of the paravertebral muscles.[252]

The risk for scoliotic curve progression increases with linear (height) growth in idiopathic adolescent scoliosis[172]; data on curve progression in PWS or other forms of neuromuscular scoliosis are lacking, although anecdotal information suggests a similar association. Back curvature should be carefully monitored during treatment with growth-promoting therapies, such as growth hormone and anabolic steroids.

The impact of scoliosis on health is not well defined in PWS. Back and lower extremity pain are frequent complaints in scoliotic individuals without PWS. However, experience indicates that these complaints are rarely encountered in those with PWS, even in cases with severely abnormal curvatures, possibly because of the known deficit in pain sensation.[105]

In infants without PWS, severe curves have been associated with respiratory compromise, but these degrees of curvature are not often observed in infants with PWS. Mild to moderate curvatures (>30 degrees) in adolescents and adults without PWS are thought to cause inefficient coordination of respiratory muscle function and the thoracic cage, resulting in decreased maximal inspiratory and expiratory pressures and reduction in exercise capacity.[121] Compensatory mechanisms involve recruitment of thoracic and abdominal musculature, and this is likely to be deficient in the presence of hypotonia and decreased muscle mass, as is observed in PWS. In non-PWS scoliosis, respiratory symptomatology progressively increases with curves of >70 degrees, with significant risks for respiratory failure at >100 degrees. Experience suggests that a contribution to respiratory compromise is observed with lesser degrees of curvature in PWS; however, published data are lacking.

Excessive kyphosis or lordosis can also be a concern.[134] The normal back has a balanced thoracic kyphosis and lumbar lordosis. Extreme accentuation (hyperkyphosis) or extension of the kyphotic curve into the lumbar region can lead to spinal cord stretching, injury, and paraplegia in non-PWS individuals. Extreme lordosis can be associated with respiratory compromise. There are no published reports of kyphosis- or lordosis-related morbidities in PWS, although clinical experience indicates that these types of curvature abnormalities are not uncommon.[82]

Given the high rate of occurrence and the potential associated morbidities, all individuals with PWS should be regularly screened for abnormal back curvature. Some PWS experts have recommended spine radiographs at regular intervals for all patients with PWS, while others recommend regular clinical screening, with radiographs obtained only if indicated by abnormal findings. The Adams forward bending test is the standard clinical screening procedure for scoliosis, performed with the patient standing with his/her back to the observer and feet together, bending forward with knees flexed and arms extended (hanging). Abnormal spinal curvature, asymmetric elevation of one shoulder, and/or pelvic tilt should all be indications for further investigation. A similar examination with the patient standing upright, combined with examination of the gait, can also be a useful screening tool. A lateral examination should be performed to detect abnormal kyphotic and lordotic curves. Clinical detection of scoliosis in individuals with PWS may be hindered by obesity, hypotonia, and lack of cooperation; in such cases, radiographic screening may be indicated. Diagnosis is based on the standing radiograph, with anteroposterior (for scoliosis) and lateral (for kyphosis/lordosis) views.

Consultation with an experienced orthopedist is essential as soon as an abnormal curvature is identified, and the treatment plan should be carefully coordinated with other therapies. Treatment of scoliosis in PWS requires conscientious medical observation to determine appropriate timing of treatment: a spinal orthosis (brace) for mild to moderate curves and surgical intervention for severe, rapidly-progressing curves. While intermittent spine radiographs may or may not be indicated for routine screening, they are definitely a mainstay of monitoring for progression. In neuromuscular scoliosis, bracing is typically used to improve the functional position of the trunk and may permit delay of surgical procedures in a younger patient. In non-PWS patients, progression of mild degrees of curvature associated with hypotonia can be slowed with bracing[20]; specific data in PWS are lacking.

Surgical stabilization of scoliosis can be difficult in PWS. Problems with rod fixation may be due to poor bone and tissue quality.[82,173] In addition, there can be significant morbidity due to respiratory compromise and infection. In a review of 33 cases of individuals with PWS treated with posterior spinal instrumentation (rod placement), excess bleeding (55%), soft bone (27%), fragile soft tissues (9%), sepsis (15%), and respiratory problems (15%) were the most commonly reported complications.[113] Growth hormone therapy has been anecdotally reported to improve surgical outcome, perhaps by improving bone and tissue quality and respiratory function.

Finally, there are limited data regarding efficacy of scoliosis therapy in PWS. In non-PWS individuals with scoliosis, surgical therapy can prevent respiratory compromise; however, similar therapy can be detrimental in cases where lung function is already impaired.[121]

Osteoporosis

Osteoporosis has been defined as a "skeletal disorder characterized by compromised bone strength predisposing a person to an increased risk of fracture."[3,155] Bone strength is dependent on both bone mineral and bone quality. However, since overall bone quality and strength are difficult to measure, bone mineral, which accounts for about two thirds of total bone strength, is commonly used to define osteoporosis.

A complete discussion of normal bone physiology is beyond the scope of this text. However, a few basic principles are essential to understanding bone pathology in PWS. Throughout life, there is continuous loss and gain of bone mineral. During childhood, there is progressive net accretion of bone matrix and subsequent mineralization. Bone mineral density, the amount of bone mineral per unit of bone mass, is relatively low in early life and gradually increases throughout childhood and adolescent growth. Peak bone mass is reached at completion of linear growth, and peak bone mineral density occurs in the third decade of life. Thereafter, there is progressive net loss of bone mineral.

Optimal net accretion of bone mass and mineralization during childhood and adolescence is dependent upon several factors, including hormones (growth hormone, gonadal steroids, vitamin D), nutrition (particularly calcium), and neuromuscular stimulation (exercise). Deficiencies in any of these areas could lead to inadequate bone mass accretion and osteoporosis, either in childhood/adolescence or later in life, as the normal loss of bone mineral occurs. Deficiencies of these factors after attainment of peak bone mass will lead to accelerated loss of bone mineral and osteoporosis. In addition, several factors may inhibit or reverse bone mineralization, including chronic use of glucocorticoids, chronic systemic illness, and immobility.

As discussed in other sections, individuals with PWS have low growth hormone, low gonadal steroid levels, and low neuromuscular activity. In addition, vitamin D levels may be low due to lack of sun exposure, and there may be deficient intakes of both calcium and vitamin D for individuals on calorie-restricted diets. Therefore, it is not particularly surprising that low bone mineral has been frequently reported in PWS.[27,33,41,108,127,223] Biochemical markers of bone turnover may also be elevated,[223] as might be expected from the probable etiologies. On the other hand, there may be a protective effect(s) of weight, weight-bearing exercise, and/or body fat on bone mineralization,[148,174] although these have not been demonstrated in PWS. Anecdotal experience suggests that decreased bone mineral density in PWS is associated with increased fracture risk and, perhaps, abnormal progression of spinal curvature; there are no published data supporting these associations.

For many reasons, dual-energy X-ray absorptiometry (DXA) has become the standard diagnostic method for assessment of bone mineral density[70]; principles of this method are discussed in the next chapter. Most research related to osteoporosis has focused on women, since osteoporosis and fracture risk in postmenopausal women are major public and individual health concerns. Osteoporosis-related fragility fractures in this population are primarily located in the hip and lumbar spine. (A fragility fracture can be defined as a fracture that occurs during normal activity or minor trauma that would not normally cause a fracture in a healthy young adult.) Because of this, standards for diagnosis of osteoporosis are based on DXA analysis of bone mineral density in the femoral neck and the upper lumbar vertebrae (L1–4).

DXA facilities are readily available in most medical institutions. A DXA scan is noninvasive, painless, involves minimal radiation exposure (less than a chest X-ray) and takes only a few minutes. When ordered for assessment of bone mineral status (e.g., to diagnose or monitor osteoporosis), the report typically includes the measured bone mineral parameters and a T-score for the bone mineral density at the hip and spine. The T-score is the standard deviations from the average measurement for a healthy, young, white female at peak bone mass (the reference standard). Therefore, a lumbar T-score of –2.0 means that the L1–4 bone mineral density is 2 standard deviations below the mean for a healthy young, white female. For adult women, fracture risk and definition of osteoporosis is based on the T-score for bone mineral density (or bone mineral content), as follows[3]:

Normal: T-score between –1 and +1
Osteopenia: T-score less than –1 but greater than –2.5
Osteoporosis: T-score less than –2.5
Severe osteoporosis: T-score less than –2.5 *and* one or more fragility fractures

It should be noted that osteopenia is not considered to be a disorder, and it is not currently defined as a risk factor for fragility fractures.

By definition, all children and adolescents can be expected to have a low T-score since peak bone mineralization has not been attained. Therefore, DXA measurements for children often refer to standard deviation (Z) scores, based on norms established by age, sex, and ethnicity,[70] using the same guidelines referred to for T-scores (osteoporosis defined as Z-score less than –2.5). Adult male standards are also not well defined and results are usually interpreted relative to the female T-score standard. However, fracture risk has not been defined for T- or Z-score levels in children or adult men.

Another caveat with DXA procedures is that results vary considerably according to instrumentation.[70,124] Technologies vary among manufacturers and direct comparisons of results are not feasible. Alternate sites (e.g., finger, ankle, heel) and alternate technologies, such as ultrasound bone density measurements, are also not directly comparable to traditional DXA results and may not be appropriately standardized in relation to age, sex, and fracture risk. Therefore, physicians should be careful to obtain sequential measurements using a single type of device (e.g., the same body regions measured using a device from the same manufacturer) if the intent is to compare results between procedures.

Although a relationship between bone mineral status and fracture risk has not been established for PWS, many specialists include DXA analyses in the routine care plan for individuals with PWS.[127] Because prevention of osteoporosis and fracture risk is more effective than secondary treatment, DXA screening should begin in childhood or early adolescence, during the period of bone accretion. Even in the presence of a normal DXA analysis, a repeat scan may be advisable if risk factors are present (e.g., hypogonadism, growth hormone deficiency, low activity) to assess the direction and rate of bone mineral changes.

Preventive treatment of low bone mineral density in PWS should include nutritional therapy (vitamin D and calcium supplementation, if needed), hormonal replacement (discussed in other sections of this chapter and in Chapter 7), and exercise. Caretakers should also be aware that individuals with PWS may have decreased deep pain sensation; therefore, bone fractures may not be accompanied by complaints of pain. Unusual gait, posture, or limb movements may signal a need for further investigation.

If there is inadequate accretion of bone mineral, excessive loss of bone mineral, or continuing osteoporosis with or without fragility fractures, bisphosphonate treatment may be advised, particularly for adolescents and adults. Typical regimens involve weekly oral doses of alendronate or risendronate, or monthly ibandronate. These medications must be administered on an empty stomach with water, with a sufficient subsequent interval before food intake to prevent reflux and esophageal complications. If this type of administration is not feasible, intermittent intravenous bisphosphonate preparations, such as pamidronate, may be used.[56] Since osteoporosis-associated morbidity has not been well studied in PWS, careful monitoring of individual therapy is necessary. A trial off bisphosphonate, with continued annual monitoring of bone density, may be attempted after an adequate bone mineral density has been achieved.

Although not specifically related to osteoporosis, it should be noted that one survey study has reported a possible 10-fold increase in the prevalence of hip dysplasia in PWS relative to the general population

(10% of 565 respondents).[241] This potentially important finding requires more objective confirmation, and possible etiologies have not been identified. Hip dysplasia can lead to limited mobility and degenerative hip disease.

Respiratory and Vascular Systems

Respiratory Disorders

Physiology

As reviewed above, respiratory disorders, including apnea, sleep-related disorders, and hypoventilation, are a major cause of morbidity in PWS, affecting at least 50% of individuals. In addition, respiratory failure is the major identified cause of mortality.

The control of normal respiration involves complex, interrelated mechanisms under both involuntary and voluntary control. The entire process of respiration involves three primary components[26]: (1) thoracic movement and ventilation, or exchange of air between the lungs and atmosphere; (2) exchange of gases (oxygen, CO_2, etc.) across the alveolar membranes; and (3) transport in the circulation, delivery, and exchange in the tissues. The end result is maintenance of tissue oxygen and CO_2 homeostasis. Respiratory disorders in PWS exclusively involve the first component of respiration. No primary abnormalities of the lung tissue or oxygen/CO_2 transport have been identified in PWS.

The first component of respiration is dependent on coordinated neural control of the respiratory musculature, including the upper airway, thoracic muscles, diaphragm, and abdominal muscles. During inspiration, the diaphragm moves down into the abdomen, creating negative pressure in the lungs and influx of air. Additional chest wall muscles may be recruited, especially during exertion. Air entry through nasal and oral paths require coordinated actions of upper airway musculature, including pharyngeal elevation, laryngeal opening, and protective closure (e.g., during swallowing and vocalization). Expiration is primarily a passive event due to elastic recoil of the chest wall and lungs; however, chest wall and abdominal muscles may be recruited during exertion. Due to the high content of Type 1 (slow twitch) and Type 2A muscle fibers, the normal diaphragm is relatively resistant to fatigue. However, the upper airway musculature has a high proportion of fast-twitch fibers and, therefore, a greater susceptibility to fatigue.

The control of breathing is via chemoreceptors that send signals to the central nervous system. Peripheral chemoreceptors in the carotid bodies sense the partial pressure of oxygen (not oxygen content per se), as well as increased PCO_2, acidemia, and low perfusion; these chemoreceptors are not essential for maintenance of respiration. Central chemoreceptors are distributed along the brainstem in several groups; these chemoreceptors respond primarily to increased PCO_2 and decreased pH in the cerebrospinal fluid, both changes resulting in increased respiration.

The automatic rhythmic nature of breathing is controlled by three respiratory centers, located in the brainstem pons, solitary nucleus (dorsal medulla), and ventral medulla. The pontine respiratory group has connections to the hypothalamus and cerebral cortex, has inspiratory and expiratory functions, and may be involved in establishing respiratory patterns. The medullary centers, and particularly the pre-Bötzinger region of the ventral respiratory group, appear to be essential for establishing respiratory rhythmicity.

Pathophysiology

Using a modification of the classification included in a recent review,[156] reported sleep and breathing disorders and representative studies in PWS are as follows:

1. Abnormalities of Daytime Breathing
   a. Restrictive lung disease[85]
   b. Abnormal ventilatory response to hypercapnia and/or hypoxia[13]

2. Abnormalities of Sleep and Wakefulness
   a. Excessive daytime sleepiness[43,47,48,78,97,100,101,115,125,138,224]
   b. Disorders of REM sleep
      i. Respiratory abnormalities during rapid-eye movement (REM) sleep[100,138,221]
      ii. Disordered timing of REM sleep[115,138,221,226]
   c. Abnormalities of arousal—reduced arousal during sleep in response to hypoxic or hypercapnic stimuli[12,132]
   d. Sleep-disordered breathing
      i. Obstructive sleep apnea[90,183]
      ii. Alveolar hypoventilation[156]
      iii. Decreased oxygen saturation[100,115,221]
      iv. Reduced ventilatory response to hypoxia and/or hypercapnia[132,183]

The etiology and PWS-specific occurrence of these respiratory disorders is controversial. For instance, some studies show a surprisingly low occurrence of sleep apnea in PWS.[97,100,115] The excessive daytime sleepiness bears some resemblance to narcolepsy, a multifactorial genetic condition characterized by excessive, disabling daytime sleepiness, cataplexy and sleep paralysis, and severe disturbances in REM sleep. However, the bulk of information suggests that PWS and narcolepsy are distinct.[156] In addition, non-PWS narcolepsy has not been associated with genes within the PWS locus.

A hypothalamic disturbance has been proposed as a common link for the observed disorders. However, as reviewed above, the major control of resting automatic respiration is at the brainstem level. Ablation of higher centers, including the hypothalamus and cortex, has minimal effect on resting respiration.[107] On the other hand, the hypothalamus is involved in adjusting respiration in response to hypoxia, hypercapnia, and other sensory input, modulating cortical and brainstem outputs to affect ventilation. The extent of this influence in humans is not completely defined. In lower species, the hypothalamic response to hypercapnia is mediated by GABA, which may be of interest since several non-imprinted GABA-receptor subunit genes are located in the PWS region of chromosome 15q. In rodents, disruption of *Ndn*, an imprinted gene within the PWS region, results in deficiency of oxytocin and gonadotropin-releasing hormone neurons in the hypothalamus,[149] which is of interest in relation to the observed deficiency of hypothalamic oxytocin neurons in human PWS[202] (although CSF oxytocin levels are reported to be high in PWS[139]) and the variable hypogonadotropic hypogonadism, as reviewed in a previous section. Disruption of *Ndn* in rodents also leads to a disorder of respiratory drive similar to that observed in PWS; this effect may be due to a defect at the brainstem level.[175] However, as of this writing, no specific hypothalamic abnormalities have been identified in human PWS or animal models to account for the clinically observed respiratory disorders.

Obesity has been proposed as a possible mechanism for obstructive sleep apnea in PWS. However, as recently reviewed,[156] the data are controversial with some studies indicating no relationship. Obesity or, more specifically, overweight has also been postulated to contribute to restrictive lung disease in PWS, as has been described in morbidly overweight individuals without PWS.[137] However, data suggest that respiratory muscle hypotonia may be a more important contributory factor,[85,156] especially since respiratory problems can be observed without excess body weight, as in neonates with PWS.

Sleep and respiratory disorders similar to those in PWS have been described in individuals with non-PWS neuromuscular disorders.[18,55,121,166,192] In myotonic dystrophy, a condition often compared to PWS, a decreased number of neurons in the medullary arcuate nucleus (a postulated brainstem respiratory control center) was observed in 3 individuals with hypoventilation, as compared with 5 with myotonic dystrophy without hypoventilation and 18 controls.[161] Although these data were interpreted to be consistent with a primary central respiratory defect in myotonic dystrophy, it may also be consistent with retrograde loss of neurons due to peripheral muscle degeneration, weakness, and disuse, a situation that may also theoretically occur in PWS. Central, obstructive, and combined apnea, similar to the apnea described in PWS, have been characterized in non-PWS individuals with neuromuscular disease causing weakness of the respiratory musculature.[121]

Individuals with neuromuscular disorders may also be more prone to respiratory compromise due to normal reductions in accessory respiratory and upper airway muscle tone that occur during REM sleep,[143] a sleep period that appears to be particularly affected in PWS. Additional upper airway problems in PWS (e.g., decreased airway diameter, thick oral secretions, and possible adenotonsillar hypertrophy[156,183]) may further compound this problem.

Finally, individuals with PWS are notably deficient in growth hormone and insulin-like growth factor-I (IGF-I, also known as somatomedin-C), a growth hormone-dependent peptide (see Chapter 7). Both growth hormone and IGF-I have been directly or indirectly shown to be respiratory stimulants.[181] Withdrawal of growth hormone therapy in adults with non-PWS GH deficiency has been associated with a shift from obstructive to central apnea and an increase in slow-wave sleep time.[157]

The consequences of respiratory disorders in PWS have not been well defined, although this category of disorder is the most commonly identified cause of demise. As with other disorders involving respiratory muscle weakness, a decreased ability to compensate during acute lung disease or hypermetabolic states may account for the relatively high occurrence of pneumonia- and fever-related mortality. The decreased respiratory capacity can also lead to hypertrophy and eventual failure of the right heart (cor pulmonale), another commonly specified cause of death in PWS (see next section).

Excessive daytime sleepiness has been correlated with severity of behavior disorders in PWS.[121] In non-PWS individuals, obstructive sleep apnea has been correlated with risks for cardiovascular and cerebrovascular diseases[165,168] and a variety of other comorbidities, including diabetes mellitus[248] and decreased cognition and school performance.[76,164,216]

Evaluation

The evaluation of respiratory and sleep disorders has gained recent prominence due to reports of respiratory-related mortality in children with PWS who received GH treatment[158,228] (see Chapter 7). An initial review by the manufacturer of Genotropin®, the GH brand labeled for treatment of PWS, revealed seven cases, all of whom had evidence of respiratory compromise (five with pneumonia) either preceding or at the time of demise ("Prader-Willi Syndrome and Death," Pfizer Global Pharmaceuticals, October 2, 2003). This led to the placement of an additional label in the package insert on April 30, 2003, including the following advisory:

Growth hormone is contraindicated in patients with Prader-Willi syndrome who . . . have severe respiratory impairment. . . . Patients with Prader-Willi syndrome should be evaluated for upper airway obstruction before initiation of treatment with growth hormone. If during treatment with growth hormone patients show signs of upper airway obstruction (including onset of or increased snoring), treatment should be interrupted. All patients with Prader-Willi syndrome should be evaluated for sleep apnea and monitored if sleep apnea is suspected.

Subsequent and previous case reviews of individuals with PWS who were *not* treated with growth hormone demonstrated that respiratory disorders are the major cause of morbidity and mortality. Consideration by the Clinical Advisory Board of PWSA (USA) recognized the overall risk for respiratory disorders in PWS, resulting in a statement that partially contradicts the Genotropin labeling (Figure 5.1).

Of these two statements, the PWSA (USA) guidelines are more consistent with clinical experience and published data. However, although these recommendations provide valuable guidance, there are limitations that prevent a completely evidence-based approach to respiratory disorders in PWS.

Most of the attention regarding respiratory disorders has focused on identification of obstructive sleep apnea (OSA). This is based largely on data in non-PWS populations showing an association of OSA with morbidity (see above); it should be noted, however, that these data are not considered to be definitive.[25] A recent review[156] found that the reported prevalence of obstructive sleep apnea in PWS ranged from 0 to 100%. Available data are limited by small sample size, biased sampling, and differences in screening and diagnostic criteria. Nonetheless, the overall impression is that a significant proportion of individuals with PWS will be diagnosed with OSA if tested, i.e., that PWS is itself a risk factor for OSA. However, it is currently unclear whether all individuals with PWS benefit from knowledge of this diagnosis in the absence of associated morbidity.

Clinical screening for OSA has been proposed in both PWS and non-PWS patients to improve the diagnostic yield of testing and/or to provide a basis for treatment. Snoring has been proposed as an indicator of clinically significant OSA in non-PWS populations. However, the diagnostic sensitivity and specificity of snoring for morbidity-associated OSA (i.e., OSA associated with cardiac, neuropsychologic, or other sequelae) have not been conclusively demonstrated in either PWS or non-PWS populations.[136,210] In non-PWS children, it has been stated that OSA "is very unlikely in the absence of snoring." Snoring is dependent on both relaxation of the pharyngeal muscles and airflow; therefore, hypotonia-related hypoventilation in PWS could decrease the occurrence of snoring even in the presence of OSA. Other proposed clinical indicators of significant morbidity-associated OSA, including obesity (or overweight), sleep history, and daytime somnolence, have not been validated in PWS and are controversial in non-PWS populations.

The "gold standard" for diagnosis of OSA is nocturnal polysomnography, commonly referred to as a "sleep study," in which various parameters (oxygen saturation, carbon dioxide content, brain wave activity, sleep/wake status) are monitored during sleep. However, the utility of this procedure is limited by cost (currently ~US$1,000–$1,500) and availability (particularly for pediatric testing). In addition, there is controversy over the interpretation of polysomnographic data that is only partly resolved by consensus criteria, and outcome-based normative data are not available. Abbreviated polysomnography, alternate procedures, and home monitors have even more serious limitations and cannot be recommended in PWS.

Polysomnography prior to surgery, particularly tonsillectomy/adenoidectomy, has been recommended for PWS (see above) and for children without PWS who have adenotonsillar hypertrophy and evidence of OSA. However, it is not clear how these data impact the performance of the surgical procedures. Artificial respiratory support is required

PWSA (USA) Clinical Advisory Board, 2003 Consensus Statement

Problems with sleep and sleep-disordered breathing have been long known to affect individuals with Prader-Willi Syndrome (PWS). The problems have been frequently diagnosed as sleep apnea (obstructive-OSA, central, or mixed) or hypoventilation with hypoxia. Disturbances in sleep architecture (delayed sleep onset, frequent arousals, and increased time of wakefulness after sleep onset) are also frequently common. Although prior studies have shown that many patients with PWS have relatively mild abnormalities in ventilation during sleep, it has been known for some time that certain individuals may experience severe obstructive events that may be unpredictable.

Factors that seem to increase the risk of sleep-disordered breathing include young age, severe hypotonia, narrow airway, morbid obesity, and prior respiratory problems requiring intervention, such as respiratory failure, reactive airway disease, and hypoventilation with hypoxia. Due to a few recent fatalities reported in individuals with PWS who were on growth hormone therapy (GH) some physicians have also added this as an additional risk factor. One possibility (that is currently unproven) is that GH could increase the growth of lymphoid tissue in the airway thus worsening already existing hypoventilation or OSA. Nonetheless, it must be emphasized that there are currently no definitive data demonstrating GH causes or worsens sleep-disordered breathing. However, to address this new concern, as well as the historically well documented increased risk of sleep-related breathing abnormalities in PWS, the Clinical Advisory Board of the PWSA (USA) makes the following recommendations:

1. A sleep study or a polysomnogram that includes measurement of oxygen saturation and carbon dioxide for evaluation of hypoventilation, upper airway obstruction, obstructive sleep apnea and central apnea should be contemplated for all individuals with Prader-Willi syndrome. These studies should include sleep staging and be evaluated by experts with sufficient expertise for the age of the patient being studied.
2. Risk factors that should be considered to expedite the scheduling of a sleep study should include:
   a. Severe obesity—weight over 200% of ideal body weight (IBW).
   b. History of chronic respiratory infections or reactive airway disease (asthma).
   c. History of snoring, sleep apnea, or frequent awakenings from sleep.
   d. History of excessive daytime sleepiness, especially if this is getting worse.
   e. Before major surgery, including tonsillectomy and adenoidectomy.
   f. Prior to sedation for procedures, imaging scans, and dental work.
   g. Prior to starting growth hormone or if currently receiving growth hormone therapy.

Additional sleep studies should be considered if patients have the onset of one of these risk factors, especially a sudden increase in weight or change in exercise tolerance. If a patient is being treated with growth hormone, it is not necessary to stop the growth hormone before obtaining a sleep study unless there has been a new onset of significant respiratory problems.

Any abnormalities in sleep studies should be discussed with the ordering physician and a pulmonary specialist knowledgeable about treating sleep disturbances to ensure that a detailed plan for treatment and management is made. Referral to a pediatric or adult pulmonologist with experience in treating sleep apnea is strongly encouraged for management of the respiratory care.

Figure 5.1. Recommendations for Evaluation of Breathing Abnormalities Associated with Sleep in Prader-Willi Syndrome.

In addition to a calorically-restricted diet to ensure weight loss or maintenance of an appropriate weight, a management plan may include modalities such as supplemental oxygen, continuous positive airway pressure (CPAP), or BiPAP. Oxygen should be used with care as some individuals may have hypoxemia as their only ventilatory drive and oxygen therapy may actually worsen their breathing at night. Behavior training is sometimes needed to gain acceptance of CPAP or BiPAP. Medications to treat behavior may be required to ensure adherence to the treatment plan.

If sleep studies are abnormal in the morbidly obese child or adult (IBW > 200%) the primary problem of weight should be addressed with an intensive intervention—specifically, an increase in exercise and dietary restriction. Both are far preferable to surgical interventions of all kinds. Techniques for achieving this are available from clinics and centers that provide care for PWS and from the national parent support organization (PWSA-USA). Behavioral problems interfering with diet and exercise may need to be addressed simultaneously by persons experienced with PWS.

If airway-related surgery is considered, the treating surgeon and anesthesiologist should be knowledgeable about the unique pre- and postoperative problems found in individuals affected by Prader-Willi syndrome (see "Medical News" article regarding "Anesthesia and PWS" written by Drs. Loker and Rosenfeld in *The Gathered View*, Vol. 26, Nov.– Dec. 2001). Tracheostomy surgery and management presents unique problems for people with PWS and should be avoided in all but the most extreme cases. Tracheostomy is typically not warranted in the compromised, morbidly obese individual because the fundamental defect is virtually always hypoventilation, not obstruction. Self-endangerment and injury to the site are common in individuals with PWS who have tracheostomies placed.

At this time there is no direct evidence of a causative link between growth hormone and the respiratory problems seen in PWS. Growth hormone has been shown to have many beneficial effects in most individuals with PWS including improvement in the respiratory system. Decisions in the management of abnormal sleep studies should include a risk/benefit ratio of growth hormone therapy. It may be reassuring for the family and the treating physician to obtain a sleep study prior to the initiation of growth hormone therapy and after 6-8 weeks of therapy to assess the difference that growth hormone therapy may make. A follow-up study after one year of treatment with growth hormone may also be indicated.

Members of the Clinical Advisory Board are available for consultation with physicians and families through the Prader-Willi Syndrome Association (USA).

Figure 5.1. *Continued.*

during such procedures regardless of test results and the supposition that test results may be predictive of recovery or surgical outcome has not been proven.[164]

In contrast to OSA and sleep studies, there has been little attention paid to routine pulmonary function testing in PWS (e.g., measurement of lung volumes, inspiratory and expiratory pressures, oxygen and carbon dioxide while awake).[156] Pulmonary function testing provides important measures of ventilatory capacity and respiratory muscle function, both of which are known to be abnormal in PWS. The relative lack of attention to such testing may, in part, be due to the usual need for active cooperation from the patient and consequent perceived difficulties with PWS. Nonetheless, modified pulmonary function testing data in children with PWS have been reported[85,131] with no significant reports of problems with test feasibility.

In summary, published data do not allow for evidence-based specific recommendations for evaluation of respiratory disorders in PWS. Experience-based recommendations, such as those proposed by PWSA (USA) (see Figure 5.1) and by Nixon and Brouillette,[156] provide reasonable guidance. However, consultation with clinicians with experience in the medical management of PWS is recommended before and after testing.

At the current time, it does not seem advisable to *routinely* test all individuals with PWS for OSA or pulmonary function abnormalities outside of a research protocol. A large proportion will be diagnosed with a respiratory abnormality regardless of symptomatology, and there are no data correlating the results of such testing with risks for morbidity or mortality. In the absence of this information, routine testing could cause undue cost and concern. Polysomnography and/or pulmonary function testing in PWS would appear to be most useful in two circumstances:

1. To assist with therapeutic planning for patients with definite, persistent evidence of clinically significant respiratory compromise or sleep disorder, e.g., 2a–2d of the PWSA (USA) risk factors—hypertension, evidence of pulmonary hypertension, or cor pulmonale.

2. To monitor therapies that may affect respiratory function, e.g., continuous positive airway pressure (CPAP), growth hormone.

Once a decision has been made to proceed with testing, it may be reasonable to routinely obtain pulmonary function testing and to limit the performance of polysomnography to those individuals with clinical evidence of sleep-disordered breathing or excessive daytime sleepiness, as has been recommended by Nixon and Brouillette.[156]

Treatment
Given the high probability of finding a respiratory abnormality in an individual with PWS and the potentially invasive nature of some treat-

ment modalities, the decision to initiate treatment should be carefully considered, preferably in consultation with specialists in PWS and cardiorespiratory care. Among the points to be considered are the following:

1. The indications for the evaluation—e.g., the signs and symptoms that prompted testing. If the testing was routine and the patient is asymptomatic, is treatment indicated? Or is continued monitoring preferable?

2. The feasibility and efficacy of treatment options. In most cases, this will be based on the consultant's own knowledge and experience, limited published data in PWS, and a larger published experience in non-PWS populations (which may or may not be directly relevant).

3. Overall clinical status and expected benefits for the individual patient.

Evaluation and treatment of contributory co-morbidities should be first and foremost, even in individuals with PWS who do not have documented respiratory abnormalities. Exercise and physical therapy may help improve respiratory and general muscle tone, thereby improving ventilatory effort. Exercise and weight control may also reduce the restrictive component of respiratory impairment.[34,61,101,225] Bracing and, if indicated, surgical correction of significant scoliosis (see section on scoliosis) may improve pulmonary function, particularly if the curvature is severe.

In non-PWS individuals with obstructive sleep apnea, particularly those with evidence of tonsillar hypertrophy, adenotonsillectomy has been reported to improve respiratory status, snoring, and neurocognitive and other disease measures, especially over the short term. However, there is limited published experience with adenotonsillectomy and a related procedure, uvulo-palato-pharyngoplasty, in PWS,[16,102,156] and concerns have been raised regarding pre- and postoperative risks even in non-PWS populations.[164]

Nasal mask or cannula delivery of oxygen and/or ventilatory support via continuous positive airway pressure or intermittent positive pressure ventilation, usually as a nocturnal therapy, may be beneficial in patients with chronic hypoxemia, although patient cooperation can be problematic.[48,156,194] In the current view of these authors and according to the recommendations of the PWSA (USA) Clinical Advisory Board, tracheostomy is not a preferable treatment option in PWS.

Central respiratory stimulants, such as medroxyprogesterone, caffeine, and theophylline, have not been adequately studied in PWS and may potentially have adverse effects on behavior. Modafinil, a relatively new medication for treatment of narcolepsy and obstructive sleep apnea, has also not been studied in PWS.[4]

Finally, growth hormone treatment has been associated with significant improvements in pulmonary function, including resting ventilation, airway occlusion pressure, and ventilatory response to carbon dioxide in nine children with PWS treated for 12 months.[131] Less significant improvements were noted with 6 months of therapy.[89]

Cardiovascular and Cerebrovascular Systems

Congenital cardiac, vascular, or cerebrovascular malformations and disorders have not been reported to occur with increased frequency in PWS. In addition, hypertension appears to have a surprisingly low occurrence in PWS in view of the obesity/overweight; this could be related to the unexpectedly low occurrence of hyperinsulinemia[147] (see Chapter 6).

Cor Pulmonale

Cor pulmonale (from Latin: *heart of the lungs*) can be generally defined as right heart dysfunction resulting from pulmonary disease. This condition is the most commonly reported cardiovascular complication in PWS and is often cited as a cause of mortality,[125,200,211,222] although careful examination of the literature reveals few well-documented cases.

Clinical criteria for cor pulmonale have been controversial. Weitzenblum[239] proposed that cor pulmonale be defined as "pulmonary arterial hypertension resulting from diseases affecting the structure and/or the function of the lungs; pulmonary arterial hypertension results in right ventricular enlargement (hypertrophy or dilatation) and may lead with time to right heart failure." A consensus panel sponsored by the World Health Organization[142,239] lists five categories of pulmonary hypertension. Categories 1 (pulmonary arterial hypertension), 2 (pulmonary venous hypertension), 4 (caused by chronic thrombotic and/or embolic disease), and 5 (caused by disorders directly affecting the pulmonary vasculature) do not appear to be relevant to PWS.

Category 3 is defined as "pulmonary hypertension associated with disorders of the respiratory system and/or hypoxemia." Eight subcategories are included: 3.1 chronic obstructive pulmonary disease, 3.2 interstitial lung disease, 3.3 sleep-disordered breathing, 3.4 alveolar hypoventilation disorders, 3.5 chronic exposure to high altitude, 3.6 neonatal lung disease, 3.7 alveolar capillary dysplasia, and 3.8 other. As described in the preceding section, categories 3.1, 3.3, and 3.4 are most relevant to PWS.

Significant pulmonary hypertension is caused by *chronic* alveolar hypoxemia ($PaO_2 < 55$–$60\,mmHg$), as may occur with neuromuscular disease and resultant hypoventilation, or with severe obstructive apnea. Alveolar hypoxemia causes pulmonary vasoconstriction, a normal physiologic response. When this occurs chronically over a long period of time, it can lead to structural remodeling of the pulmonary vasculature and chronic pulmonary hypertension. The increased pressure causes resistance to the outflow of the right heart ventricle, leading to right heart hypertrophy or dilatation, with a risk for eventual cardiac failure. In non-PWS populations, chronic obstructive sleep apnea can be associated with a 17% to 41% occurrence of pulmonary hypertension. It has been stated that in this group of patients without intrinsic lung disease, pulmonary hypertension is rarely observed in the absence of daytime (awake) hypoxemia.[142]

In non-PWS patients, clinical signs and non-invasive testing for pulmonary hypertension are considered to have variable specificity (i.e., indicative of the disorder if present) and relatively low sensitivity (does not rule out pulmonary hypertension if not present) and are usually seen late in the course of disease.[239] These findings include peripheral edema (especially ankle edema), a typical tricuspid regurgitation murmur, accentuation of the second heart sound, electrocardiogram evidence for right ventricular hypertrophy, and increased diameter of the pulmonary artery on chest X-ray. Echocardiographic methods are considered to be the best noninvasive tests for pulmonary hypertension and right heart status, correlating relatively well with more invasive methods. Magnetic resonance imaging (MRI) and other methods may provide additional information.

The treatment of pulmonary hypertension and, particularly, cor pulmonale involves (1) provision of oxygen to prevent pulmonary hypoxemia and vasoconstriction and (2) vasodilators. Surgical intervention may be necessary in severe cases. Evaluation and treatment of these disorders are best handled by an experienced cardiologist.

As of this writing, the percentage of PWS individuals who meet stringent criteria for diagnosis of pulmonary hypertension, with or without right heart dysfunction, is not known. In addition, risk factors, clinical signs and symptoms, natural history, treatment indications, and outcomes are not defined. Peripheral edema is a not uncommon finding, particularly in adults, but personal experience suggests a low correlation with a confirmed diagnosis of right heart failure. Since hypoxemia is the major cause of pulmonary hypertension and consequent right heart failure, documentation of decreased pulmonary function, chronic alveolar hypoventilation and low PaO_2 (e.g., by oximetry), particularly with daytime occurrence, may be indications for further cardiac evaluation and/or referral to a cardiologist. Cor pulmonale appears to be a frequently reported cause of mortality in adults with PWS; a recommendation could be made for routine cardiac evaluation of all adults with PWS.

Other Vascular Conditions

Only a few well-documented cases of coronary artery disease have been reported in PWS. A 26-year-old male with PWS and Type 2 diabetes mellitus was found to have a myocardial infarction and inoperable 3-vessel coronary artery disease after presenting with bilateral lower limb swelling.[123] A 28-year-old woman with PWS was found to have stenosis of the left anterior descending artery with probable myocardial infarction after presenting with chest pain.[163] In addition, there are scattered postmortem reports of atherosclerotic heart disease in PWS, usually as an incidental finding; this is also a common incidental finding in the general population. Biochemical risk markers for atherosclerotic heart disease (e.g., C-reactive protein, homocysteine) have not been reported in PWS.

As with ischemic heart disease, there are only scattered reports of cerebrovascular accident (stroke) in PWS. A case of apparent vaso-occlusive stroke with consequent moyamoya syndrome (revascularization of the middle cerebral arteries) was reported in a teenager with PWS and Type 2 diabetes mellitus.[120]

Since ischemic vascular disease is commonly linked to obesity and sleep-disordered breathing in the general population,[165,168,248] the low prevalence of these conditions in PWS is surprising and may be due to the lower-than-expected occurrence of dyslipidemia and insulin-resistance (see section on Obesity). In this respect, it is notable that 2 of the 3 documented cases occurred in individuals with PWS and concurrent Type 2 diabetes mellitus.

Miscellaneous Medical Concerns

Thermoregulatory Disorders, Autonomic Dysfunction, and Anesthesia Risk

The ability to maintain core body temperature within a narrow range is a major characteristic of normal mammalian physiology.[37,247] Ambient (environmental) temperature sensation occurs via peripheral nerves that signal to a hypothalamic regulatory center, resulting in appropriate signaling back to the periphery through the autonomic and sympathetic nervous systems. A similar mechanism may be triggered with exposure to endogenous pyrogens (i.e., with febrile illness), although body temperature may be less efficiently regulated in these conditions.

During the usual resting state, body temperature (36.5°C) is higher than in the environment (~25°C); the temperature difference is largely accounted for by heat generation resulting from routine metabolic processes. During ambient temperature fluctuations, both voluntary behavioral reactions (e.g., clothing and activity) and physiologic mechanisms are important elements of thermoregulation. When ambient temperature rises, the body dissipates heat via (1) vasodilation, which increases environmental heat exchange via radiation and convection (~75% of heat loss), and (2) sweating with evaporative cooling (~25%). When ambient temperature falls, the thermogenic response includes (1) norepinephrine-stimulated vasoconstriction, leading to decreased transfer of core body heat to the skin, and (2) shivering thermogenesis, in which there is progressive recruitment of muscle fibers that undergo involuntary, rhythmic contractions resulting in heat generation. Muscle thermogenesis can be further augmented by voluntary physical exertion; however, prolonged cold exposure and exertion can lead to fatigue and declines in both exertional and shivering thermogenesis. In small mammals and human infants, nonshivering thermogenesis from brown fat is a crucial mechanism for thermoregulation; however, this mechanism appears to be minimally active in human adults.[37]

Clinically, thermoregulatory disorders can be defined as (1) an inability to normally regulate core body temperature in response to environmental temperature fluctuations, (2) extreme high fevers during illness, and/or (3) risk for hypo- or hyperthermia during anesthesia.

A survey study of 27 PWS individuals (2 to 46 years old) reported at a scientific meeting of PWSA (USA) found that "clinically important abnormal temperature control is not a common event in PWS."[39] At the same meeting, five infants with PWS and unexplained hyperthermia (temperatures above 101°F) were reported.[244] In a compilation of cases used for derivation of clinical diagnostic criteria for PWS, "abnormal temperature sensitivity" was reported in 4.6% or 3 of 65 PWS subjects and none of 11 non-PWS individuals (this feature was included as a "supportive finding").[105] A survey study of 85 children with PWS found an increased occurrence of febrile convulsions and hypothermia (temperature less than 94°F) as compared with 85 non-PWS siblings and 118 well children[243]; however, there were no differences from a group of 105 non-PWS neurodevelopmentally-impaired children. Hypothermia (rectal temperature 30°C) with complete recovery has been reported in a 5-month-old infant with PWS who may have been exposed to a very low ambient temperature.[238] Hyperthermia during illness was also postulated to contribute to mortality in several infants with PWS, possibly due to increased respiratory demand and compromise.[200] Together, these reports suggest but do not confirm a possible risk for temperature dysregulation in infants with PWS.

A cold-exposure study was conducted in 6 PWS subjects and 3 non-PWS individuals placed into a cold-room with temperature set at 4°C with light clothing for 1 hour.[29] Rectal (core) temperatures were noted to rise in all 3 non-PWS and 2 PWS individuals. Rectal temperature showed a small decrease in 3 of 6 PWS subjects, implying a defect in cold-adaptation.

Temperature dysregulation has also been reported in a 37-year-old woman with PWS who had three separate episodes of hypothermia (27°C–32°C) associated with infections.[167] The risk for hypothermia was attributed to the use of risperidone, a dopamine and serotonin type 2 receptor antagonist. Hypothermia resolved after cessation of risperidone but recurred with use of olanzapine, a similar medication. Hypothermia has also been reported with these medications in non-PWS individuals[24] and with other psychotropic medications that are often used in PWS[118,119]; therefore, it is not clear whether the risk is greater for individuals with PWS.

Autonomic dysfunction has been proposed as a mechanism for the possible disorders of thermoregulation in PWS. The autonomic nervous system regulates smooth muscle, exocrine glands, and cardiac muscle and is largely responsible for maintaining somatic homeostasis through involuntary mechanisms. This is distinguished from the somatic motor system, which is responsible for primarily voluntary control of skeletal muscles. The hypothalamus is considered to be the control center for the autonomic system; however, multiple areas of the cerebral cortex, brain stem, cerebellum, and spinal cord also play roles. The autonomic system is divided into the sympathetic and parasympathetic systems. The sympathetic system arises from spinal cord neurons, giving rise to segmental innervation of smooth muscle in the eye, heart, GI tract, and reproductive organs. The parasympathetic system arises from both brainstem neurons, including cranial nerves III, VII, IX, and X (vagus nerve) and sacral neurons, and is largely responsible for peristalsis and exocrine secretion.

Evidence for a generalized autonomic (parasympathetic) dysfunction in PWS is based primarily on a study of 14 subjects with PWS (4 to 40 years old, 8 female) and a comparison group of individuals without PWS.[65] Within the PWS cohort, 50% had abnormal pupillary constriction to pilocarpine and, on EKG, 6 of 14 had an abnormal ratio of the 30th to 15th R-R interval. A higher resting pulse and lower orthostatic pulse response was correlated with body mass index (BMI). Decreased frequency of respiratory sinus arrhythmia was also reported.[63] A later study in 32 individuals with PWS in a thermoneutral environment (72°F) was interpreted to show that resting cutaneous temperature and capillary blood flow were not different from controls, thereby supporting a theory of central as opposed to peripheral autonomic dysfunction,[64] although the data presented seem to show higher capillary blood flow and mean cutaneous temperature in the group with uniparental disomy (n = 9) as compared with the deletion cases (n = 23) and controls (n = 5). Contrary to these findings, a detailed study in 26 individuals with PWS showed no differences in autonomic regulation of cardiac function as compared with age-, gender-, and BMI-matched controls.[231]

Deficient sweating in response to environmental warming has been anecdotally mentioned as evidence for thermoregulatory and autonomic dysfunction in PWS. However, there are no published studies characterizing this condition. Other evidence for autonomic dysfunction, such as decreased ability to vomit and decreased salivation, are discussed in the next chapter.

Therefore, there is no current evidence that autonomic dysfunction accounts for thermoregulatory defects in PWS. In view of the normal physiologic dependence on shivering thermogenesis during exposure to cold, a risk for core hypothermia may be related to the decreased muscle mass.

Disordered thermoregulation during general anesthesia has also been raised as a concern in PWS, particularly the risk for malignant hyperthermia. This anecdotal concern arises in part from the known increased risk for malignant hyperthermia in congenital myopathies, such as congenital myotonia and muscular dystrophy. Malignant hyperthermia is caused by a disorder of calcium transport in skeletal muscle, leading to increased intracellular calcium, uncoupling of mitochondrial oxidative phosphorylation, and excess heat production to several times greater than normal.[87,179] Clinically, the condition may be triggered by several general anesthetic agents and muscle relaxants, and is initially characterized by muscle rigidity, hyperthermia, and acidosis. To date, there are no documented reports of malignant hyperthermia or defects of muscle calcium transport in PWS.

A similar but pathologically unrelated condition, neuroleptic malignant syndrome, has been reported as an idiosyncratic reaction to a variety of psychotropic agents,[71,87] including several that are often prescribed in PWS. These agents are thought to trigger an inappropriate hypothalamic response by inhibiting dopamine or enhancing serotonin action, leading to muscle rigidity and hyperthermia. In addition, use of serotonin reuptake inhibitors, particularly in combination with other agents affecting serotonin action, has been associated with "serotonin syndrome" in non-PWS individuals.[87] Serotonin syndrome is characterized by hyper-reflexia, tremor, muscle rigidity, autonomic dysfunction, mental confusion, and variable occurrence of hyperthermia, resulting from the muscle contractions. There are no published reports of either neuroleptic malignant syndrome or serotonin syndrome in PWS, despite the widespread use of potential triggering agents.

Finally, there have been no reports of postoperative core hypothermia with postanesthetic shivering in PWS. In non-PWS patients, this condition is considered to be an important adverse event during postoperative recovery and is thought to be due to inhibition of thermoregulation by general anesthetic agents.[67]

A review of eight cases of children with PWS undergoing general anesthesia showed no particular complications that were not related to obesity,[60] e.g., difficult intravenous access and sleep apnea. Similar data in adults with PWS have not been presented.

In summary, clinical experience suggests that a proportion of individuals with PWS may be at risk for thermoregulatory disorders, particularly environment-related hypothermia, which could be related to a diminished capacity for shivering thermogenesis. A clinically significant risk for temperature dysregulation may be particularly noted for infants with PWS, who may be prone to both hyperthermia with illness and hypothermia with environmental cold exposure. Unfortunately, however, there are insufficient data to answer the question of whether specific thermoregulatory abnormalities are more common in PWS and no conclusive proof of autonomic dysfunction. In addition, there are no data to indicate an increased risk for malignant hyperthermia, neuroleptic malignant syndrome, or serotonin syndrome. Nevertheless, in view of the relative paucity of published data and the potentially serious nature of these conditions, practitioners should be aware of the possible risks when confronted with specific situations.

Ophthalmologic Disorders

Esotropia has been reported in 43% of 15 infants and 66% of 65 older individuals with PWS, and was included as a supportive finding for clinical diagnosis.[105] However, there are surprisingly few published studies characterizing ocular abnormalities and their treatment in PWS, and documentation of ophthalmologic findings is lacking in many descriptive series of the syndrome.

Esotropia is a type of strabismus (dysconjugate eye movement) in which one eye turns inward (toward the nose) while the other is fixated in a forward gaze. In a series of 46 patients clinically diagnosed with PWS, 22 were found to have esotropia, while 7 had exotropia, or outward turning of the eye.[98] Additional defects included myopia (15%), significant astigmatism (oblong deformation of the eye leading to blurred vision, 41%), amblyopia (decreased visual acuity in one eye, 24%), and iris transillumination defects (33%). In a detailed study of 12 patients with PWS, Roy et al.[180] found only 1 patient with esotropia, 2 with exotropia, and 9 with orthophoria (normal conjugate eye movements). One patient had myopic chorioretinal degeneration, 1 had disc atrophy and foveal hypolasia, and 5 (42%) were noted to have telecanthus (increased distance between the inner corners of the eyes without orbit displacement).

A later series by Fox et al.[74] compared 20 PWS patients with deletion and 7 with uniparental maternal disomy with 16 controls matched for age, body composition, and intelligence. Strabismus was found in 6 deletion subjects (30%), 4 disomy subjects (57%), and 4 controls (25%)—not significantly different between groups. Abnormal iris transillumination was found in 4 deletion and none of the disomy or control subjects. Foveal hypoplasia was not identified in any of the groups, while findings of astigmatism, amblyopia, and amisometropia were not significantly different between the PWS and control groups. Stereopsis differed (p < 0.002) between the PWS (45% deletion, 43% disomy) and controls (81%), indicating an increased defect in 3-D perception in the PWS group. The mean level of myopia was also different (p = 0.01) between control (refractive error −2.33 ± 1.65SD) and PWS (−5.56 ± 5.38) subjects.

Ocular hypopigmentation has also been noted in PWS.[103,129,234] Generalized skin hypopigmentation is a well-known feature of PWS (~50% of cases) and may be associated with deletion of the *P* (pink-eyed dilution) gene, a nonimprinted gene located within the PWS deletion boundaries.[30,198] Mutations of the *P* gene are the cause of oculocutaneous albinism type 2 (OCA2) in humans, in which there is hypopigmentation of the eyes, skin, and hair. Associated findings include strabismus, loss of stereopsis, nystagmus, and foveal hypoplasia, apparently due to abnormal routing of the optic nerve pathways.[30] The similarity of these eye findings along with skin hypopigmentation to features of PWS suggested a similar pathogenesis. An initial study of 6 patients with PWS and strabismus found abnormal visual-evoked potentials (VEP) in 3 of 4 hypopigmented patients and in neither of the 2 normally-pigmented patients.[52] The VEP abnormality was indistinguishable from that observed in OCA2, suggesting a similar abnormality in optic nerve pathways in PWS. However, the presence of strabismus in those patients with normal VEP was unexplained and subsequent studies failed to confirm these initial findings.[11,180,234]

Overall, the data suggest that strabismus, both esotropia and exotropia, is not uncommon in PWS, and may be related to ocular muscle hypotonia.[98] A variety of other ophthalmologic disorders may also be present, including ocular hypopigmentation. These disorders could be related to deletion of the *P* gene, which is also associated with skin hypopigmentation. However, this link is far from certain because misrouting of optic nerve pathways, such as observed in OCA2, has not been confirmed in PWS.

Evaluation and treatment of eye conditions in PWS requires consultation with an experienced ophthalmologist. Treatment of congenital strabismus has been noted to have variable outcome in non-PWS cases[81,83] and there are limited data in PWS.[93]

Sensory Function

Decreased peripheral and deep pain sensation is a commonly reported feature of PWS and, in the authors' clinical experience, a definite characteristic of the condition. However, published data relating to pain sensation in PWS is remarkably scant. Decreased pain sensation has been reported in 18% of individuals with PWS, with other reports indicating decreased pinprick sensitivity and decreased dermal and epidermal innervation.[40,103] In the formulation of diagnostic criteria for PWS, a high pain threshold was reported in 9% of 65 typical PWS, 9% of 11 atypical PWS, and 9% of 11 control subjects.[104] Easy bruising has also been anecdotally reported in PWS[40] and could, in part, be related to the decreased ability to feel pain; no other contributory factors have been identified.

Interestingly, disruption of the *Ndn* gene (an imprinted gene in the human PWS locus) in mice leads to increased skin-scraping behavior,[149] which may be analogous to the skin-picking behavior in human PWS that has been theoretically related to aberrant sensory input.

Detailed studies in five children with PWS showed normal tactile perception (stereognosis), normal sensory nerve conduction (normal myelinization), and a 50% reduction in sensory nerve action potentials.[28] These latter results suggest that decreased density of peripheral nerve fibers may account for the decreased pain sensation.

A possible etiology for decreased somatic pain sensation was recently proposed by Lucignani et al.,[135] who found a reduction in gamma-aminobutyric acid type-A (GABA-A) receptors in areas of the cerebral cortex that are related to pain response in six young adults with PWS (all deletion). The GABA-A receptor is composed of several subunits, three of which are coded by nonimprinted genes in the PWS-region of chromosome 15q.

In view of the decreased pain sensation, it is important for caretakers to be vigilant for signs of infection and injury that might otherwise be signaled by pain or discomfort. Acute or unexpected changes in behavior, mobility, or gait may warrant further investigation.

Mitochondrial DNA

The mitochondria are intracellular organelles that are ancestrally derived from bacteria and live in the mutually supportive environment of the host cells.[10] Mitochondria-dependent metabolic pathways, involving electron transport and oxidative phosphorylation, are the major source of energy in human cells. Some of the proteins involved in these processes are coded in the mitochondrial DNA while others are coded in the nuclear DNA of the cell, demonstrating an evolved symbiotic relationship. A variety of human disorders, primarily involving muscle and central nervous system abnormalities, have been associated with defects in these genes.

As of this writing, none of the genes involved in the Prader-Willi region of chromosome 15q have been associated with mitochondrial function or cell energetics. However, the similarities of the muscle hypotonia in PWS to some types of mitochondrial myopathy have raised suspicions that there may be a relationship. Thus far, three cases of mitochondrial myopathy have been identified in children with PWS. A 2-year-old child with PWS and mitochondrial myopathy (complex I/IV respiratory chain deficiency diagnosed by muscle biopsy) was reported to have had open heart surgery for atrial septal defect under general anesthesia without complications.[187] Two cases of PWS (maternal uniparental disomy) with mitochondrial myopathy (complex I deficiency) were reported in an abstract[237]; the diagnosis of mitochondrial myopathy was apparently pursued because of the neonatal hypotonia and the delayed clinical suspicion of PWS. The authors concluded that the mitochondrial myopathy was "likely a secondary rather than a primary event." However, the co-occurrence of these two relatively uncommon conditions suggests the possibility of a common pathogenesis and a need for larger-scale screening studies. A recent case of fatal metabolic acidosis in an infant with PWS is also suggestive of mitochondrial DNA disease.[249]

There has been recent interest in the relationship of coenzyme Q10 (CoQ10) and PWS. CoQ10 is a protein encoded by the nuclear, rather than mitochondrial, DNA that plays an essential role in mitochondrial and extra-mitochondrial electron transport, as well as a variety of other vital functions.[213] The CoQ10 gene is not located in the PWS region. Low blood levels of CoQ10 have been identified in several disorders, including Alzheimer's, Huntington's, and Parkinson's diseases, and in some studies there is a suggestion that CoQ10 therapy may be beneficial for some of these conditions.[189] Low blood levels of CoQ10 in children with PWS were reported in an abstract[114] and beneficial effects on neurodevelopment were reported for one infant. However, a recent report found that although plasma CoQ10 levels were lower in 16 individuals with PWS as compared with controls, the levels were not different from a group of obese individuals.[32] The authors postulated that the lower plasma CoQ10 levels were due to the lower muscle mass and decreased energy expenditure in PWS. In view of these latter data and a lack of studies showing treatment efficacy, CoQ10 is not currently recommended as a medical therapy for PWS.

Epilepsy

Epilepsy is defined as a disorder of the central nervous system that causes convulsions (seizures). The physical manifestations of epilepsy range from subtle changes in behavior to major episodes of generalized involuntary motor activity. Various environmental factors, including increased body temperature, may increase the propensity or threshold for generation of the abnormal electrical impulses in the brain that lead to seizures.

There have been occasional reports of epilepsy or seizures in individuals with PWS[230] and a suggestion that infants with PWS may be susceptible to febrile convulsions (see above); however, several systematic population surveys have failed to mention epilepsy as a specific feature of PWS.[104,133,170] On the other hand, epilepsy is a well-characterized feature of Angelman syndrome[79] and is also associated with other chromosome 15q11–q13 defects, including interstitial duplications of maternal origin within the imprinted region.[46,205,208,233] A recent study of children referred from general pediatrics and neurology clinics reported a history of febrile seizures in 18 of 40 individuals (45%) with PWS and deletion, as compared with 1 of 14 patients (7%) with maternal uniparental disomy.[219] The same group reported seizures in 42 of 47 patients (89%) with Angelman syndrome and deletion, as compared with 4 of 9 patients (44%) with paternal uniparental disomy.[220] As noted by the authors, these results suggest that a lowering of the seizure threshold, leading to a risk for febrile seizures in PWS and all seizures in Angelman syndrome, could be due to haploinsufficiency of a nonimprinted seizure-related gene, such as those that code for subunits of the gamma-butyric acid receptor. Additional studies to clarify these relationships are necessary.

Cancer

Concerns have been raised regarding the possible association of PWS with increased risk for cancer. In 1985, Hall[86] published three cases of leukemia in individuals with PWS. No additional cases were presented until 1999, when Cassidy et al.[42,57] reported results from a survey study of the PWSA (USA) membership. Of 1,077 valid responses (from 1,852 members surveyed), there were 32 confirmed cases of cancer, of which 19 were benign; 3 of 53 deaths were attributed to cancer. Eight cases involved cancer types that are monitored by the Surveillance, Epidemiology and End Results (SEER) Program, a project of the U.S. National Cancer Institute, including 3 myeloid leukemia (1 acute, 2 chronic, probably the same cases previously published by Hall). Other SEER-listed tumors in PWS included single cases of seminoma, dermatofibrosarcoma, ovarian teratoma, Hodgkin's lymphoma, and cardiac lymphoma. The 8 observed cases were compared with the calculated expected number of 4.8 (based on data from SEER), indicating that there may be an increased overall occurrence of cancer in PWS.

Other cases of cancer reported in PWS include single cases of poorly differentiated hepatoblastoma in a 16-month-old boy (deletion 15q),[92] hepatocyte adenoma in an 11-year-old girl with known hepatic steatosis (fatty liver),[203] "pseudo-Kaposi" sarcoma in a 25-year-old male,[66] and multiple endocrine neoplasia type 1 (MEN 1).[151]

Abnormalities of chromosome 15q, which is the location of the PWS region, are frequently found in tumor tissues from patients without PWS. Such tumors include breast cancers, bladder carcinoma, lymphoblastic leukemia, and myeloid leukemia.[45,96,153,246] The 15q abnormalities identified in these conditions include deletions, translocations, loss of heterozygosity, and duplications. In detailed studies, it appears that there may be tumor suppressor genes located distal to the Prader-Willi region,[96,153] loss of which may lead to tumorigenesis. The relevance of these findings to PWS is currently unknown and tumor karyotype information is lacking for most cases of cancer in PWS.

Infections

Individuals with PWS may be prone to a variety of infections. However, any increased risk of infection appears to be primarily related to other characteristics of the syndrome. Primary immunodeficiency disorders have not been reported in PWS.

As described in previous sections, respiratory infections are a frequent cause of morbidity and mortality in PWS. The risk for serious respiratory infection is probably due to hypotonia of the respiratory musculature and consequent poor ventilatory capacity and reserve rather than inherent susceptibility to particular pathogenic organisms. Given the known risks for respiratory decompensation, judicious use of antibiotics should be considered if there is evidence of infection, even if a specific agent cannot be identified.

Dental infections are also common, as described in the next chapter, and are likely related to defects in salivation, high intake of sugar, and poor hygiene.

Skin infections are frequently observed in PWS. Contributory factors include obsessive skin picking (discussed elsewhere in this text) and decreased pain sensation, allowing traumatic skin injuries to progress to infection. All individuals with PWS should be examined regularly for cutaneous infections, particularly at pressure points (feet, buttocks) and in less-visible areas (scalp, perianal area, groin, axillae). In most cases, infected lesions can be successfully managed with topical antisepsis and topical or oral antibiotics. Culture and sensitivity studies should be considered for persistent or unusual infections, and attention should be given to infectious agents endemic to the community, such as methicillin-resistant staph aureus. Finally, the presence of systemic symptoms or acute onset of unusual behavior, body temperature instability, respiratory compromise, or uncharacteristic body movements may signal the presence of a deep tissue or systemic infection, such as osteomyelitis or pneumonia.

References

1. Conference report: First International Scientific Workshop on Prader-Willi Syndrome and Other Chromosome 15q Deletion Disorders. May 2–3, 1991, DeLeeuwenhorst, The Netherlands. Abstracts. *American Journal of Medical Genetics*. 1992;42(2):220–269.
2. Greenswag LR, Alexander RC, eds. *Management of Prader-Willi Syndrome*. 2nd ed. New York, NY: Springer-Verlag; 1995.
3. Osteoporosis prevention, diagnosis, and therapy. NIH Consensus Statement. 2000;17(1):1–45.
4. New indications for modafinil (Provigil). *The Medical Letter on Drugs and Therapeutics*. 2004;46(1181):34–35.
5. Afifi AK. Histology and cytology of muscle in Prader Willi syndrome. *Developmental Medicine and Child Neurology*. 1969;11(3):363–364.
6. Afifi AK, Zellweger H. Pathology of muscular hypotonia in the Prader-Willi syndrome. Light and electron microscopic study. *Journal of the Neurological Sciences*. 1969;9(1):49–61.
7. Ahmed SF, Hughes IA. The genetics of male undermasculinization. *Clinical Endocrinology (Oxf)*. 2002;56(1):1–18.

8. Akefeldt A, Tornhage CJ, Gillberg C. A woman with Prader-Willi syndrome gives birth to a healthy baby girl. *Developmental Medicine and Child Neurology.* 1999;41(11):789–790.

9. Ammini AC, Pandey J, Vijyaraghavan M, Sabherwal U. Human female phenotypic development: role of fetal ovaries. *Journal of Clinical Endocrinology and Metabolism.* 1994;79(2):604–608.

10. Andreu AL, DiMauro S. Current classification of mitochondrial disorders. *Journal of Neurology.* 2003;250(12):1403–1406.

11. Apkarian P, Spekreijse H, van Swaay E, van Schooneveld M. Visual evoked potentials in Prader-Willi syndrome. *Documenta Ophthalmologica.* 1989; 71(4):355–367.

12. Arens R, Gozal D, Burrell BC, et al. Arousal and cardiorespiratory responses to hypoxia in Prader-Willi syndrome. *American Journal of Respiratory and Critical Care Medicine.* 1996;153(1):283–287.

13. Arens R, Gozal D, Omlin KJ, et al. Hypoxic and hypercapnic ventilatory responses in Prader-Willi syndrome. *Journal of Applied Physiology.* 1994; 77(5):2224–2230.

14. Argov Z, Gardner-Medwin D, Johnson MA, Mastaglia FL. Patterns of muscle fiber-type disproportion in hypotonic infants. *Archives of Neurology.* 1984;41(1):53–57.

15. Arlet V, Odent T, Aebi M. Congenital scoliosis. *European Spine Journal.* 2003;12(5):456–463.

16. Attal P, Lepajolec C, Harboun-Cohen E, Gaultier C, Bobin S. [Obstructive sleep apnea-hypopnea syndromes in children. Therapeutic results]. *Annales d'Oto-laryngologie et de Chirurgie Cervico Faciale.* 1990;107(3): 174–179.

17. Aughton DJ, Cassidy SB. Physical features of Prader-Willi syndrome in neonates. *American Journal of Diseases of Children.* 1990;144(11):1251–1254.

18. Bandla H, Splaingard M. Sleep problems in children with common medical disorders. *Pediatric Clinics of North America.* 2004;51(1):203–227.

19. Barthold JS, Gonzalez R. The epidemiology of congenital cryptorchidism, testicular ascent and orchiopexy. *Journal of Urology.* 2003;170(6 Pt 1): 2396–2401.

20. Berven S, Bradford DS. Neuromuscular scoliosis: causes of deformity and principles for evaluation and management. *Seminars in Neurology.* 2002;22(2):167–178.

21. Bhasin S, Bremner WJ. Emerging issues in androgen replacement therapy. *Journal of Clinical Endocrinology and Metabolism.* 1997;82(1):3–8.

22. Bier DM, Kaplan SL, Havel RJ. The Prader-Willi syndrome: regulation of fat transport. *Diabetes.* 1977;26(9):874–881.

23. Bin-Abbas B, Conte FA, Grumbach MM, Kaplan SL. Congenital hypogonadotropic hypogonadism and micropenis: effect of testosterone treatment on adult penile size. Why sex reversal is not indicated. *Journal of Pediatrics.* 1999;134(5):579–583.

24. Blass DM, Chuen M. Olanzapine-associated hypothermia. *Psychosomatics.* 2004;45(2):135–139.

25. Boehlecke B. Controversies in monitoring and testing for sleep-disordered breathing. *Current Opinion in Pulmonary Medicine.* 2001;7(6):372–380.

26. Bolton CF, Chen R, Wijdicks EFM, Zifko U. *Neurology of Breathing.* Philadelphia, PA: Butterworth Heinemann; 2004.

27. Brambilla P, Bosio L, Manzoni P, Pietrobelli A, Beccaria L, Chiumello G. Peculiar body composition in patients with Prader-Labhart-Willi syndrome. *American Journal of Clinical Nutrition.* 1997;65(5):1369–1374.

28. Brandt BR, Rosen I. Impaired peripheral somatosensory function in children with Prader-Willi syndrome. *Neuropediatrics*. 1998;29(3):124–126.

29. Bray GA, Dahms WT, Swerdloff RS, Fiser RH, Atkinson RL, Carrel RE. The Prader-Willi syndrome: a study of 40 patients and a review of the literature. *Medicine (Baltimore)*. 1983;62(2):59–80.

30. Brilliant MH. The mouse p (pink-eyed dilution) and human P genes, oculocutaneous albinism type 2 (OCA2), and melanosomal pH. *Pigment Cell Research*. 2001;14(2):86–93.

31. Butler JV, Whittington JE, Holland AJ, Boer H, Clarke D, Webb T. Prevalence of, and risk factors for, physical ill-health in people with Prader-Willi syndrome: a population-based study. *Developmental Medicine and Child Neurology*. 2002;44(4):248–255.

32. Butler MG, Dasouki M, Bittel D, Hunter S, Naini A, DiMauro S. Coenzyme Q10 levels in Prader-Willi syndrome: comparison with obese and non-obese subjects. *American Journal of Medical Genetics*. 2003;119A(2): 168–171.

33. Butler MG, Haber L, Mernaugh R, Carlson MG, Price R, Feurer ID. Decreased bone mineral density in Prader-Willi syndrome: comparison with obese subjects. *American Journal of Medical Genetics*. 2001;103(3): 216–222.

34. Bye AM, Vines R, Fronzek K. The obesity hypoventilation syndrome and the Prader-Willi syndrome. *Australian Paediatric Journal*. 1983;19(4): 251–255.

35. Cacciari E, Zucchini S, Carla G, et al. Endocrine function and morphological findings in patients with disorders of the hypothalamo-pituitary area: a study with magnetic resonance. *Archives of Disease in Childhood*. 1990;65(11):1199–1202.

36. Calisti L, Giannessi N, Cesaretti G, Saggese G. [Endocrine study in the Prader-Willi syndrome. Apropos of 5 cases]. *Minerva Pediatrica*. 1991; 43(9):587–593.

37. Cannon B, Nedergaard J. Brown adipose tissue: function and physiological significance. *Physiological Reviews*. 2004;84(1):277–359.

38. Carroll JL, McColley SA, Marcus CL, Curtis S, Loughlin GM. Inability of clinical history to distinguish primary snoring from obstructive sleep apnea syndrome in children. *Chest*. 1995;108(3):610–618.

39. Cassidy S, McKillop J. Termperature regulation in Prader-Willi syndrome. *American Journal of Medical Genetics*. 1991;41:528.

40. Cassidy SB. Prader-Willi syndrome. *Current Problems in Pediatrics*. 1984; 14(1):1–55.

41. Cassidy SB. Prader-Willi syndrome. Characteristics, management, and etiology. *The Alabama Journal of Medical Sciences*. 1987;24(2):169–175.

42. Cassidy SB, Dele Davis H, Rose S, et al. Cancer in Prader-Willi syndrome. Paper presented at PWSA (USA) 14th Annual Scientific Conference, San Diego, CA, July 1999.

43. Cassidy SB, McKillop JA, Morgan WJ. Sleep disorders in Prader-Willi syndrome. *Dysmorphology and Clinical Genetics*. 1990;4:13–17.

44. Cassidy SB, Rubin KG, Mukaida CS. Genital abnormalities and hypogonadism in 105 patients with Prader-Willi syndrome. *American Journal of Medical Genetics*. 1987;28:922–923.

45. Chen SN, Xue YQ, Wu YF, Pan JL. [Cytogenetic and molecular genetic studies on a variant of t(15;17), ins(17;15)(q21;q14q22) in an acute promyelocytic leukemia patient]. *Zhonghua Yi Xue Yi Chuan Xue Za Zhi*. 2004; 21(1):77–79.

46. Chifari R, Guerrini R, Pierluigi M, et al. Mild generalized epilepsy and developmental disorder associated with large inv dup(15). *Epilepsia.* 2002; 43(9):1096–1100.

47. Clarke DJ, Waters J, Corbett JA. Adults with Prader-Willi syndrome: abnormalities of sleep and behaviour. *Journal of the Royal Society of Medicine.* 1989;82(1):21–24.

48. Clift S, Dahlitz M, Parkes JD. Sleep apnoea in the Prader-Willi syndrome. *Journal of Sleep Research.* 1994;3(2):121–126.

49. Cook DM, Ludlam WH, Cook MB. Route of estrogen administration helps to determine growth hormone (GH) replacement dose in GH-deficient adults. *Journal of Clinical Endocrinology and Metabolism.* 1999;84(11): 3956–3960.

50. Cortes D, Thorup JM, Visfeldt J. Cryptorchidism: aspects of fertility and neoplasms. A study including data of 1,335 consecutive boys who underwent testicular biopsy simultaneously with surgery for cryptorchidism. *Hormone Research.* 2001;55(1):21–27.

51. Costeff H, Holm VA, Ruvalcaba R, Shaver J. Growth hormone secretion in Prader-Willi syndrome. *Acta Paediatrica Scandinavica.* 1990;79(11): 1059–1062.

52. Creel DJ, Bendel CM, Wiesner GL, Wirtschafter JD, Arthur DC, King RA. Abnormalities of the central visual pathways in Prader-Willi syndrome associated with hypopigmentation. *New England Journal of Medicine.* 1986; 314(25):1606–1609.

53. Crino A, Greggio NA, Beccaria L, Schiaffini R, Pietrobelli A, Maffeis C. [Diagnosis and differential diagnosis of obesity in childhood]. *Minerva Pediatrica.* 2003;55(5):461–470.

54. Crino A, Schiaffini R, Ciampalini P, et al. Hypogonadism and pubertal development in Prader-Willi syndrome. *European Journal of Pediatrics.* 2003;162(5):327–333.

55. Culebras A. Sleep and neuromuscular disorders. *Neurologic Clinics.* 1996; 14(4):791–805.

56. Daragon A, Pouplin S. Potential benefits of intermittent bisphosphonate therapy in osteoporosis. *Joint Bone Spine.* 2004;71(1):2–3.

57. Davies HD, Leusink GL, McConnell A, et al. Myeloid leukemia in Prader-Willi syndrome. *Journal of Pediatrics.* 2003;142(2):174–178.

58. De Filippo RE, Barthold JS, Gonzalez R. The application of magnetic resonance imaging for the preoperative localization of nonpalpable testis in obese children: an alternative to laparoscopy. *Journal of Urology.* 2000; 164(1):154–155.

59. de Zwaan M, Mitchell JE. Opiate antagonists and eating behavior in humans: a review. *Journal of Clinical Pharmacology.* 1992;32(12):1060–1072.

60. Dearlove OR, Dobson A, Super M. Anaesthesia and Prader-Willi syndrome. *Paediatric Anaesthesia.* 1998;8(3):267–271.

61. Deschildre A, Martinot A, Fourier C, et al. [Effects of hypocaloric diet on respiratory manifestations in Willi-Prader syndrome]. *Archives of Pediatrics.* 1995;2(11):1075–1079.

62. Dickey RP, Holtkamp DE. Development, pharmacology and clinical experience with clomiphene citrate. *Human Reproduction Update.* 1996;2(6): 483–506.

63. DiMario FJ, Jr., Bauer L, Volpe J, Cassidy SB. Respiratory sinus arrhythmia in patients with Prader-Willi syndrome. *Journal of Child Neurology.* 1996; 11(2):121–125.

64. DiMario FJ, Jr., Burleson JA. Cutaneous blood flow and thermoregulation in Prader-Willi syndrome patients. *Pediatric Neurology.* 2002;26(2):130–133.

65. DiMario FJ, Jr., Dunham B, Burleson JA, Moskovitz J, Cassidy SB. An evaluation of autonomic nervous system function in patients with Prader-Willi syndrome. *Pediatrics.* 1994;93(1):76–81.

66. Donhauser G, Eckert F, Landthaler M, Braun-Falco O. [Pseudo-Kaposi sarcoma in Prader-Labhart-Willi syndrome]. *Hautarzt.* 1991;42(7):467–470.

67. Doufas AG. Consequences of inadvertent perioperative hypothermia. *Best Practice and Research. Clinical Anaesthesiology.* 2003;17(4):535–549.

68. Eiholzer U, l'Allemand D, van der Sluis I, Steinert H, Gasser T, Ellis K. Body composition abnormalities in children with Prader-Willi syndrome and long-term effects of growth hormone therapy. *Hormone Research.* 2000;53(4):200–206.

69. Eiholzer U, Nordmann Y, l'Allemand D, Schlumpf M, Schmid S, Kromeyer-Hauschild K. Improving body composition and physical activity in Prader-Willi syndrome. *Journal of Pediatrics.* 2003;142(1):73–78.

70. Ellis KJ. Human body composition: in vivo methods. *Physiological Reviews.* 2000;80(2):649–680.

71. Farver DK. Neuroleptic malignant syndrome induced by atypical antipsychotics. *Expert Opin Drug Saf.* 2003;2(1):21–35.

72. Fieldstone A, Zipf WB, Schwartz HC, Berntson GG. Food preferences in Prader-Willi syndrome, normal weight and obese controls. *International Journal of Obesity and Related Metabolic Disorders.* 1997;21(11):1046–1052.

73. Flemons WW, Littner MR, Rowley JA, et al. Home diagnosis of sleep apnea: a systematic review of the literature. An evidence review cosponsored by the American Academy of Sleep Medicine, the American College of Chest Physicians, and the American Thoracic Society. *Chest.* 2003; 124(4):1543–1579.

74. Fox R, Sinatra RB, Mooney MA, Feurer ID, Butler MG. Visual capacity and Prader-Willi syndrome. *Journal of Pediatric Ophthalmology and Strabismus.* 1999;36(6):331–336.

75. Friedman BC, Hendeles-Amitai A, Kozminsky E, et al. Adenotonsillectomy improves neurocognitive function in children with obstructive sleep apnea syndrome. *Sleep.* 2003;26(8):999–1005.

76. Gale SD, Hopkins RO. Effects of hypoxia on the brain: neuroimaging and neuropsychological findings following carbon monoxide poisoning and obstructive sleep apnea. *Journal of the International Neuropsychological Society.* 2004;10(1):60–71.

77. Greenberg F, Elder FF, Ledbetter DH. Neonatal diagnosis of Prader-Willi syndrome and its implications. *American Journal of Medical Genetics.* 1987; 28(4):845–856.

78. Greenswag LR. Adults with Prader-Willi syndrome: a survey of 232 cases. *Developmental Medicine and Child Neurology.* 1987;29(2):145–52.

79. Guerrini R, Carrozzo R, Rinaldi R, Bonanni P. Angelman syndrome: etiology, clinical features, diagnosis, and management of symptoms. *Paediatric Drugs.* 2003;5(10):647–661.

80. Gunay-Aygun M, Schwartz S, Heeger S, O'Riordan MA, Cassidy SB. The changing purpose of Prader-Willi syndrome clinical diagnostic criteria and proposed revised criteria. *Pediatrics.* 2001;108(5):E92.

81. Gunton KB, Nelson BA. Evidence-based medicine in congenital esotropia. *Journal of Pediatric Ophthalmology and Strabismus.* 2003;40(2):70–73.

82. Gurd AR, Thompson TR. Scoliosis in Prader-Willi syndrome. *Journal of Pediatric Orthopedics.* 1981;1(3):317–320.

83. Guthrie ME, Wright KW. Congenital esotropia. *Ophthalmology Clinics of North America*. 2001;14(3):419–424.

84. Guthrie RD, Smith DW, Graham CB. Testosterone treatment for micropenis during early childhood. *Journal of Pediatrics*. 1973;83(2):247–252.

85. Hakonarson H, Moskovitz J, Daigle KL, Cassidy SB, Cloutier MM. Pulmonary function abnormalities in Prader-Willi syndrome. *Journal of Pediatrics*. 1995;126(4):565–570.

86. Hall BD. Leukaemia and the Prader-Willi syndrome. *Lancet*. 1985; 1(8419):46.

87. Halloran LL, Bernard DW. Management of drug-induced hyperthermia. *Current Opinion in Pediatrics*. 2004;16(2):211–215.

88. Hamilton CR, Jr., Scully RE, Kliman B. Hypogonadotropinism in Prader-Willi syndrome. Induction of puberty and spermatogenesis by clomiphene citrate. *American Journal of Medicine*. 1972;52(3):322–329.

89. Haqq AM, Stadler DD, Jackson RH, Rosenfeld RG, Purnell JQ, LaFranchi SH. Effects of growth hormone on pulmonary function, sleep quality, behavior, cognition, growth velocity, body composition, and resting energy expenditure in Prader-Willi syndrome. *Journal of Clinical Endocrinology and Metabolism*. 2003;88(5):2206–2212.

90. Harris JC, Allen RP. Is excessive daytime sleepiness characteristic of Prader-Willi syndrome? The effects of weight change. *Archives of Pediatrics and Adolescent Medicine*. 1996;150(12):1288–1293.

91. Haschke F, Hohenauer L. Endocrine studies in four patients with Prader-Labhart-Willi syndrome during early infancy and childhood. *Pediatric Research*. 1978;12:1100.

92. Hashizume K, Nakajo T, Kawarasaki H, et al. Prader-Willi syndrome with del(15)(q11,q13) associated with hepatoblastoma. *Acta Paediatrica Japonica*. 1991;33(6):718–722.

93. Hatsukawa Y, Ishizaka M, Nihmi A, Mitarai K, Furukawa A, Yamagishi T. Treatment of A-pattern esotropia with marked mongoloid slanting palpebral fissures. *Japanese Journal of Ophthalmology*. 2001;45(5):482–486.

94. Hattori S, Mochio S, Kageyama A, Nakajima T, Akima M, Fukunaga N. [An autopsy case of Prader-Labhart-Willi syndrome]. *No To Shinkei*. 1985;37(11):1059–1066.

95. Hayashi M, Itoh M, Kabasawa Y, Hayashi H, Satoh J, Morimatsu Y. A neuropathological study of a case of the Prader-Willi syndrome with an interstitial deletion of the proximal long arm of chromosome 15. *Brain and Development*. 1992;14(1):58–62.

96. Heerema NA, Sather HN, Sensel MG, et al. Abnormalities of chromosome bands 15q13–15 in childhood acute lymphoblastic leukemia. *Cancer*. 2002; 94(4):1102–1110.

97. Helbing-Zwanenburg B, Kamphuisen HA, Mourtazaev MS. The origin of excessive daytime sleepiness in the Prader-Willi syndrome. *Journal of Intellectual Disabilities Research*. 1993;37 (Pt 6):533–541.

98. Hered RW, Rogers S, Zang YF, Biglan AW. Ophthalmologic features of Prader-Willi syndrome. *Journal of Pediatric Ophthalmology and Strabismus*. 1988;25(3):145–150.

99. Herrinton LJ, Zhao W, Husson G. Management of cryptorchidism and risk of testicular cancer. *American Journal of Epidemiology*. 2003;157(7):602–605.

100. Hertz G, Cataletto M, Feinsilver SH, Angulo M. Sleep and breathing patterns in patients with Prader-Willi syndrome (PWS): effects of age and gender. *Sleep*. 1993;16(4):366–371.

101. Hertz G, Cataletto M, Feinsilver SH, Angulo M. Developmental trends of sleep-disordered breathing in Prader-Willi syndrome: the role of obesity. *American Journal of Medical Genetics*. 1995;56(2):188–190.

102. Hiroe Y, Inoue Y, Higami S, Suto Y, Kawahara R. Relationship between hypersomnia and respiratory disorder during sleep in Prader-Willi syndrome. *Psychiatry and Clinical Neurosciences.* 2000;54(3):323–325.

103. Hittner HM, King RA, Riccardi VM, et al. Oculocutaneous albinoidism as a manifestation of reduced neural crest derivatives in the Prader-Willi syndrome. *American Journal of Ophthalmology.* 1982;94(3):328–337.

104. Holm VA, Cassidy SB, Butler MG, et al. Diagnostic criteria for Prader-Willi syndrome. In: Cassidy SB, ed. *Prader-Willi Syndrome and Other Chromosome 15q Deletion Disorders.* New York, NY: Springer-Verlag; 1992.

105. Holm VA, Cassidy SB, Butler MG, et al. Prader-Willi syndrome: consensus diagnostic criteria. *Pediatrics.* 1993;91(2):398–402.

106. Holm VA, Laurnen EL. Prader-Willi syndrome and scoliosis. *Developmental Medicine and Child Neurology.* 1981;23(2):192–201.

107. Horn EM, Waldrop TG. Suprapontine control of respiration. *Respiration Physiology.* 1998;114(3):201–211.

108. Hoybye C, Hilding A, Jacobsson H, Thoren M. Metabolic profile and body composition in adults with Prader-Willi syndrome and severe obesity. *Journal of Clinical Endocrinology and Metabolism.* 2002;87(8):3590–3597.

109. Husmann DA, Levy JB. Current concepts in the pathophysiology of testicular undescent. *Urology.* 1995;46(2):267–276.

110. Hutson JM, Hasthorpe S, Heyns CF. Anatomical and functional aspects of testicular descent and cryptorchidism. *Endocrine Reviews.* 1997;18(2):259–280.

111. Jaffray B, Moore L, Dickson AP. Prader-Willi syndrome and intratubular germ cell neoplasia. *Medical and Pediatric Oncology.* 1999;32(1):73–74.

112. Jaskulsky SR, Stone NN. Hypogonadism in Prader-Willi syndrome. *Urology.* 1987;29(2):207–208.

113. Jones MW. Scoliosis and its treatment in the Prader-Willi syndrome. In: Cassidy SB, ed. *Prader-Willi Syndrome and Other Chromosome 15q Deletion Disorders.* New York: Springer-Verlag; 1992:199–209.

114. Judy WV, Stogsdill WW. Coenzyme Q10 and mitochondria function. Paper presented at PWSA (USA) 15th Annual Scientific Day, Pittsburgh, PA, July 2000.

115. Kaplan J, Fredrickson PA, Richardson JW. Sleep and breathing in patients with the Prader-Willi syndrome. *Mayo Clinic Proceedings.* 1991;66(11):1124–1126.

116. Katcher ML, Bargman GJ, Gilbert EF, Opitz JM. Absence of spermatogonia in the Prader-Willi syndrome. *European Journal of Pediatrics.* 1977;124(4):257–260.

117. Kauli R, Prager-Lewin R, Laron Z. Pubertal development in the Prader-Labhart-Willi syndrome. *Acta Paediatrica Scandinavica.* 1978;67(6):763–767.

118. Kudoh A, Takase H, Takazawa T. Chronic treatment with antidepressants decreases intraoperative core hypothermia. *Anesthesia and Analgesia.* 2003;97(1):275–279.

119. Kudoh A, Takase H, Takazawa T. Chronic treatment with antipsychotics enhances intraoperative core hypothermia. *Anesthesia and Analgesia.* 2004;98(1):111–115.

120. Kusuhara T, Ayabe M, Hino H, Shoji H, Neshige R. [A case of Prader-Willi syndrome with bilateral middle cerebral artery occlusion and moyamoya phenomenon]. *Rinsho Shinkeigaku.* 1996;36(6):770–773.

121. Laghi F, Tobin MJ. Disorders of the respiratory muscles. *American Journal of Respiratory and Critical Care Medicine.* 2003;168(1):10–48.

122. L'Allemand D, Eiholzer U, Rousson V, et al. Increased adrenal androgen levels in patients with Prader-Willi syndrome are associated with insulin, IGF-I, and leptin, but not with measures of obesity. *Hormone Research*. 2002;58(5):215–222.

123. Lamb AS, Johnson WM. Premature coronary artery atherosclerosis in a patient with Prader-Willi syndrome. *American Journal of Medical Genetics*. 1987;28(4):873–880.

124. Laskey MA. Dual-energy X-ray absorptiometry and body composition. *Nutrition*. 1996;12(1):45–51.

125. Laurance BM, Brito A, Wilkinson J. Prader-Willi syndrome after age 15 years. *Archives of Disease in Childhood*. 1981;56(3):181–186.

126. Lee PD, Hwu K, Henson H, et al. Body composition studies in Prader-Willi syndrome: effects of growth hormone therapy. *Basic Life Sciences*. 1993;60:201–205.

127. Lee PDK. Endocrine and metabolic aspects of Prader-Willi syndrome. In: Greenswag LR, Alexander RC, eds. *Management of Prader-Willi Syndrome*. 2nd ed. New York, NY: Springer-Verlag; 1995:32–57.

128. Leissner, Filipas, Wolf, Fisch. The undescended testis: considerations and impact on fertility. *BJU International*. 1999;83(8):885–892.

129. Libov AJ, Maino DM. Prader-Willi syndrome. *Journal of the American Optometric Association*. 1994;65(5):355–359.

130. Linde R, McNeil L, Rabin D. Induction of menarche by clomiphene citrate in a fifteen-year-old girl with the Prader-Labhart-Willi syndrome. *Fertility and Sterility*. 1982;37(1):118–120.

131. Lindgren AC, Hellstrom LG, Ritzen EM, Milerad J. Growth hormone treatment increases CO_2 response, ventilation and central inspiratory drive in children with Prader-Willi syndrome. *European Journal of Pediatrics*. 1999;158(11):936–940.

132. Livingston FR, Arens R, Bailey SL, Keens TG, Ward SL. Hypercapnic arousal responses in Prader-Willi syndrome. *Chest*. 1995;108(6):1627–1631.

133. Lofterod B. Prader-Willi syndrome in Norway. In: Cassidy SB, ed. *Prader-Willi Syndrome*. Berlin: Springer-Verlag; 1992:131–136.

134. Lonstein JE. Congenital spine deformities: scoliosis, kyphosis, and lordosis. *The Orthopedic Clinics of North America*. 1999;30(3):387–405.

135. Lucignani G, Panzacchi A, Bosio L, et al. GABA(A) receptor abnormalities in Prader-Willi syndrome assessed with positron emission tomography and [(11)C] flumazenil. *NeuroImage*. 2004;22(1):22–28.

136. Lysdahl M, Haraldsson PO. Long-term survival after uvulopalatopharyngoplasty in nonobese heavy snorers: a 5- to 9-year follow-up of 400 consecutive patients. *Archives of Otolaryngology—Head and Neck Surgery*. 2000;126(9):1136–1140.

137. Mallory GB, Jr., Fiser DH, Jackson R. Sleep-associated breathing disorders in morbidly obese children and adolescents. *Journal of Pediatrics*. 1989;115(6):892–897.

138. Manni R, Politini L, Nobili L, et al. Hypersomnia in the Prader Willi syndrome: clinical-electrophysiological features and underlying factors. *Clinical Neurophysiology*. 2001;112(5):800–805.

139. Martin A, State M, Anderson GM, et al. Cerebrospinal fluid levels of oxytocin in Prader-Willi syndrome: a preliminary report. *Biological Psychiatry*. 1998;44(12):1349–1352.

140. McCarthy RE. Management of neuromuscular scoliosis. *The Orthopedic Clinics of North America*. 1999;30(3):435–449.

141. McGuffin WL, Jr., Rogol AD. Response to LH-RH and clomiphene citrate in two women with the Prader-Labhart-Willi syndrome. *Journal of Clinical Endocrinology and Metabolism.* 1975;41(2):325–331.

142. McLaughlin VV, Rich S. Severe pulmonary hypertension: critical care clinics. *Critical Care Clinics.* 2001;17(2):453–467.

143. McNicholas WT. Impact of sleep on respiratory muscle function. *Monaldi Archives for Chest Disease.* 2002;57(5–6):277–280.

144. Merchant-Larios H, Moreno-Mendoza N. Onset of sex differentiation: dialog between genes and cells. *Archives of Medical Research.* 2001; 32(6):553–558.

145. Mitchell RB, Kelly J, Call E, Yao N. Long-term changes in quality of life after surgery for pediatric obstructive sleep apnea. *Archives of Otolaryngology—Head and Neck Surgery.* 2004;130(4):409–412.

146. Mitchell RB, Kelly J, Call E, Yao N. Quality of life after adenotonsillectomy for obstructive sleep apnea in children. *Archives of Otolaryngology—Head and Neck Surgery.* 2004;130(2):190–194.

147. Mogul HR, Lee PD, Whitman B, et al. Preservation of insulin sensitivity and paucity of metabolic syndrome symptoms in Prader-Willi syndrome adults: preliminary data from the U.S. multi-center study. *Obesity Research.* 2004;12(1):171, HT-08.

148. Mora S, Gilsanz V. Establishment of peak bone mass. *Endocrinology and Metabolism Clinics of North America.* 2003;32(1):39–63.

149. Muscatelli F, Abrous DN, Massacrier A, et al. Disruption of the mouse Necdin gene results in hypothalamic and behavioral alterations reminiscent of the human Prader-Willi syndrome. *Human Molecular Genetics.* 2000;9(20):3101–3110.

150. Nagai T, Mimura N, Tomizawa T, Monden T, Mori M. Prader-Willi syndrome with elevated follicle stimulating hormone levels and diabetes mellitus. *Internal Medicine.* 1998;37(12):1039–1041.

151. Nakajima K, Sakurai A, Kubota T, et al. Multiple endocrine neoplasia type 1 concomitant with Prader-Willi syndrome: case report and genetic diagnosis. *American Journal of the Medical Sciences.* 1999;317(5): 346–349.

152. Nardella MT, Sulzbacher SI, Worthington-Roberts BS. Activity levels of persons with Prader-Willi syndrome. *American Journal of Mental Deficiency.* 1983;87(5):498–505.

153. Natrajan R, Louhelainen J, Williams S, Laye J, Knowles MA. High-resolution deletion mapping of 15q13.2–q21.1 in transitional cell carcinoma of the bladder. *Cancer Research.* 2003;63(22):7657–7662.

154. Nieman LK. Management of surgically hypogonadal patients unable to take sex hormone replacement therapy. *Endocrinology and Metabolism Clinics of North America.* 2003;32(2):325–336.

155. NIH Consensus Development Panel on Osteoporosis Prevention, Diagnosis, and Therapy. Osteoporosis prevention, diagnosis, and therapy. *Journal of the American Medical Association.* 2001;285(6):785–795.

156. Nixon GM, Brouillette RT. Sleep and breathing in Prader-Willi syndrome. *Pediatric Pulmonology.* 2002;34(3):209–217.

157. Nolte W, Radisch C, Rodenbeck A, Wiltfang J, Hufner M. Polysomnographic findings in five adult patients with pituitary insufficiency before and after cessation of human growth hormone replacement therapy. *Clinical Endocrinology (Oxf).* 2002;56(6):805–810.

158. Nordmann Y, Eiholzer U, l'Allemand D, Mirjanic S, Markwalder C. Sudden death of an infant with Prader-Willi syndrome–not a unique case? *Biology of the Neonate.* 2002;82(2):139–141.

159. O'Connor DB, Archer J, Hair WM, Wu FCW. Exogenous testosterone, aggression, and mood in eugonadal and hypogonadal men. *Physiology and Behavior.* 2002;75(4):557–566.

160. Ohashi T, Takeda K, Morioka M, et al. [Endocrinological study on Prader-Willi syndrome: report of four cases and review of literature (author's translation)]. *Nippon Hinyokika Gakkai Zasshi.* 1980;71(9): 999–1009.

161. Ono S, Takahashi K, Kanda F, et al. Decrease of neurons in the medullary arcuate nucleus in myotonic dystrophy. *Acta Neuropathologica (Berl).* 2001;102(1):89–93.

162. O'Sullivan AJ, Crampton LJ, Freund J, Ho KK. The route of estrogen replacement therapy confers divergent effects on substrate oxidation and body composition in postmenopausal women. *Journal of Clinical Investigation.* 1998;102(5):1035–1040.

163. Page SR, Nussey SS, Haywood GA, Jenkins JS. Premature coronary artery disease and the Prader-Willi syndrome. *Postgraduate Medical Journal.* 1990; 66(773):232–234.

164. Pang KP, Balakrishnan A. Paediatric obstructive sleep apnoea: is a polysomnogram always necessary? *Journal of Laryngology and Otology.* 2004; 118(4):275–278.

165. Parra Ordaz O. Sleep-disordered breathing and cerebrovascular disease. *Archivos de Bronconeumologia.* 2004;40(1):34–38.

166. Perrin C, Unterborn JN, Ambrosio CD, Hill NS. Pulmonary complications of chronic neuromuscular diseases and their management. *Muscle and Nerve.* 2004;29(1):5–27.

167. Phan TG, Yu RY, Hersch MI. Hypothermia induced by risperidone and olanzapine in a patient with Prader-Willi syndrome. *Medical Journal of Australia.* 1998;169(4):230–231.

168. Phillips BG, Somers VK. Sleep disordered breathing and risk factors for cardiovascular disease. *Current Opinion in Pulmonary Medicine.* 2002; 8(6):516–520.

169. Pinilla L, Gonzalez LC, Tena-Sempere M, Aguilar E. 5-HT1 and 5-HT2 receptor activation reduces N-methyl-D-aspartate (NMDA)-stimulated LH secretion in prepubertal male and female rats. *European Journal of Endocrinology.* 2003;148(1):121–127.

170. Pozzan GB, Cerruti F, Corrias A, et al. A multicenter Italian study on Prader-Willi syndrome. In: Cassidy SB, ed. *Prader-Willi Syndrome and Other Chromosome 15q Deletion Disorders.* New York, NY: Springer-Verlag; 1992:137–151.

171. Rankin KM, Mattes RD. Role of food familiarity and taste quality in food preferences of individuals with Prader-Willi syndrome. *International Journal of Obesity and Related Metabolic Disorders.* 1996;20(8):759–762.

172. Reamy BV, Slakey JB. Adolescent idiopathic scoliosis: review and current concepts. *American Family Physician.* 2001;64(1):111–116.

173. Rees D, Jones MW, Owen R, Dorgan JC. Scoliosis surgery in the Prader-Willi syndrome. *Journal of Bone and Joint Surgery, British volume.* 1989; 71(4):685–688.

174. Reid IR. Relationships among body mass, its components, and bone. *Bone.* 2002;31(5):547–555.

175. Ren J, Lee S, Pagliardini S, et al. Absence of Ndn, encoding the Prader-Willi syndrome-deleted gene necdin, results in congenital deficiency of central respiratory drive in neonatal mice. *Journal of Neuroscience.* 2003;23(5):1569–1573.

176. Riegler HC. Torsion of intra-abdominal testis: an unusual problem in diagnosis of the acute surgical abdomen. *Surgical Clinics of North America.* 1972;52(2):371–374.

177. Ritzen EM, Bolme P, Hall K. Endocrine physiology and therapy in Prader-Willi syndrome. In: Cassidy SB, ed. *Prader-Willi Syndrome and Other Chromosome 15q Deletion Disorders.* New York, NY: Springer-Verlag; 1992:153–169.

178. Robinson AC, Jones WG. Prader Willi syndrome and testicular tumour. *Clinical Oncology (Royal College of Radiologists [Great Britain]).* 1990;2(2): 117.

179. Rosenbaum HK, Miller JD. Malignant hyperthermia and myotonic disorders. *Anesthesiology Clinics of North America.* 2002;20(3):623–664.

180. Roy MS, Milot JA, Polomeno RC, Barsoum-Homsy M. Ocular findings and visual evoked potential response in the Prader-Willi syndrome. *Canadian Journal of Ophthalmology.* 1992;27(6):307–312.

181. Saaresranta T, Polo O. Hormones and breathing. *Chest.* 2002;122(6): 2165–2182.

182. Saenger P, Dimartino-Nardi J. Premature adrenarche. *Journal of Endocrinological Investigation.* 2001;24(9):724–733.

183. Schluter B, Buschatz D, Trowitzsch E, Aksu F, Andler W. Respiratory control in children with Prader-Willi syndrome. *European Journal of Pediatrics.* 1997;156(1):65–68.

184. Schmidt H, Schwarz HP. Premature adrenarche, increased growth velocity and accelerated bone age in male patients with Prader-Labhart-Willi syndrome. *European Journal of Pediatrics.* 2001;160(1):69–70.

185. Schrander-Stumpel CT, Curfs LM, Sastrowijoto P, Cassidy SB, Schrander JJ, Fryns JP. Prader-Willi syndrome: causes of death in an international series of 27 cases. *American Journal of Medical Genetics.* 2004;124A(4): 333–338.

186. Seyler LE, Jr., Arulanantham K, O'Connor CF. Hypergonadotropic-hypogonadism in the Prader-Labhart-Willi syndrome. *Journal of Pediatrics.* 1979;94(3):435–437.

187. Sharma AD, Erb T, Schulman SR, Sreeram G, Slaughter TF. Anaesthetic considerations for a child with combined Prader-Willi syndrome and mitochondrial myopathy. *Paediatric Anaesthesia.* 2001; 11(4):488–490.

188. Shimizu H, Negishi M, Takahashi M, et al. Dexamethasone suppressible hypergonadotropism in an adolescent patient with Prader-Willi syndrome. *Endocrinologia Japonica.* 1990;37(1):165–169.

189. Shults CW. Coenzyme Q10 in neurodegenerative diseases. *Current Medicinal Chemistry.* 2003;10(19):1917–1921.

190. Silverthorn KH, Hornak JE. Beneficial effects of exercise on aerobic capacity and body composition in adults with Prader-Willi syndrome. *American Journal of Mental Retardation.* 1993;97(6):654–658.

191. Sinisi AA, Pasquali D, Notaro A, Bellastella A. Sexual differentiation. *Journal of Endocrinological Investigation.* 2003;26(3 Suppl):23–28.

192. Sivak ED, Shefner JM, Sexton J. Neuromuscular disease and hypoventilation. *Current Opinion in Pulmonary Medicine.* 1999;5(6):355–362.

193. Smith A, Loughnan G, Steinbeck K. Death in adults with Prader-Willi syndrome may be correlated with maternal uniparental disomy. *Journal of Medical Genetics.* 2003;40(5):e63.

194. Smith IE, King MA, Siklos PW, Shneerson JM. Treatment of ventilatory failure in the Prader-Willi syndrome. *European Respiratory Journal.* 1998; 11(5):1150–1152.

195. Smith JD, Neeman J, Wulff J, Seely JR. Clinical-metabolic study of the Prader-Willi syndrome. *Journal, Oklahoma State Medical Association.* 1970;63(6):234–238.

196. Sone S. Muscle histochemistry in the Prader-Willi syndrome. *Brain and Development.* 1994;16(3):183–188.

197. Soriano RM, Weisz I, Houghton GR. Scoliosis in the Prader-Willi syndrome. *Spine.* 1988;13(2):209–211.

198. Spritz RA, Bailin T, Nicholls RD, et al. Hypopigmentation in the Prader-Willi syndrome correlates with P gene deletion but not with haplotype of the hemizygous P allele. *American Journal of Medical Genetics.* 1997; 71(1):57–62.

199. Stauder JE, Brinkman MJ, Curfs LM. Multi-modal P3 deflation of event-related brain activity in Prader-Willi syndrome. *Neuroscience Letters.* 2002;327(2):99–102.

200. Stevenson DA, Anaya TM, Clayton-Smith J, et al. Unexpected death and critical illness in Prader-Willi syndrome: report of ten individuals. *American Journal of Medical Genetics.* 2004;124A(2):158–164.

201. Swaab DF, Hofman MA, Lucassen PJ, Purba JS, Raadsheer FC, Van de Nes JA. Functional neuroanatomy and neuropathology of the human hypothalamus. *Anatomy and Embryology (Berl).* 1993;187(4): 317–330.

202. Swaab DF, Purba JS, Hofman MA. Alterations in the hypothalamic paraventricular nucleus and its oxytocin neurons (putative satiety cells) in Prader-Willi syndrome: a study of five cases. *Journal of Clinical Endocrinology and Metabolism.* 1995;80(2):573–579.

203. Takayasu H, Motoi T, Kanamori Y, et al. Two case reports of childhood liver cell adenomas harboring beta-catenin abnormalities. *Human Pathology.* 2002;33(8):852–855.

204. Theodoridis CG, Albutt EC, Chance GW. Blood lipids in children with the Prader-Willi syndrome. A comparison with simple obesity. *Australian Paediatric Journal.* 1971;7(1):20–23.

205. Thomas NS, Browne CE, Oley C, Healey S, Crolla JA. Investigation of a cryptic interstitial duplication involving the Prader-Willi/Angelman syndrome critical region. *Human Genetics.* 1999;105(5):384–387.

206. Tolis G, Lewis W, Verdy M, et al. Anterior pituitary function in the Prader-Labhart-Willi (PLW) syndrome. *Journal of Clinical Endocrinology and Metabolism.* 1974;39(6):1061–1066.

207. Toppari J. Physiology and disorders of testicular descent. *Endocrine Development.* 2003;5:104–109.

208. Torrisi L, Sangiorgi E, Russo L, Gurrieri F. Rearrangements of chromosome 15 in epilepsy. *American Journal of Medical Genetics.* 2001;106(2): 125–128.

209. Trifiro G, Livieri C, Bosio L, et al. Neonatal hypotonia: don't forget the Prader-Willi syndrome. *Acta Paediatrica.* 2003;92(9):1085–1089.

210. Trotter MI, D'Souza AR, Morgan DW. Simple snoring: current practice. *Journal of Laryngology and Otology.* 2003;117(3):164–168.

211. Tseng CH, Chen C, Wong CH, Wong SY, Wong KM. Anesthesia for pediatric patients with Prader-Willi syndrome: report of two cases. *Chang Gung Medical Journal.* 2003;26(6):453–457.

212. Tu JB, Hartridge C, Izawa J. Psychopharmacogenetic aspects of Prader-Willi syndrome. *Journal of the American Academy of Child and Adolescent Psychiatry.* 1992;31(6):1137–1140.

213. Turunen M, Olsson J, Dallner G. Metabolism and function of coenzyme Q. *Biochimica et Biophysica Acta.* 2004;1660(1–2):171–199.

214. Uehling D. Cryptorchidism in the Prader-Willi syndrome. *Journal of Urology.* 1980;124(1):103–104.
215. Ulrich U, Nowara I, Rossmanith WG. Serotoninergic control of gonado-trophin and prolactin secretion in women. *Clinical Endocrinology (Oxf).* 1994;41(6):779–785.
216. Urschitz MS, Guenther A, Eggebrecht E, et al. Snoring, intermittent hypoxia and academic performance in primary school children. *American Journal of Respiratory and Critical Care Medicine.* 2003;168(4):464–468.
217. van Mil EG, Westerterp KR, Gerver WJ, Van Marken Lichtenbelt WD, Kester AD, Saris WH. Body composition in Prader-Willi syndrome compared with nonsyndromal obesity: Relationship to physical activity and growth hormone function. *Journal of Pediatrics.* 2001;139(5):708–714.
218. Vanderschueren D, Vandenput L. Androgens and osteoporosis. *Andrologia.* 2000;32(3):125–130.
219. Varela M, Kok F, Setian N, Kim C, Koiffmann C. Impact of molecular mechanisms, including deletion size, on Prader-Willi syndrome pheno-type: study of 75 patients. *Clinical Genetics.* 2005;67(1):47–52.
220. Varela MC, Kok F, Otto PA, Koiffmann CP. Phenotypic variability in Angelman syndrome: comparison among different deletion classes and between deletion and UPD subjects. *European Journal of Human Genetics.* 2004;12(12):987–992.
221. Vela-Bueno A, Kales A, Soldatos CR, et al. Sleep in the Prader-Willi syn-drome. Clinical and polygraphic findings. *Archives of Neurology.* 1984; 41(3):294–296.
222. Vellayappan K, Ngiam TE, Low PS. Tonsillar hypertrophy, cor pulmonale and cardiac failure. *Annals of the Academy of Medicine, Singapore.* 1981;10(4):461–465.
223. Vestergaard P, Kristensen K, Bruun JM, et al. Reduced bone mineral density and increased bone turnover in Prader-Willi syndrome compared with controls matched for sex and body mass indexa cross-sectional study. *The Journal of Pediatrics.* 2004;144(5):614–619.
224. Vgontzas AN, Bixler EO, Kales A, et al. Daytime sleepiness and REM abnormalities in Prader-Willi syndrome: evidence of generalized hypo-arousal. *International Journal of Neuroscience.* 1996;87(3–4):127–139.
225. Vgontzas AN, Bixler EO, Kales A, Vela-Bueno A. Prader-Willi syndrome: effects of weight loss on sleep-disordered breathing, daytime sleepiness and REM sleep disturbance. *Acta Paediatrica.* 1995;84(7):813–814.
226. Vgontzas AN, Kales A, Seip J, et al. Relationship of sleep abnormalities to patient genotypes in Prader-Willi syndrome. *American Journal of Medical Genetics.* 1996;67(5):478–482.
227. Vitale ML, Chiocchio SR. Serotonin, a neurotransmitter involved in the regulation of luteinizing hormone release. *Endocrine Reviews.* 1993;14(4): 480–493.
228. Vliet GV, Deal CL, Crock PA, Robitaille Y, Oligny LL. Sudden death in growth hormone-treated children with Prader-Willi syndrome. *Journal of Pediatrics.* 2004;144(1):129–131.
229. Vogels A, Van Den Ende J, Keymolen K, et al. Minimum prevalence, birth incidence and cause of death for Prader-Willi syndrome in Flanders. *European Journal of Human Genetics.* 2004;12(3):238–240.
230. Vranjesevic D, Jovic N, Brankovic S. [Case report of a boy with Prader-Willi syndrome and focal epilepsy]. *Srpski Arhiv Za Celokupno Lekarstvo.* 1989;117(5–6):351–359.
231. Wade CK, De Meersman RE, Angulo M, Lieberman JS, Downey JA. Prader-Willi syndrome fails to alter cardiac autonomic modulation. *Clinical Autonomic Research.* 2000;10(4):203–206.

232. Walterspiel JN, Wolff J, Heinze E. [Prader-Labhart-Willi syndrome with precocious puberty. (author's transl)]. *Klinische Padiatrie*. 1981;193(2): 120–121.

233. Wang NJ, Liu D, Parokonny AS, Schanen NC. High-resolution molecular characterization of 15q11-q13 rearrangements by array comparative genomic hybridization (array CGH) with detection of gene dosage. *American Journal of Human Genetics*. 2004;75(2):267–281.

234. Wang XC, Norose K, Kiyosawa K, Segawa K. Ocular findings in a patient with Prader-Willi syndrome. *Japanese Journal of Ophthalmology*. 1995;39(3): 284–289.

235. Wannarachue N, Ruvalcaba RH. Hypogonadism in Prader-Willi syndrome. *American Journal of Mental Deficicency*. 1975;79(5):592–603.

236. Warnock JK, Clayton AH, Shaw HA, O'Donnell T. Onset of menses in two adult patients with Prader-Willi syndrome treated with fluoxetine. *Psychopharmacology Bulletin*. 1995;31(2):239–242.

237. Wassmer E, Robinson BH, Tein I. Dual pathology in two hypotonic children with Prader-Willi syndrome and muscle mitochondrial complex deficiency. *Archives of Disease in Childhood*. 2003;88:A70.

238. Watanabe T, Iwabuchi H, Oishi M. Accidental hypothermia in an infant with Prader-Willi syndrome. *European Journal of Pediatrics*. 2003;162(7–8): 550–551.

239. Weitzenblum E. Chronic cor pulmonale. *Heart*. 2003;89(2):225–230.

240. Weninger M, Frisch H, Widhalm K, Schernthaner G. [Endocrine studies on the Prader-Labhart-Willi syndrome: puberty induction in a 19-year-old boy after long-term treatment with an LHRH analog]. *Experimental and Clinical Endocrinology*. 1983;82(1):8–14.

241. West LA, Ballock RT. High incidence of hip dysplasia but not slipped capital femoral epiphysis in patients with Prader-Willi syndrome. *Journal of Pediatric Orthopedics*. 2004;24(5):1–3.

242. Whittington JE, Holland AJ, Webb T, Butler J, Clarke D, Boer H. Population prevalence and estimated birth incidence and mortality rate for people with Prader-Willi syndrome in one UK Health Region. *Journal of Medical Genetics*. 2001;38(11):792–798.

243. Williams MS, Rooney BL, Williams J, Josephson K, Pauli R. Investigation of thermoregulatory characteristics in patients with Prader-Willi syndrome. *American Journal of Medical Genetics*. 1994;49(3):302–307.

244. Wise M, Zoghbi H, Edwards M, Byrd L, Guttmacher A, Greenberg F. Hyperthermia in infants with Prader-Willi syndrome. *American Journal of Medical Genetics*. 1991;41:528.

245. Wu RH, Hasen J, Warburton D. Primary hypogonadism and 13/15 chromosome translocation in Prader-Labhart-Willi syndrome. *Hormone Research*. 1981;15(3):148–158.

246. Yahata N, Ohyashiki K, Iwase O, et al. A sole del(15q) anomaly in post-myelodysplasia acute myeloid leukemia. *Leukemia Research*. 1998;22(9): 845–847.

247. Young AJ, Castellani JW. Exertion-induced fatigue and thermoregulation in the cold. *Comparative Biochemistry and Physiology. Part A, Molecular and Integrative Physiology*. 2001;128(4):769–776.

248. Young T, Skatrud J, Peppard PE. Risk factors for obstructive sleep apnea in adults. *Journal of the American Medical Association*. 2004;291(16): 2013–2016.

249. Zaglia F, Zaffanello M, Biban P. Unexpected death due to refractory metabolic acidosis and massive hemolysis in a young infant with Prader-Willi syndrome. *American Journal of Medical Genetics*. 2005;132(2): 219–221.

250. Zappulla F, Salardi S, Tassinari D, et al. [Hypothalamo-hypophyseal-gonadal axis in the Prader-Labhart-Willi syndrome]. *Minerva Pediatrica.* 1981;33(5):201–204.
251. Zarate A, Soria J, Canales ES, Kastin AJ, Schally AV, Guzman Toledano R. Pituitary response to synthetic luteinizing hormone-releasing hormone in Prader-Willi syndrome, prepubertal and pubertal children. *Neuroendocrinology.* 1974;13(6):321–326.
252. Zellweger H, Schneider HJ. Syndrome of hypotonia-hypomentia-hypogonadism-obesity (HHHO) or Prader-Willi syndrome. *American Journal of Diseases of Children.* 1968;115(5):588–598.

6

Gastrointestinal System, Obesity, and Body Composition

Ann O. Scheimann, Phillip D.K. Lee, and Kenneth J. Ellis

Obesity or overweight is arguably the most obvious physical feature of PWS. Perhaps for this reason, more than 80% of the over 1,600 publications to date regarding PWS mention, discuss, and/or explore the topic of obesity. Despite this attention, the diagnosis, pathogenesis, optimal treatment, monitoring, and outcome of this condition in PWS remain undefined. In addition, the paradox of the underweight infant with PWS evolving into an overweight child and adult has led to considerable speculation regarding pathophysiology.

In addition to the obvious physical feature of obesity/overweight and questions regarding optimal nutritional management, a variety of gastrointestinal (GI) disorders have been identified in PWS, with frequencies similar to respiratory disorders. A description of the GI system and related disorders in PWS will start this chapter, followed by a discussion of obesity and related nutrition and medical issues. Finally, a discussion of analytical methods for body composition analysis and their application in PWS is presented (see Chapter 5 for a complete outline of Part II).

Gastrointestinal System and Disorders

The primary function of the gastrointestinal system is to facilitate intake, digestion and absorption of nutrients. The elements of the system and their functions are as follows:

1. Oropharynx and Esophagus: sucking, mastication (chewing and softening of food), salivation, swallowing (deglutition), transfer of food to the stomach
2. Stomach and Intestines: digestion, production of regulatory hormones, absorption of nutrients and elimination of waste

Elements of these processes may be relevant to the pathophysiology of PWS. Therefore, each of the following sections begins with a very brief description of the relevant physiology, followed by a review of PWS-related pathophysiology and treatment.

Oropharynx

Physiology

In the resting state, in which a person is not eating or vocalizing, the oral cavity undergoes continual involuntary salivation and swallowing. Saliva is derived primarily from the parotid, submandibular, and sublingual glands in the oral cavity (90%); another 10% is derived from scattered salivary glands. An average adult produces 0.5 to 1.5 L per day. Saliva is composed primarily of water with dissolved proteins, enzymes, and minerals; the composition differs according to the source gland. The major functions of saliva include lubrication and cleansing of the oral cavity, neutralization of acids, inhibition of microbial growth, and protection of dentition.[133,157]

Taste, smell, and palatability are among the initial food qualities that determine intake and retention of substances in the oral cavity. After entry of solid or semisolid food into the oral cavity, a complex neuromuscular process of chewing (mastication), salivation, and swallowing (deglutition) follows. In the neonatal period, ingestion of liquid substances normally involves primarily deglutition (without mastication), and the physical mechanisms differ somewhat from that which occurs with solid food ingestion.

Mastication combines a number of processes, including reduction of food to smaller pieces suitable for deglutition. The teeth serve as passive tools for this process and are controlled by coordinated activity of the powerful jaw muscles.[102,187] The muscles of the tongue participate in churning and mixing of food in the oral cavity and movement of food toward the esophagus. Coordinated movement of the jaw, laryngeal, and pharyngeal muscles occurs both voluntarily and involuntarily, with neurosensory input and feedback.[119] Mastication may also play a role in feedback regulation of appetite,[163] although this has yet to be demonstrated in humans.

Increased salivation occurs in anticipation of food intake (e.g., via visual and olfactory inputs), during taste (gustatory stimulus), and during mastication. The neural inputs for this process have been summarized.[157] Gustatory input via cranial nerves VII (facial), IX (glossopharyngeal), and X (vagus) and masticatory input via cranial nerve V (trigeminal) to the brainstem salivary center is then relayed back to the salivary glands via cranial nerves VII (to the submandibular and sublingual glands) and IX (to the parotid glands), resulting in increased salivation. The stimulated parotid gland, which produces an amylase-rich saliva, can reach 50% of total production during mastication. During mastication, saliva has key roles in enhancing taste perception, solubilizing food products, initiation of starch and lipid digestion, and preparation of food boluses for swallowing.

The process of swallowing starts with movement of processed food toward the back of the oral cavity, accomplished primarily by voluntary movement of the tongue. The second stage involves a reflex elevation of the pharynx and peristaltic movement of the food into the esophagus. The larynx also elevates, with closure of the epiglottis, thereby protecting the airway during swallowing.[171] The final stage of swallowing involves anterograde esophageal peristalsis, resulting in movement of the food bolus into the stomach. Swallowing is mediated by input via cranial nerves IX and X to the brainstem swallowing center, with output via cranial nerves V, VII, IX, X, and XII.[64,157] The entire process of swallowing is facilitated by saliva, which provides the necessary lubrication. In addition, saliva buffers the oropharynx and esophagus against acidic food and back leakage of gastric acid.

The human neonate relies exclusively upon the sucking reflex for nutrient ingestion.[70] The suck reflex develops relatively early during fetal life and involves intra- and peri-oral stimulation, leading to activation of brainstem centers interacting with the motor cortex, leading to rhythmic motor activity mediated by cranial nerves V, VII, and XII. Nonnutritive sucking (e.g., use of a pacifier) may have somewhat different dynamics and regulation from nutritive sucking (i.e., breast- or bottle-feeding),[70] the latter presumably involving additional coordination with swallowing. Although a sucking motor pattern can be identified in the 10- to 12-weeks' gestation fetus, coordination of sucking, swallowing, and respiration does not occur until after 35 weeks.[89,144] As with oral food intake after the neonatal period, the suck reflex is highly dependent upon oropharyngeal muscle tone and function.

Pathology in PWS

Generalized hypotonia in the neonate with PWS is manifested by an extremely weak suck reflex, lacking in both strength and endurance.[87] In addition, apparent lack of coordination between suck/swallow and breathing has been anecdotally observed in some infants. Although not yet studied in detail, hypotonia of the laryngeal, pharyngeal, and esophageal musculature could lead to further problems with swallowing, airway protection, and efficient movement and retention of liquid in the stomach.

Although the oropharyngeal hypotonia usually improves sufficiently to allow adequate oral nutrition by 6 to 12 months of age, the underlying problem probably continues throughout the life span. Older individuals with PWS are often noted to avoid meat and other foods that require a relatively high oromotor effort, which may in part account for the noted preference for carbohydrates over protein.[69] Micrognathia and microdontia (small lower jaw and teeth), noted in some individuals with PWS, may further compound the problem by providing less muscle bulk and surface area.

Perhaps even more problematic is the lack of adequate salivation. Unusually viscous saliva has been noted in the majority of patients with PWS,[22,105] and decreased volume of saliva is virtually universal. Saliva collection and analysis from 25 individuals with PWS (1 to 53 years old) showed an unstimulated salivary flow rate of 0.16 g/min, as compared with 0.54 g/min in controls.[94] Stimulation of salivary flow by mastication of paraffin increased the flow to only 0.38 g/min, as compared with 2.38 g/min in controls (it should also be noted that saliva could not be collected from an additional 15 PWS subjects due to inadequate flow and/or viscosity). PWS saliva was noted to be extremely viscous, with increased concentrations of all measured solutes, including fluoride (165%), calcium (226%), phosphorous (157%), chloride (124%), sodium (154%), and protein (126%). No differences were noted between the 3 uniparental disomy and 17 deletion subjects (5 did not have detailed genetic studies). A similar, but not entirely identical, condition of decreased salivation and hyperconcentration of salivary fluid has been noted in non-PWS patients with denervation of the parotid gland.[130] However, a normal but prominent appearance of the three major salivary glands, despite decreased salivation, was noted in one subject with PWS.[204]

Decreased salivary secretion, or xerostomia ("dry mouth"), leads to decreased natural cleansing of the oral cavity, severe dental caries, enamel erosion, infection, and tooth loss.[7,11,94] These disorders are similar to those observed with xerostomia associated with other conditions.[133] Enamel erosion is due to inadequate salivary buffering of food-derived acids from citrus, acidic substances (including carbonated sodas, both regular and diet), and bacterial metabolism of dietary sugar and starch,[18,204] resulting in resorption of bone mineral in the acidic milieu.

In non-PWS patients, xerostomia has been associated with speech abnormalities (dysphonia), a sensation of thirst resulting in frequent sipping, oral discomfort, difficulty with mastication and swallowing, taste disturbances (dysgeusia), heartburn, and halitosis.[133,157] Several of these features are also noted in PWS, although direct cause/effect relationships have not been systematically studied.

Treatment

In neonates with PWS, hypotonic suck and lack of a coordinated feeding mechanism can often lead to a severe failure-to-thrive. Nasogastric tube-feeding is often used to meet nutritional needs,[79] and many infants require gastrostomy tube placement to facilitate feeding for the first few months of life.

The use of treatment strategies to improve oromotor strength and coordination of swallowing can significantly enhance feeding success. Such strategies may include early introduction of occupational and speech therapies, use of adaptive devices, positioning strategies, jaw-strengthening exercises, thickening agents for liquids, and use of low-calorie binding agents. These therapies reduce the need for parenteral (tube) feedings, as shown by our experience at Texas Children's Hospital (Figure 6.1). Infants with PWS who received supplemental oromotor therapy required nasogastric feedings for a mean of 40 days, as compared with 234 days for infants who did not receive this therapy (p = 0.003).

Occupational and speech therapy are often utilized after the neonatal period (see separate chapters); the efficacy of these treatments in relation to feeding behavior and food preferences has not been studied in PWS, and there is poor documentation of results in other forms of dysphagia associated with muscle disease.[101] Nonetheless, these therapies are generally recommended for individuals with PWS.

Nonpharmacologic treatment of xerostomia in non-PWS patients often involves the use of natural secretagogues (e.g., sour lozenges,

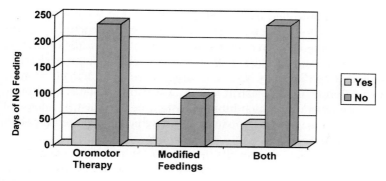

Figure 6.1. Impact of supplemental therapies on the duration of nasogastric tube feedings in infants with PWS (total N = 15). The y-axis is the mean number of days that nasogastric feedings were required. The x-axis shows supplemental therapies that were administered. (*Source:* A. Scheimann, unpublished data.)

sugarless chewing gum) to provide continuous salivary gland stimulation. This approach has been reported in some individuals with PWS,[204] but larger scale studies have not been conducted. Pharmacologic treatment of xerostomia has relied primarily upon topical application of artificial saliva, although the use of salivary secretagogues is increasing[161]; these modalities have not been studied in PWS.

Given the high risks for severe dental caries and infection, regular dental care must be established for all individuals with PWS. Oral hygiene should be instituted in infancy, even if dental eruption has not yet occurred, using soft foam toothbrushes and wetting solutions. This may be particularly important for infants on parenteral feedings, where there is virtually no stimulation of salivation or wetting from oral feeds. Fluoride treatment may also be recommended. For treatment of caries, adhesive dental techniques have been recommended.[18]

Avoidance of sugars, thick starchy foods, citrus, carbonated drinks, and other acidic foods is advisable. In addition, individuals with PWS should be encouraged to include copious amounts of water with each meal to facilitate mastication, swallowing, and oral cleansing.

Stomach and Intestines

Physiology[50]

After swallowing, esophageal peristalsis results in delivery of food boluses into the antrum of the stomach. In the resting condition, the lower esophageal sphincter, composed of specialized smooth muscle cells (not a true sphincter) maintains a positive pressure to prevent gastric contents from moving back into the esophagus. During swallowing, this pressure is released in response to local stimuli, including vasoactive intestinal peptide (VIP) and nitric oxide, thereby allowing food to enter the antrum of the stomach.

The stomach, which begins at the lower esophageal sphincter and ends at the pyloric sphincter, is composed of inner circular and outer longitudinal layers of smooth muscle which undergo rhythmic contractions controlled by a pacemaker located in the main portion of the stomach. The stomach is lined by secretory cells, including parietal, chief, and mucus cells. In response to food-related stimuli (smell, visual, cerebral), the vagus nerve signals release of acid from the parietal cells; this effect is mediated by histamine and gastrin. The acid environment, in turn, inhibits further gastrin release and also stimulates local production of somatostatin, which inhibits histamine and gastrin release. The parietal cells also produce intrinsic factor, which is required for absorption of vitamin B_{12}.

Vagal stimulation also results in release of pepsinogen from the chief cells of the stomach. In the presence of gastric acid, pepsinogen is processed to pepsin, which is the major enzyme involved in the initial steps of protein digestion.

Neurons within the stomach produce a number of substances that participate in regulation of gastrin, histamine, acid, and somatostatin release. These include acetylcholine and calcitonin gene-related peptide from the vagus nerve and pituitary adenylate cyclase-activating peptide, VIP, gastrin-releasing peptide, galanin, and nitric oxide from enteric neurons. Some of these peptides are also postulated to affect normal eating behavior and are discussed in the section on obesity.

When liquid substances enter the stomach, vagal stimulation results in relaxation of the proximal stomach, where the liquid is retained until gastric emptying. Solid foods are mixed, digested, and reduced to small particles in the distal stomach. Gastric emptying is accomplished by both muscular contractions of the stomach and by alternate opening and closure of the pyloric sphincter, which is under sympathetic and vagal control, respectively.

After passing through the pyloric sphincter, partially digested material enters the duodenum, where the fat, protein, and gastric acid stimulate duodenal production of cholecystokinin (CCK) and secretin. These peptides are absorbed into the bloodstream and travel to the pancreas, activating vagal stimulation of pancreatic enzyme secretion into the intestinal lumen. Duodenal distention also leads to pancreatic enzyme secretion through direct vagal stimulation (enteropancreatic reflex). These enzymes include (1) pancreatic amylase, which digests carbohydrates to oligosaccharides; (2) lipase and colipase, which digest fat (triglycerides) to monoglycerides and fatty acids; and (3) trypsin, chymotrypsin, and elastase, which digest peptides (resulting from pepsin digestion of proteins in the stomach) to oligopeptides.

Further digestion into amino acids, fatty acids, monoglycerides, and monosaccharides occurs in the small intestine. These nutrients are absorbed into the bloodstream from the intestinal lumen. In addition, the small and large intestines are responsible for absorption of water and electrolytes. Finally, the large intestine and anal sphincter are responsible for the process of solid waste elimination, or defecation.

A number of peptides are released from the gastrointestinal tract and pancreas into the bloodstream during the process of food intake, digestion, and absorption. These include insulin, glucagon, and postulated appetite-regulatory hormones. Some of these peptides are discussed in the section on obesity.

Pathology and Treatment in PWS

Retrograde Movement of Ingested Substances: Normally, ingested nutrients move in an anterograde (forward) fashion from the mouth to stomach to intestines with minimal backflow. Retrograde movement of food from the stomach into the oropharynx may occur under two primary circumstances: (1) gastroesophageal reflux and (2) emesis. Voluntary regurgitation and reprocessing of food from the stomach, or rumination, may also occur. Retrograde movement of intestinal contents has been observed in cases of severe obstruction and constipation in patients without PWS, but these problems have not been reported to be a particular concern in PWS.

Gastroesophageal reflux is a passive phenomenon in which liquid and partially digested food moves up the esophagus from the stomach into the oropharynx. Gastroesophageal reflux is a relatively common finding in otherwise normal (non-PWS) infants, occurring in the majority of infants under 4 months of age[153] and usually resolving within the first year of life.[190] The severity may be increased in infants with hypotonia, prematurity, or other predisposing conditions. Contributory factors may include transient relaxation of the lower esophageal sphincter pressure and positioning after feeds. It has been postulated that reflux may trigger arousal, thereby being protective against sudden death in infants.[190] In addition, in non-PWS infants, it has been found that a nasogastric tube increases the frequency of reflux episodes.[160]

In severe cases, chronic reflux of gastric acid can cause esophagitis, esophageal stricture, and cellular dysplasia and carcinoma of the distal esophagus. In older children and adults, gastroesophageal reflux may be associated with symptoms of "acid reflux" or heartburn. Endoscopic evaluation and surgical treatment may be necessary.

Possible gastroesophageal reflux has been occasionally reported in PWS,[18] but systematic documentation of prevalence has not been performed. Anecdotal experience suggests that clinically significant reflux is not as commonly observed in infants with PWS as might be expected. In addition, typical symptoms and complications due to gastroesophageal reflux have not been reported. This may, in part, be due to the lower volumes per oral feed that are usually ingested by PWS infants.

There are concerns regarding the possibility that due to hypotonia, infants with PWS may be unable to adequately protect the airway during reflux episodes, thereby increasing the risks for aspiration pneumonia and respiratory compromise. As a safety measure, reflux precautions should be taken for all infants with PWS, especially if taking substantial volumes of oral bolus feeds, and continued until the child is ambulatory. Optimal precautions have not, however, been defined in infants with PWS. In non-PWS infants, a 30-degree incline post feeds (e.g., in an infant seat) has been traditionally recommended. Given the concerns that this may worsen reflux in some infants due to increased intra-abdominal pressure, the prone or left-lateral position has been recommended,[160,190] but the relative advantages of this positioning have not been tested in infants with PWS. In any case, a supine position should be avoided.

Gastroesophageal reflux and/or complications related to this condition have not been reported in older individuals with PWS. Since individuals with PWS have decreased pain sensation (see Chapter 5), typical symptoms of heartburn may not be a reliable indicator of acid reflux. In one case of a 27-year-old man with PWS, heartburn was reported, but no abnormalities were noted on endoscopic examination.[204]

Emesis is an active process that may be considered to be a normal protective reflex. When emetic agents and toxins enter the gastrointestinal lumen, mucosal chemoreceptors are triggered which then signal through the vagus nerve back to a brainstem emetic center.[107] Emetic toxins in the bloodstream signal directly to this same area through the area postrema of the brainstem. Processing of these signals results in sequential signaling through vagal and other motor neurons, resulting in retching (simultaneous, forceful contractions of the diaphragm and abdominal muscles) and expulsion (prolonged forceful contraction of the abdominal muscles in coordination with the rib cage and pharyngeal and laryngeal muscles). Retrograde intestinal contraction occurs with gastric relaxation. Emesis then results from sequential and coordinated increases in intra-abdominal and intrathoracic pressure. Active retrograde peristalsis of the stomach or esophagus is not thought to be involved in emesis. Hypothalamic release of vasopressin and oxytocin may also be essential elements of emesis.

A commonly reported feature of PWS is a decreased ability to vomit, with a complete absence of "natural" or induced (e.g., with syrup of ipecac) vomiting noted in a large proportion of individuals.[3,104] The reasons for this are not completely known. Hypotonia of the diaphragmatic, abdominal, and intercostal muscles may be contributory since forceful contractions of these muscles are required for emesis. A deficiency of oxytocin neurons[179] could play a role, although CSF oxytocin levels are reportedly elevated in PWS.[140] Vagal autonomic dysfunction is also a possibility, although, as reviewed in Chapter 5, the evidence for autonomic dysfunction in PWS is limited.

Caution has been advised regarding reliance on emetic agents, particularly syrup of ipecac, in the treatment of accidental poisoning for individuals with PWS since the response may be inadequate. The American Academy of Pediatrics no longer recommends routine supply or use of syrup of ipecac for home treatment of childhood ingestion[46]; therefore, this issue may be a moot point, at least in the U.S. Instead, parents and guardians are advised to call the local poison control center for guidance. However, healthcare practitioners and guardians should be aware of the decreased ability to vomit in the event that an emetic therapy is considered in the emergency room or other medical care facility.

Although most individuals with PWS have a decreased ability to vomit, others may have rumination, a condition characterized by *voluntary* regurgitation of gastric contents that are then rechewed and reswallowed. A survey study found that 10% to 17% of 313 individuals with PWS reported a history of rumination and that approximately half of this group had a history of emesis.[3] Rumination was also suspected in a 17-year-old who was found to have gastric secretions in her pharynx despite fasting during preparation for general anesthesia.[174] Rumination may be a form of self-stimulation and, in the case of PWS, a means of obtaining food, albeit reprocessed.

Regurgitated food usually contains gastric acid, which may add to problems with dental enamel erosion. Therefore, in addition to behavioral treatment, the use of pharmacologic agents to block stomach acid secretion should be considered in patients who have rumination.

Gastric Dilatation: In 1997, Wharton et al.[197] reported six females with massive gastric dilatation; two died of gastric necrosis, one died of cardiac arrest, and three survived. Fever, abdominal pain, and distention were presenting signs, and vomiting was reported in two cases. These individuals had all had strict dietary control; the authors postulated that gastric muscular atony and atrophy may have occurred as a result of the dietary limitations, resulting in dilatation and necrosis following sudden ingestion of a large quantity of food. No additional cases have been published and the prevalence of this condition in PWS is unknown. However, aside from the usual recommendations for prevention of binge eating, caretakers should be vigilant for signs of acute onset of unusual abdominal distention, fever, and emesis.

Bowel Complaints: As indicated in Figure 6.2, complaints related to bowel function are frequently reported by individuals with PWS (data summarized from various sources). For the most part, these appear to

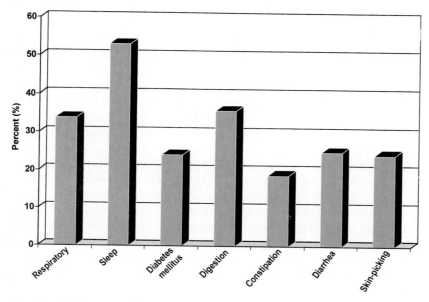

Figure 6.2. Symptom prevalence in adults with PWS (*Sources:* Butler et al.,[27] 2002, Holm et al.,[104] 1992, and personal communications from S.B. Cassidy and B.Y. Whitman.)

be secondary to factors related to eating. As of this writing, no intrinsic abnormalities in bowel anatomy or function have been associated with PWS.

Composite data indicate that 20% of individuals with PWS report constipation. Constipation can be defined as hard, dry bowel movements, usually occurring less than three times a week. Abdominal and rectal pain, rectal fissures, hemorrhoids, and rectal bleeding (bright red blood) may occur in association with the disordered defecation. In addition, affected individuals may have abdominal distention, bloating, and a general feeling of discomfort. Lack of dietary fiber and inadequate liquid ingestion may be contributory factors. Hypotonia of the pelvic floor and abdominal muscles can increase the difficulty of defecation. General physical immobility is also associated with constipation. Although hypothyroidism does not occur with increased frequency in PWS, it is fairly common in the general population and often presents with constipation.

Treatment of constipation involves provision of adequate dietary water and fiber, encouragement of physical activity, and thyroid hormone replacement, if deficient. In some cases, rectal stimulation, irrigation, suppositories, or enemas may be necessary to clear the rectal ampulla. Laxatives may be helpful in some cases; however, chronic use is not recommended.

Diarrhea, often thought of as the opposite of constipation, can be defined as loose, watery, and frequent stools. Diarrhea is reported somewhat more frequently than constipation in PWS. Noninfectious diarrhea can be caused by consumption of large amounts of poorly absorbed dietary sweeteners (e.g, sorbitol or fructose) or fat substitutes (olestra), use of antiabsorptive agents (e.g., orlistat), and food intolerance (e.g., lactose intolerance). Consumption of contaminated foods, not uncommon in PWS due to the foraging behavior, enhances the likelihood of acquiring infectious diarrhea. A careful history, examination of the stool and blood count, stool cultures (including cultures for *C. difficile*, if indicated), and parasite studies (including *giardia lamblia*) may be included in the evaluation.

Treatment of diarrhea is dependent upon the cause. Diarrhea, especially if copious, watery, and acidic, can cause perianal irritation, bleeding, and infection. These secondary conditions also require attention and treatment.

Rectal ulcers may occur as a result of a regional skin picking sometimes termed "rectal digging."[14,38,143] This behavior is often exacerbated by rectal irritation from constipation, diarrhea, or large stools. Symptoms may include mucoid rectal discharge, bloody stools, constipation, rectal pain, abdominal pain, and tenesmus. Behavioral and possibly psychotropic agent therapies are the recommended therapy for skin picking (see Chapter 12). Stool softeners and treatment of constipation and other contributory factors are also necessary to avoid further rectal irritation.

Treatment of bowel disorders in individuals with PWS often requires ongoing specialized treatment and monitoring. A multidisciplinary approach is often necessary to optimize therapy.

Other Gastrointestinal Organs

No intrinsic structural or functional abnormalities in pancreatic exocrine or endocrine function have been identified in PWS. Disordered serum levels of pancreatic secretions (e.g., insulin, pancreatic polypeptide) have been reported, as discussed elsewhere, but are not postulated to be due to primary pancreatic disease.

Nonalcoholic fatty liver disease (NAFLD) is a condition that typically occurs in obese individuals, particularly those with insulin resistance, characterized by lipid accumulation in the liver.[67] NAFLD may progress through the stages of fat accumulation (steatosis), fibrosis and inflammatory necrosis (nonalcoholic steatohepatitis), cirrhosis, and liver failure. The first stage is fairly common in the general obese population; subsequent progression is less common, but NAFLD is a major cause of nonalcoholic liver failure in the overall population. Diagnostic signs include elevated liver enzymes, liver enlargement, and ultrasound evidence of steatosis. Hepatic steatosis has been occasionally reported in cases of PWS,[99,203] but the overall frequency is not known. The treatment of NAFLD has not yet been delineated, although metformin may have beneficial effects in the early stages of disease.[112]

There are two case reports of childhood liver tumor in PWS, an adenoma and a hepatoblastoma.[95,180] It is not known whether these were specific or chance associations.

Obesity and Nutrition

Definitions

Obesity or overweight is both a major feature of PWS and a burgeoning problem in the non-PWS population. However, the recognition, diagnosis, and characteristics of obesity in PWS are very different from virtually every other population. In addition, while the terms *obesity* and *overweight* may be interchangeable in the general population, this may not be the case in PWS. This is not merely a semantic argument; distinguishing obesity from overweight may have important pathophysiologic and treatment implications for PWS.

The derivation of the word obesity is somewhat obscure. An online dictionary of etymology (www.etymonline.com) has the following listing:

obesity—1611, from Fr. *obésité*, from L. *obesitas* "fatness, corpulence," from *obesus* "that has eaten itself fat," pp. of *obdere* "to eat all over, devour," from *ob* "over" + *edere* "eat." The adj. *obese* is attested from 1651.

Other sources trace the word to *ob* (Eng: toward) *ese* (Ger: eating).

In modern medical terminology, obesity is often defined as a condition characterized by excessive body fat. While some definitions include reference to overweight resulting from excess body fat, a more standard scientific definition refers to an excess proportion of fat to nonfat mass. This latter definition is probably most relevant to PWS.

In normal human physiology and with usual diet composition, as weight is gained, fat and nonfat (bone, muscle, water) mass increase in a nearly linear relationship, with fat increasing at a faster rate than nonfat mass.[75] This "companionship of lean and fat"[75] holds for a variety of human conditions in which total body weight is altered, including diabetes mellitus, anorexia nervosa, and "normal" obesity. In an underweight individual, there is a relatively higher proportion of lean mass. As body weight increases, the ratio of fat to nonfat mass gradually increases, but both compartments increase in a linearly correlated fashion.

Because of this relationship, a ratio of weight to height provides a reasonable estimate of total body fat, with weight (kg) divided by height squared (m^2), also known as body mass index (BMI), providing the best approximation of more direct measures of body fat (see Body Composition section in this chapter). For non-PWS populations, BMI provides a convenient noninvasive indicator of body fat. As defined by the Centers for Disease Control, obesity in adults is equated with a BMI of ≥ 25. However, it is possible to have a BMI in the obese range without actually being obese, as in the case of a body builder who has increased muscle mass.

Since the proportions of body fat to height change during human growth, childhood *overweight* is defined as a BMI $\geq$95th percentile as compared with the age-/sex-related norms; *obesity* is not defined by BMI alone. As with adults, overweight usually but not always equates with obesity. For instance, athletic adolescent males may have increased muscle mass, decreased fat mass, and a relatively high BMI.

PWS is one of the few conditions in which the companionship between fat and lean does not hold. Forbes noted that while individuals with a number of conditions showed a linear, superimposable relationship of lean mass and weight, individuals with PWS were clear outliers, with a marked deficit in lean mass for weight.[75,76] Other studies have shown an excess of body fat in individuals with PWS, including underweight infants.[20,40,54,55]

Therefore, in the natural history of PWS, while overweight invariably equates to obesity, normal and underweight are also accompanied by increased body fat. At least by the definition of increased fat to lean, all individuals with PWS, regardless of weight, may be considered obese. Distinction of obesity and overweight (total body mass) may be important in terms of defining related morbidity risks and treatment. In addition, total fat mass and fat distribution may have different pathophysiologic implications from obesity and overweight.

Pathogenesis

Given the above discussion, the pathogenesis of the increased fat mass in PWS can be separated into two considerations, which may or may not be related to one another:

1. What causes the apparent violation of the "companionship rule"; i.e., why does fat mass increase in an abnormal proportion to lean mass as weight increases?

2. What causes the apparently insatiable appetite and the consequent unlimited weight (primarily fat) gain?

Knowledge of the pathogenesis of these conditions could be crucial to designing adequate treatment protocols.

The inappropriate proportion of fat to nonfat mass probably begins *in utero* and is accompanied by an absolute deficit in muscle and bone mass. This could result from either (1) inappropriate preferential shuttling of nutrients into fat, thereby causing a deficit in lean (bone and muscle) mass, and/or (2) inappropriately low utilization of nutrients by bone and/or muscle, leading to a default deposition of fat.

The mechanisms by which the human body normally shuttles ingested substrate (glucose, amino acids, fatty acids) into fat, muscle, and bone have not been completely defined[72,98,115] and a discussion of this topic is beyond the scope of this chapter. There is evidence that hormones and neuropeptides that may regulate appetite (e.g., leptin, neuropeptide-Y (NPY), adiponectin, Agouti-related protein) may also affect substrate partitioning[98,115]; however, much of the data has been collected in rodents and may not apply directly to humans. In addition, no specific defects in the physiology or action of these substances have been reported in PWS, and none are coded within the PWS gene region. Therefore, it is not known at this time whether the primary body composition abnormality in PWS is excess lipogenesis, decreased formation of muscle and/or bone, or a concomitant dysregulation of both fat and nonfat mass accretion.

The propensity for insatiable appetite and weight gain in PWS is likewise not completely understood. Normal eating behavior has been separated into three components: (1) an initial, relatively acute "drive to eat," or *hunger*; (2) an immediate postmeal feeling of fullness, or *satiety*; and (3) a longer-lasting feeling of satisfaction, or *satiation*. The control mechanisms for each of these stages are likely to be different. On a theoretical basis, environmental stimuli and voluntary control are more likely to influence hunger, whereas satiety and satiation may be more dependent on intrinsic physiologic control. Studies indicate that the primary deficit in PWS involves the third stage, satiation.[131]

Satiation may be controlled by endogenous appetite suppressants and stimulants (orexins). Theoretically, the balance between these two components is maintained by peripheral metabolic and/or neurogenic signals. Many of the characterized appetite-regulatory pathways in rodents and humans have been localized to the arcuate nucleus of the hypothalamus. A primary hypothalamic defect has been postulated to drive the hyperphagia in PWS, although no relevant functional or structural abnormalities have been identified thus far.[81,82,85,86] The possibility exists that the primary disorder may involve a defect in peripheral satiety signaling, perhaps related to the defective nutrient cycling and body composition abnormality. This latter model would agree with the clinical observation that the eating behavior in PWS more closely resembles nutritional deprivation or starvation rather than normal hunger.[103,127]

As of this writing, the study of appetite regulatory peptides in the general population and in PWS is actively developing.[50,56,199,201] The following list summarizes current knowledge of several postulated appetite regulatory peptides in relation to PWS. In general, the current model of appetite regulation includes peptides generated within the gastrointestinal system or body tissues that are released into the bloodstream (endocrine) or nervous system (neurocrine). These peptides then feed back to specific receptors in the hypothalamus, resulting in generation of signaling within the central nervous system and back to the body tissues.

1. Cholecystokinin (CCK)

CCK is produced by the endocrine I cells in the duodenum and jejunum in response to fat, amino acids, and gastric acid. Actions include stimulation of pancreatic enzyme secretion, gallbladder contraction, intestinal motility, insulin secretion, and pancreatic polypeptide secretion. CCK delays gastric emptying and inhibits release of gastric acid. In the brain, high concentrations of CCK are present in the cerebral cortex. CCK is postulated to be responsible for initiating satiety during a meal.[201] In PWS, fasting levels of CCK are normal but unlike in non-PWS controls, fasting CCK is not correlated with free fatty acid levels.[31] In response to a protein meal, CCK levels rise normally in individuals with PWS.[183]

2. Pancreatic Polypeptides

The pancreatic polypeptides PP, PYY (peptide YY), and NPY (neuropeptide Y) are produced in the pancreatic islets of Langerhans (PP), the large and small bowel (PYY), and peptidergic neurons of the stomach, small intestine, and central nervous system (NPY). PP secretion is primarily stimulated by protein intake and cholinergic activity. PYY secretion is stimulated by mixed meals and oleic acid. NPY functions as a neurotransmitter, with high concentrations in the arcuate nucleus of the hypothalamus. PP and PYY are postulated to play a role in satiety.

PP secretion in response to a protein meal has been reported to be deficient in PWS.[183,209,210] Short-term infusion of PP was initially reported to cause a mild inhibition of food intake in females with PWS[13]; however, a more detailed investigation indicated no effect.[208]

PYY levels have been reported to be low in non-PWS obese individuals, and PYY infusion causes a reduction in food intake.[9] PYY levels are reportedly low in PWS[29]; however, infusion studies have not been reported.

NPY activity in the hypothalamus is postulated to stimulate food intake.[199] Hypothalamic NPY neurons appear to be normal in PWS.[85] Serum levels of NPY are reported to be low-normal in adults with PWS and do not change with GH therapy.[108,109]

3. Hypocretins (orexins)

Hypocretins are neurocrine peptides that are postulated to stimulate feeding. Hypocretin-containing neurons in the lateral hypothalamus are stimulated in response to hypoglycemia. Hypocretins are also postulated to play roles in regulating energy expenditure and sleep/wake cycles. In a study of hypocretin levels in relation to sleep disorders,[142] cerebrospinal fluid hypocretin levels were found to be normal in a 16-year-old with PWS. However, cerebrospinal fluid hypocretin levels were reported to be low in four patients with PWS, in association with daytime sleepiness.[154]

4. Agouti-Gene-Related Protein (AGRP)

AGRP production is co-localized with NPY in the hypothalamus. AGRP stimulates food intake by inhibiting the actions of melanocortin, an anorexigenic neurocrine peptide. AGRP expression was reported to be decreased in the neonatal mouse model of PWS, in which there is failure to gain weight (as with human PWS).[80] However, AGRP neurons appear to be normal in older individuals with PWS.[85]

5. Ghrelin

Ghrelin is a peptide released from cells in the stomach. The normal function of ghrelin is not completely defined. In both rodents and humans, serum ghrelin levels rise progressively during fasting and it is postulated that ghrelin provides a signal to initiate food intake. In agreement with this theory, pharmacologic administration of ghrelin to mice results in increased food intake. However, the ghrelin knockout (deficient) mouse has no apparent defect in appetite or any other body function.[178] Recent studies indicate that the hyperphagic effect of exogenously administered ghrelin in mice is mediated by NPY/AGRP neurons.[44]

Serum ghrelin levels have been reported to be elevated in individuals with PWS[29,48,51,84,92,109] and are postulated to contribute to the hyperphagia. However, examination of the data reveals that levels are not increased in all individuals with PWS (although hyperphagia is virtually universal) and mean levels are not significantly different from normal in some studies.[16] In addition, serum ghrelin levels appear to decrease appropriately in response to meals and somatostatin administration in PWS.[16,93] Therefore, as of this writing, the role of ghrelin in PWS pathophysiology is not known.

6. Opioids (endorphins)

Opioid peptides produced within the central nervous system function as neurotransmitters. In some cases, opioids may enhance intake of foods, particularly sugary foods, perhaps by inducing positive sensations.[199] Opioids have been postulated to play a role in various types of eating disorders, including PWS.[116] However, serum levels of beta-endorphin were found to be normal in children with PWS,[138] and administration of opioid inhibitors had no effect on food intake.[207,211]

7. Leptin

Leptin is produced by fat and appears to signal adequacy of energy storage back to the brain.[26,37,39,41,132] Except in those conditions involving genetic defects in leptin or leptin receptor expression, serum leptin levels correlate directly with body fat.[202] In cases of genetic defects in leptin synthesis, the hyperphagia and other physiologic abnormalities are alleviated by leptin administration. However, it is not clear at this point whether pharmacologic administration of leptin to normal obese individuals, who have high leptin levels, will lead to significantly decreased food intake. It has been postulated that leptin functions primarily to signal energy deficiency rather than adequacy.

In PWS, leptin levels are increased and appear to be directly correlated with body fat.[26,37,39,158] Molecular defects of the leptin gene have not been identified in PWS.[33] Growth hormone therapy may decrease leptin levels in PWS; this effect is probably related to decreased absolute body fat.[59,111,148,205]

Energy Expenditure

Previous studies have described alterations in metabolic rate in individuals with PWS. Schoeller et al.[169] noted problems with use of common mathematical formulae to predict the basal metabolic rate (BMR) in adults with PWS and advocated use of the Cunningham BMR formula to adjust for the deficit in lean mass (FFM). Subsequent studies[100,188] have demonstrated a low basal metabolic rate with varied interpretation of data dependent upon the technique of body composition analysis. However, although resting energy expenditure may be normal or near-normal for lean mass, lean mass is deficient, leading to deficient expenditure for total body mass.[12,82,169]

Despite differences in body composition, energy expenditure during physical activity is similar to that of controls[151,152] when corrected for lean mass. However, individuals with PWS are less active than controls.[49,189] The combination of diminished BMR and activity level necessitates lower caloric intake or a significant increase in physical activity to avoid excessive weight gain.

Associated Morbidities

Body fat itself is rarely a direct cause of morbidity or mortality. Fat embolism is an example of direct fat-related morbidity, but this condition is not reported to occur with increased frequency in PWS. In PWS, the increased *proportion* of fat to lean mass (regardless of weight) is largely the result of decreased muscle mass. The resultant hypotonia is a major contributor to morbidity and mortality, as discussed in previous sections. As body weight increases above normal in PWS, the fat mass itself becomes problematic.

Body fat can contribute indirectly to pathophysiology in two ways: (1) complications due to a mass effect, i.e., in morbidly obese individuals, and (2) via metabolic complications related to fat.

The detrimental effects of excess body mass are well recognized. In particular, increased fat is associated with respiratory impairment and obstructive apnea. The sheer weight of excess fat in the presence of low muscle mass also contributes to impaired physical mobility and difficulty with daily tasks. Many adults with PWS adopt a typical hypotonic posturing, with both arms folded across the upper abdomen, while others become wheelchair-dependent in young adult life. Increased fat mass may also theoretically exacerbate scoliosis and fragility fractures involving weight-bearing bones and joints (vertebrae and hips), but, on the other hand, increased weight may augment bone mineral density, as shown in non-PWS populations.[129]

Metabolic effects of increased body fat have been well defined in the general population. In non-PWS children, early puberty (particularly adrenarche),[163] accelerated linear growth, and advanced bone age can occur; final adult height is usually not adversely affected despite the accelerated physical development. Adrenarche is the portion of sexual development characterized by increased production of adrenal androgen precursors, which, in the peripheral tissues, are converted to testosterone and dihydrotestosterone. In normal puberty, adrenarche is responsible for secondary sexual hair growth in females; the effect of adrenarche is usually less notable in boys due to testicular production of testosterone. The hormones produced during adrenarche also cause acceleration of bone growth and epiphyseal closure. The physiologic control of the timing of adrenarche has not been defined; however, insulin resistance and hyperinsulinemia are associated with earlier onset.

In children with PWS, premature adrenarche occurs in a relatively small subset of cases[121,126,167] and is manifested by early (before age 8 to 9 years) appearance of pubic hair. Anecdotal experience suggests that the frequency of premature adrenarche is less than might be expected in a similarly overweight non-PWS population. In affected children, the increased linear growth rate due to adrenarche often replicates a normal, non-PWS growth rate (which is actually accelerated for PWS). Unfortunately, the end result is often extreme short stature due to premature epiphyseal closure, which is quite different from the normal stature attained by non-PWS children with obesity-related premature adrenarche. Therefore, "idiopathic" premature adrenarche cannot be considered to be a benign condition in PWS.

In older children, adolescents, and adults, obesity can lead to metabolic syndrome, also known as dysmetabolic or insulin-resistance syndrome (defined by various criteria, but generally including insulin resistance, overweight/obesity, dyslipidemia, and hypertension; all associated with increased cardiovascular risk), polycystic ovary syndrome, and Type 2 diabetes mellitus.[58] In non-PWS populations, insulin resistance and consequent morbidities have been specifically associated with increased abdominal visceral fat (as opposed to subcutaneous fat).

Glucose intolerance and Type 2 (also known as non-insulin-dependent) diabetes mellitus (T2DM) can occur in patients with PWS.[27,42,78,91,113,123,150,162,170,172,203] The usual case of T2DM in PWS is indistinguishable from non-PWS obesity-related diabetes, which is occurring with increasing frequency worldwide.[198] Unlike Down and Turner syndromes, there does not appear to be any unique risk for Type 1 (insulin-dependent) diabetes mellitus (T1DM) in PWS. In addition, no specific metabolic abnormalities have been identified in PWS to suggest a unique predisposition to T2DM except for obesity.[21,136] A reduced amount of insulin receptors was noted in one report, but the clinical significance of this finding is uncertain.[120]

Recent data indicate that individuals with PWS may have a lower risk for insulin resistance and T2DM than would be expected based on the degree of overweight. Average fasting insulin levels are reported to be low in children and adults with PWS and there is a relative lack of clinical signs consistent with insulin resistance,[57,111,145,156,181,206] although some reports have found elevated fasting insulin,[125,155] and others have reported elevated fasting but decreased 2-hour postprandial insulin.[128] However, the bulk of evidence suggests that insulin levels are relatively low in most individuals with PWS, arguing against an increased frequency of insulin resistance. In addition, serum levels of adiponectin, a protein secreted by adipocytes that is thought to increase insulin sensitivity, are unexpectedly high in PWS,[110] whereas low levels are usually observed in non-PWS patients with obesity and insulin resistance.

In non-PWS individuals, insulin resistance syndromes and T2DM are part of a spectrum of disorders related to increased body fat and, in particular, intra-abdominal visceral fat[73,165] (as distinguished from subcutaneous fat). Lipid deposition in muscle and other body tissues may also be contributory.[114] However, in obese individuals with PWS, subcutaneous but not visceral fat has been found to be increased[83,182]; the mechanisms for this occurrence have not been defined. In addition, visceral fat characteristics for individuals with PWS and insulin resistance have not yet been reported.

Monitoring for signs and symptoms of insulin resistance and T2DM should be a routine element of care for all individuals with PWS. Some cases of T2DM may present with the classic symptoms of diabetes mellitus: polyuria, polydipsia, and, in some cases, unexpected weight loss despite continued hyperphagia. Ketoacidosis, obtundation, and disordered consciousness may occur in the most severe cases.[203] However, most individuals with insulin resistance or T2DM will be asymptomatic. A classic, but not universal, physical sign of insulin resistance is acanthosis nigricans: hyperpigmented, velvet-textured skin in the nuchal, axillary, inguinal, and other folds of the body thought to be due to direct or indirect effects of hyperinsulinemia on keratinocytes.[184]

Individuals suspected of having insulin resistance should be screened for associated morbidities, including dyslipidemia (fasting lipid panel) and hypertension. Diagnosis of impaired glucose tolerance (IGT) and T2DM should be made according to criteria of the American Diabetes Association[2]:

1. Symptoms of diabetes plus casual plasma glucose concentration ≥200 mg/dl (11.1 mmol/l). Casual is defined as any time of day without regard to time since last meal. The classic symptoms of diabetes include polyuria, polydipsia, and unexplained weight loss.

or

2. FPG ≥ 126 mg/dl (7.0 mmol/l). Fasting is defined as no caloric intake for at least 8 hours.

or

3. Two-hour post-load glucose ≥200 mg/dl (11.1 mmol/l) during an OGTT. The test should be performed as described by the World Health Organization (WHO), using a glucose load containing the equivalent of 75-g anhydrous glucose dissolved in water.

Since criteria 1 and 3 do not require an elevated fasting glucose, and a 2-hour oral glucose tolerance test may not always be feasible, some practitioners utilize a random glycated hemoglobin or hemoglobin A1c measurement to screen for clinically significant glucose intolerance and diabetes mellitus.

Treatment

Nutritional management of PWS can be separated into four major areas of concern:

1. Control of under- and overweight
2. Optimization and conservation of lean mass
3. Special nutritional considerations
4. Treatment of obesity-related morbidities

Evolving clinical needs for the individual with PWS requires adaptation of nutritional support. During infancy, diminished muscle tone affects the volume of caloric intake during feedings. A variety of techniques are available for nutritional support of the infant with PWS including adaptive feeding bottles and nipples (e.g., Haberman feeder, cleft palate nurser, adaptive nipples), thickening agents (Thick-It, cereal), formula concentration, and nasogastric tubes. The feeding therapy utilized is determined by the adequacy of swallowing skills, and nutritional status.

Nasogastric and gastrostomy feedings are commonly used to meet the nutritional needs of infants with PWS. Gavranich et al.,[79] reported use of tube feedings for a mean of 8.6 weeks among 67% of infants with PWS in New South Wales. Since gastrointestinal absorption and motility are essentially normal, intravenous total parenteral nutrition is usually not required. In infants receiving feedings primarily through a non-oral route, oral feedings as tolerated and non-nutritive sucking should be continued to encourage development of oromotor strength and coordination.

Intake parameters during infancy can be patterned along guidelines from the Nutrition Committee of the American Academy of Pediatrics.[47] During the first 6 months of life, breast milk and infant formula should serve as the primary nutritional source and should be given in usual amounts. Solids are generally introduced at 5 to 6 months of age and advanced in texture, dependent upon oral motor skills. Higher-calorie solids, desserts, and juices are commonly avoided. Through close monitoring of growth data over the first 2 years, oral intake can be appropriately adjusted to maintain weight for height between the

25th and 80th percentiles. Caloric restriction under the guidance of an experienced nutritionist or other health care provider is required, with supplementation of deficient vitamins and minerals, if weight gain becomes excessive.

Nutritional strategies beyond the toddler years focus on avoidance of overweight. Limited studies[106,159] have evaluated the caloric requirements for individuals with PWS. Weight maintenance has been reported with intakes of 8 to 11 kcal/cm/day (non-PWS children require 11–14 kcal/cm/day; cm = height); weight loss has been documented with intakes of 7 kcal/cm/day. Sample calorie guidelines, adapted from guidelines published by PWSA (USA)[17] are listed in Table 6.1. It should be noted that while these guidelines are based on logical criteria, there are no prospective data regarding efficacy. In addition, these guidelines may be excessive for children who are unusually inactive and, conversely, inadequate for children with PWS receiving growth hormone therapy.

Similar guidelines have not been specifically formulated for adolescents and adults with PWS, although a general recommendation has been 800 to 1,000 kcal/day for weight loss. These calorie guidelines are a significant reduction from usual food intake in the general population; therefore, the individual with PWS will probably need to have different food preparation and provision from the rest of the household.

A typical approach to a calorie-restricted weight control treatment plan is to introduce a "balanced" calorie reduction, with maintenance of the usual carbohydrate-protein-fat proportions (i.e., 60%-15%-25%, respectively). Emphasis on low-glycemic-index carbohydrates (i.e., slowly-absorbed complex carbohydrates rather than, for instance, high sugar foods) may also reduce insulin secretion, facilitate optimal nutrient utilization and have a positive effect on satiety,[135] although these effects have not been studied in individuals with PWS.

Adherence to a calorie-restricted diet requires intensive and continuous monitoring of intake and regular dietary counseling, including analysis of food histories and attention to possible associated nutrient

Table 6.1. Sample Calorie Guidelines

Age (yr)	Average Height (cm)	Weight Maintenance (kcal/d)	Weight Loss (kcal/d)
Female			
3	89	700–980	630
5	102	800–1120	720
7	112	880–1232	790
9	135	1060–1484	954
Male			
3	94	740–1036	660
5	107	840–1176	760
7	119	940–1316	845
9	122	960–1344	864

deficiencies. Behavioral aspects of this plan require attention to all potential sources for intake including cafeterias, school buses, classroom activities ("life skills"), vending machine access, neighbors, and convenience stores as well as home access (e.g., pantry, garbage cans, refrigerator, tabletop). Locks on kitchen doors and refrigerators are often recommended elements of this plan. More detailed discussions of behavioral and environmental management strategies appear elsewhere (see Chapters 12 and 18).

There is no doubt that total calorie restriction will achieve weight maintenance or loss if completely implemented; however, it remains unresolved as to whether this approach is justified in view of the physiology. As mentioned above, there is growing consensus that the food foraging and apparently insatiable appetite in PWS may be triggered by internal mechanisms that more closely resemble true starvation than non-PWS pre-meal hunger or eating behavior. If this is true, then the intentional restriction of all intake could have detrimental effects on overall behavior and well-being and may, in fact, augment foraging and food-sneaking behavior. This hypothesis has not been tested, although it appears to hold some validity on review of anecdotal patient experience.

An equally important issue relates to adverse effects on body composition. Although increased body fat and overweight is a major visible morbidity in PWS, a more crucial functional morbidity is the lack of lean mass. In non-PWS individuals, induction of an energy deficit (e.g., fasting, total calorie restriction) and weight loss using a balanced nutrient intake results in loss of not only fat mass, but also lean mass. In addition, the lower the total calorie intake, the higher the proportion of lean mass lost. In non-PWS individuals, excess body fat will provide a relative protection against loss of lean mass (i.e., thinner individuals lose proportionately more lean mass than obese individuals during weight loss).[77]

As stated by Dr. Gilbert Forbes, a pioneer in this field of research, "There is no level of reduced energy intake that will completely spare LBM [lean body mass] when significant amounts of body weight are lost."[75] In individuals with PWS, where total lean mass is deficient even in the presence of overweight, there is no reason to suspect that a balanced calorie restriction diet will result in preservation of absolute lean mass.

However, lean mass can be at least relatively preserved during calorie deficit by preferentially preserving protein intake. The initial observations of this phenomenon preceded the currently popular low-carbohydrate, increased protein diets.[75] In a short-term metabolic unit study in four patients with PWS, a protein-sparing (1.5 gm of meat protein per kg body weight), ketogenic, modified fast preserved positive nitrogen balance and lean body mass in the presence of significant weight reduction.[15] A similar nutritional approach in an obese ventilator-dependent adolescent with PWS apparently facilitated weight management and weaning from the ventilator.[45]

In addition to potential effects on preservation of lean mass, protein may have a greater positive effect on satiety than carbohydrates or fat, as demonstrated in both short- and long-term human studies.[5,25] Although this effect has not been investigated in individuals with PWS, it was postulated to occur in outpatient follow-up of patients involved in the previously mentioned study using a protein-sparing fast.[15]

A ketogenic protein-sparing fast is probably not practical for most individuals with PWS. However, several other approaches to high-protein, lower carbohydrate meal planning are available. Popular diets using this approach typically result in 28% to 35% calories from protein, 8% to 40% from carbohydrate, and the remainder from fat.[5] One of the authors (PDKL) has prescribed modified lower-carbohydrate, lower-fat guidelines in his PWS clinics for several years, an approach which is similar to other, perhaps more stringent regimens.[5,146,147] The basic five-step approach is as follows: (1) elimination of sugar and all packaged foods containing >5 g sugar per serving, (2) limitation of complex carbohydrates to one to two servings three times daily, with one serving = 15 gm of carbohydrate, (3) avoidance of fried and fatty foods, (4) encouragement of lean protein intake, and (5) provision of "free foods"—i.e., carb-free, low-fat, low-calorie—for ad lib snacking and/or foraging. In addition, an exercise guideline of 30 minutes sustained activity, three to five times weekly is recommended. This regimen is provided as a one-page guideline and reviewed during clinic visits. For most individuals, adherence to this simple plan results in a substantial reduction in total calories. This approach has the advantages of being easy to learn (requires teaching of food label readings and basic carbohydrate counting), relatively straightforward implementation, and possible integration into usual household eating patterns. More structured dietary regimens can be added to this basic program for individual patients.

Whatever dietary approach to weight control is taken, it is important that it be logical, consistent, easily implemented, emphasized at each clinic visit, and carefully monitored. Modifications should be made for individual patients, and in children developmental changes should be taken into account.

Pharmacologic agents for weight management should be considered as adjuncts and not primary treatment modalities. None of the appetite suppressant or antiabsorptive agents marketed for obesity treatment have shown efficacy in PWS, and systematic studies have not been published. Some of the psychotropic agents commonly used in PWS have been anecdotally reported to control food-foraging behavior, but exacerbation of overeating has also been observed.[118] Anecdotal reports indicate that growth hormone therapy may have a beneficial effect on eating behavior, but this has not been demonstrable in objective studies. In a short-term uncontrolled trial, the anti-epileptic medication, topiramate, was reported to improve eating and other behaviors in seven patients with PWS, resulting in weight loss in three[175]; however, several of the subjects were on concomitant medications and the overall results were not entirely conclusive.

Personal experience and one report[42] indicate that metformin may have efficacy for weight control in PWS when coupled with dietary management, particularly with carbohydrate limitation. Similar results have been reported in treatment of non-PWS obesity.[6,146,147] The mechanism(s) of action for metformin in weight management have not been completely elucidated, although an anorectic effect has been demonstrated in animal studies.[112]

Surgical options for weight management are not generally recommended in PWS. Bariatric surgery causes weight loss through either a diminished capacity for food intake and/or via reduced digestion and absorption of food. In non-PWS obesity, studies in adolescents and adults show efficacy in promoting weight loss, although surgical risks, gastrointestinal problems, and malabsorption are a concern.[71]

Experience with bariatric surgery in PWS has been less encouraging, as summarized in Table 6.2, although relatively long-term success has been reported in selected patients.

There is no apparent effect of bariatric surgery on hyperphagia in PWS. Therefore, the need for dietary intervention and monitoring is not eliminated. Over the long-term, weight gain may recur after the patient develops compensatory dietary strategies. At the current time, bariatric surgery should be considered only in severe cases in which serious obesity-related morbidities are present and rapid weight loss is considered to be potentially beneficial.

Special Nutritional Considerations

Vitamin and mineral supplementation is highly recommended for individuals with PWS and particularly for those on a balanced, calorie-restriction diet. For example, the sample calorie guidelines from PWSA (USA)[17] listed in Table 6.1 are deficient in calcium, iron, vitamin D, vitamin E, biotin, pantothenic acid, magnesium, zinc, and copper; multivitamin and mineral supplementation is recommended in the publication. In addition, many individuals with PWS have limited sun exposure, especially those affected with hypopigmentation. Since a large proportion of the body stores of vitamin D are synthesized in response to sunlight, lack of sun exposure can result in vitamin D deficiency and an increased risk for osteoporosis.

As a general rule, it is recommended that all patients with PWS receive daily multivitamin and mineral supplementation in consideration of their individualized meal plan and sun exposure. An over-the-counter preparation may be adequate for many patients. Others may require additional monitored supplementation with calcium, trace minerals, and vitamins. Trace mineral, iron, and fat-soluble vitamin supplementation should be carefully monitored to avoid overload.

Obesity-Related Morbidities

Premature Adrenarche

In children without PWS, premature adrenarche is usually a benign condition that does not require specific therapy; indeed, specific therapies have not been proven to have efficacy.[163] In some cases, premature adrenarche may be associated with early central puberty, which may require treatment. As mentioned previously, premature adrenarche in

Table 6.2. Bariatric Surgery in Prader-Willi Syndrome

Reference	Type of Surgery	No. of Patients	Median Age	Median Weight	Success Rate	Complications
Soper et al. (1975)[176]	Gastroplasty	7	15 yrs	92.5 kg	43%?	• 57% required revisions due to inadequate weight (wt) loss
Anderson et al. (1980)[4]	91% Gastric bypass 9% Gastroplasty	11	13 yrs	85 kg		• 54% required revision due to inadequate wt loss • 1 (9%) wound infection • 1 dumping/diarrhea • 1 death from uncontrolled wt gain
Fonkalsrud and Bray (1981)[74]	Vagotomy	1	17 yrs	120 kg	29 kg initial wt loss, followed by 20 kg gain	• 20-kg weight gain
Touquet et al. (1983)[185]	Jejunoileal bypass	1	24 yrs	181 kg	62 kg (1 yr)	• Postoperative wound infection • DVT/pulmonary embolus • 4–5 stools/day
Laurent-Jacard et al. (1991)[124]	Biliopancreatic diversion	3	27.6 yrs	84.5 kg	Significant wt loss 1st year, followed by wt gain (2½–6 yrs)	• Diarrhea • Vitamin D, vitamin B12, folate, and iron deficiency

Study	Procedure	N	Age	Weight/BMI	Outcome	Complications
Dousei et al. (1992)[53]	Vertical banded gastroplasty	1	21 yrs	57.4 kg	Initial improved DM control	Short-term wt loss followed by break of staple line and wt gain
Chelala et al. (1997)[43]	Laparoscopic adjustable gastric band	1	?	?		Death 45 days postoperatively from GI bleeding
Grugni et al. (2000)[88]	Biliopancreatic diversion	1	24 yrs	80 kg	Initial wt loss, but wt gain without restriction	Diarrhea, severe osteopenia, anemia, hypoproteinemia
Marinari et al. (2001)[139]	Biliopancreatic diversion	15	21 yrs	127 kg	56%–59% wt loss at 2–3 yrs, then regained 10%–20% of wt lost	2 deaths from unrelated causes (no vitamin levels or bone density data provided)
Braghetto et al. (2003)[19]	95% gastrectomy with Roux-en-Y, hypocaloric diet	1	15 yrs	BMI = 57.7	70 kg weight loss over 1 yr; BMI = 30	No surgical complications reported
Kobayashi et al. (2003)[117]	Roux-en-Y gastric bypass	1	30 yrs	146 kg	54-kg weight loss over 18 mos, improved lipid profile	No complications

children with PWS is not a benign condition since it is associated with severe short stature and an inadequate increase in height velocity.[126,167] Although obesity is thought to be pathogenetic for this condition, there is no evidence that intensive weight control after onset will slow the progress of the adrenarche. Prevention of predisposing factors for premature adrenarche via weight management beginning in very early childhood is the best recommendation at this point. In cases where the process has already started, growth hormone therapy should be considered even if current height is normal in order to optimize final adult height.

Insulin Resistance and Type 2 Diabetes Mellitus

The treatment of insulin resistance and T2DM in PWS should follow current standard-of-care guidelines for these conditions in the non-PWS population. A comprehensive discussion of this topic is beyond the scope of this chapter; the reader is referred to the current literature and healthcare organizations for more detailed protocols (e.g., American Diabetes Association, American Association of Clinical Endocrinologists).

In general, the first-line approach should include diet and exercise, as described above for treatment of obesity and overweight; in milder cases, these therapies may lead to complete resolution of the disorder. Metformin should be considered a first-line pharmacotherapy for insulin resistance, especially if T2DM is present.[42,112] Insulin may be necessary in cases where there is evidence of insulin deficiency (keto-acidosis, unexplained weight loss) but should be avoided in all other cases since it may augment increases in body fat. Sulfonylureas and PPAR-agonists also have a tendency to increase body fat, and the efficacy of these agents in individuals with PWS and T2DM has not been shown. The authors' anecdotal experience suggests that diet, exercise, and metformin are sufficient for treatment of most individuals with PWS and insulin resistance and/or T2DM.

Monitoring of patients with insulin resistance and T2DM should include periodic evaluation of fasting lipid profiles and blood pressure. Fasting insulin levels can be checked periodically, but the clinical utility of this measurement, which can be highly variable, is not defined. Clinically, weight, calculated body mass index, waist circumference, and status of acanthosis nigricans (if present) can be useful.

Individuals treated with metformin should have routine annual monitoring of liver and kidney function tests; an elevated serum creatinine level is a contraindication to therapy. With T2DM, routine home monitoring of fasting and postprandial glucose levels and periodic glycated hemoglobin measurements are necessary, with an optimal goal of achieving normal levels for both parameters.

There have been surprisingly few reports of diabetes-related complications in PWS.[8,192] However, individuals with diabetes mellitus should be routinely monitored for evidence of retinopathy, nephropathy, hypertension, and cardiovascular disease, with institution of appropriate therapy as needed as per general standard-of-care practice guidelines.

Measurement of Body Composition

The measurement and monitoring of body composition has become an essential element in clinical research and management of individuals with PWS. In particular, body composition measurements may be used to diagnose decreased bone mineral density (osteopenia, osteoporosis) and monitor changes in total body fat and nonfat mass.

Anthropometry

Anthropometry refers to body measurements, such as length or height, weight, circumferences, and skinfold thickness. Although anthropometric techniques (weight-for-height indices, skinfold thickness, waist-to-hip ratio) have been used for many years to indirectly estimate body composition, these techniques are less commonly used for that purpose today. Instead, if detailed body composition analysis is needed, more sophisticated models and techniques are used, as discussed below.

Height growth is an essential feature of human development and should be measured for all children at regular intervals. The measurements should be plotted to the nearest fractional age on charts compiled from the background normal population. Such charts are available for the U.S. pediatric population from the U.S. Centers for Disease Control (www.cdc.gov). Standards for most other industrialized countries are available. Children under 2 years of age should be measured using recumbent length, which is the basis for these standards. Height velocity charts are also available. Procedures for accurate measurement and calculation of height and velocity are also described on the CDC Web site.

Height growth is basically a measure of long bone (leg, spine) growth, which in turn is dependent on a number of genetic, hormonal, structural, and mechanical factors. Abnormal height growth in children can be the first sign of a systemic abnormality. As mentioned previously, height growth in children with PWS is highly variable but usually abnormally low starting at or before the toddler stage. As shown in several population studies, the average final adult height in PWS is at approximately 2 standard deviations below the mean for the background population.[1,22,36,97,123,149] Scoliosis, if present, may also account for loss of height growth.

Weight is basically a measure of total body mass, regardless of composition. As a stand-alone measurement, it has little intrinsic utility, even if plotted on normative curves. However, the clinical value of a weight measurement is increased when analyzed in conjunction with height, that is, in determination of weight-for-height or body mass index. Assuming that fat and lean mass are present in a predictable proportion (see previous discussion regarding "companionship rule" above), these ratios can provide a measure of body fatness (normative charts available on www.cdc.gov). However, in conditions where there is an excess proportion of lean mass (e.g., body building) or deficient lean mass/excess fat mass (e.g., PWS), these ratios do not provide an accurate estimation of body fat.

Anthropometric measurements for head circumference, hand and foot length, and other body parts have been published for PWS[28,32,34–36,65,141,196] (see Appendix C). Head circumference measurements are primarily used as an indicator of brain growth in infants and toddlers under 3 years of age. Careful monitoring of head growth is also important for detection of craniosynostosis, premature closure of the cranial sutures resulting in severe neurologic sequelae. One such case in PWS has been published,[66] and the authors are aware of other cases.

Waist circumference and waist to hip circumference ratios have been used to indicate visceral fat and consequent risks for T2DM and cardiovascular disease in non-PWS adults. However, this correlation is likely to be less reliable in PWS since visceral fat may not be increased.

Skinfold thicknesses, measured using calipers at selected body sites, are sometimes used to estimate body fat and non-fat mass. Skinfold thickness is basically a measure of subcutaneous fat. Using assumptions and validated algorithms regarding the proportions of subcutaneous fat to other body compartments, fat and lean mass can be estimated. Although skinfold thickness measurements have been used in several key studies of PWS,[30,54,96] inter-observer variability and lack of validated disease-specific (including PWS) standards and algorithms are major limitations to routine clinical use.[193]

Body Composition Modeling

The first step in determining body composition is to select a model that provides clinically relevant measurements (Figure 6.3). The most extreme form of body composition analysis, elemental or chemical

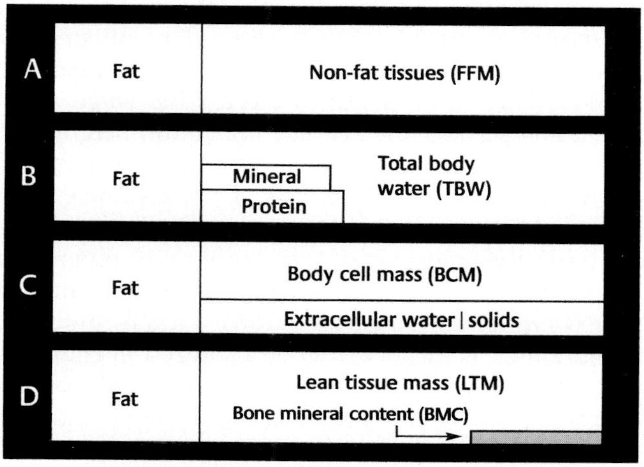

Figure 6.3. Multicompartment models of body composition. **A.** Basic 2-compartment model, weight = Fat + FFM (fat-free mass). **B.** Water, protein, and mineral subcompartments of FFM. **C.** Body cell mass and extracellular water and solids subcompartments of FFM. **D.** 3-compartment dual-energy X-ray absorptiometry (DXA) model.

analysis after ashing, is primarily of research interest since it is not feasible in an individual living organism. Elemental analysis can be also performed *in vivo* using isotope counting and neutron activation methods[62]; these measurements currently have limited clinical utility and will not be discussed.

The simplest clinically useful model of body composition, the 2-compartment (2-C) model, divides total body mass or weight (Wt) into fat mass (FM) and fat-free mass (FFM). Body fatness, in turn, can be defined as the FM/Wt ratio, expressed as a percentage. The basic 2-C model requires only one measurement to be made and is the easiest to use when the assessment of body fatness is the primary aim.

More detailed clinical models,[194,195] particularly for considering issues related to nutrition or growth, separate the FFM into components, creating a multicompartment model (Figure 6.3). A useful multicompartment model separates FFM into water and/or protein (muscle) and mineral (bone) subcomponents. Direct assay of FFM components is difficult but has been facilitated by recently developed imaging techniques. These sophisticated techniques, such as dual-energy X-ray absorptiometry (DXA), computed tomography (CT), and magnetic resonance imaging (MRI), allow us to view the components of the living body.[52]

In the following sections of this chapter, the various methods that are available will be described in the context of their application in 2-, 3-, and 4-compartment models of body composition.

Methods Based on the 2-Compartment (2-C) Model

Underwater Weighing: For the 2-C model, Wt = FM + FFM, and $Vol_{TB} = Vol_{FM} + Vol_{FFM}$ (Total body volume = FM + FFM volumes). An accurate measurement of body weight is relatively easy to achieve, i.e., using a weight scale; the measurement of body volume is more challenging.

The classic technique of measuring body volume, underwater weighing (UWW, or hydrostatic weighing) relies on Archimedes' principle, which states that a body immersed in a fluid is buoyed up by a force equal to the weight of the displaced fluid. The procedure involves measuring body weight while totally submerged underwater, after exhalation of air from the lungs. The body volume is calculated by subtracting the underwater weight from the regularly measured body weight, the weight differential being equal to the weight of displaced water, and dividing the difference by the density of water. This number is adjusted for the measured residual lung volume.

The density of the body (ρ_{TB}) is the ratio of body weight (Wt) to body volume (Vol_{TB}). The classic UWW relationship between body fatness (%FM) and body density (ρ_{TB}) for a 2-compartment model is

$$\%FM = 100 \times \{(k_1/\rho_{TB}) + k_2\}$$

where the constants (k_1, k_2) are determined by the values selected for ρ_{FM} and ρ_{FFM} (densities of the fat and fat-free mass).[24,173] The density of fat can be assumed to be constant (0. 9004 g/cc), whereas the density of FFM is not constant, changes with growth,[134] and is altered by diseases and medications. Since body water is the major contributor to the mass and volume of the FFM, changes in the relative water content (hydration) will have significant effects on ρ_{FFM}. Other components of the FFM can also change, but these have a secondary impact compared with hydration.

For short-term longitudinal studies in healthy subjects, using age- and gender-adjusted constants, the 2-C model is adequate for the assessment of changes in body fatness. However, it does not provide information about components of the FFM.

Air-Displacement Plethysmography: There can be several reasons why the underwater weighing technique cannot be used, e. g., lack of necessary equipment, fear of water, difficulty breathing underwater, too buoyant to be easily submerged. An alternative technique called air-displacement plethysmography (ADP) can also be used to measure body volume. The advantage of ADP is that the problems related to water are eliminated, although the subject still needs to wear a tight-fitting bathing suit and cap to cover scalp hair. The objective is to determine the volume of air that is displaced by the subject's body.[68,134] At present, there are only two commercial ADP instruments available (BODPOD for adults and PEAPOD for infants, Life Measurements Inc., Concord, Calif.).

The ADP technique is based on Boyle's Law for gases: pressure multiplied by volume is constant if temperature does not change. Poisson's Law for gases is also used in order to adjust for the subject's breathing (heated moist air coming from the lungs) and for the isothermal effects of air in contact with the subject's skin and body hair.

For the BODPOD measurement, the subject sits on a bench in a small test chamber that is about the size of a telephone booth. This chamber is connected via a diaphragm to a reference chamber behind the bench. The door (which has a large window) is closed, the diaphragm is oscillated at a low frequency, and the pressure difference between the two chambers is measured. The BODPOD instrument also has a built-in spirometer system that can be used to measure the subject's residual lung volume. It should be noted that without preliminary training, the spirometer procedure can be difficult for some subjects to perform correctly.

ADP estimates of body volume are highly correlated with those for UWW, and the two results are virtually interchangeable for healthy adults.[68] Additional studies with children may be needed, especially where body composition may be abnormal. However, it is reasonable to expect that the ADP technology may replace the underwater weighing technique for 2-C analysis in adult and pediatric populations.

Total Body Water and Potassium: Two alternate methods can be used in a 2-C model to estimate body fatness by measuring body water and cellular components of the FFM. The body composition parameters that are measured for these methods are total body water (TBW) or total body potassium (TBK), respectively.[60,62]

For these two assays, the simple 2-C equations for body fatness are as follows:

$$\%FM_{TBW} = 100 \times (Wt - TBW/k_3)$$

$$\%FM_{TBK} = 100 \times (Wt - TBK/k_4)$$

where the values for k_3 and k_4 are assumed to be relatively constant at a given age for a healthy subject. However, during infancy and childhood, the relative hydration and potassium content of the total FFM are not constant, and they may also be altered by disease and/or medications.[60] That is, the same limitations that were possible for the 2-C UWW or ADP models are also present for the 2-C TBW and TBK models. This is an inherent limitation of the 2-C model and not the methods. An advantage of the TBW and TBK assays, compared with UWW or ADP, is that these assays provide useful information on their own about the composition of FFM.

For the TBW assay, the subject drinks a small amount of isotope-labeled (non-radioactive) water. Several hours later, a body fluid sample (blood, urine, or saliva) is collected and stored for later analysis using isotope-ratio mass spectroscopy (MS) or Fourier-transformed infrared spectroscopy (FT-IRS). Since the amount of the tracer given to the subject is known, and its concentration in the water part of the fluid sample is measured, the total volume of the dilution space can be easily calculated.[168] The value for the conversion constant (k_3) that is used most often is 0.732 for older children and adults, gradually increasing to 0.83 for infants. That is, FFM contains, on average, about 73.2% water for healthy children and adolescents. This percentage, however, may be altered with diseases, such as severe malnutrition or edema, and by some drugs. A clear advantage of the TBW technique is that bulky instrumentation is not needed at the times of isotope administration and sample collection, thus making it a suitable choice for field studies. The MS or FT-IRS analysis could be performed immediately, but the usual practice is to collect multiple samples for batched analysis, thus the TBW results are often delayed until some time after the actual procedure.

The results of the TBK assay, on the other hand, can be obtained immediately. This assay takes advantage of a natural signal that is being constantly emitted from the potassium in the human body. A small fraction of natural potassium is radioactive (^{40}K), emitting characteristic gamma rays (1.46 MeV) at the rate of about 200 gammas per minute per gram of potassium. This signal can be detected external to the body using a whole-body counter,[63] usually a whole-room shielded counting chamber. Based on numerous studies over the past 40 years, the values for the conversion constant (k_4) are estimated at 59 to 61 mEq/kg for adult females and 62 to 64 mEq/kg for adult males. For infants, the ratio is reduced to ~43 mEq/kg because of increased hydration.[60,61] A major limitation with this assay is that the measurement instruments are not portable, and the number of available instruments and facilities is very limited. However, it is well recognized that the TBK assay is the best choice for monitoring body cell mass (BCM), the active metabolizing tissues of the FFM.[60] In many diseases, knowledge of the patient's BCM status has a higher priority than knowledge of body fatness.

Bioelectrical Impedance: The bioelectrical techniques for assaying body composition were developed as alternatives to the isotope dilution assay of total body water. The attractiveness of this technology is that the instruments are small in size, relatively inexpensive, don't require extensive operator training, and the results are immediately available. The principle of the bioelectrical impedance technique is that the body has general electrical properties, which primarily reflect the volume of the FFM and its electrolyte content.[90] Three approaches have been developed for human use: (1) single frequency (50 kHz) bioelectrical impedance analysis (BIA), (2) multifrequency (5–1000 kHz) bioelectrical impedance spectroscopy (BIS), and (3) total body electrical conductivity (TOBEC).

For the BIA and BIS procedures, pairs of electrodes are attached at the hand and foot. A weak electrical current is passed between the electrodes, and resistance (R) and reactance (Xc) are measured. The BIA assay uses a single frequency (50 kHz), while the BIS technique varies the frequency (5–1000 kHz). The basic BIA and BIS theory results in the assumption that the Ht^2/R ratio is directly proportional to TBW or FFM.[10] Some BIA instruments have been designed to measure only the upper body (electrodes placed on the hands) or lower body (subject stands in the electrodes), while other investigators have chosen to perform segmental BIA measurements (placing multiple electrodes at many sites on the body).

There are at least 30 single-frequency BIA devices commercially available. Unfortunately, there are almost an equal number of algorithms to choose from for the calculation of TBW, FFM, or %FM. Furthermore, some investigators have suggested that disease-specific calibrations of BIA should be used.[191] Although this approach may, at first, appear attractive, this type of "work around" simply avoids the more difficult issues related to the limited accuracy of the basic BIA theory and algorithms.

For the total body electrical conductivity (TOBEC) assay, no electrodes are placed on the body. Instead, the body passes through a large diameter electrical coil. The free charge particles in the body will attempt to align with the external magnetic field within the bore of the coil, causing a small measured perturbation in the coil current. The procedure takes only a few minutes to perform, and can be repeated as frequently as needed without risk to the subject.

Similar to the BIA and BIS assays, the TOBEC technique is a secondary assay, which means that the measured value (called the TOBEC number) must be calibrated with a more direct assay of FFM or TBW. A limitation with the TOBEC instrument is that it is very large compared with the BIA and BIS devices; hence it is not portable, and its initial cost is substantially higher. Thus, the number of TOBEC instruments worldwide is extremely limited, as with those based on the TBK technique.

Methods Based on the 3-Compartment (3-C) Model

Body Density + TBW: As pointed out for the 2-C density models (UWW and ADP), the major limitation was the assumption that TBW was a fixed percentage of the total FFM. This constraint can be overcome by using a 3-compartment (3-C) density model where the measurement of TBW is included. In this case, the %FM equation becomes:

$$\%FM = 100 \times \{(2.118rTB) - 0.78\ TBW/Wt - 1.354\}$$

where the density of body fat and solids are $0.9004\,g/cc$ and $1.565\,g/cc$, respectively. The advantage of this model is that the hydration state of the FFM can be variable, without introducing additional error in the estimate for body fatness. The disadvantage is that two assays (UWW or ADP *and* TBW) are needed, which increases the complexity and decreases feasibility of the procedure.

Dual-Energy X-ray Absorptiometry: Absorptiometric techniques were developed in the 1960s to examine bone because of the concerns that astronauts would experience significant bone loss during space flight. Over the last 40 years, significant advances have been made with this technology, evolving into the current technique, called dual-energy X-ray absorptiometry (DXA).[62] DXA assays of the hip and spine have become the clinical standards for the assessment of bone mineral density, used to screen for the increased risk of bone fractures, especially in postmenopausal women. In addition, a whole body DXA can be obtained in approximately 3 minutes with a very low total radiation dose ($<10\,\mu Sv$).

During a DXA procedure, the subject lies supine on the exam table, clothed but with removal of metal objects. An X-ray scanning arm passes over the selected body parts or the whole body. The X-rays are attenuated during passage through body tissues. The net signal is detected, converted into pixels, and analyzed using algorithms. For diagnosis and monitoring of osteoporosis, the scan is usually limited to the hip and/or lumbar vertebral spine, which are sites that are prone to osteoporotic fragility fractures. For more complete body composition analysis, a whole body scan is obtained.

In order for DXA to provide a quantitative measure of bone, the physics requires that the density of the overlying soft tissue must be known. This is accomplished by analyzing the nonbone pixels in the image next to the bone-containing pixels for their relative fat-to-lean content. Thus, a whole-body DXA scan can be used to produce a quasi-3-compartment (3-C) model: bone mineral content (BMC), fat mass (FM), and nonbone lean tissue mass (LTM). The sum of the bone-containing pixels provides a measurement called the bone area (BA), and areal bone mineral density (BMD) is defined as BMC/BA. It is to be noted that the DXA-derived BMD (g/cm^2) value is not true bone density (g/cm^3) but a projection of the total body mass onto a 2-dimensional or plainer image of the body.

A clear advantage of whole-body DXA is that the body scan image is divided into 10 general regions (head, arms, legs, trunk, pelvic, spine, etc.) with BMC, BMD, fat, and LTM calculated for each region. Thus, not only can the body FM be examined but also its regional distribution. Although this information is useful, it does not, for example, distinguish between subcutaneous fat and visceral fat stores in the body.

Methods Based on Body Imaging Techniques

To obtain precise body composition information, anatomical imaging techniques, such as computed tomography (CT) and magnetic resonance imaging (MRI), are used. The CT technique uses collimated beams of X-rays that are passed through the body, and an array of detectors is positioned on the opposite side of the body to detect the transmitted signal.[62] The X-ray source and detector assembly are rotated as a single unit around the body, and the data are collected and reconstructed to generate a cross-sectional image or "slice" of the body for each rotation. The physics of the CT procedure makes it a quantitative assay, i.e., the relative density (g/cc) of each pixel in the cross-sectional image can be obtained. Thus, anatomical regions such as the subcutaneous adipose tissue (SAT) layer, muscle, skin, internal organs, bone, and visceral fat deposits (VAT) can be identified. A disadvantage with routine CT imaging is that the radiation exposure is higher than that needed for a DXA scan, and whole body or extensive regional scans are not feasible.[23] The image requirements for a clinical CT scan can be relaxed when it is used for body composition analysis, such as for determination of VAT and SAT, which reduces the dose significantly.[177]

Magnetic resonance imaging (MRI) also provides internal images of the body, which tend to be superior in anatomical quality to those obtained using CT. One can easily identify the subcutaneous and visceral fat areas, for example, in an abdominal MRI image. However, the quantitative quality of the MRI image is much less than that obtained with the CT scan, and a whole body MRI can take up to 10 times longer to perform than a DXA. On the other hand, the MRI technique has several advantages, including the ability to distinguish VAT and SAT, and the system can be tuned to respond specifically to lipids contained within the lean tissues.[137,166] This may help to better understand the association of excess adiposity with some chronic diseases, such as T2DM.[114]

Methods Selection

The pathogenesis of PWS involves alterations in body composition that are atypical for human physiology and pathophysiology. In particular, there is an inherent increase in fat mass that is disproportionate to lean mass regardless of weight. The distribution of this fat mass could be relevant to defining cardiovascular risk. Lean mass is extremely deficient, as reflected in the hypotonia and decreased spontaneous activity. Finally, bone mineral density tends to be low, particularly in older individuals with PWS, increasing the risks for osteoporosis and fragility fractures. Given these considerations, body composition monitoring is an important element of clinical care for individuals with PWS. However, the selection of methods can be confusing.

As discussed above, there are a number of techniques that can be used to examine human body composition,[62] each with its own sets of advantages and disadvantages. Two-compartment (fat, nonfat) and 3-compartment (fat, bone, nonbone lean) models have been the most popular and clinically feasible. Higher-level models may provide a more complete picture of body composition, but multiple assays are required, some of which are not routinely found in most institutions.

In terms of 2-compartment models, the BIA technique has become widely available partly because it is relatively easy to perform, requires minimum investment, and is portable (for field studies or bedside use). Unfortunately, the body composition results are often not much better than those obtained using anthropometric measurements, such as body mass index calculations or skinfold measurements.

For more accurate 2-compartment modeling, air displacement plethysmography (i.e., using the BODPOD instrument) offers an attractive alternative to the more difficult underwater weighing technique.[68] However, the feasibility of this procedure in PWS has not been tested. In addition, although the methods for 2-compartment modeling are designed to measure fat-free mass (with fat mass calculated as the residual), they have the disadvantage of not providing additional information regarding the nonfat component (i.e., bone and nonbone lean mass).

Table 6.3 summarizes key considerations for body composition measurement methods, classified by the measured component. Costs are based on 2004 U.S. estimates. Precision and accuracy of each method are listed, as well as the minimal detectable change.

For practical purposes, the DXA procedure has become the reference method for the clinical assessment of bone mineral. This clinical acceptance of DXA, and its accessibility at many institutions, is also driving the use of this technique as a reference for the measurement of body fatness and lean mass. For patients with PWS, the DXA procedure has the additional advantages of rapidity, minimal patient cooperation requirements, and provision of measurements for all three clinically relevant body compartments. The 1% to 3% analytical precision of the DXA method allows detection of relatively small longitudinal changes in bone, fat, and nonbone lean tissues. However, there are significant differences in the calculation of %FM by DXA versus more sophisticated techniques; therefore, further improvements in DXA may be needed before a consensus can be reached[186] regarding the utility of DXA in estimating FM and nonbone lean mass. In addition, DXA methodologies differ according to manufacturers; results are not directly interchangeable; and pediatric, age-related, and ethnic normal ranges have not been widely accepted.[9,62,122,200] A minor disadvantage of DXA is that an exposure to X-rays is required; however, the dose is very small ($>10\,\mu Sv$) and carries minimal risk. From a practical standpoint, the DXA platform has body weight limitations; scans cannot be performed on individuals over 300 pounds on some commonly used DXA machines.

Single-slice abdominal CT and MRI methods are excellent choices when information about the distribution of body fat in the abdominal region is important. This information could have relevance to defining cardiovascular risk; however, these methods have not yet been validated for monitoring of individual patients, PWS or otherwise. In addition, these methods do not provide information about whole body tissue status.

Table 6.3. Comparison of Body Composition Methodologies

Measured Compartment	Method	Instrument Cost (US$)	Procedure Charge (US$)	Precision %	Accuracy %	MDC[i] Amt. (%)
Total Body Water	Deuterium dilution[a]	25–70 K	50–100	1–2	2–3	2 L (5)
	BIA/BIS[b]	3–7 K	30	2–4	3–7	4 L (10)
	QMR[c]	100 K	—	1	—	1 L (3)
Fat-Free Mass	UWW[d]	25–35 K	45	1–2	2–3	2 kg (4)
	ADP[e]	30–80 K	45	1–2	2–3	2 kg (4)
	BIA/BIS	3–7 K	30	2–4	2–8	4 kg (7)
	DXA[f]	125 K	150	2	1–2	1 kg (2)
Fat Mass	DXA	125 K	150	2–3	3–5	2 kg (11)
	CT[g]	800 K	350	3–4	3–4	— (10)
	MRI[h]	1500 K	600	3–4	3–4	— (10)
Bone Mineral Density	DXA	125 K	150	1	2–3	0.04 g/cm^2 (4)[j]
	CT	800 K	175	1	1	1.2 mg/cc (1)[k]

[a] Requires baseline body fluid sample, such as plasma, and 2–4 hours post-dose sample. Can be assayed using infrared spectroscopy or mass spec.; results are not "immediate," 2nd assay must be delayed 15–30 days to allow for initial dose to clear. Use of ^{18}O instead of deuterium results in 5–10-fold increase in assay cost.

[b] Single (BIA) and multifrequency (BIS) bioelectrical impedance analysis. At least 30 commercial instruments, with equal number of prediction equations. Can be repeated as needed.

[c] Quantitative magnetic resonance has only been used with small animals, but has been shown to be more accurate than dilution method and can be repeated as needed without harm. Magnetic field strength is about 1/200 of routine MRI instruments.

[d] Underwater weighing, requires subject to be totally submerged in water while air is exhaled from lungs. Classic method in use for more than 50 years, limited use for infants and older adults.

[e] Air-displacement plethysmography replaces underwater weighing method. Easily tolerated by subjects, can be repeated as frequently as needed, and both adult and infant-sized instruments are available.

[f] Dual-energy X-ray absorptiometry. Often considered as "gold standard" for *in vivo* bone mineral measurement. Gives regional information about body fat distribution.

[g] Computed tomography gives information about internal distribution of fat. Most frequently used to assay subcutaneous and visceral fat components of abdominal fat. Scan times in seconds, but frequency of repeat scans is limited by the radiation dose.

[h] Magnetic resonance imaging gives information and image of internal distribution of fat. Can be repeated as needed, but requires substantially longer time than CT.

[i] Minimum detectable change (MDC) for an individual. Value in parenthesis is the change expressed as a percent based on body composition of a 79-kg male with 25% fat. For CT and MRI, a 10% change in either fat subcompartment should be detectable.

[j] Values are for areal bone mineral density; similar values for total body, spine, or hip DXA.

[k] True density of bone, usually performed only for the spine.

In summary, with the various techniques that are available today, it is possible to obtain an accurate *in vivo* assay of the human body and to monitor the changes with growth or treatment of diseases. DXA is currently the most useful procedure for assessment of body composition in individuals with PWS; however care should be exercised in the proper interpretation of results.

References

1. Guidelines for the use of growth hormone in children with short stature. A report by the Drug and Therapeutics Committee of the Lawson Wilkins Pediatric Endocrine Society. *Journal of Pediatrics.* 1995;127(6):857–867.
2. Report of the expert committee on the diagnosis and classification of diabetes mellitus. *Diabetes Care.* 2003;26 Suppl 1:S5–S20.
3. Alexander RC, Greenswag LR, Nowak AJ. Rumination and vomiting in Prader-Willi syndrome. *American Journal of Medical Genetics.* 1987;28(4): 889–895.
4. Anderson AE, Soper RT, Scott DH. Gastric bypass for morbid obesity in children and adolescents. *Journal of Pediatric Surgery.* 1980;15(6):876–881.
5. Anderson GH, Moore SE. Dietary proteins in the regulation of food intake and body weight in humans. *Journal of Nutrition.* 2004;134(4):974S–979S.
6. Avenell A, Broom J, Brown TJ, et al. Systematic review of the long-term effects and economic consequences of treatments for obesity and implications for health improvement. *Health Technology Assessment.* 2004;8(21): 1–194.
7. Banks PA, Bradley JC, Smith A. Prader-Willi syndrome–a case report of the multidisciplinary management of the orofacial problems. *British Journal of Orthodontics.* 1996;23(4):299–304.
8. Bassali R, Hoffman WH, Chen H, Tuck-Muller CM. Hyperlipidemia, insulin-dependent diabetes mellitus, and rapidly progressive diabetic retinopathy and nephropathy in Prader-Willi syndrome with del(15)(q11.2q13). *American Journal of Medical Genetics.* 1997;71(3):267–270.
9. Batterham RL, Cohen MA, Ellis SM, et al. Inhibition of food intake in obese subjects by peptide YY3–36. *New England Journal of Medicine.* 2003;349(10):941–948.
10. Baumgartner R. Electrical impedance and total body electrical conductivity. In: Roche AF, Heymsfield SB, Lohman T, eds. *Human Body Composition.* Champaign, IL: Human Kinetics; 1996:79–108.
11. Bazopoulou-Kyrkanidou E, Papagiannoulis L. Prader-Willi syndrome: report of a case with special emphasis on oral problems. *Journal of Clinical Pediatric Dentistry.* 1992;17(1):37–40.
12. Bekx MT, Carrel AL, Shriver TC, Li Z, Allen DB. Decreased energy expenditure is caused by abnormal body composition in infants with Prader-Willi syndrome. *Journal of Pediatrics.* 2003;143(3):372–376.
13. Berntson GG, Zipf WB, O'Dorisio TM, Hoffman JA, Chance RE. Pancreatic polypeptide infusions reduce food intake in Prader-Willi syndrome. *Peptides.* 1993;14(3):497–503.
14. Bhargava SA, Putnam PE, Kocoshis SA, Rowe M, Hanchett JM. Rectal bleeding in Prader-Willi syndrome. *Pediatrics.* 1996;97(2):265–267.
15. Bistrian BR, Blackburn GL, Stanbury JB. Metabolic aspects of a protein-sparing modified fast in the dietary management of Prader-Willi obesity. *New England Journal of Medicine.* 1977;296(14):774–779.
16. Bizzarri C, Rigamonti AE, Giannone G, et al. Maintenance of a normal meal-induced decrease in plasma ghrelin levels in children with Prader-Willi syndrome. *Hormone and Metabolic Research.* 2004;36(3):164–169.
17. Borgie K. *Nutrition Care for Children with Prader-Willi Syndrome: A Nutrition Guide for Parents of Children with Prader-Willi Syndrome Ages 3–9 Years.* Sarasota, FL: PWSA (USA); 1994.
18. Bots CP, Schueler YT, Brand HS, van Nieuw Amerongen A. [A patient with Prader-Willi syndrome. Characteristics, oral consequences and treatment options]. *Nederlands Tijdschrift Voor Tandheelkunde.* 2004;111(2):55–58.

19. Braghetto I, Rodriguez A, Debandi A, et al. [Prader-Willi syndrome (PWS) associated to morbid obesity: surgical treatment]. *Revista Medica de Chile.* 2003;131(4):427–431.

20. Brambilla P, Bosio L, Manzoni P, Pietrobelli A, Beccaria L, Chiumello G. Peculiar body composition in patients with Prader-Labhart-Willi syndrome. *American Journal of Clinical Nutrition.* 1997;65(5):1369–1374.

21. Bray GA. Genetic, hypothalamic and endocrine features of clinical and experimental obesity. *Progress in Brain Research.* 1992;93:333–40; discussion 340–341.

22. Bray GA, Dahms WT, Swerdloff RS, Fiser RH, Atkinson RL, Carrel RE. The Prader-Willi syndrome: a study of 40 patients and a review of the literature. *Medicine (Baltimore).* 1983;62(2):59–80.

23. Brenner DJ, Elliston CD. Estimated radiation risks potentially associated with full-body CT screening. *Radiology.* 2004;232(3):735–738.

24. Brozek J, Grande F, Anderson JT, Keys A. Densitometric analysis of body composition: revision of some quantitative assumptions. *Annals of the New York Academy of Sciences.* 1963;110:113–140.

25. Buchholz AC, Schoeller DA. Is a calorie a calorie? *American Journal of Clinical Nutrition.* 2004;79(5):899S-906.

26. Bueno G, Moreno LA, Pineda I, et al. Serum leptin concentrations in children with Prader-Willi syndrome and non-syndromal obesity. *Journal of Pediatric Endocrinology and Metabolism.* 2000;13(4):425–430.

27. Butler JV, Whittington JE, Holland AJ, Boer H, Clarke D, Webb T. Prevalence of, and risk factors for, physical ill-health in people with Prader-Willi syndrome: a population-based study. *Developmental Medicine and Child Neurology.* 2002;44(4):248–255.

28. Butler MG. Prader-Willi syndrome: current understanding of cause and diagnosis. *American Journal of Medical Genetics.* 1990;35(3):319–332.

29. Butler MG, Bittel DC, Talebizadeh Z. Plasma peptide YY and ghrelin levels in infants and children with Prader-Willi syndrome. *Journal of Pediatric Endocrinology and Metabolism.* 2004;17(9):1177–1184.

30. Butler MG, Butler RI, Meaney FJ. The use of skinfold measurements to judge obesity during the early phase of Prader-Labhart-Willi syndrome. *International Journal of Obesity.* 1988;12(5):417–422.

31. Butler MG, Carlson MG, Schmidt DE, Feurer ID, Thompson T. Plasma cholecystokinin levels in Prader-Willi syndrome and obese subjects. *American Journal of Medical Genetics.* 2000;95(1):67–70.

32. Butler MG, Haynes JL, Meaney FJ. Anthropometric study with emphasis on hand and foot measurements in the Prader-Willi syndrome: sex, age and chromosome effects. *Clinical Genetics.* 1991;39(1):39–47.

33. Butler MG, Hedges L, Hovis CL, Feurer ID. Genetic variants of the human obesity (OB) gene in subjects with and without Prader-Willi syndrome: comparison with body mass index and weight. *Clinical Genetics.* 1998;54(5):385–393.

34. Butler MG, Levine GJ, Le JY, Hall BD, Cassidy SB. Photoanthropometric study of craniofacial traits of individuals with Prader-Willi syndrome. *American Journal of Medical Genetics.* 1995;58(1):38–45.

35. Butler MG, Meaney FJ. An anthropometric study of 38 individuals with Prader-Labhart-Willi syndrome. *American Journal of Medical Genetics.* 1987;26(2):445–455.

36. Butler MG, Meaney FJ. Standards for selected anthropometric measurements in Prader-Willi syndrome. *Pediatrics.* 1991;88(4):853–860.

37. Butler MG, Moore J, Morawiecki A, Nicolson M. Comparison of leptin protein levels in Prader-Willi syndrome and control individuals. *American Journal of Medical Genetics.* 1998;75(1):7–12.

38. Campeotto F, Naudin C, Viot G, Dupont C. [Rectal self-mutilation, rectal bleeding and Prader-Willi syndrome]. *Archives of Pediatrics.* 2001;8(10): 1075–1077.

39. Carlson MG, Snead WL, Oeser AM, Butler MG. Plasma leptin concentrations in lean and obese human subjects and Prader-Willi syndrome: comparison of RIA and ELISA methods. *Journal of Laboratory and Clinical Medicine.* 1999;133(1):75–80.

40. Carrel AL, Allen DB. Effects of growth hormone on body composition and bone metabolism. *Endocrine.* 2000;12(2):163–172.

41. Cento RM, Proto C, Spada RS, et al. Serum leptin concentrations in obese women with Down syndrome and Prader-Willi syndrome. *Gynecological Endocrinology.* 1999;13(1):36–41.

42. Chan NN, Feher MD, Bridges NA. Metformin therapy for diabetes in Prader-Willi syndrome. *Journal of the Royal Society of Medicine.* 1998; 91(11):598.

43. Chelala E, Cadiere GB, Favretti F, et al. Conversions and complications in 185 laparoscopic adjustable silicone gastric banding cases. *Surgical Endoscopy.* 1997;11(3):268–271.

44. Chen HY, Trumbauer ME, Chen AS, et al. Orexigenic action of peripheral ghrelin is mediated by neuropeptide Y and agouti-related protein. *Endocrinology.* 2004;145(6):2607–2612.

45. Collier SB, Walker WA. Parenteral protein-sparing modified fast in an obese adolescent with Prader-Willi syndrome. *Nutrition Reviews.* 1991;49(8): 235–238.

46. Committee on Injury, Violence, and Poison Prevention (American Academy of Pediatrics). Poison treatment in the home. *Pediatrics.* 2003; 112(5):1182–1185.

47. Committee on Nutrition, American Academy of Pediatrics, ed. *Pediatric Nutrition Handbook.* 5th ed. Elk Grove Village, IL: American Academy of Pediatrics; 2003.

48. Cummings DE, Clement K, Purnell JQ, et al. Elevated plasma ghrelin levels in Prader Willi syndrome. *Nature Medicine.* 2002;8(7): 643–644.

49. Davies PS, Joughin C. Using stable isotopes to assess reduced physical activity of individuals with Prader-Willi syndrome. *American Journal on Mental Retardation.* 1993;98(3):349–353.

50. Debas HT. *Gastrointestinal Surgery: Pathophysiology and Management.* New York, NY: Springer; 2004.

51. DelParigi A, Tschop M, Heiman ML, et al. High circulating ghrelin: a potential cause for hyperphagia and obesity in Prader-Willi syndrome. *Journal of Clinical Endocrinology and Metabolism.* 2002;87(12):5461–5464.

52. Despres J, Ross D, Lemieux S. Imaging techniques applied to the measurement of human body composition. In: Roche AF, Heymsfield SB, Lohman T, eds. *Human Body Composition.* Champaign, IL: Human Kinetics; 1996:149–166.

53. Dousei T, Miyata M, Izukura M, Harada T, Kitagawa T, Matsuda H. Long-term follow-up of gastroplasty in a patient with Prader-Willi syndrome. *Obesity Surgery.* 1992;2(2):189–193.

54. Eiholzer U, Blum WF, Molinari L. Body fat determined by skinfold measurements is elevated despite underweight in infants with Prader-Labhart-Willi syndrome. *Journal of Pediatrics.* 1999;134(2):222–225.

55. Eiholzer U, l'Allemand D, van der Sluis I, Steinert H, Gasser T, Ellis K. Body composition abnormalities in children with Prader-Willi syndrome and long-term effects of growth hormone therapy. *Hormone Research.* 2000;53(4):200–206.

56. Eiholzer U, l'Allemand D, Zipf WB, eds. *Prader-Willi Syndrome as a Model for Obesity.* Basel: Karger; 2003.

57. Eiholzer U, Stutz K, Weinmann C, Torresani T, Molinari L, Prader A. Low insulin, IGF-I and IGFBP-3 levels in children with Prader-Labhart-Willi syndrome. *European Journal of Pediatrics.* 1998;157(11):890–893.

58. Einhorn D, Reaven GM, Cobin RH, et al. American College of Endocrinology position statement on the insulin resistance syndrome. *Endocrine Practice.* 2003;9(3):237–252.

59. Elimam A, Lindgren AC, Norgren S, et al. Growth hormone treatment downregulates serum leptin levels in children independent of changes in body mass index. *Hormone Research.* 1999;52(2):66–72.

60. Ellis K. Total body potassium: A reference measurement for the body cell mass. In: Pierson R, Jr., ed. *Quality of the Body Cell Mass.* New York, NY: Springer; 2000:119–129.

61. Ellis KJ. Whole-body counting and neutron activation analysis. In: Roche AF, Heymsfield SB, Lohman T, eds. *Human Body Composition.* Champaign, IL: Human Kinetics; 1996:79–108.

62. Ellis KJ. Human body composition: in vivo methods. *Physiology Reviews.* 2000;80(2):649–680.

63. Ellis KJ, Shypailo R. Whole body potassium measurements independent of body size. In: Ellis KJ, Eastman J, eds. *Human Body Composition: In Vivo Methods, Models and Assessment.* New York, NY: Plenum Press; 1993: 371–376.

64. Ertekin C, Aydogdu I. Neurophysiology of swallowing. *Clinical Neurophysiology.* 2003;114(12):2226–2244.

65. Farag HM, Kotb SM, Sweify GA, Fawzy RK, Ismail SR. A diagnostic clinical genetic study of craniofacial dysmorphism. *Eastern Mediterranean Health Journal.* 1999;5(3):470–477.

66. Fehlow P, Miosge W, Scholl U, Walther F. [Prader-Willi syndrome in association with self-injury behavior and premature craniosynostosis]. *Klinische Padiatrie.* 1996;208(3):135–136.

67. Festi D, Colecchia A, Sacco T, Bondi M, Roda E, Marchesini G. Hepatic steatosis in obese patients: clinical aspects and prognostic significance. *Obesity Reviews.* 2004;5(1):27–42.

68. Fields DA, Goran MI, McCrory MA. Body-composition assessment via air-displacement plethysmography in adults and children: a review. *American Journal of Clinical Nutrition.* 2002;75(3):453–467.

69. Fieldstone A, Zipf WB, Schwartz HC, Berntson GG. Food preferences in Prader-Willi syndrome, normal weight and obese controls. *International Journal of Obesity and Related Metabolic Disorders.* 1997;21(11): 1046–1052.

70. Finan DS, Barlow SM. Intrinsic dynamics and mechanosensory modulation of non-nutritive sucking in human infants. *Early Human Development.* 1998;52(2):181–197.

71. Fisher BL, Schauer P. Medical and surgical options in the treatment of severe obesity. *The American Journal of Surgery.* 2002;184(6 Suppl 2): S9–S16.

72. Flier JS. Physiology: is brain sympathetic to bone?*Nature.* 2002;420(6916): 619–622.

73. Fliers E, Kreier F, Voshol PJ, et al. White adipose tissue: getting nervous. *Journal of Neuroendocrinology.* 2003;15(11):1005–1010.

74. Fonkalsrud EW, Bray G. Vagotomy for treatment of obesity in childhood due to Prader-Willi syndrome. *Journal of Pediatric Surgery.* 1981;16(6): 888–889.

75. Forbes GB. The companionship of lean and fat. *Basic Life Sciences.* 1993;60:1–14.

76. Forbes GB. A distinctive obesity: body composition provides the clue. *American Journal of Clinical Nutrition.* 1997;65(5):1540–1541.

77. Forbes GB. Body fat content influences the body composition response to nutrition and exercise. *Annals of the New York Academy of Sciences.* 2000;904:359–365.

78. Forssman H, Hagberg B. Prader-Willi syndrome in boy of ten with prediabetes. *Acta Paediatrica.* 1964;53:70–78.

79. Gavranich J, Selikowitz M. A survey of 22 individuals with Prader-Willi syndrome in New South Wales. *Australian Paediatric Journal.* 1989;25(1): 43–46.

80. Ge Y, Ohta T, Driscoll DJ, Nicholls RD, Kalra SP. Anorexigenic melanocortin signaling in the hypothalamus is augmented in association with failure-to-thrive in a transgenic mouse model for Prader-Willi syndrome. *Brain Research.* 2002;957(1):42–45.

81. Goldstone AP. Prader-Willi syndrome: advances in genetics, pathophysiology and treatment. *Trends in Endocrinology and Metabolism.* 2004;15(1): 12–20.

82. Goldstone AP, Brynes AE, Thomas EL, et al. Resting metabolic rate, plasma leptin concentrations, leptin receptor expression, and adipose tissue measured by whole-body magnetic resonance imaging in women with Prader-Willi syndrome. *American Journal of Clinical Nutrition.* 2002;75(3):468–475.

83. Goldstone AP, Thomas EL, Brynes AE, et al. Visceral adipose tissue and metabolic complications of obesity are reduced in Prader-Willi syndrome female adults: evidence for novel influences on body fat distribution. *Journal of Clinical Endocrinology and Metabolism.* 2001;86(9):4330–4338.

84. Goldstone AP, Thomas EL, Brynes AE, et al. Elevated fasting plasma ghrelin in Prader-Willi syndrome adults is not solely explained by their reduced visceral adiposity and insulin resistance. *Journal of Clinical Endocrinology and Metabolism.* 2004;89(4):1718–1726.

85. Goldstone AP, Unmehopa UA, Bloom SR, Swaab DF. Hypothalamic NPY and agouti-related protein are increased in human illness but not in Prader-Willi syndrome and other obese subjects. *Journal of Clinical Endocrinology and Metabolism.* 2002;87(2):927–937.

86. Goldstone AP, Unmehopa UA, Swaab DF. Hypothalamic growth hormone-releasing hormone (GHRH) cell number is increased in human illness, but is not reduced in Prader-Willi syndrome or obesity. *Clinical Endocrinology (Oxf).* 2003;58(6):743–755.

87. Greenberg F, Elder FF, Ledbetter DH. Neonatal diagnosis of Prader-Willi syndrome and its implications. *American Journal of Medical Genetics.* 1987;28(4):845–856.

88. Grugni G, Guzzaloni G, Morabito F. Failure of biliopancreatic diversion in Prader-Willi syndrome. *Obesity Surgery.* 2000;10(2):179–181; discussion 182.

89. Gryboski JD. Gastrointestinal function in the infant and young child. *Clinical Gastroenterology.* 1977;6(2):253–265.

90. Guo SM, Roche AF, Chumlea WC, Miles DS, Pohlman RL. Body composition predictions from bioelectric impedance. *Human Biology.* 1987;59(2): 221–233.

91. Hall BD, Smith DW. Prader-Willi syndrome: a resume of 32 cases including an instance of affected first cousins, one of whom is of normal stature and intelligence. *Journal of Pediatrics.* 1972;81(2):286–293.

92. Haqq AM, Farooqi IS, O'Rahilly S, et al. Serum ghrelin levels are inversely correlated with body mass index, age, and insulin concentrations in normal children and are markedly increased in Prader-Willi syndrome. *Journal of Clinical Endocrinology and Metabolism.* 2003;88(1):174–178.

93. Haqq AM, Stadler DD, Rosenfeld RG, et al. Circulating ghrelin levels are suppressed by meals and octreotide therapy in children with Prader-Willi syndrome. *Journal of Clinical Endocrinology and Metabolism.* 2003;88(8): 3573–3576.

94. Hart PS. Salivary abnormalities in Prader-Willi syndrome. *Annals of the New York Academy of Sciences.* 1998;842:125–131.

95. Hashizume K, Nakajo T, Kawarasaki H, et al. Prader-Willi syndrome with del(15)(q11,q13) associated with hepatoblastoma. *Acta Paediatrica Japanica.* 1991;33(6):718–722.

96. Hauffa BP, Schlippe G, Gillessen-Kaesbach G. Adiposity indices in German children and adolescents with genetically confirmed Prader-Willi syndrome (PWS). *International Journal of Obesity and Related Metabolic Disorders.* 2001;25 Suppl 1:S22–S25.

97. Hauffa BP, Schlippe G, Roos M, Gillessen-Kaesbach G, Gasser T. Spontaneous growth in German children and adolescents with genetically confirmed Prader-Willi syndrome. *Acta Paediatrica.* 2000;89(11):1302–1311.

98. Havel PJ. Update on adipocyte hormones: regulation of energy balance and carbohydrate/lipid metabolism. *Diabetes.* 2004;53(90001): S143–S151.

99. Hayashi M, Itoh M, Kabasawa Y, Hayashi H, Satoh J, Morimatsu Y. A neuropathological study of a case of the Prader-Willi syndrome with an interstitial deletion of the proximal long arm of chromosome 15. *Brain Development.* 1992;14(1):58–62.

100. Hill J, Kaler M, Spetalnick B, Reed G, Butler M. Resting metabolic rate in Prader-Willi syndrome. *Dysmorphology and Clinical Genetics.* 1990;4: 27–32.

101. Hill M, Hughes T, Milford C. Treatment for swallowing difficulties (dysphagia) in chronic muscle disease. *Cochrane Database of Systematic Reviews* (Online). 2004(2):CD004303.

102. Hoh JF. "Superfast" or masticatory myosin and the evolution of jaw-closing muscles of vertebrates. *Journal of Experimental Biology.* 2002;205(Pt 15):2203–2210.

103. Holland A, Whittington J, Hinton E. The paradox of Prader-Willi syndrome: a genetic model of starvation. *Lancet.* 2003;362(9388):989–991.

104. Holm VA, Cassidy SB, Butler MG, et al. Diagnostic criteria for Prader-Willi syndrome. In: Cassidy SB, ed. *Prader-Willi Syndrome and Other Chromosome 15q Deletion Disorders.* New York, NY: Springer-Verlag; 1992.

105. Holm VA, Cassidy SB, Butler MG, et al. Prader-Willi syndrome: consensus diagnostic criteria. *Pediatrics.* 1993;91(2):398–402.

106. Holm VA, Pipes PL. Food and children with Prader-Willi syndrome. *American Journal of Diseases of Children.* 1976;130(10):1063–1067.

107. Hornby PJ. Central neurocircuitry associated with emesis. *American Journal of Medicine.* 2001;111 Suppl 8A:106S–112S.

108. Hoybye C. Endocrine and metabolic aspects of adult Prader-Willi syndrome with special emphasis on the effect of growth hormone treatment. *Growth Hormone and IGF Research.* 2004;14(1):1–15.

109. Hoybye C, Barkeling B, Espelund U, Petersson M, Thoren M. Peptides associated with hyperphagia in adults with Prader-Willi syndrome before and during GH treatment. *Growth Hormone and IGF Research.* 2003;13(6): 322–327.

110. Hoybye C, Bruun JM, Richelsen B, Flyvbjerg A, Frystyk J. Serum adipo-nectin levels in adults with Prader-Willi syndrome are independent of anthropometrical parameters and do not change with GH treatment. *European Journal of Endocrinology.* 2004;151(4):457–461.

111. Hoybye C, Hilding A, Jacobsson H, Thoren M. Metabolic profile and body composition in adults with Prader-Willi syndrome and severe obesity. *Journal of Clinical Endocrinology and Metabolism.* 2002;87(8):3590–3597.

112. Hundal RS, Inzucchi SE. Metformin: new understandings, new uses. *Drugs.* 2003;63(18):1879–1894.

113. Illig R. [Immunologically determinable insulin and glucose tolerance in the Prader-Labhart-Willi syndrome]. *Schweizerische Medizinische Wochen-schrift.* 1968;98(19):723–724.

114. Jacob S, Machann J, Rett K, et al. Association of increased intramyocellular lipid content with insulin resistance in lean nondiabetic offspring of type 2 diabetic subjects. *Diabetes.* 1999;48(5):1113–1119.

115. Jeanrenaud B, Rohner-Jeanrenaud F. Effects of neuropeptides and leptin on nutrient partitioning: dysregulations in obesity. *Annual Review of Medicine.* 2001;52(1):339–351.

116. Johnson RD. Opioid involvement in feeding behaviour and the pathogen-esis of certain eating disorders. *Medical Hypotheses.* 1995;45(5):491–497.

117. Kobayashi J, Kodama M, Yamazaki K, et al. Gastric bypass in a Japanese man with Prader-Willi syndrome and morbid obesity. *Obesity Surgery.* 2003;13(5):803–805.

118. Kohn Y, Weizman A, Apter A. Aggravation of food-related behavior in an adolescent with Prader-Willi syndrome treated with fluvoxamine and fluoxetine. *International Journal of Eating Disorders.* 2001;30(1):113–117.

119. Koolstra JH. Dynamics of the human masticatory system. *Critical Reviews in Oral Biology and Medicine.* 2002;13(4):366–376.

120. Kousholt AM, Beck-Nielsen H, Lund HT. A reduced number of insulin receptors in patients with Prader-Willi syndrome. *Acta Endocrinologica (Copenh).* 1983;104(3):345–351.

121. L'Allemand D, Eiholzer U, Rousson V, et al. Increased adrenal androgen levels in patients with Prader-Willi syndrome are associated with insulin, IGF-I, and leptin, but not with measures of obesity. *Hormone Research.* 2002;58(5):215–222.

122. Laskey MA. Dual-energy X-ray absorptiometry and body composition. *Nutrition.* 1996;12(1):45–51.

123. Laurance BM, Brito A, Wilkinson J. Prader-Willi syndrome after age 15 years. *Archives of Disease in Childhood.* 1981;56(3):181–186.

124. Laurent-Jaccard A, Hofstetter JR, Saegesser F, Chapuis Germain G. Long-term result of treatment of Prader-Willi syndrome by Scopinaro's bilio-pancreatic diversion: study of three cases and the effect of dextrofenfluramine on the postoperative evolution. *Obesity Surgery.* 1991;1(1):83–87.

125. Lautala P, Knip M, Akerblom HK, Kouvalainen K, Martin JM. Serum insulin-releasing activity and the Prader-Willi syndrome. *Acta Endocrino-logica Supplementum (Copenh).* 1986;279:416–421.

126. Lee PD. Effects of growth hormone treatment in children with Prader-Willi syndrome. *Growth Hormone and IGF Research.* 2000;10 Suppl B:S75–S79.

127. Lee PDK. Model for a peripheral signalling defect in Prader-Willi syn-drome. In: Eiholzer U, L'Allemand D, Zipf WB, eds. *Prader-Willi Syndrome as a Model for Obesity.* Basel: Karger; 2003:70–81.

128. Lee PDK, Hwu K, Brown BT, Greenberg F, Klish WJ. Endocrine investiga-tions in children with Prader-Willi syndrome. *Dysmorphology and Clinical Genetics.* 1992;6:27–28.

129. Leonard MB, Shults J, Wilson BA, Tershakovec AM, Zemel BS. Obesity during childhood and adolescence augments bone mass and bone dimensions. *American Journal of Clinical Nutrition.* 2004;80(2):514–523.

130. Levin SL, Khaikina LI. Is there neural control over electrolyte reabsorption in the human salivary gland? *Clinical Science (Lond).* 1987;72(5):541–548.

131. Lindgren AC, Barkeling B, Hagg A, Ritzen EM, Marcus C, Rossner S. Eating behavior in Prader-Willi syndrome, normal weight, and obese control groups. *Journal of Pediatrics.* 2000;137(1):50–55.

132. Lindgren AC, Marcus C, Skwirut C, et al. Increased leptin messenger RNA and serum leptin levels in children with Prader-Willi syndrome and nonsyndromal obesity. *Pediatric Research.* 1997;42(5):593–596.

133. Lingstrom P, Moynihan P. Nutrition, saliva, and oral health. *Nutrition.* 2003;19(6):567–569.

134. Lohman T. Assessment of body composition in children. *Pediatric Exercise Science.* 1989;1:19–30.

135. Ludwig DS. Dietary glycemic index and obesity. *Journal of Nutrition.* 2000;130(2):280S–283S.

136. Mace J, Gotlin RW, Dubois R. Letter: Pathogenesis of obesity in Prader-Willi syndrome. *Journal of Pediatrics.* 1974;84(6):927–928.

137. Machann J, Bachmann OP, Brechtel K, et al. Lipid content in the musculature of the lower leg assessed by fat selective MRI: intra- and interindividual differences and correlation with anthropometric and metabolic data. *Journal of Magnetic Resonance Imaging.* 2003;17(3):350–357.

138. Margules DL, Inturrisi CE. Beta-endorphin immunoreactivity in the plasma of patients with the Prader-Labhart-Willi syndrome and their normal siblings. *Experientia.* 1983;39(7):766–767.

139. Marinari GM, Camerini G, Novelli GB, et al. Outcome of biliopancreatic diversion in subjects with Prader-Willi Syndrome. *Obesity Surgery.* 2001;11(4):491–495.

140. Martin A, State M, Anderson GM, et al. Cerebrospinal fluid levels of oxytocin in Prader-Willi syndrome: a preliminary report. *Biological Psychiatry.* 1998;44(12):1349–1352.

141. Meaney FJ, Butler MG. Craniofacial variation and growth in the Prader-Labhart-Willi syndrome. *American Journal of Physical Anthropology.* 1987;74(4):459–464.

142. Mignot E, Lammers GJ, Ripley B, et al. The role of cerebrospinal fluid hypocretin measurement in the diagnosis of narcolepsy and other hypersomnias. *Archives of Neurology.* 2002;59(10):1553–1562.

143. Misumi A, Sera Y, Matsuda M, et al. Solitary ulcer of the rectum: report of a case and review of the literature. *Gastroenterologia Japonica.* 1981;16(3):286–294.

144. Mizuno K, Ueda A. The maturation and coordination of sucking, swallowing, and respiration in preterm infants. *Journal of Pediatrics.* 2003; 142(1):36–40.

145. Mogul HR, Lee PD, Whitman B, et al. Preservation of insulin sensitivity and paucity of metabolic syndrome symptoms in Prader Willi syndrome adults: preliminary data from the U. S. multi-center study. *Obesity Research.* 2004;12(1):171, HT-08.

146. Mogul HR, Peterson SJ, Weinstein BI, Li J, Southren AL. Long-term (2-4 year) weight reduction with metformin plus carbohydrate-modified diet in euglycemic, hyperinsulinemic, midlife women (Syndrome W). *Heart Disease.* 2003;5(6):384–392.

147. Mogul HR, Peterson SJ, Weinstein BI, Zhang S, Southren AL. Metformin and carbohydrate-modified diet: a novel obesity treatment protocol: pre-

liminary findings from a case series of nondiabetic women with midlife weight gain and hyperinsulinemia. *Heart Disease.* 2001;3(5):285–292.

148. Myers SE, Davis A, Whitman BY, Santiago JV, Landt M. Leptin concentrations in Prader-Willi syndrome before and after growth hormone replacement. *Clinical Endocrinology (Oxf).* 2000;52(1):101–105.

149. Nagai T, Matsuo N, Kayanuma Y, et al. Standard growth curves for Japanese patients with Prader-Willi syndrome. *American Journal of Medical Genetics.* 2000;95(2):130–134.

150. Nagai T, Mori M. Prader-Willi syndrome, diabetes mellitus and hypogonadism. *Biomedicine and Pharmacotherapy.* 1999;53(10):452–454.

151. Nardella MT, Sulzbacher SI, Worthington-Roberts BS. Activity levels of persons with Prader-Willi syndrome. *American Journal of Mental Deficiency.* 1983;87(5):498–505.

152. Nelson RA, Anderson LF, Gastineau CF, Hayles AB, Stamnes CL. Physiology and natural history of obesity. *Journal of the American Medical Association.* 1973;223(6):627–630.

153. Nelson SP, Chen EH, Syniar GM, Christoffel KK. Prevalence of symptoms of gastroesophageal reflux during infancy: a pediatric practice-based survey. Pediatric Practice Research Group. *Archives of Pediatric and Adolescent Medicine.* 1997;151(6):569–572.

154. Nevsimalova S, Vankova J, Stepanova I, Seemanova E, Mignot E, Nishino S. Hypocretin deficiency in Prader-Willi syndrome. *European Journal of Neurology.* 2005;12(1):70–72.

155. Parra A, Cervantes C, Schultz RB. Immunoreactive insulin and growth hormone responses in patients with Prader-Willi syndrome. *Journal of Pediatrics.* 1973;83(4):587–593.

156. Partsch CJ, Lammer C, Gillessen-Kaesbach G, Pankau R. Adult patients with Prader-Willi syndrome: clinical characteristics, life circumstances and growth hormone secretion. *Growth Hormone and IGF Research.* 2000;10 Suppl B:S81–S85.

157. Pedersen AM, Bardow A, Jensen SB, Nauntofte B. Saliva and gastrointestinal functions of taste, mastication, swallowing and digestion. *Oral Diseases.* 2002;8(3):117–129.

158. Pietrobelli A, Allison DB, Faith MS, et al. Prader-Willi syndrome: relationship of adiposity to plasma leptin levels. *Obesity Research.* 1998;6(3): 196–201.

159. Pipes PL, Holm VA. Weight control of children with Prader-Willi syndrome. *Journal of the American Dietetic Association.* 1973;62(5):520–524.

160. Poets CF. Gastroesophageal reflux: a critical review of its role in preterm infants. *Pediatrics.* 2004;113(2):e128–e132.

161. Porter SR, Scully C, Hegarty AM. An update of the etiology and management of xerostomia. *Oral Surgery, Oral Medicine, Oral Pathology, Oral Radiology & Endodontics.* 2004;97(1):28–46.

162. Reed WB, Ragsdale W, Jr., Curtis AC, Richards HJ. Acanthosis nigricans in association with various genodermatoses. With emphasis on lipodystrophic diabetes and Prader-Willi syndrome. *Acta Dermato-Venereologica.* 1968;48(5):465–473.

163. Saenger P, Dimartino-Nardi J. Premature adrenarche. *Journal of Endocrinological Investigation.* 2001;24(9):724–733.

164. Sakata T, Yoshimatsu H, Masaki T, Tsuda K. Anti-obesity actions of mastication driven by histamine neurons in rats. *Experimental Biology and Medicine (Maywood).* 2003;228(10):1106–1110.

165. Scheen AJ. Pathophysiology of type 2 diabetes. *Acta Clinica Belgica.* 2003;58(6):335–341.

166. Schick F, Machann J, Brechtel K, et al. MRI of muscular fat. *Magnetic Resonance in Medicine.* 2002;47(4):720–727.

167. Schmidt H, Schwarz HP. Premature adrenarche, increased growth velocity and accelerated bone age in male patients with Prader-Labhart-Willi syndrome. *European Journal of Pediatrics.* 2001;160(1):69–70.

168. Schoeller D. Hydrometry. In: Roche AF, Heymsfield SB, Lohman T, eds. *Human Body Composition.* Champaign, IL: Human Kinetics; 1996:25–44.

169. Schoeller DA, Levitsky LL, Bandini LG, Dietz WW, Walczak A. Energy expenditure and body composition in Prader-Willi syndrome. *Metabolism.* 1988;37(2):115–120.

170. Schuster DP, Osei K, Zipf WB. Characterization of alterations in glucose and insulin metabolism in Prader-Willi subjects. *Metabolism.* 1996;45(12):1514–1520.

171. Shaker R, Hogan WJ. Normal physiology of the aerodigestive tract and its effect on the upper gut. *American Journal of Medicine.* 2003;115 Suppl 3A:2S–9S.

172. Sills IN, Rapaport R. Non-insulin dependent diabetes mellitus in a prepubertal child with Prader-Willi syndrome. *Journal of Pediatric Endocrinology and Metabolism.* 1998;11(2):281–282.

173. Siri W. Body composition from fluid spaces and density: analysis of methods. In: Brozek J, Henschel A, eds. *Techniques for Measuring Body Composition.* Washington, DC: National Academy of Sciences; 1961:223–224.

174. Sloan TB, Kaye CI. Rumination risk of aspiration of gastric contents in the Prader-Willi syndrome. *Anesthesia and Analgesia.* 1991;73(4):492–495.

175. Smathers SA, Wilson JG, Nigro MA. Topiramate effectiveness in Prader-Willi syndrome. *Pediatric Neurology.* 2003;28(2):130–133.

176. Soper RT, Mason EE, Printen KJ, Zellweger H. Gastric bypass for morbid obesity in children and adolescents. *Journal of Pediatric Surgery.* 1975;10(1):51–58.

177. Starck G, Lonn L, Cederblad A, Alpsten M, Sjostrom L, Ekholm S. Dose reduction for body composition measurements with CT. *Applied Radiation and Isotopes.* 1998;49(5–6):561–563.

178. Sun Y, Ahmed S, Smith RG. Deletion of ghrelin impairs neither growth nor appetite. *Molecular and Cellular Biology.* 2003;23(22):7973–7981.

179. Swaab DF, Purba JS, Hofman MA. Alterations in the hypothalamic paraventricular nucleus and its oxytocin neurons (putative satiety cells) in Prader-Willi syndrome: a study of five cases. *Journal of Clinical Endocrinology and Metabolism.* 1995;80(2):573–579.

180. Takayasu H, Motoi T, Kanamori Y, et al. Two case reports of childhood liver cell adenomas harboring beta-catenin abnormalities. *Human Pathology.* 2002;33(8):852–855.

181. Tauber M, Barbeau C, Jouret B, et al. Auxological and endocrine evolution of 28 children with Prader-Willi syndrome: effect of GH therapy in 14 children. *Hormone Research.* 2000;53(6):279–287.

182. Thomas EL, Saeed N, Hajnal JV, et al. Magnetic resonance imaging of total body fat. *Journal of Applied Physiology.* 1998;85(5):1778–1785.

183. Tomita T, Greeley G, Jr., Watt L, Doull V, Chance R. Protein meal-stimulated pancreatic polypeptide secretion in Prader-Willi syndrome of adults. *Pancreas.* 1989;4(4):395–400.

184. Torley D, Bellus GA, Munro CS. Genes, growth factors and acanthosis nigricans. *British Journal of Dermatology.* 2002;147(6):1096–1101.

185. Touquet VL, Ward MW, Clark CG. Obesity surgery in a patient with the Prader-Willi syndrome. *British Journal of Surgery.* 1983;70(3):180–181.

186. Treuth MS, Butte NF, Wong WW, Ellis KJ. Body composition in prepubertal girls: comparison of six methods. *International Journal of Obesity and Related Metabolic Disorders.* 2001;25(9):1352–1359.

187. van der Bilt A, Bosman F, van der Glas HW. [Masticatory muscles. Part VII. Masticatory muscles and mastication. How do we get small pieces of food?]. *Nederlands Tijdschrift Voor Tandheelkunde.* 1998;105(1): 4–6.

188. van Mil EA, Westerterp KR, Gerver WJ, et al. Energy expenditure at rest and during sleep in children with Prader-Willi syndrome is explained by body composition. *American Journal of Clinical Nutrition.* 2000;71(3): 752–756.

189. van Mil EG, Westerterp KR, Kester AD, et al. Activity related energy expenditure in children and adolescents with Prader-Willi syndrome. *International Journal of Obesity and Related Metabolic Disorders.* 2000;24(4): 429–434.

190. Vandenplas Y, Hauser B. Gastro-oesophageal reflux, sleep pattern, apparent life threatening event and sudden infant death. The point of view of a gastro-enterologist. *European Journal of Pediatrics.* 2000;159(10): 726–729.

191. VanderJagt DJ, Harmatz P, Scott-Emuakpor AB, Vichinsky E, Glew RH. Bioelectrical impedance analysis of the body composition of children and adolescents with sickle cell disease. *Journal of Pediatrics.* 2002;140(6): 681–687.

192. Walker JD, Warren RE. Necrobiosis lipoidica in Prader-Willi–associated diabetes mellitus. *Diabetic Medicine.* 2002;19(10):884–885.

193. Wang J, Thornton JC, Kolesnik S, Pierson RN Jr. Anthropometry in body composition: an overview. *Annals of the New York Academy of Sciences.* 2000;904(1):317–326.

194. Wang ZM, Heshka S, Pierson RN, Jr., Heymsfield SB. Systematic organization of body-composition methodology: an overview with emphasis on component-based methods. *American Journal of Clinical Nutrition.* 1995;61(3):457–465.

195. Wang ZM, Pierson RN, Jr., Heymsfield SB. The five-level model: a new approach to organizing body-composition research. *American Journal of Clinical Nutrition.* 1992;56(1):19–28.

196. Ward RE, Jamison PL, Allanson JE. Quantitative approach to identifying abnormal variation in the human face exemplified by a study of 278 individuals with five craniofacial syndromes. *American Journal of Medical Genetics.* 2000;91(1):8–17.

197. Wharton RH, Wang T, Graeme-Cook F, Briggs S, Cole RE. Acute idiopathic gastric dilation with gastric necrosis in individuals with Prader-Willi syndrome. *American Journal of Medical Genetics.* 1997;73(4):437–441.

198. Wild S, Roglic G, Green A, Sicree R, King H. Global prevalence of diabetes: estimates for the year 2000 and projections for 2030. *Diabetes Care.* 2004;27(5):1047–1053.

199. Williams G, Cai XJ, Elliott JC, Harrold JA. Anabolic neuropeptides. *Physiology and Behavior.* 2004;81(2):211–222.

200. Wong WW, Hergenroeder AC, Stuff JE, Butte NF, Smith EO, Ellis KJ. Evaluating body fat in girls and female adolescents: advantages and disadvantages of dual-energy X-ray absorptiometry. *American Journal of Clinical Nutrition.* 2002;76(2):384–389.

201. Woods SC. Gastrointestinal satiety signals I. An overview of gastrointestinal signals that influence food intake. *American Journal of Physiology. Gastrointestinal and Liver Physiology.* 2004;286(1):G7–13.

202. Woods SC, Schwartz MW, Baskin DG, Seeley RJ. Food intake and the regulation of body weight. *Annual Review of Psychology.* 2000;51(1): 255–277.

203. Yigit S, Estrada E, Bucci K, Hyams J, Rosengren S. Diabetic ketoacidosis secondary to growth hormone treatment in a boy with Prader-Willi syndrome and steatohepatitis. *Journal of Pediatric Endocrinology and Metabolism.* 2004;17(3):361–364.

204. Young W, Khan F, Brandt R, Savage N, Razek AA, Huang Q. Syndromes with salivary dysfunction predispose to tooth wear: case reports of congenital dysfunction of major salivary glands, Prader-Willi, congenital rubella, and Sjogren's syndromes. *Oral Surgery, Oral Medicine, Oral Pathology, Oral Radiology and Endodontics.* 2001;92(1):38–48.

205. Zadik Z, Wittenberg I, Segal N, et al. Interrelationship between insulin, leptin and growth hormone in growth hormone-treated children. *International Journal of Obesity and Related Metabolic Disorders.* 2001;25(4): 538–542.

206. Zipf WB. Glucose homeostasis in Prader-Willi syndrome and potential implications of growth hormone therapy. *Acta Paediatrica Supplement.* 1999;88(433):115–117.

207. Zipf WB, Berntson GG. Characteristics of abnormal food-intake patterns in children with Prader-Willi syndrome and study of effects of naloxone. *American Journal of Clinical Nutrition.* 1987;46(2):277–281.

208. Zipf WB, O'Dorisio TM, Berntson GG. Short-term infusion of pancreatic polypeptide: effect on children with Prader-Willi syndrome. *American Journal of Clinical Nutrition.* 1990;51(2):162–166.

209. Zipf WB, O'Dorisio TM, Cataland S, Dixon K. Pancreatic polypeptide responses to protein meal challenges in obese but otherwise normal children and obese children with Prader-Willi syndrome. *Journal of Clinical Endocrinology and Metabolism.* 1983;57(5):1074–1080.

210. Zipf WB, O'Dorisio TM, Cataland S, Sotos J. Blunted pancreatic polypeptide responses in children with obesity of Prader-Willi syndrome. *Journal of Clinical Endocrinology and Metabolism.* 1981;52(6):1264–1266.

211. Zlotkin SH, Fettes IM, Stallings VA. The effects of naltrexone, an oral beta-endorphin antagonist, in children with the Prader-Willi syndrome. *Journal of Clinical Endocrinology and Metabolism.* 1986;63(5):1229–1232.

Growth Hormone and Prader-Willi Syndrome

Aaron L. Carrel, Phillip D.K. Lee, and Harriette R. Mogul

In the previous edition of this textbook, various lines of evidence for the occurrence of a deficiency in the growth hormone (GH) axis were reviewed as one of several sections in a chapter entitled "Endocrine and Metabolic Aspects of Prader-Willi Syndrome."[76] At that time, it was suggested that GH therapy may have beneficial effects on growth and body composition in individuals with PWS. Over the succeeding decade, a number of scientific investigations have provided definitive evidence in support of these suggestions.[16,37,72,73] In 2000, biosynthetic GH became the first and, to date, only medication to receive regulatory approval for treatment of children with PWS. In addition, it appears that GH may also have potential utility in adults with PWS.[61,86] This chapter presents a comprehensive review of relevant physiology and pathophysiology of the GH system and GH treatment efficacy and safety in PWS.

GH/IGF Axis Pathophysiology

Short stature is one of the cardinal features of PWS and was included in the initial description[98] of the condition in 1956. Numerous studies have shown that growth is usually compromised during childhood and that the average final adult height is approximately 2 standard deviations below the mean for the normal population[1,15,19,34,48,57,59,60,70,89,91,124] (see Appendix C). In individual cases, final adult height has been noted to be related to midparental height—i.e., PWS individuals with taller parents will be relatively taller than other PWS individuals[18,97]; however, final height is virtually always significantly lower than the actual midparental height in the usual case of PWS. Although a proportion of children with PWS will have accelerated growth and normal stature during childhood, often in association with premature adrenarche, final adult height in these cases may be further compromised by the accelerated bone maturation (see previous chapter). Thus, PWS is one of only a few conditions in which obesity is associated with short stature and is clearly distinguished from exogenous obesity, in which childhood linear growth is accelerated and final height is normal or increased.[46]

Early investigators suspected that growth hormone (GH) deficiency might be involved in the pathogenesis of the short stature, and several studies have shown deficient serum GH levels in response to standard stimulation tests in children[4,15,20,21,29,48,78,91,94,114] and adults[62,95] with PWS. However, these results may not be entirely convincing since GH levels are low in non-PWS obesity that is not associated with short stature.[67]

GH is synthesized in the anterior pituitary gland and released into the bloodstream. GH itself is thought to have minimal, if any, effect on somatic growth. Instead, GH stimulates synthesis of other growth factors in liver and other tissues. These growth factors then act to stimulate growth of body tissues, including bone and muscle.

One of the primary growth factors that mediate the GH effects is insulin-like growth factor-I (IGF-I). In true GH deficiency, IGF-I levels are very low, and this is associated with decreased linear growth. On the other hand, in common exogenous obesity, GH levels are low but IGF levels are normal or elevated,[67] and this is associated with normal linear growth. IGF-I is carried in the bloodstream by several specific binding proteins, including IGFBP-3. Therefore, measurement of either serum IGF or serum IGFBP-3 levels provides an indirect measure of GH secretion and a direct measure of the GH/IGF axis.

Reliable laboratory assays were developed for IGF-I in the mid-1970s and for IGFBP-3 in the mid-1980s. Since then, IGF-I and IGFBP-3 levels have been reported to be abnormally low in both children and adults with PWS[3,28,29,45,56,62,64,75,78,81,99] The combination of low GH and low IGF-I argues for the existence of a true GH/IGF axis deficiency in PWS.[38,76]

GH/IGF axis deficiency is also suggested by the body composition characteristics in PWS. As discussed in the preceding chapter, both fat and lean mass compartments are increased in common exogenous obesity. However, in PWS, fat mass is preferentially increased,[14,74] a condition that is also found in other forms of GH deficiency.[27]

Although the composite data strongly support the existence of a GH/IGF axis deficiency, a small but significant proportion of individuals with PWS will have normal GH responses on standard testing. In many of these cases, the IGF-I levels are low and growth and body composition are characteristic for PWS and GH deficiency. This paradoxical situation may be explained by the physiology of the GH/IGF system and limitations of GH testing.

GH (somatotropin) is secreted by specialized cells (somatotropes) in the anterior pituitary gland in an episodic manner; bursts of GH are released into the bloodstream in an intermittent and somewhat unpredictable pattern, with each burst having a relatively short half-life in serum. The pattern of these bursts is regulated by a complex system, involving inhibition of GH release by intermittent secretion of somatostatin (inhibitory) and GH-releasing factor (stimulatory) from the arcuate nucleus of the hypothalamus, coupled with negative feedback signals from the periphery to the CNS.[47] Larger bursts of GH release are observed during fasting, exercise, sleep, and during puberty. Between these bursts, GH levels are normally low. The total daily secretion, or total area-under-the-curve, for GH is presumably proportionate to the levels of IGF-I.

Because of this episodic secretion, random GH levels have little utility in the assessment of GH adequacy since these levels are likely to be low. Therefore, various testing protocols are used to predictably stimulate GH secretion, using either physiologic (fasting, monitored exercise) and/or pharmacologic (insulin-induced hypoglycemia, clonidine, arginine, ornithine, L-DOPA, glucagons) stimuli followed by repeated, timed blood sampling.

Although provocative GH testing has been a widely accepted diagnostic procedure for many decades, there is considerable controversy regarding reliability and clinical usefulness.[51] A large proportion of individuals will fail to achieve a normal GH peak on one test but not another, resulting in a requirement for two or more tests. In addition, individuals who fail testing in childhood may have a normal response when tested at a later time. Furthermore, there is lack of agreement on the definition of a "normal" response level, with published criteria ranging from a peak GH level of 3–15 ng/mL, and this is further complicated by considerable variability between assay methods.

Finally, there are questions regarding the physiologic relevance of GH secretion. GH is undoubtedly the major factor that stimulates secretion of growth-promoting factors such as IGF-I. However, other factors are also involved. For instance, in some individuals who have GH deficiency due to craniopharyngioma, IGF-I levels and linear growth are normal. This latter situation is thought to be due to hyperphagia and obesity, with high insulin levels stimulating IGF-I synthesis. In addition, children with short stature and growth failure may have low IGF-I levels despite normal or high GH levels; the extreme example of this situation involves defects in the GH receptor. Therefore, although very low GH levels are indicative of pituitary dysfunction, neither high nor low levels are necessarily predictive of IGF-I levels or linear growth patterns.

Returning to PWS, it is evident that the majority of individuals have a deficiency in IGF-I and characteristics of GH deficiency, although many patients will have apparently normal stimulated GH levels. The specific reasons for this discordance in PWS are not entirely known. However, several points should be considered:

1. There is no evidence for GH resistance as might be observed, for instance, with GH receptor defects. GH levels are not high and a therapeutic response is seen with usual GH replacement therapy.

2. There is no evidence for an IGF-I synthetic defect in response to GH; IGF-I levels increase during GH replacement therapy (see below).

3. The low GH levels could be related to obesity, as is seen with exogenous obesity. However, in this latter situation, inhibition of GH

secretion is thought to be due to excess free fatty acids[85]; free fatty acid levels are not elevated in PWS.[12,85,113]

4. GH deficiency in PWS has been attributed to intrinsic hypothalamic dysfunction; however, as discussed in previous sections, there are no definitive data in support of this hypothesis, nor for the alternate hypothesis of a defect in peripheral signaling.[77]

5. The fact that stimulated GH levels are normal in a significant proportion of individuals with PWS implies that there is no intrinsic defect in pituitary GH synthesis associated with the syndrome. However, GH testing is acknowledged to be nonphysiologic;[51] therefore, the possibility of a defect in the daily regulation of pituitary GH secretion cannot be completely excluded. Low levels of IGF-I could be due to low total daily secretion of GH despite normal stimulated levels. This latter hypothesis is supported by the positive response to GH therapy in PWS.

6. The lack of normal IGF-I levels despite hyperphagia and obesity, as might be expected from the experience with craniopharyngioma patients,[96] could in part be due to the lack of insulin resistance and hyperinsulinemia in PWS (see preceding chapter).

In summary, short stature and body composition abnormalities in PWS are consistent with a defect in the GH/IGF axis.[38,76] Although the exact mechanisms have not been delineated, an insufficient production of IGF-I is evident in most cases. The composite data indicate that a complete defect in pituitary GH synthesis may not be involved, although stimulated GH levels are low in a considerable proportion of cases.

As a final comment to this discussion, although GH therapy was historically specified for use in GH deficiency, increased availability of biosynthetic GH has led to its use in a number of conditions that do not necessarily involve GH deficiency. In addition to PWS, current regulatory labeling for GH use in the U.S. and Europe includes childhood growth failure associated with several conditions in which GH and/or IGF-I levels are typically normal (chronic renal failure, Turner syndrome, small for gestational age, idiopathic short stature) and adult HIV-associated wasting. Therefore, while the evidence for GH/IGF axis deficiency provided impetus for investigations of GH therapy in PWS, current therapeutic recommendations are guided primarily by overall treatment efficacy.

GH Therapy

GH was first identified in the 1920s, with purification of primate GH achieved in the 1950s.[50] Therapeutic use in a child with GH deficiency was first reported in the late 1950s. Until 1985, GH was available only as a purified preparation from animal sources and human cadaver donors, and non-primate GH was found to be relatively ineffective in humans. A U.S. national program had been initiated in 1963 (and shortly thereafter in other countries) to collect donated human cadaver pituitary glands and purify GH for therapeutic use. This very limited, sporadic supply was distributed gratis to GH-deficient children through an application and approval process. In 1978, commercial supplies of cadaver-derived human GH became available.

In 1985, a case of Creutzfeldt-Jakob disease, a fatal neurodegenerative disorder, was reported in a patient who had received cadaver-derived human GH.[58] Other cases were identified worldwide and were suspected to be due to transmission of an infectious agent in the cadaver-derived GH preparations. Cadaver-derived GH was removed from clinical use and replaced by biosynthetic GH, manufactured through recombinant DNA technology. No cases of Creutzfeldt-Jakob disease have been reported in patients treated with synthetic GH. In addition, the cause of this disease was recently confirmed to be an unusual infectious agent, called a prion, which was probably transmitted via impurities in the cadaver-derived GH preparations.[121]

At the time of its precipitous release into the marketplace in 1985, synthetic GH had not yet gone through complete safety and efficacy trials. Therefore, a system of postmarketing surveillance was established by agreement between the GH manufacturers and the U.S. Food and Drug Administration. In 1987, Genentech Inc. established the National Cooperative Growth Study (NCGS) in the U.S.[123] and Kabi Pharmaceuticals (now Pfizer) established the Kabi International Growth Database (KIGS).[100] Over the subsequent years, additional post-marketing studies have been initiated for GH and other pharmaceuticals in the U.S. (www.fda.gov/cder/pmc) and throughout the industrialized world, providing information regarding safety and efficacy of prescription medications.

In addition, the increased supply of GH has opened the doors for treating conditions other than childhood GH deficiency. Other U.S. FDA-approved indications and the year of approval include: childhood Turner syndrome (1996), childhood growth failure due to chronic renal failure (1997), HIV-associated wasting in adults (1998), adult GH deficiency (1998), childhood PWS with growth failure (2000), childhood growth failure in children who were small for gestational age at birth (2001), childhood severe idiopathic short stature with abnormally low predicted adult height (2003), and in patients with short bowel syndrome who require nutritional support (2003). It should be noted that the European regulatory labeling for GH in PWS is "for improvement of growth and body composition in children," while the U.S. labeling specifies "long-term treatment of pediatric patients who have growth failure due to Prader-Willi syndrome" without mention of the important beneficial effects of GH (Genotropin and Genotropin/Genotonorm package inserts, Pfizer). Labeling in other countries tends to follow the European indications.

The first preparations of synthetic GH were provided as lyophilized powder, which was then reconstituted with liquid diluent by the user and injected using an insulin syringe. This type of preparation is still available. However, pen injection devices, using GH cartridges and disposable needles, are perhaps the most popular modality in use today. Disposable single- and multi-dose and nondisposable needle-free air injection devices are also available. All commercial GH preparations in use today are biochemically identical to natural human GH and are highly purified.

Effects of GH Treatment in Prader-Willi Syndrome

Linear and Other Skeletal Growth

Initial studies of the effect of GH treatment of children with PWS focused on growth rate acceleration and improvement in stature as primary therapeutic goals. Early reports showing that GH treatment increases growth rate in PWS children did not include control subjects and were relatively short-term.[4,32,56,74,75]

Recent longer-term studies (2–5 years) have provided additional evidence supporting a significant and sustained growth response to daily GH administration.[3,26] Favorable effects on growth were noted in a Swedish study of 18 children with PWS treated with GH for 5 years.[83] A Swiss study of 23 PWS children treated with GH for a median of 3.5 years showed an increase in mean height standard deviation score (SDS) of 1.8, with adult height predictions approaching midparental target height.[41]

In a treatment/no treatment control study, height velocity increased from −1.9 to +6.0 SDS during the first year of GH administration (0.1 IU/kg/day) in 15 children with PWS, compared with a decrease from −0.1 to −1.4 SDS in the no-treatment group (n = 12). For the treatment group, this corresponded to a mean growth rate of approximately 12 cm/year, which is greater than that observed for virtually all other conditions treated with GH.[81]

Small hands and feet (acromicria) were also positively affected by GH therapy, tending toward normalization in children with PWS.[37,41,88] Arm span and sitting height showed similar trends.[37]

Body Composition and Metabolism

GH treatment of GH-deficient children without PWS not only restores linear growth, but also promotes growth of lean body mass, decreases fat mass by increasing fat oxidation and total body energy expenditure, increases bone mineral density following an initial period of increased bone resorption, and improves cardiovascular risk factors.[68,122] Children with PWS respond to GH therapy with similar improvements in body composition and metabolism, and these effects are arguably more valuable than change in growth velocity.

Increased lean body mass and bone density have been reported in several uncontrolled studies of GH therapy in children with PWS.[31,40,74,80] Favorable effects to reduce absolute or percent body fat have also been reported, although these effects are less consistent than for lean body mass and bone density,[3,40,74] as may be expected due to the overall variability in calorie intake, energy output, and body weight in the general population.

In a recent trial in children with PWS, dramatic improvements in lean body mass, bone density, and linear growth were demonstrated in the GH-treated group ($1.0\,mg/m^2/day$) as compared with a nontreated group over the first year.[23] Fat utilization was also increased, while percent body fat stabilized or decreased in most subjects. Over subsequent years of the study, in which the nontreated group was crossed over into treatment, progressive increases in lean body mass and linear growth and continued beneficial effects on body fat were noted.[11,22,23,24,88]

In the same study, a dose response was noted during the dose-ranging study at 24 to 48 months.[26] During this study period, continued progressive increases in lean body mass and height improvement and stabilization of percent fat occurred with administration of either $1.0\,mg/m^2/day$ or $1.5\,mg/m^2/day$ of GH, but not with a lower dose of $0.3\,mg/m^2/day$. Prior improvements in bone mineral density were sustained regardless of dose. The GH treatment responses in children with PWS are greatest during the first 12 months, as is typically observed for other types of childhood GH treatment. The rates of positive changes in lean body mass and bone mineral density slowed but did not regress during more prolonged GH therapy at doses $\geq 1.0\,mg/m^2/day$, and continued to be higher than expected compared with reference data for healthy children without PWS.

Given their reduced lean body mass, children with PWS would be expected to demonstrate markedly reduced resting energy expenditure (REE). Prior to GH treatment, children with PWS showed reduced REE compared with predicted values for non-PWS children matched for surface area ($22.4 \pm 4.4\,kcal/m^2/hour$ versus $43.6 \pm 3.2\,kcal/m^2/hour$; $p < 0.0001$).[26] GH therapy increased REE in children with PWS in parallel with changes in lean body mass, with a similar dose-dependence; i.e., effects were noted at both $1.0\,mg/m^2/day$ and $1.5\,mg/m^2/day$ of GH but not with a lower dose of $0.3\,mg/m^2/day$.

GH deficiency is associated with lipogenesis and fat storage predominating over the accretion of lean mass, even in the absence of overt obesity. Preference for fat utilization as an energy source is reflected in a reduction of respiratory quotient (RQ). The RQ normally ranges from 0.7 (strong predominance of fatty acid oxidation) to 1.0 (exclusive oxidation of carbohydrate) to >1.0 (indicating lipogenesis from carbohydrate). GH treatment in PWS children was associated with a decrease in RQ values, indicating increased utilization of fat for energy.[26] Thus, compared with non-GH-treated PWS controls, GH-treated PWS patients demonstrated a shift in energy derived from oxidation of fat, coincident with reductions in fat mass.

Muscle and Respiratory Function

As reviewed in a previous chapter, decreased muscle mass, hypotonia, and respiratory dysfunction are major contributors to morbidity and mortality in PWS. It is clear that GH therapy improves lean body mass;

however, perhaps of greatest importance to children with PWS and their families is the hope that these changes will be accompanied by improvements in physical strength and function.

Over the years, anecdotal reports of improvements in physical stamina, strength, and agility have been reported in association with GH treatment of children with PWS.[76] Parental reports include a variety of new gross motor skills—e.g., independently climbing up the school bus steps, carrying a gallon carton of milk at the grocery store, participating in a normal gym class without restrictions, being able to join a karate class.[37] However, few studies have been conducted to quantify changes in physical performance.

In an uncontrolled interview study of 12 children with PWS treated with GH, all parents reported improvements in increased physical performance and activity; this effect was judged to be the most important clinical result of treatment.[37] In four children with PWS treated for 1 year, significant improvements in peak and mean power were demonstrated using an ergometer.[40]

Improvements in physical strength and agility were carefully documented in a previously described study[26] using objective measures, including a timed run, sit-ups, and weight lifting. Significant improvements in running speed, broad jump distance, number of sit-ups, and number of arm curls were documented after 12 months of GH treatment compared with controls. During 48 months of GH treatment, improvements in broad jumping and sit-ups were maintained, while further improvements were noted in running speed and arm curls. Despite the fact that the PWS children still scored well below the non-PWS children for all parameters studied, the improvements in strength and agility were associated with "real-life" functional benefits.

Similar positive results on body composition and resting energy expenditure have been reported in a 12-month double-blind, placebo-controlled, randomized crossover GH treatment study of 12 children with PWS.[55]

As reviewed in a previous chapter, the etiology of respiratory disorders in PWS has not been completely defined; however chest wall and oropharyngeal hypotonia are likely to be major contributing factors. Significant improvements in minute ventilation, airway occlusion pressure, and ventilatory response to CO^2 were reported after 6 to 9 months of GH treatment in nine children with PWS.

In the randomized, crossover study mentioned above,[55] significant improvements in peak flow rate, vital capacity, and forced expiratory flow rate were observed after 6 months of GH treatment. In addition, the number and duration of apneic events tended to decline.

In 20 children with PWS involved in a longitudinal study of GH therapy,[26] significant improvements in respiratory function were seen after 1 year of therapy and were sustained at retesting after 24 months of therapy.[88]

Figures 7.1 through 7.9 show the physiological effects of GH deficiency, GH therapy, and weight control successes and failures in various individuals with PWS (photos courtesy of their families).

Behavioral and Psychosocial Effects

Few studies have addressed the effects of GH therapy on behavior and quality of life in patients with PWS, although parents and guardians often offer anecdotal reports of improvement, particularly in relation to eating behavior, alertness, and ability to concentrate. In addition, one might suspect that improved motor function would be associated with improvements in school performance and social functioning. The impression of increased alertness and activity has been reported in uncontrolled studies of GH therapy in PWS.[40]

Whitman et al.[120] have presented the most comprehensive study of behavioral parameters during GH therapy, as part of a controlled study of GH therapy in children with PWS. The behavioral component of this study involved 27 children receiving GH treatment and 14 control subjects studied using a modified Offord Survey Diagnostic Instrument. This instrument includes a behavioral checklist section and a parental questionnaire section. On the checklist section, no statistically significant between-group effects, positive or negative, were noted for multiple measures of psychological and behavioral function. However, within-group analysis showed a significant reduction of depressive symptoms only in the GH treatment group. In the under-11-year-old group, a significant increase in "attention-deficit/hyperactivity" symptoms was observed, perhaps in agreement with the anecdotal reports of increased alertness. In the questionnaire section, parents reported improved school performance, memory, and family/social relationships with GH therapy.

Infants

Infants with PWS often suffer considerable morbidity related to hypotonia and respiratory disorders. Increased body fat and decreased lean mass have also been demonstrated, even in infants who are underweight.[39,119] However, deficits in linear growth may not be evident until after 12 to 24 months of age. In addition, questions regarding the safety and efficacy of GH therapy in severely hypotonic infants have been raised and gained further prominence with two reports of deaths of infants treated with GH.[44,90] For these reasons, studies of GH therapy in infants with PWS have received particular interest.

In a study of 11 children with PWS under 2 years of age, Eiholzer et al.[42] reported significant improvements in lean body mass (estimated by stable isotope dilution) and stabilization of fat mass compared with a group of six infants receiving only CoQ10 therapy over a 1-year period. Continued improvements in lean mass were observed over a 30-month GH treatment period. The same investigators have reported accelerated psychomotor development in infants treated with GH,

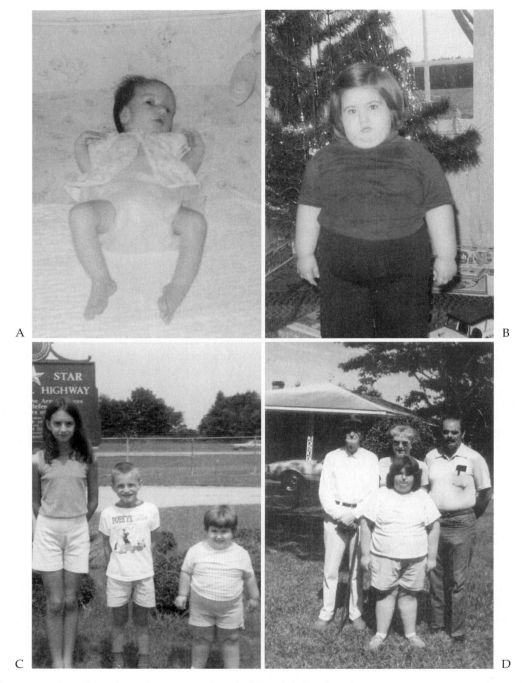

Figure 7.1. Female with PWS not treated with GH: adult height 4 ft. 11 ins.; maximum weight 310 lbs. **A.** Age 1.5 months. **B.** Age 3.5 years, before being diagnosed with congestive heart failure and pneumonia. **C.** Age 4, with older siblings. **D.** Age 16, with mother and maternal relatives.

Figure 7.1. E. Age 21, with male with PWS not treated with GH. **F.** Age 24. **G.** Age 28, with parents and older brother. **H.** Age 31, 1.5 years before her death.

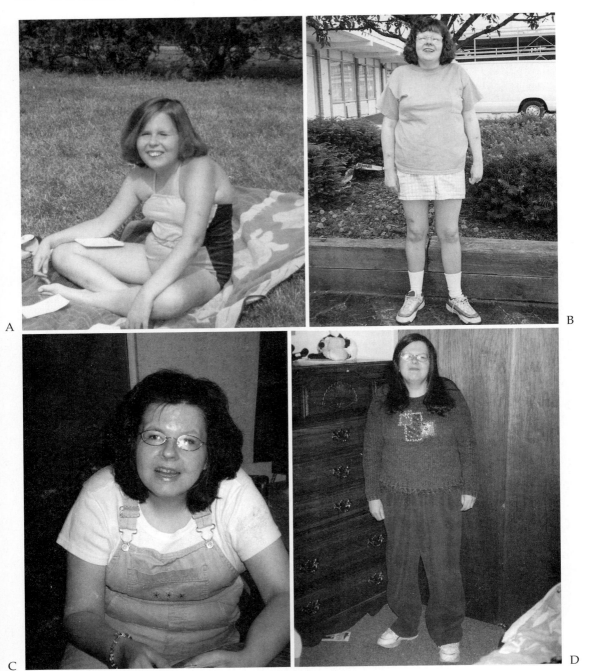

Figure 7.2. Female with PWS not treated with GH: adult height 4 ft. 10 ins.; maximum weight 118 lbs. **A.** Age ~14 years. **B.** Age 28 years. **C.** Age 32 years. **D.** Age 34 years.

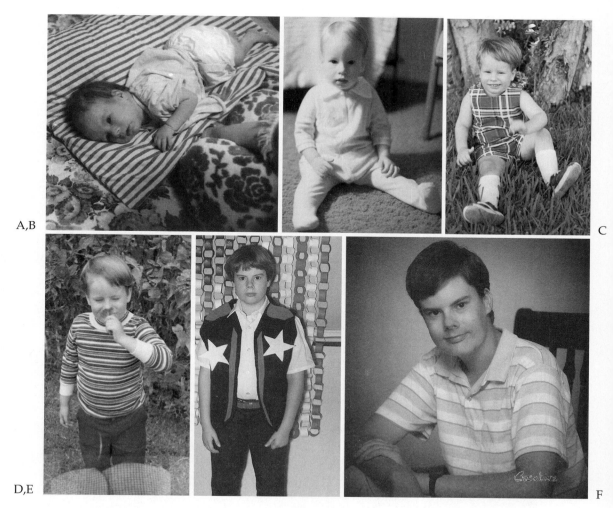

Figure 7.3. Male with PWS, not treated with GH but with controlled weight over lifetime. **A.** Age 9 weeks. **B.** Age 17 months. **C.** Age 2.5 years. **D.** Age 4.5 years. **E.** Age 12 years. **F.** Age 19 years.

Figure 7.4. Male with PWS, treated with GH from age 9 months. **A.** Age 4, height ~42 ins., weight 37 lbs. **B.** Age 4, playing soccer with nondisabled children.

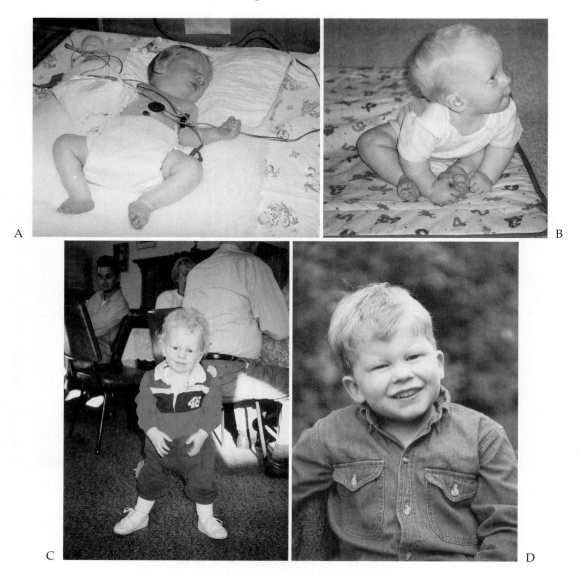

Figure 7.5. Male with PWS, treated with GH from age 8 years. **A.** Newborn. **B.** Age 10 months. **C.** Age 2 years. **D.** Age 5 years.

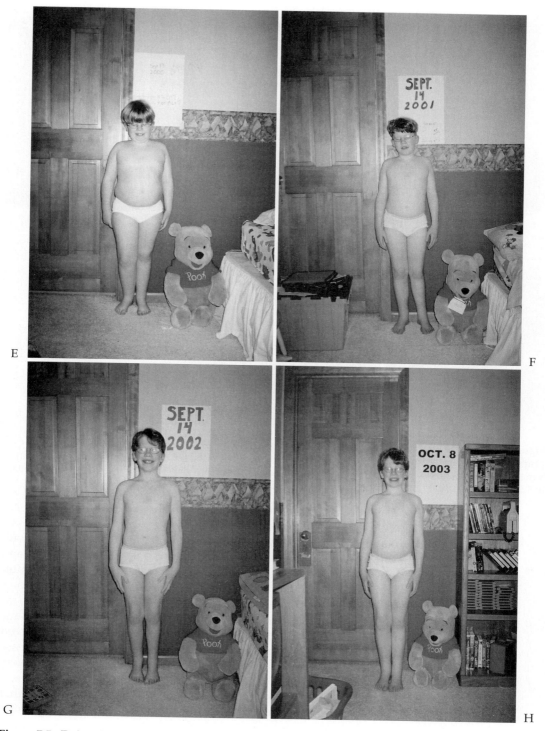

Figure 7.5. E. Age 8 years, at start of GH treatment; height 46.5 ins., weight 69.5 lbs. **F.** Age 9 years, after 12 months of GH treatment; height 51 ins., weight 73.5 lbs. **G.** Age 10 years, after 25 months of GH treatment; height 55.5 ins., weight 81 lbs. **H.** Age 11 years, after 36 months of GH treatment; height 58.5 ins., weight 90 lbs.

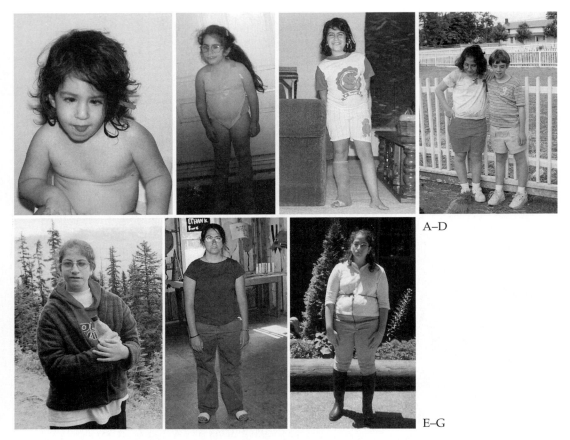

A–D

E–G

Figure 7.6. Female with PWS, treated with GH from age 7.5 to 14 years. **A.** Age 2.5 years. **B.** Age 5 years. **C.** Age 8 years, after start of GH. **D.** Age 12 (left), after 4+ years of GH, with 10-year-old treated female with PWS (shown also in Figure 7.7). **E.** Age 13, after 6+ years of GH. **F.** Age 14, after 7+ years of GH. **G.** Age 15 (in riding outfit), after stopping GH treatment.

which was attributed to improvements in muscle function.[37,43] No adverse effects of GH were noted during these studies.

In a randomized study (GH versus no-treatment) of 25 infants with PWS (mean age 15.5 months), a significant increase in lean mass and decrease in fat mass, both measured by DXA scan, was observed after 6 months of treatment.[119] In addition, age-equivalent motor scores increased approximately twice as fast in the treatment group, as measured by the Toddler Infant Motor Evaluation (TIME). Dramatic increases in head circumference within the normal range were noted in the treatment group with no evidence of CNS pathology; this increase presumably represents increased brain growth. No adverse effects were noted.

Based on these studies and composite experience, it is evident that GH therapy in infants with PWS is efficacious in improving growth, body composition, and neurodevelopment. No specific safety problems have been observed in the published studies. As discussed below, the cases of death reported during GH therapy are likely to be unrelated to GH therapy.

Figure 7.7. Female with PWS, treated with GH from age 9 years. **A.** Age ~11 months. **B.** Age 4.5 years. **C.** Age 6 years. **D.** Age 8.5 years, 6 months before start of GH treatment. **E.** Age 10 years, after ~2 years of GH treatment, wearing scoliosis brace. **F.** Age 13 years, after 4+ years of GH treatment. **G.** Age 15 years, height 5 ft. 5.5 ins., weight ~170 lbs., temporarily off GH before restarting at adult level.

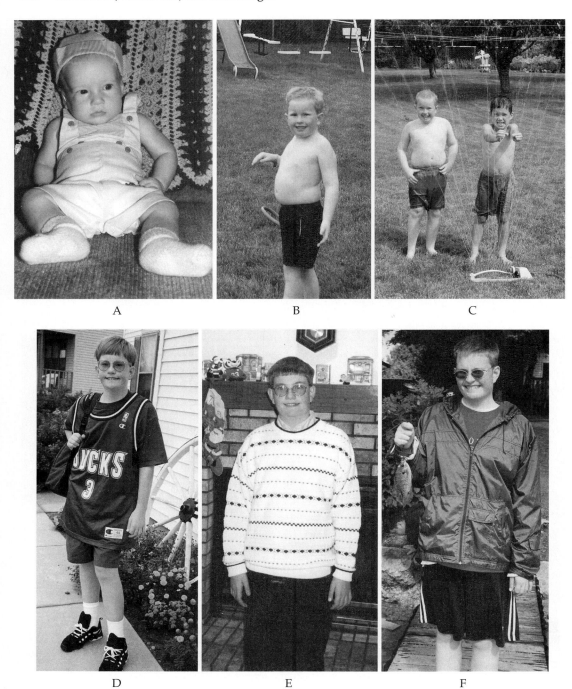

Figure 7.8. Male with PWS, treated with GH from age 11 years: **A.** Age 3 months. **B.** Age 3 years. **C.** Age 9 years (left), with 7-year-old sibling. **D.** Age 13 years, after 1.5 years of GH treatment. **E.** Age 15 years, after 3.5 years of GH (treatment then was stopped for 2 years). **F.** Age 18 years, after restarting GH at adult level 8 months earlier.

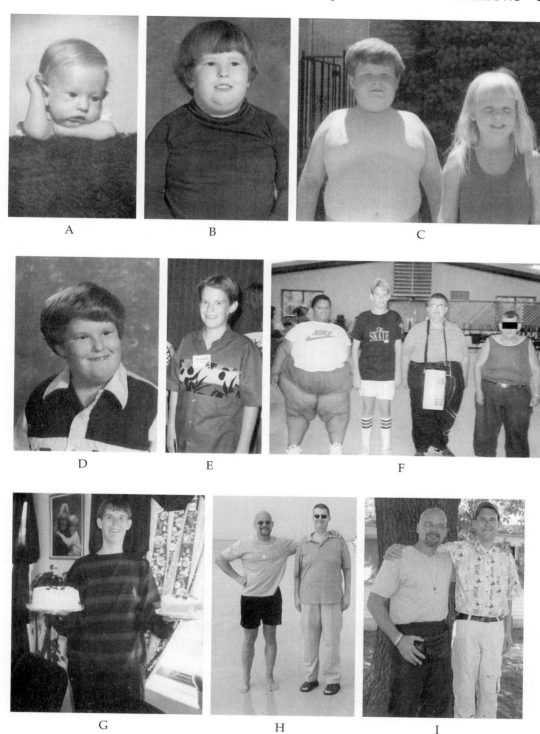

Figure 7.9. Male with PWS, treated with GH from age 14 years: **A.** Age 5 months. **B.** Age 4 years. **C.** Age 7 years, with sister age 6. **D.** Age 8 years. **E.** Age ~12 years, showing weight loss achieved through food management. **F.** Age 16 years (second from left), after 2 years of GH treatment, shown with three untreated males with PWS. **G.** Age 18, after 4 years of GH treatment. **H.** Age 30 (at right) with father, height 5 ft. 10 ins., weight 186 lbs., after discontinuing GH. **I.** Age 32 (at right) with father, weight 139 lbs., after restarting GH treatment.

Adults

Knowledge of the natural medical course of adults with PWS is an evolving area of interest. Over the past decade, there has been an emphasis on earlier initiation of preventive interventions, more intensive and coordinated management of medical and psychosocial problems, and recognition and replacement therapy of GH deficiency.

Although linear bone growth is not a concern during adult life, detrimental effects of GH deficiency have been well documented in non-PWS adults with GH deficiency, including osteoporosis, increased body fat, decreased lean mass, increased risks for cardiovascular disease, fatigue, and psychological depression. In 1995, GH replacement therapy in adults with either childhood or adult-onset GH deficiency received regulatory approval from the U.S. Food and Drug Administration and in Europe. Numerous reports documenting the safety and efficacy of GH replacement in non-PWS GH-deficient adults have been published in the last decade[115,116] including large placebo-controlled, randomized clinical trials[30,104] and a growing number of multinational observational studies from surveillance databases (see below). As demonstrated in placebo-controlled randomized clinical trials, GH therapy improves body composition,[6,35,65] lipid profiles,[52,65,115,116] bone density,[7,102] and quality of life[103] in GH-deficient adults.

Recent studies indicate that adults with PWS have GH/IGF axis deficiency, as a continuation of the condition that begins in childhood.[62,95] In addition, several clinical features of adult PWS are similar to those observed in GH-deficient adults without PWS, and initial studies of therapy indicate similar treatment efficacy.[61,63]

The distinct psychosocial and clinical risk profiles in PWS suggest that GH replacement therapy could have unique benefits in this subpopulation of GH-deficient adults. In addition, there may be special considerations involved with GH replacement therapy in adults with PWS:

1. In contrast to most other individuals with childhood-onset GH deficiency, the majority of current PWS adults did not receive GH therapy as children. This raises concerns regarding potential side effects of initiating GH replacement therapy in treatment-naïve adults.
2. The presence of concomitant lifetime sex-steroid deficiencies could exacerbate the manifestations of GH deficiency.
3. The co-existence of significant medical conditions, including morbid obesity, scoliosis, insulin resistance, Type 2 diabetes mellitus (T2DM), and cardiovascular disease may occur at a relatively young age in PWS and require additional surveillance during GH treatment.
4. The effects of GH treatment on sleep apnea and other pulmonary problems in adults with PWS are uncertain.
5. Cognitive, behavioral, and social considerations could complicate both initiation and long-term administration of GH injections. The specialized living situations of many PWS individuals may be a limiting factor.

Thus, from a theoretical perspective, adults with PWS have a wider spectrum for both potential benefit and risk than do other adults with GH deficiency.

Pretherapy Testing

Since most children with PWS will meet regulatory requirements for GH use, formal testing of GH secretion is not usually necessary prior to therapy. However, testing may be indicated in a few instances:

1. In the U.S., GH therapy is specifically labeled for treatment of PWS *with growth failure*, rather than for body composition abnormalities associated with PWS. In some cases, growth failure per se may be difficult to document. For instance, as previously described, children with PWS and premature adrenarche may have relatively accelerated linear growth, but compromised final height. In such cases, GH testing may be useful to document the alternate therapeutic indication of GH deficiency.

2. *In utero* and neonatal problems, including prematurity and hypoxia, are known risk factors for panhypopituitarism (i.e., GH deficiency plus other pituitary hormone deficiencies, such as TSH and ACTH). Infants with PWS who suffer these types of problems should be considered for a complete pituitary evaluation, including direct measurement of GH levels, since replacement of GH and other hormones may be necessary before the clinical development of "growth failure." Unexpected neonatal hypoglycemia may be a warning sign of hypopituitarism; low GH and/or cortisol levels measured during hypoglycemia can be diagnostic for pituitary deficiency.

Another problematic situation is the neonate with PWS. Since evidence suggests that GH replacement may be beneficial for brain growth and neuromotor development, it may be clinically unwise to withhold therapy while awaiting documentation of growth failure. Consultation with a pediatric endocrinologist with experience in the treatment of neonatal PWS is highly recommended.

Other specific testing that may be considered for children with PWS prior to therapy includes the following:

1. Thyroid function testing (T4, TSH). Although thyroid abnormalities do not occur with increased frequency in PWS, thyroid problems are common in the general population and, if untreated, may have detrimental effects on neonatal development as well as limit the efficacy of GH therapy at all ages.

2. Fasting blood glucose and/or glycated hemoglobin (hemoglobin A1c). Elevations in either of these levels may indicate impaired glucose tolerance or Type 2 diabetes mellitus, either of which may be exacerbated by GH therapy.

3. Bone age radiograph may be useful to assess final height potential, particularly if the bone age is >6 years.

4. Liver enzymes to assess for hepatic steatosis and renal function tests (urea nitrogen and creatinine), if clinically indicated.

5. DXA scan to assess hip and spine density (standards generally not available below 8 years of age at the time of this writing), preferably in combination with a whole-body DXA scan to assess body composition.

6. Scoliosis radiographs, if clinically indicated (see discussion in Chapter 5).

Whenever feasible, formal baseline testing of motor function should be documented for comparison purposes. Video documentation may also be considered.

Since GH replacement therapy does not yet have specific regulatory labeling for treatment of adult PWS, formal testing of the GH/IGF axis GH secretion is necessary, despite the controversies regarding the physiologic relevance of stimulatory testing (as discussed in a preceding section). Results of population-based studies suggest that the measurement of IGF-I might be a useful initial screening test to detect abnormalities in the GH/IGF axis in young adults.[2,84] While the presence of a normal IGF-I does not rule out GH deficiency, the finding of a level that is below the age-specific normal range strongly suggests the possibility of this diagnosis. Because IGF-I levels naturally decline with age, a low level is more likely to indicate GH deficiency in younger, as compared with older, adults. Therefore, a two-stage screening procedure using an initial IGF-I determination followed by formal provocative GH testing could be a useful and cost-effective method to identify GH deficiency in young adults with PWS. The IGF-I screening procedure in older adults with PWS may be less sensitive and specific; however, since GH therapy in adults is guided by monitoring of IGF-I levels, measurement of this level prior to GH testing and/or GH therapy is recommended.

At the current time, there is considerable controversy regarding the optimal agent to use for provocative GH testing in adults.[8,13] Pro and con arguments relating to safety, validity, reliability (reproducibility), and cost can be given for virtually all secretagogues commonly used, including clonidine, L-DOPA, propranolol, pyridostigmine, arginine, GH-releasing factor and analogues, and insulin. Secretagogue selection is further complicated in adults with PWS in view of the potential interaction of obesity, gonadal steroid deficiency, and the presumed hypothalamic level of the GH secretory defect.[112] Insights gleaned from an increasing body of information based on studies of secretagogues in normal adults,[9] non-GH-deficient obese adults,[107] individuals with PWS,[10,21,53,54,110] and adults with presumed hypothalamic GH deficiency[110] suggest that L-DOPA may be the most effective and, perhaps, safest GH provocative agent in adult PWS.

The exact mechanism of GH deficiency in PWS, while not definitively ascertained, is often attributed to a hypothalamic defect, perhaps via a deficiency of GH-releasing factor (see preceding sections for further discussion). Recent publications suggest that L-DOPA and other dopaminergic stimuli may act to stimulate pituitary GH secretion by augmenting hypothalamic production or release of GH-releasing factor, in contradistinction to another commonly used secretagogue, arginine.[105] The data further suggest that arginine increases GH secretion as a consequence of inhibiting hypothalamic somatostatin release in addition to having a direct stimulatory action on pituitary somatotrophs. This information, collected in non-PWS adults, suggests that L-DOPA and similar secretagogues that act at the hypothalamic level may be preferable to agents such as arginine, which have actions below the hypothalamic level, if the intent is to detect a defect in GH-releasing factor, as may be the case with PWS.

Arguing against the use of L-DOPA is the possibility of false positive (i.e., false low) results. In a study of 22 non-PWS, normal volunteers, one subject failed to respond to L-DOPA.[9] Several studies have shown blunted GH secretory response in association with increased adiposity in non-PWS adults.[71,107] Therefore, a second stimulation test or sequential testing might be useful in documenting or confirming GH deficiency in obese adults, such as those with PWS. However, an argument can be made for single-agent testing in adults who have a low IGF-I level and other clinical evidence consistent with GH deficiency.

In a typical provocative GH test, the subject is studied after an overnight fast (recent carbohydrate intake may inhibit GH secretion). A baseline sample is drawn, followed by administration of the provocative agent and timed serum sampling for a period thereafter.[51] A normal response is defined as a peak GH level above a selected level. Unfortunately, there is considerable disagreement regarding a diagnostic cutoff level in both children and adults. While a peak level of less than 3 to 5 mcg/L is generally accepted as an indication of GH deficiency, some experts have advocated higher boundary levels of up to 15 mcg/L. Therefore, the interpretation of results from a provocative GH test should be interpreted in conjunction with IGF-I and/or IGFBP-3 levels and clinical signs and symptoms.

For adults with PWS, other recommended tests prior to GH therapy are similar to those listed above for children with PWS, with these exceptions/additions:

1. Bone age radiographs are not indicated in adults since they have already achieved epiphyseal closure.
2. On the other hand, since scoliosis may progress during adult life, appropriate radiographs should be obtained, if clinically indicated.
3. IGF-I levels, used for therapeutic monitoring, should be obtained at baseline.
4. Because height growth cannot be used as a somatic efficacy measure, a baseline DXA scan, preferably with measurement of whole body composition in addition to hip/spine bone mineral density, is highly recommended.
5. Adults may be at an increased risk for glucose intolerance and Type 2 diabetes mellitus. Therefore, consideration should be given to inclusion of an oral glucose test prior to initiation of GH therapy.

6. Consideration should be given to a baseline prostate-specific antigen (PSA) level in males and PAP smears and mammograms in females.
7. A baseline quality-of-life assessment may also be useful.

Childhood GH Therapy

In a general sense, GH therapy for children with PWS follows the same course as for other conditions. Monitoring of linear growth, growth potential (e.g., via bone age radiographs), and weight serve as the primary indicators of therapeutic effect and the need for dose adjustment. However, while somatic growth is an important effect of therapy, the beneficial effects of GH on body composition and physical function are of greater importance in children with PWS. In this sense, height growth serves primarily as an obvious surrogate marker for overall therapeutic efficacy, including body composition.

Currently, labeling for GH use in children with PWS states that "Generally, a dose of 0.24 mg/kg body weight/week is recommended" (full prescribing information at www.genotropin.com), or 0.034 mg/kg/day. The European labeling for Genotropin/Genotonorm includes the alternate dose calculation of 1.0 mg/m^2/day based on body surface area. This recommendation is apparently based on data from Europe, where GH doses used are generally lower than in the U.S. for all treated conditions. In a review of 15 sets of studies relating to GH use in PWS, Eiholzer[37] found doses ranging from 0.02 to 0.05 mg/kg/day. The highest reported doses (0.04–0.05 mg/kg/day) were those from some of the earliest studies,[3,4,56,74] and one of the authors (PDKL) has continued to use a standard dose of 0.05 mg/kg/day for all pediatric patients with PWS. Another author (ALC) has co-authored studies regarding GH dose-response in PWS,[26] as described in a preceding section of this chapter.

Controversy over GH dosage in PWS is mirrored by similar discussions in non-PWS conditions that have endured for many decades.[49,66] Although GH therapy has been in use for over 40 years and synthetic GH was released in 1985, surprisingly few data have been published regarding optimal dosage for any condition. In most studies related to dose response, the maximal measured response occurs at the highest dose tested, which is the case in relation to GH effects on body composition in children with PWS.[26] Even for a straightforward endpoint such as height growth in non-PWS GH-deficient children, optimal dosage has not been defined, although it is clear that for a given weekly dosage, daily administration is more efficacious than a less frequent dose interval and that increased efficacy is not necessarily proportionate to increased dose.[49,66] Indeed, the average dose used for treatment of childhood GH deficiency, 0.3 mg/kg/week (0.043 mg/kg/day), seems to have evolved primarily from composite practice experience and is higher than at least one manufacturer's recommendation (0.16–0.24 mg/kg/week, www.genotropin.com). In addition, this dose is approximately three times the weekly dose used prior to the availability of synthetic GH and three times the estimated physiologic GH secretion rate for normal children. Higher weight-based doses have been recommended for children with Turner syndrome (0.375 mg/kg/week), growth failure associated with chronic renal failure (0.35 mg/kg/week), small-for-gestational-age children with failure of catch-up growth (0.48 mg/kg/week), and for "pubertal dosing" of GH-deficient children (0.7 mg/kg/week).

Efforts to define individualized upper-dose limits based on IGF-I levels are complicated by high interassay and interlab variability, intra-individual physiologic variability, age- and sex-dependent normal ranges, and extremely wide normal ranges, especially during puberty. In addition, long-term efficacy data based on IGF-I or other biochemical response markers are not available. There are virtually no data relating adverse effects to doses within the usual prescribed ranges.[66] For PWS, an argument has been made that dosing should be based on lean mass rather than on total body weight. However, this would require accurate determination of lean mass at each visit and, more importantly, there are no data to support the effectiveness of such an approach.

A prudent approach to GH therapy in children with PWS may be to use a dose that is at least at the labeled recommendation of 0.24 mg/kg/week (0.034 mg/kg/day, based on total body weight) but not to exceed the highest published efficacious dose (0.35 mg/kg/week, or 0.05 mg/kg/day). The Genotropin/Genotonorm package labeling includes a warning that "Daily doses of 2.7 mg should not be exceeded." The authors recommend consultation with a pediatric endocrinologist with experience in PWS treatment when initiating GH therapy for an individual with PWS.

As mentioned above, height is a reasonable measure of clinical efficacy in children with PWS. As observed in other GH-treated populations, for a given weight-based dose the height velocity will be greatest during the first year of therapy, typically into supranormal ranges; this may be due to "catch-up" growth. This is often followed by a decrease into essentially normal height velocity ranges, transient augmentation with initial sex-steroid replacement, and eventual attainment of a final adult height.

Other clinical parameters that should be monitored during childhood GH therapy include the following:

1. Weight, body mass index, body fat: Although numerous studies indicate that GH therapy may significantly reduce body fat in children with PWS, this effect is not universal. GH therapy may actually augment accretion of body fat in children who do not have a controlled diet.

2. Head circumference: Infants with PWS may have an initial preferential growth in head circumference. Failure of head growth should raise suspicions of craniosynostosis, especially if the fontanels are prematurely closed; this disorder requires immediate evaluation. In cases with an exuberant increase in head growth, hydrocephalus can usually be excluded by careful clinical examination. Widely-spaced sutures and/or a bulging anterior fontanel should be cause for concern and further evaluation.

3. Back curvature: As discussed in Chapter 5, scoliosis is a common occurrence in PWS, and an abnormal curve may worsen during rapid growth. It should be emphasized that GH therapy is not thought to cause scoliosis in PWS, and there is no current evidence for stopping GH therapy because of this condition. Close coordination with orthopedic care is needed. A significant scoliotic curve should also be taken into account when monitoring height.

4. Bone age radiograph: An annual hand/wrist radiograph for determination of bone age was formerly a standard practice during childhood GH treatment. The primary utility of a bone age is for estimation of epiphyseal closure and further growth potential (i.e., final adult height potential). To avoid unnecessary radiation exposure, it may be best to limit bone age radiographs to situations where there is a concern regarding further growth potential. It should also be remembered that height is determined by growth of the vertebrae, hips, and legs—these epiphyses typically close after those of the hand. Therefore, bone age radiographs should always be interpreted together with actual measurements of height velocity.

5. DXA: Although DXA provides useful research information regarding changes in body composition, its role in clinical monitoring of childhood GH therapy is not clear. Hip and vertebral measurements are useful for diagnosing and monitoring osteoporosis, particularly for adolescents and young adults. The clinical utility of whole body DXA is currently under investigation.

6. Laboratory testing: At the current time, there is no evidence to support *routine* monitoring of IGF-I, IGFBP-3, thyroid function tests, glucose tolerance, hemoglobin A1c, or other laboratory tests during GH therapy of children with PWS. Of course, such tests may be indicated in individual cases based on signs and symptoms, and an individual physician may have his or her own routine monitoring guidelines based on personal experience and opinion.

7. Whenever possible, muscle function, physical activity, school performance, and quality-of-life measures should be monitored.

Childhood GH therapy is usually continued until attainment of final adult height. It should be noted that height growth may continue after apparent closure of hand and wrist epiphyseal centers (i.e., on a typical bone age radiograph), since the hip, knee, and vertebral epiphyses tend to mature more slowly than the distal centers. Patients are usually seen in clinic at 3- to 4-month intervals for dose adjustments and monitoring. Lack of significant height gain (after correcting for the effects of scoliosis) over two visit intervals (or approximately 6 to 8 months) may be an indication for cessation of childhood GH therapy and consideration for transitioning to the adult GH therapy regimen. The issue of how to optimally transition GH-deficient individuals from childhood to adult therapy is somewhat controversial, with some practitioners recommending immediate transitioning and others advocating cessation of GH therapy and periodic reevaluation of GH/IGF status and body composition.[106] Limited experience suggests that individuals with PWS have a rapid reversal of body composition benefits with GH cessation; therefore, long-term GH withdrawal cannot be recommended. In all cases, careful monitoring during the transition period is essential.

Adult GH Therapy

As with non-PWS adults,[33] the goals of GH therapy in adults with PWS include improvements in (1) body composition, with reduction of fat mass and increase in lean mass; (2) lipid profiles; (3) bone density; and (4) quality of life. In addition, as with treatment of children with PWS, improvement in muscle function is also a desirable outcome.

In contrast to childhood GH therapy, where GH dosing is based primarily on body size (weight or surface area) and growth response, GH dosing in adults is more typically administered in a fixed amount (mg/day) with adjustments based on laboratory measures of response. In adult patients with adequate nutritional status and normal liver function, the serum IGF-I level is considered to be the best clinical surrogate measure of GH adequacy.

For adults with GH deficiency, a typical starting dose is 0.1 to 0.2 mg/day, without adjustment for weight or surface area. (Note that when corrected for a typical adult weight, this dose is several-fold lower than in children). Subsequent dose changes are often based on routine measurements of IGF-I at 1- to 6-month intervals, with GH dose changes in 0.1- to 0.2-mg/day increments/decrements to achieve an IGF-I level within 0 to +1 standard deviation of age- and sex-specific normal ranges. Experience suggests that a typical stable dose to maintain normal IGF-I levels is 0.6 to 1.0 mg/day for adults with PWS.

In general, IGF-I should be measured approximately 1 month after a GH dose change. It should be noted that IGF-I levels show considerable variability based on assay methodology and laboratory. Therefore, it is important to select a single laboratory with good reference ranges and quality control procedures.

Other parameters to be monitored during adult GH therapy include the following:

1. Weight, BMI, body fat—For the same reasons as stated for children. Bioelectrical impedance assessment can provide a reasonably objective estimate of body fat, as discussed in the preceding chapter.

2. Back curvature.

3. DXA—Osteoporosis is a major cause of morbidity in adults with PWS; therefore, regular measurements (every 1 to 2 years) of hip and spine bone mineral density are recommended. Whole body DXA may provide useful estimates of body composition, as described in the preceding chapter.

4. Polysomnography and/or pulmonary function testing may be indicated in individual cases, as discussed in Chapter 5.

5. Laboratory testing—Annual fasting lipid panel, fasting glucose, hemoglobin A1c, and general chemistry profile (including liver enzymes) may be recommended, especially for very obese patients and/or those with uncontrolled excessive weight gain. More frequent

monitoring may be needed in the first few months of therapy and for those patients with known T2DM, dyslipidemia, or hepatic steatosis.

6. Hemoglobin and hematocrit may also be included in the routine monitoring for menstruating women.

7. As mentioned above, IGF-I levels should be monitored as a surrogate measure of GH adequacy. Whenever possible, physical activity, job performance, psychological status and quality of life should be monitored. In selected patients, cardiology evaluations (echocardiogram, treadmill stress test) may be indicated.

Treatment of adult GH deficiency is presumably a lifelong necessity. However, there are no long-term, randomized, placebo-controlled studies of GH replacement in adults; therefore, clinical experience is an essential element for guiding therapy. Adult patients receiving GH therapy can be followed less frequently than children, perhaps at 4- to 6-month intervals.

Safety and Contraindications

There are no long-term, randomized, placebo-controlled studies of GH therapy in any population. Safety and efficacy monitoring of adult GH replacement therapy is based primarily on observational data from large post-marketing surveillance databases,[5,87,100] which together contain tens of thousands of patient years of treatment experience. Despite the intrinsic methodological limitations of such studies due to the absence of a comparison (control) arm, these databases suggest that GH replacement is safe and confers distinct clinical benefits when administered in appropriate doses. Nevertheless, health care professionals should be aware of potential adverse events during GH therapy.

As described in preceding sections, GH therapy during childhood may be associated with changes in physical features and body proportions (faces, hands, feet) due to bone growth and remodeling. In children with PWS, these changes generally lead to a "normalization" of body features and are not considered adverse effects. However, after epiphyseal closure, continued treatment at childhood dose levels can lead to a coarsening of features and bone overgrowth, as is observed in acromegaly. Therefore, it is important to monitor the response to GH during childhood and transition to adult dosage when long bone growth has ceased.

Scoliosis is a common, progressive condition in PWS, regardless of GH therapy. Therefore, all children and adults with PWS should have regular monitoring of back curvature and appropriate orthopedic monitoring and treatment. Progression of scoliosis may worsen during rapid growth, such as during a pubertal growth spurt or, theoretically, with childhood GH or gonadal steroid therapy. However, studies have not shown an independent effect of GH therapy on progression in PWS.[79,81] Anecdotal reports from experienced orthopedists indicate that GH therapy may improve bone quality and efficacy of surgical treatment for scoliosis.

Slipped capital femoral epiphysis (SCFE) is a childhood condition in which the femoral epiphyseal plate is dislocated posteriorly, resulting in hip (and referred) pain and an abnormal gait. This condition is known to be associated with non-PWS obesity and may also be associated with hypopituitarism, hypothyroidism, and GH therapy.[123] A survey study of 898 members of PWSA (USA) revealed only one case of SCFE among 565 respondents, suggesting a much lower than expected incidence.[118] However, since pain is a major presenting sign of SCFE and individuals with PWS are known to have an increased pain threshold, this could be an underestimate. Clinicians and other caretakers should be vigilant for abnormalities in gait that could signal a need for further evaluation.

GH deficiency is associated with decreased body water[101] and GH replacement therapy can be associated with acute retention of salt and water.[122] In both children and adults, this can manifest as pseudotumor cerebri or idiopathic intracranial hypertension,[123] which is associated with acute onset of headache, visual changes, gait disturbance, nausea, and dizziness. This condition is accompanied by papilledema and could lead to visual loss, although this complication would be unusual for GH-associated cases. Treatment involves cessation and/or lowering of the GH dose. As of this writing, there has been 1 reported case of pseudotumor cerebri in PWS,[56] which may be comparable to the observed risk of 1.6 cases per 1,000 patient years in GH-deficient non-PWS children.[123]

Another manifestation of GH-related salt and water retention is peripheral edema, which occurs almost exclusively in adult GH-treated patients. This complication is said to be the most common emergent complaint during GH replacement therapy in adults, and it was particularly frequent in early investigations when higher doses were administered. As with pseudotumor cerebri, treatment involves a lowering of the GH dose. Mild to moderate peripheral edema has been anecdotally observed in adults with PWS receiving GH therapy; however, the frequency of this adverse event has not been defined. Peripheral edema also was reported in 3 of 12 pediatric patients with PWS who received higher-dose GH therapy.[79,81]

GH therapy of non-PWS, pituitary GH deficiency often unmasks an underlying TSH deficiency, leading to deficiencies in T4 (thyroxine) and T3 (triiodothyronine). Therefore, measurement of thyroid hormone levels (T4, T3, and/or the free forms of these hormones) is recommended for true pituitary GH deficiency. However, this co-occurrence of GH and TSH deficiency has not been reported in PWS, and repeated studies have not shown an increased risk for hypothyroidism, either central (TSH deficiency) or peripheral (primary hypothyroidism) (see Medical Considerations chapter).

GH therapy can cause a decreased T4 level by increasing the conversion[122] of T4 to T3; in these cases, T4 will be low or low-normal, while TSH and T3 will be normal. Some clinicians choose to treat this condition with thyroxine, although there are no data to suggest treatment efficacy. This condition is often transient, even with continuation of GH therapy, and invariably resolves after cessation of GH therapy. One suspected case of this occurrence has been reported in PWS.[79]

GH therapy can acutely decrease insulin sensitivity by counteracting endogenous insulin action. In cases where there is obesity preceding insulin resistance, and/or a genetic predisposition to insulin resistance, the addition of GH therapy can trigger the onset of T2DM. Furthermore, GH therapy could potentially worsen glycemic control in patients with poorly-controlled T2DM, although the longer-term effects on lean and fat mass may increase insulin sensitivity. Post-marketing surveillance databases indicate that this type of complication may be particularly frequent in GH-treated patients with Turner syndrome or intracranial tumors.[122] Increased frequency of GH-induced T2DM has not been formally reported in PWS; however, one of the authors (PDKL) has observed a few such cases, all of whom were very obese and had preceding evidence of insulin resistance or glucose intolerance. Cessation of GH treatment led to reversal of the T2DM, and subsequent weight control allowed restarting therapy at a lower dose without recurrence of T2DM. A worsening of well-controlled, pre-existing T2DM has not been observed with GH therapy. Some clinicians have anecdotally recommended that individuals with PWS not be given GH therapy if they have uncontrolled weight gain and/or are >200% of ideal body weight. GH therapy does not adversely affect carbohydrate metabolism in children with PWS who do not have preceding evidence for insulin resistance and/or glucose intolerance.[69]

The question of whether GH therapy may increase cancer risk has been raised based on limited data showing possible increases in risk for leukemia and colorectal cancer in GH-treated populations.[92,122] This concern is accentuated in PWS due to suggestions, albeit not proven, that there may be an increased overall risk for cancer (see Medical Considerations chapter). Ongoing analyses of post-marketing surveillance databases have not shown an increased risk for primary, secondary, or recurrent cancers in GH-treated individuals, including those with PWS.[123] GH is relatively contraindicated in conditions where there is a known risk for cancer, such as Bloom syndrome,[122] although it is not entirely clear that GH contributes to the risk. This contraindication cannot be logically applied to PWS, where the suggestion of an increased cancer risk has not been proven.

In 2002, Eiholzer and colleagues reported two cases of sudden death in children with PWS during GH treatment.[44,90] These cases and subsequent published reports[117] raised an international concern that GH therapy may increase mortality risk in children with PWS. A worldwide survey of GH-treated PWS patients through mid-2004 revealed a total of 13 cases of death in children, aged 0.7 to 16 years, treated for 2 weeks to 1.5 years. Although the cause of death was not definitely determined in all cases, respiratory compromise, often with concurrent pulmonary infection, seems to be a common link.

These cases spurred a worldwide interest in mortality in PWS. Published reports indicate that over the same approximate time period, there have been more than 30 deaths in children with PWS who were *not* treated with GH,[17,44,90,93,108,109,111,117] and the authors are aware of additional unpublished cases. The cases with known cause of death are remarkably similar to those reported during GH therapy. A publication dated prior to regulatory labeling of GH for use in PWS advocated for collaboration to investigate the phenomenon of sudden death in PWS.[108]

Additional arguments against a relationship of GH therapy and risk of sudden death include (1) the absence of evidence for GH as a cause for increased contributory morbidity in PWS and (2) published studies showing that GH therapy improves respiratory function in PWS.[55,82] Moreover, the dose of GH is known for 11 of the 13 cases of death during GH therapy and all were relatively low, with only 2 individuals treated at or above the recommended level (0.24 and 0.26 mg/kg/week) and the remainder treated at >0.10 to 0.23 mg/kg/week. As of this writing, the relationship between GH treatment and mortality in PWS continues to be a subject of intense investigation. However, a logical conclusion based on current information might be that which was suggested by a statement in the first published report[90]: "The boy reported here . . . thus died before the effects of GH therapy could manifest themselves . . ."

Finally, there have been anecdotal concerns that GH therapy could exacerbate behavioral problems in children with PWS. However, detailed, objective behavioral studies conducted during a 2-year, controlled study of GH therapy in 41 children (27 treatment, 14 control) found no differences for measures of attentional symptoms, anxiety, obsessive-compulsive complex, violence, or psychotic symptoms.[120]

In summary, GH therapy in general has had a remarkable safety record. Adverse effects specifically observed in PWS are either reversible with cessation or reduction of GH therapy (water and salt retention, exacerbation of insulin resistance) or related to pre-existing conditions that require separate attention (scoliosis). Caretakers should be vigilant for the possibility of rare events, such as malignancy, and for complications that are usually signaled by pain in non-PWS conditions (SCFE, pseudotumor cerebri), although none of these complications have been reported to be specifically associated with GH treatment of PWS. Current concerns about the risk for sudden death in PWS are under investigation and a relationship to GH therapy is not supported by available data.

Comprehensive Care

Traditional care guidelines for individuals with PWS, including many of those included in this text, are based on experience in non-GH-treated individuals. At the risk of oversimplifying, the major guidelines can be summarized as follows:

1. Strict dietary limitations to prevent overweight
2. Developmental, occupational, and educational therapy
3. Intensive physical therapy and exercise to counteract the hypotonia
4. Psychological/psychiatric evaluation and therapy

Although the above therapies have provided immeasurable benefits for countless individuals with PWS and their caretakers, they produce no substantial changes in the characteristic body habitus, limited muscle mass and function, or behavior. At each stage of life, a typical, distinctive "look" could be expected, as clearly depicted in the life-sequence pictures of non-GH-treated individuals included in this book and other references.[37,70]

As is also depicted in life-sequence photographs in this text and elsewhere,[24,37] GH therapy has resulted in a remarkable change in the appearance and physical function of individuals with PWS, alleviating many but not all of the associated morbidities as summarized in Table 7.1.[72,73] These changes are most dramatic for individuals treated with GH from a young age, but may also be important for adolescents and adults.

Table 7.1. Morbidity and Treatment in Prader-Willi Syndrome

Morbidity	Contributory Factors	Therapy
Genital hypoplasia, cryptorchidism, hypogonadism	Gonadotropin deficiency	Surgery, gonadal steroids
Hypotonia	GH/IGF deficiency, other factors not yet defined	Physical therapy, GH
Scoliosis	? hypotonia	Orthopedic care
Osteoporosis	GH/IGF deficiency, hypogonadism	GH, gonadal steroids, bisphosphonates
Respiratory dysfunction	Hypotonia, ?central component	GH, surgery, and supportive care if clinically indicated
Obesity (increased body fat, independent of weight)	Not fully defined	Diet, exercise, GH
Hyperphagia, overweight	Not fully defined	Diet, exercise ? GH effect
Acromicria	GH/IGF deficiency	GH
Short stature	GH/IGF deficiency	GH
Insulin resistance, T2DM, related conditions	Obesity	Diet, exercise ? GH effect
Cognitive defect	Not fully defined	Behavioral and educational therapy
OCD-like behavior, psychosis	Not fully defined	Psychotherapy ± psychotropic agents
Low academic and social functioning	Composite effect of physical and mental limitations	Behavioral, educational, and psycho-therapy ? GH effect

For many of the morbidities, the full therapeutic effects of GH are not yet completely defined. GH may have a dual effect on the spectrum of disorders related to obesity, an acute exacerbation of insulin resistance, followed by increased insulin sensitivity as muscle mass increases; this possibility is currently under investigation. Although initial data indicate potential beneficial effects on psychosocial and behavioral function, long-term treatment outcomes have not been reported.

The improvements in physical appearance and function inevitably facilitate psychosocial integration and interactions. In addition, there may be increased resting and active energy expenditure related to increased muscle mass. Additionally, improved medical attention to gonadal steroid replacement may further enhance physical appearance and function. The implications of these changes for comprehensive care of PWS are not completely known, and are likely to be fully defined only with long-term follow-up of patients receiving GH and gonadal steroid replacement. However, as stated by an international consensus panel in 2001,[36] a comprehensive team approach to the individual with PWS is the most logical avenue for provision of optimal care (see Appendix B).

We end this chapter with a few of the additional medical care considerations that are introduced by hormone replacement recommendations:

- *Diet:* It is not clear at this time whether GH has a significant effect on hyperphagia, although experience suggests that dietary control remains a crucial element of care. However, calorie-guided diets must now take into account the increased requirements associated with GH therapy. Strict application of traditional PWS calorie limitations may lead to undernutrition. Weight, body composition, and, for children, growth should be carefully monitored.
- *Physical therapy and exercise:* Traditional expectations for improvement need to be raised and therapy should be individualized. For school-age children, mainstreaming into usual age-related physical education and selected other activities may be possible.
- *Cognitive function and behavior:* There is no current evidence that GH therapy improves cognitive abilities or behavior, although parents reported improvements in memory and school performance[120] in a controlled study of school-aged children. Therapists and educators should be aware that adherence to traditional assumptions of cognitive limitations may not be appropriate. On the other hand, normalization of physical appearance does not eliminate the need for proper attention to typical PWS-related behavior disorders.
- *Sexual behavior:* Normalization of physical function and appearance increases the likelihood of social interactions that may result in sexual activity and relationships. Age-appropriate counseling, caretaker education, birth control precautions, and minimization of risks for sexual abuse and sexually transmitted disease are important elements of a comprehensive team approach.

References

1. Guidelines for the use of growth hormone in children with short stature. A report by the Drug and Therapeutics Committee of the Lawson Wilkins Pediatric Endocrine Society. *Journal of Pediatrics.* 1995;127(6): 857–867.

2. Aimaretti G, Corneli G, Razzore P, et al. Usefulness of IGF-I assay for the diagnosis of GH deficiency in adults. *Journal of Endocrinological Investigation.* 1998;21(8):506–511.

3. Angulo M, Castro-Magana M, Mazur B, Canas JA, Vitollo PM, Sarrantonio M. Growth hormone secretion and effects of growth hormone therapy on growth velocity and weight gain in children with Prader-Willi syndrome. *Journal of Pediatric Endocrinology and Metabolism.* 1996;9(3):393–400.

4. Angulo M, Castro-Magana M, Uy J. Pituitary evaluation and growth hormone treatment in Prader-Willi syndrome. *Journal of Pediatric Endocrinology.* 1991;4:167–173.

5. Attanasio AF, Bates PC, Ho KK, et al. Human growth hormone replacement in adult hypopituitary patients: long-term effects on body composition and lipid status—3-year results from the HypoCCS Database. *Journal of Clinical Endocrinology and Metabolism.* 2002;87(4):1600–1606.

6. Attanasio AF, Howell S, Bates PC, et al. Body composition, IGF-I and IGFBP-3 concentrations as outcome measures in severely GH-deficient (GHD) patients after childhood GH treatment: a comparison with adult onset GHD patients. *Journal of Clinical Endocrinology and Metabolism.* 2002;87(7):3368–3372.

7. Baum HB, Biller BM, Finkelstein JS, et al. Effects of physiologic growth hormone therapy on bone density and body composition in patients with adult-onset growth hormone deficiency: a randomized, placebo-controlled trial. *Annals of Internal Medicine.* 1996;125(11):883–890.

8. Baum HB, Biller BM, Katznelson L, et al. Assessment of growth hormone (GH) secretion in men with adult-onset GH deficiency compared with that in normal men–a clinical research center study. *Journal of Clinical Endocrinology and Metabolism.* 1996;81(1):84–92.

9. Bebchuk JM, Tancer ME. Growth hormone response to clonidine and L-dopa in normal volunteers. *Anxiety.* 1994;1(6):278–281.

10. Beccaria L, Benzi F, Sanzari A, Bosio L, Brambilla P, Chiumello G. Impairment of growth hormone responsiveness to growth hormone releasing hormone and pyridostigmine in patients affected by Prader-Labhardt-Willi syndrome. *Journal of Endocrinological Investigation.* 1996; 19(10):687–692.

11. Bekx MT, Carrel AL, Shriver TC, Li Z, Allen DB. Decreased energy expenditure is caused by abnormal body composition in infants with Prader-Willi Syndrome. *Journal of Pediatrics.* 2003;143(3):372–376.

12. Bier DM, Kaplan SL, Havel RJ. The Prader-Willi syndrome: regulation of fat transport. *Diabetes.* 1977;26(9):874–881.

13. Biller BM, Samuels MH, Zagar A, et al. Sensitivity and specificity of six tests for the diagnosis of adult GH deficiency. *Journal of Clinical Endocrinology and Metabolism.* 2002;87(5):2067–2079.

14. Brambilla P, Bosio L, Manzoni P, Pietrobelli A, Beccaria L, Chiumello G. Peculiar body composition in patients with Prader-Labhart-Willi syndrome. *American Journal of Clinical Nutrition.* 1997;65(5):1369–1374.

15. Bray GA, Dahms WT, Swerdloff RS, Fiser RH, Atkinson RL, Carrel RE. The Prader-Willi syndrome: a study of 40 patients and a review of the literature. *Medicine (Baltimore).* 1983;62(2):59–80.

16. Burman P, Ritzen EM, Lindgren AC. Endocrine dysfunction in Prader-Willi syndrome: a review with special reference to GH. *Endocrine Reviews.* 2001;22(6):787–799.

17. Butler JV, Whittington JE, Holland AJ, Boer H, Clarke D, Webb T. Prevalence of, and risk factors for, physical ill-health in people with Prader-Willi syndrome: a population-based study. *Developmental Medicine and Child Neurology.* 2002;44(4):248–255.

18. Butler MG, Haynes JL, Meaney FJ. Intra-familial and mid-parental child correlations in heritability estimates of anthropometric measurements in Prader-Willi families. *Dysmorphology and Clinical Genetics.* 1990;4:2–6.

19. Butler MG, Meaney FJ. Standards for selected anthropometric measurements in Prader-Willi syndrome. *Pediatrics.* 1991;88(4):853–860.

20. Cappa M, Grossi A, Borrelli P, et al. Growth hormone (GH) response to combined pyridostigmine and GH-releasing hormone administration in patients with Prader-Labhard-Willi syndrome. *Hormone Research.* 1993;39(1–2):51–55.

21. Cappa M, Raguso G, Palmiotto T, et al. The growth hormone response to hexarelin in patients with Prader-Willi syndrome. *Journal of Endocrinological Investigation.* 1998;21(8):501–505.

22. Carrel AL, Allen DB. Effects of growth hormone on body composition and bone metabolism. *Endocrine.* 2000;12(2):163–172.

23. Carrel AL, Myers SE, Whitman BY, Allen DB. Growth hormone improves body composition, fat utilization, physical strength and agility, and growth in Prader-Willi syndrome: A controlled study. *Journal of Pediatrics.* 1999;134(2):215–221.

24. Carrel AL, Myers SE, Whitman BY, Allen DB. Prader-Willi syndrome: the effect of growth hormone on childhood body composition. *The Endocrinologist.* 2000;10(4)3S–49S.

25. Carrel AL, Myers SE, Whitman BY, Allen DB. Sustained benefits of growth hormone on body composition, fat utilization, physical strength and agility, and growth in Prader-Willi syndrome are dose-dependent. *Journal of Pediatric Endocrinology and Metabolism.* 2001;14(8):1097–1105.

26. Carrel AL, Myers SE, Whitman BY, Allen DB. Benefits of long-term GH therapy in Prader-Willi syndrome: a 4-year study. *Journal of Clinical Endocrinology and Metabolism.* 2002;87(4):1581–1585.

27. Christiansen JS. Growth hormone and body composition. *Journal of Pediatric Endocrinology and Metabolism.* 1996;9 Suppl 3:365–368.

28. Corrias A, Bellone J, Beccaria L, et al. GH/IGF-I axis in Prader-Willi syndrome: evaluation of IGF-I levels and of the somatotroph responsiveness to various provocative stimuli. Genetic Obesity Study Group of Italian Society of Pediatric Endocrinology and Diabetology. *Journal of Endocrinological Investigation.* 2000;23(2):84–89.

29. Costeff H, Holm VA, Ruvalcaba R, Shaver J. Growth hormone secretion in Prader-Willi syndrome. *Acta Paediatrica Scandinavica.* 1990;79(11):1059–1062.

30. Cuneo RC, Judd S, Wallace JD, et al. The Australian multicenter trial of growth hormone (GH) treatment in GH-deficient adults. *Journal of Clinical Endocrinology and Metabolism.* 1998;83(1):107–116.

31. Davies PS. Body composition in Prader-Willi syndrome: assessment and effects of growth hormone administration. *Acta Paediatrica Supplement.* 1999;88(433):105–108.

32. Davies PS, Evans S, Broomhead S, et al. Effect of growth hormone on height, weight, and body composition in Prader-Willi syndrome. *Archives of Disease in Childhood.* 1998;78(5):474–476.

33. Drake WM, Howell SJ, Monson JP, Shalet SM. Optimizing GH therapy in adults and children. *Endocrine Reviews*. 2001;22(4):425–450.

34. Dunn HG. The Prader-Labhart-Willi syndrome: review of the literature and report of nine cases. *Acta Paediatrica Scandinavica*. 1968:Suppl 186:1+.

35. Eden S, Wiklund O, Oscarsson J, Rosen T, Bengtsson BA. Growth hormone treatment of growth hormone-deficient adults results in a marked increase in Lp(a) and HDL cholesterol concentrations. *Arteriosclerosis and Thrombosis*. 1993;13(2):296–301.

36. Eiholzer U, ed. A comprehensive team approach to the management of Prader-Willi syndrome. International Prader-Willi Syndrome Organization (IPWSO); 2001. Available at: http://www.ipwso.org.

37. Eiholzer U. *Prader-Willi Syndrome: Effects of Human Growth Hormone Treatment*. (Savage MO, ed. *Endocrine Development*, Vol. 3.) Zurich: Karger; 2001.

38. Eiholzer U, Bachmann S, L'Allemand D. Is there growth hormone deficiency in Prader-Willi syndrome? Six arguments to support the presence of hypothalamic growth hormone deficiency in Prader-Willi syndrome. *Hormone Research*. 2000;53 Suppl 3:44–52.

39. Eiholzer U, Blum WF, Molinari L. Body fat determined by skinfold measurements is elevated despite underweight in infants with Prader-Labhart-Willi syndrome. *Journal of Pediatrics*. 1999;134(2):222–225.

40. Eiholzer U, Gisin R, Weinmann C, et al. Treatment with human growth hormone in patients with Prader-Labhart-Willi syndrome reduces body fat and increases muscle mass and physical performance. *European Journal of Pediatrics*. 1998;157(5):368–377.

41. Eiholzer U, L'Allemand D. Growth hormone normalises height, prediction of final height and hand length in children with Prader-Willi syndrome after 4 years of therapy. *Hormone Research*. 2000;53(4):185–192.

42. Eiholzer U, L'Allemand D, Schlumpf M, Rousson V, Gasser T, Fusch C. Growth hormone and body composition in children younger than 2 years with Prader-Willi syndrome. *The Journal of Pediatrics*. 2004;144(6):753–758.

43. Eiholzer U, Malich S, L'Allemand D. Does growth hormone therapy improve motor development in infants with Prader-Willi syndrome? *European Journal of Pediatrics*. 2000;159(4):299.

44. Eiholzer U, Nordmann Y, L'Allemand D. Fatal outcome of sleep apnoea in PWS during the initial phase of growth hormone treatment. A case report. *Hormone Research*. 2002;58 Suppl 3:24–26.

45. Eiholzer U, Stutz K, Weinmann C, Torresani T, Molinari L, Prader A. Low insulin, IGF-I and IGFBP-3 levels in children with Prader-Labhart-Willi syndrome. *European Journal of Pediatrics*. 1998;157(11):890–893.

46. Epstein LH, McCurley J, Valoski A, Wing RR. Growth in obese children treated for obesity. *American Journal of Diseases of Children*. 1990;144(12):1360–1364.

47. Farhy LS, Veldhuis JD. Putative GH pulse renewal: periventricular somatostatinergic control of an arcuate-nuclear somatostatin and GH-releasing hormone oscillator. *American Journal of Physiology. Regulatory, Integrative and Comparative Physiology*. 2004;286(6):R1030–1042.

48. Fesseler WH, Bierich JR. [Prader-Labhart-Willi syndrome]. *Monatsschrift fur Kinderheilkunde*. 1983;131(12):844–847.

49. Frasier SD. Human pituitary growth hormone (hGH) therapy in growth hormone deficiency. *Endocrine Reviews*. 1983;4(2):155–170.

50. Frasier SD. The not-so-good old days: working with pituitary growth hormone in North America, 1956 to 1985. *Journal of Pediatrics*. 1997;131(1 Pt 2):S1–4.

51. Gandrud LM, Wilson DM. Is growth hormone stimulation testing in children still appropriate? *Growth Hormone and IGF Research.* 2004;14(3): 185–194.

52. Garry P, Collins P, Devlin JG. An open 36-month study of lipid changes with growth hormone in adults: lipid changes following replacement of growth hormone in adult acquired growth hormone deficiency. *European Journal of Endocrinology.* 1996;134(1):61–66.

53. Grosso S, Cioni M, Buoni S, Peruzzi L, Pucci L, Berardi R. Growth hormone secretion in Prader-Willi syndrome. *Journal of Endocrinological Investigation.* 1998;21(7):418–422.

54. Grugni G, Guzzaloni G, Moro D, Bettio D, De Medici C, Morabito F. Reduced growth hormone (GH) responsiveness to combined GH-releasing hormone and pyridostigmine administration in the Prader-Willi syndrome. *Clinical Endocrinology (Ox).* 1998;48(6):769–775.

55. Haqq AM, Stadler DD, Jackson RH, Rosenfeld RG, Purnell JQ, LaFranchi SH. Effects of growth hormone on pulmonary function, sleep quality, behavior, cognition, growth velocity, body composition, and resting energy expenditure in Prader-Willi syndrome. *Journal of Clinical Endocrinology and Metabolism.* 2003;88(5):2206–2212.

56. Hauffa BP. One-year results of growth hormone treatment of short stature in Prader-Willi syndrome. *Acta Paediatrica Supplement.* 1997;423: 63–65.

57. Hauffa BP, Schlippe G, Roos M, Gillessen-Kaesbach G, Gasser T. Spontaneous growth in German children and adolescents with genetically confirmed Prader-Willi syndrome. *Acta Paediatrica.* 2000;89(11):1302–1311.

58. Hintz RL. The prismatic case of Creutzfeldt-Jakob disease associated with pituitary growth hormone treatment. *Journal of Clinical Endocrinology and Metabolism.* 1995;80(8):2298–2301.

59. Holm VA, Nugent JK. Growth in the Prader-Willi syndrome. *Birth Defects Original Article Series.* 1982;18(3B):93–100.

60. Hooft C, Delire C, Casneuf J. Le syndrome de Prader-Labhart-Willi-Fanconi: etude clinique, endocrinologique et cytogenetique. *Acta Paediatrica Belgica.* 1966;20:27–50.

61. Hoybye C. Endocrine and metabolic aspects of adult Prader-Willi syndrome with special emphasis on the effect of growth hormone treatment. *Growth Hormone and IGF Research.* 2004;14(1):1–15.

62. Hoybye C, Frystyk J, Thoren M. The growth hormone-insulin-like growth factor axis in adult patients with Prader Willi syndrome. *Growth Hormone and IGF Research.* 2003;13(5):269–274.

63. Hoybye C, Hilding A, Jacobsson H, Thoren M. Metabolic profile and body composition in adults with Prader-Willi syndrome and severe obesity. *Journal of Clinical Endocrinology and Metabolism.* 2002;87(8): 3590–3597.

64. Hwu WL, Tsai WY, Lee JS, Wang PJ, Wang TR. Prader-Willi syndrome with chromosome 15 interstitial deletion: report of one case. *Zhonghua Min Guo Xiao Er Ke Yi Xue Hui Za Zhi.* 1991;32(2):105–111.

65. Johannsson G, Bjarnason R, Bramnert M, et al. The individual responsiveness to growth hormone (GH) treatment in GH-deficient adults is dependent on the level of GH-binding protein, body mass index, age, and gender. *Journal of Clinical Endocrinology and Metabolism.* 1996;81(4): 1575–1581.

66. Kemp S. Growth hormone therapeutic practice: dosing issues. *The Endocrinologist.* 1996;6:231–237.

67. Kokkoris P, Pi-Sunyer FX. Obesity and endocrine disease. *Endocrinology and Metabolism Clinics of North America.* 2003;32(4):895–914.

68. L'Allemand D, Eiholzer U, Schlumpf M, Steinert H, Riesen W. Cardiovascular risk factors improve during 3 years of growth hormone therapy in Prader-Willi syndrome. *European Journal of Pediatrics.* 2000;159(11): 835–842.

69. L'Allemand D, Eiholzer U, Schlumpf M, Torresani T, Girard J. Carbohydrate metabolism is not impaired after 3 years of growth hormone therapy in children with Prader-Willi syndrome. *Hormone Research.* 2003;59(5):239–248.

70. Laurance BM, Brito A, Wilkinson J. Prader-Willi syndrome after age 15 years. *Archives of Diseases in Childhood.* 1981;56(3):181–186.

71. Lee EJ, Kim KR, Lee KM, et al. Reduced growth hormone response to L-dopa and pyridostigmine in obesity. *International Journal of Obesity and Related Metabolic Disorders.* 1994;18(7):465–468.

72. Lee PD. Effects of growth hormone treatment in children with Prader-Willi syndrome. *Growth Hormone and IGF Research.* 2000;10 Suppl B: S75–79.

73. Lee PD. Disease management of Prader-Willi syndrome. *Expert Opinion on Pharmacotherapy.* 2002;3(10):1451–1459.

74. Lee PD, Hwu K, Henson H, et al. Body composition studies in Prader-Willi syndrome: effects of growth hormone therapy. *Basic Life Sciences.* 1993;60:201–205.

75. Lee PD, Wilson DM, Rountree L, Hintz RL, Rosenfeld RG. Linear growth response to exogenous growth hormone in Prader-Willi syndrome. *American Journal of Medical Genetics.* 1987;28(4):865–871.

76. Lee PDK. Endocrine and Metabolic Aspects of Prader-Willi Syndrome. In: Greenswag LR, Alexander RC, eds. *Management of Prader-Willi Syndrome.* 2nd ed. New York: Springer-Verlag; 1995:32–57.

77. Lee PDK. Model for a peripheral signalling defect in Prader-Willi syndrome. In: Eiholzer U, L'Allemand D, Zipf WB, eds. *Prader-Willi Syndrome as a Model for Obesity.* Basel: Karger; 2003:70–81.

78. Lee PDK, Hwu K, Brown BT, Greenberg F, Klish WJ. Endocrine investigations in children with Prader-Willi syndrome. *Dysmorphology and Clinical Genetics.* 1992;6:27–28.

79. Lindgren AC. Side effects of growth hormone treatment in Prader-Willi syndrome. *The Endocrinologist.* 2000;10:63S–64S.

80. Lindgren AC, Hagenas L, Muller J, et al. Effects of growth hormone treatment on growth and body composition in Prader-Willi syndrome: a preliminary report. The Swedish National Growth Hormone Advisory Group. *Acta Paediatrica Supplement.* 1997;423:60–62.

81. Lindgren AC, Hagenas L, Muller J, et al. Growth hormone treatment of children with Prader-Willi syndrome affects linear growth and body composition favourably. *Acta Paediatrica.* 1998;87(1):28–31.

82. Lindgren AC, Hellstrom LG, Ritzen EM, Milerad J. Growth hormone treatment increases CO_2 response, ventilation and central inspiratory drive in children with Prader-Willi syndrome. *European Journal of Pediatrics.* 1999;158(11):936–940.

83. Lindgren AC, Ritzen EM. Five years of growth hormone treatment in children with Prader-Willi syndrome. Swedish National Growth Hormone Advisory Group. *Acta Paediatrica Supplement.* 1999;88(433): 109–111.

84. Maccario M, Gauna C, Procopio M, et al. Assessment of GH/IGF-I axis in obesity by evaluation of IGF-I levels and the GH response to GHRH+arginine test. *Journal of Endocrinological Investigation.* 1999;22(6): 424–429.

85. Maccario M, Procopio M, Loche S, et al. Interaction of free fatty acids and arginine on growth hormone secretion in man. *Metabolism.* 1994;43(2):223–226.

86. Mogul HR, Medhi M, Zhang S, Southren A. Prader-Willi syndrome in adults. *The Endocrinologist.* 2000;10:65S–70S.

87. Monson JP. Long-term experience with GH replacement therapy: efficacy and safety. *European Journal of Endocrinology.* 2003;148 Suppl 2:S9–14.

88. Myers SE, Carrel AL, Whitman BY, Allen DB. Sustained benefit after 2 years of growth hormone on body composition, fat utilization, physical strength and agility, and growth in Prader-Willi syndrome. *Journal of Pediatrics.* 2000;137(1):42–49.

89. Nagai T, Matsuo N, Kayanuma Y, et al. Standard growth curves for Japanese patients with Prader-Willi syndrome. *American Journal of Medical Genetics.* 2000;95(2):130–134.

90. Nordmann Y, Eiholzer U, L'Allemand D, Mirjanic S, Markwalder C. Sudden death of an infant with Prader-Willi syndrome–not a unique case? *Biology of the Neonate.* 2002;82(2):139–141.

91. Nozaki Y, Katoh K. Endocrinological abnormalities in Prader-Willi syndrome. *Acta Paediatrica Japonica.* 1981;23:301–306.

92. Ogilvy-Stuart AL, Gleeson H. Cancer risk following growth hormone use in childhood: implications for current practice. *Drug Safety.* 2004;27(6):369–382.

93. Oiglane E, Ounap K, Bartsch O, Rein R, Talvik T. Sudden death of a girl with Prader-Willi syndrome. *Genetic Counseling.* 2002;13(4):459–464.

94. Parra A, Cervantes C, Schultz RB. Immunoreactive insulin and growth hormone responses in patients with Prader-Willi syndrome. *Journal of Pediatrics.* 1973;83(4):587–593.

95. Partsch CJ, Lammer C, Gillessen-Kaesbach G, Pankau R. Adult patients with Prader-Willi syndrome: clinical characteristics, life circumstances and growth hormone secretion. *Growth Hormone and IGF Research.* 2000;10 Suppl B:S81–85.

96. Phillip M, Moran O, Lazar L. Growth without growth hormone. *Journal of Pediatric Endocrinology and Metabolism.* 2002;15 Suppl 5:1267–1272.

97. Prader A. The Prader-Willi syndrome: an overview. *Acta Paediatrica Japonica.* 1981;23:307–311.

98. Prader A, Labhart A, Willi H. Ein syndrom von adipositas, kleinwuchs, kryptorchismus und oligophrenie nach myotonieartigem zustand im neugeborenenalter. *Schweizerische Medizinische Wochenschrift.* 1956;86:1260–1261.

99. Ranke MB. Human growth hormone therapy of non-growth hormone deficient children. *Pediatrician.* 1987;14(3):178–182.

100. Ranke MB, Dowie J. KIGS and KIMS as tools for evidence-based medicine. *Hormone Research.* 1999;51 Suppl 1:83–86.

101. Rosen T, Bosaeus I, Tolli J, Lindstedt G, Bengtsson BA. Increased body fat mass and decreased extracellular fluid volume in adults with growth hormone deficiency. *Clinical Endocrinology (Oxf).* 1993;38(1):63–71.

102. Rosen T, Hansson T, Granhed H, Szucs J, Bengtsson BA. Reduced bone mineral content in adult patients with growth hormone deficiency. *Acta Endocrinologica (Copenh).* 1993;129(3):201–206.

103. Rosilio M, Blum WF, Edwards DJ, et al. Long-term improvement of quality of life during growth hormone (GH) replacement therapy in adults with GH deficiency, as measured by questions on life satisfaction-

hypopituitarism (QLS-H). *Journal of Clinical Endocrinology and Metabolism.* 2004;89(4):1684–1693.

104. Russell-Jones DL, Watts GF, Weissberger A, et al. The effect of growth hormone replacement on serum lipids, lipoproteins, apolipoproteins and cholesterol precursors in adult growth hormone deficient patients. *Clinical Endocrinology (Oxf).*1994;41(3):345–350.

105. Saito H, Hosoi E, Yamasaki R, et al. Immunoreactive growth hormone-releasing hormone (IR-GHRH) in the feto-placental circulation and differential effects of L-dopa, L-arginine and somatostatin-14 on the plasma levels of IR-GHRH in normal adults. *Hormone and Metabolic Research.* 1997;29(4):184–189.

106. Savage MO, Drake WM, Carroll PV, Monson JP. Transitional care of GH deficiency: when to stop GH therapy. *European Journal of Endocrinology.* 2004;151 Suppl 1:S61–65.

107. Scacchi M, Pincelli AI, Cavagnini F. Growth hormone in obesity. *International Journal of Obesity and Related Metabolic Disorders.* 1999;23(3): 260–271.

108. Schrander-Stumpel C, Sijstermans H, Curfs L, Fryns JP. Sudden death in children with Prader-Willy syndrome: a call for collaboration. *Genetic Counseling.* 1998;9(3):231–232.

109. Schrander-Stumpel CT, Curfs LM, Sastrowijoto P, Cassidy SB, Schrander JJ, Fryns JP. Prader-Willi syndrome: causes of death in an international series of 27 cases. *American Journal of Medical Genetics.* 2004;124A(4): 333–338.

110. Shimokawa I, Higami Y, Okimoto T, Tomita M, Ikeda T. Effect of somatostatin-28 on growth hormone response to growth hormone-releasing hormone—impact of aging and lifelong dietary restriction. *Neuroendocrinology.* 1997;65(5):369–376.

111. Stevenson DA, Anaya TM, Clayton-Smith J, et al. Unexpected death and critical illness in Prader-Willi syndrome: report of ten individuals. *American Journal of Medical Genetics.* 2004;124A(2):158–164.

112. Swaab DF. Prader-Willi syndrome and the hypothalamus. *Acta Paediatrica Supplement.* 1997;423:50–54.

113. Theodoridis CG, Albutt EC, Chance GW. Blood lipids in children with the Prader-Willi syndrome: a comparison with simple obesity. *Australian Paediatric Journal.* 1971;7(1):20–23.

114. Theodoridis CG, Brown GA, Chance GW, Rudd BT. Plasma growth hormone levels in children with the Prader-Willi syndrome. *Australian Paediatric Journal.* 1971;7(1):24–27.

115. Vance ML. The Gordon Wilson Lecture. Growth hormone replacement in adults and other uses. *Transactions of the American Clinical and Climatological Association.* 1998;109:87–96.

116. Vance ML, Mauras N. Growth hormone therapy in adults and children. *New England Journal of Medicine.* 1999;341(16):1206–1216.

117. Vliet GV, Deal CL, Crock PA, Robitaille Y, Oligny LL. Sudden death in growth hormone-treated children with Prader-Willi syndrome. *Journal of Pediatrics.* 2004;144(1):129–131.

118. West LA, Ballock RT. High incidence of hip dysplasia but not slipped capital femoral epiphysis in patients with Prader-Willi syndrome. *Journal of Pediatric Orthopedics.* 2004;24(5):1–3.

119. Whitman B, Carrel A, Bekx T, Weber C, Allen D, Myers S. Growth hormone improves body composition and motor development in infants with Prader-Willi syndrome after six months. *Journal of Pediatric Endocrinology and Metabolism.* 2004;17(4):591–600.

120. Whitman BY, Myers S, Carrel A, Allen D. The behavioral impact of growth hormone treatment for children and adolescents with Prader-Willi syndrome: a 2-year, controlled study. *Pediatrics.* 2002;109(2):E35.
121. Will RG. Acquired prion disease: iatrogenic CJD, variant CJD, kuru. *British Medical Bulletin.* 2003;66:255–265.
122. Wit JM. Growth hormone therapy. *Best Practice and Research. Clinical Endocrinology and Metabolism.* 2002;16(3):483–503.
123. Wyatt D. Lessons from the national cooperative growth study. *European Journal of Endocrinology.* 2004;151 Suppl 1:S55–59.
124. Zellweger H, Schneider HJ. Syndrome of hypotonia-hypomentia-hypogonadism-obesity (HHHO) or Prader-Willi syndrome. *American Journal of Diseases of Children.* 1968;115(5):588–598.

Part III

Multidisciplinary Management

Editors' note: The behavioral characteristics, special educational and developmental needs, and lifelong support issues of those with Prader-Willi syndrome appear to be universal and relatively independent of cultural influence. However, the health, welfare, and educational systems that provide needed services vary widely from country to country as a function of the cultural and financial resources supporting those systems. Because the authors of this section are all from the United States, these chapters primarily reflect the U.S. cultural and economic framework for provision of these services. For example, the chapters on speech and motor development refer parents seeking services such as speech therapy and physical or occupational therapy to state-operated early intervention programs. Those in countries with a national health system that provides these services as needed may find the need to specially seek and qualify for these services quite foreign. Additionally, the educational sections are written within the framework of the U.S. system for those with "special educational needs." This education system, although mandated by federal law, is nonetheless both funded and administered by each individual state. Differential interpretation and funding of these federal mandates among states, between local educational units and the state, and between parents and local educational units can be an enormous source of stress for parents seeking an optimal educational experience for their youngster. Finally, there are differences by state in the provision of adult care services such as supported living.

We ask forbearance from those readers whose country's health, social welfare, and educational systems are significantly different from those in the United States. Nevertheless, we hope that the experience and wisdom offered in the following chapters will be helpful to all in managing some of the issues with which we have struggled in serving people with Prader-Willi syndrome.

8

Neurodevelopmental and Neuropsychological Aspects of Prader-Willi Syndrome

Barbara Y. Whitman and Travis Thompson

Almost 50 years after originally described, Prader-Willi syndrome (PWS) continues to unfold, revealing complex physiologic, genetic, and behavioral mechanisms that often seem contradictory to standard medical and clinical understandings. This is particularly true regarding the accompanying neurodevelopmental/behavioral complex (sometimes termed behavior phenotype). While still primarily noted for the distinctive food-related behavioral constellation, PWS is characterized by a far broader neurodevelopmental profile including a distinctive cluster of behavioral features along with mental retardation or learning difficulties. Even so, this profile is neither universal nor uniform; considerable phenotypic variability is observed among affected individuals and in the same individual over time.

Understanding this phenotypic variability is further complicated by several published series that report varying rates of major behavior features. Some of this reported variability is reflected in studies that either predate, or postdate, both consensus diagnostic criteria (1993)[61] and comprehensive confirmatory genetic technology (mid to late 1990s, M.G. Butler, personal communication). Thus, part of the reported differences may result from different ascertainment and diagnostic strategies.[8,25,55] Further, previously unaccounted for genetic variation (chromosome 15 deletion versus uniparental disomy [UPD]) may be the source of additional variability.

Recent studies suggest that only hyperphagia, skin picking, stubbornness, and temper tantrums are common to most affected individuals,[43] yet substantial variation is observed even among these common characteristics. A clear understanding of the genetically driven neurobehavioral profile and the impact of environmental influences on the expression of this behavioral profile is pivotal to addressing issues of intervention and remediation, yet our understanding of these aspects of the syndrome remains rudimentary and imperfect—a state of affairs that frustrates parents and caregivers alike.

In this chapter we will summarize the current understandings of the neurodevelopmental/behavioral aspects of PWS. It is first necessary, however, to discuss several historical and methodological factors that influence the status of research in this area. These include diagnostic/ genetic issues, an underappreciation of the pervasiveness of central nervous system involvement, inadequate cognitive and behavior measurement tools, and a strong bias toward measuring problematic behaviors.

Factors Influencing Current State of Research

Diagnostic/Genetic Issues

It is only in the last decade that consensus diagnostic criteria and confirmatory genetic testing for PWS have been readily available. For those with a deletion, genetically confirmed early diagnosis became possible in the early 1980s when high resolution chromosome techniques became available in most cytogenetic laboratories. These techniques, however, required experience to avoid false positive and false negative reporting. Further, these techniques were insensitive to those with uniparental disomy.

Full, reliable and valid laboratory confirmation was not certain until methylation PCR became readily available to most genetic testing laboratories in the mid to late 1990s (M.G. Butler, personal communication, 2003). Prior to that, diagnosis was often suspected but delayed until the emergence of the abnormal eating constellation and obesity. For those with a deletion, the average age at diagnosis was 6 years, and for those with uniparental disomy, age 9.[12] Even today, a small but significant number of youngsters appear to meet the majority of diagnostic criteria and match the behavioral profile for Prader-Willi syndrome yet fail diagnostic confirmation by genetic testing. Earlier study populations that had neither the benefit of standardized criteria nor genetic confirmation likely included some persons who did not have PWS; the influence of these individuals on the data is indeterminable at this point. Further, even among those with PWS, study participation may have been biased toward those with the most intense or severe behaviors.

Thus while behavior studies prior to the early 1990s can be helpful, later studies of behavior that include only those with genetic confirmation of PWS must be given greater weight. Further, since robust and reliable categorization of genetic subtype (deletion, uniparental disomy [UPD], or imprinting center defect) is even more recent, studies of the differential impact of genetic subtype on behavioral outcomes are in their infancy.

Underappreciation of the Pervasiveness of Central Nervous System Involvement

Because of the constellation of metabolic abnormalities associated with PWS, the (hypothesized) central etiologic role of hypothalamic dysfunction is frequently emphasized, often to the exclusion of other brain mechanisms. Yet the typical pattern of cognitive and learning deficits associated with PWS makes it clear that PWS is a pervasive (affecting many areas of the brain) neurodevelopmental disorder. Affected individuals evidence a wide range of cognitive (dis)abilities, often with an overlay of emotional symptoms (depression, anger dyscontrol) that derive, in part, from interactions with particular environments. Four separate sources of cognitive deficits have been identified: (1) depressed general cognitive functioning or IQ,[73] (2) general processing deficits including short-term memory,[90] (3) language processing deficits,[7,69] and (4) an inability to meld diverse and detailed internal information into relevant higher-order abstract and metacognitive concepts that guide behavior over the long term.[82,93] Each of these deficits implies the involvement of multiple neural systems, far beyond that predicted from hypothalamic dysfunction alone. Additionally, affected individuals display an evolving, multistage behavioral picture that differs somewhat between those with a deletion and those with uniparental disomy. Again, hypothalamic dysfunction alone renders an insufficient accounting of the observed complexities. Clearly the neurodevelopmental/neurocognitive picture associated with PWS reflects the dysfunction of a distributed brain system, the complexities of which have yet to be described either anatomically or neurophysiologically.

Inadequate Cognitive and Behavior Measurement Tools

Tools that measure cognition and behavior are frequently imprecise and focus on only a small segment of behavior. For instance, many IQ tests fail both to detect and to account for speech and language deficits. And while speech and language deficits can be separately tested, the interplay between speech and language deficits and uneven cognitive abilities is more difficult to demonstrate. Additionally, many earlier studies failed to use standardized behavioral measures, or used measures standardized on cognitively normal populations. The uncritical application of measures standardized in normal or non-PWS populations may inaccurately describe and overestimate the severity of behaviors noted in affected individuals. Further, some researchers measure the occurrence of specific behaviors[44,45]; others use measures that summarize several behaviors into more abstract conceptual groupings (e.g., the internalizing/externalizing factors of the Child Behavior Scale by Achenbach[25,38]) and personality traits (e.g., California Child Q Set[24,86]), the relevance of which for a PWS population has yet to be established.

An additional measurement problem is found in a number of studies. Both children and adults with PWS can be particularly difficult to test. Shortened attention spans and reduced tolerance for things that are difficult for the individual, combined with expressive language difficulties and, for many, slower-than-average processing speeds, frequently lead to behavioral refusals to complete testing protocols. As a result, many studies that use IQ scores as an independent variable rely on reports from previously administered tests for which the reliability, validity, and margin of error surrounding test administration cannot be established. Therefore results from studies using this methodology for determining IQ must be considered tentative.

A Bias Toward Measuring Problematic Behaviors

From the earliest efforts, behavioral studies have been primarily focused on describing and quantifying the development and severity of problematic behaviors and emotional/psychiatric disorders associated with PWS to the exclusion of behavioral strengths. Despite a clear call by Dykens et al.[38] in 1992 to include investigations of strengths and adaptive behaviors, this remains a rare focus in subsequent studies. Moreover, many studies include an age range encompassing infants through middle adulthood measured at a single point in time. Such a methodology makes it difficult to understand age-related behavioral issues. Further, depending on the nature of the investigation, the use of a comparison group may be appropriate, but defining an appropriate comparison group presents additional difficulties.[34] Newer brain imaging and neurophysiologic techniques offer an avenue for developing a more complete understanding of the neurobiologic substrate of behavior, but these techniques can be particularly difficult to use with this population.

Despite these difficulties, studies across time, taken together, provide a broad, general picture of behavior in PWS. Let us look at the various components of this complex behavior as we currently understand them.

Food-Related Behaviors

As indicated, the neurobehavioral picture associated with Prader-Willi syndrome is far broader than the food-related behavior constellation. Nonetheless, this constellation remains the primary defining feature of the syndrome for most family, health, and other care providers. And, when poorly or improperly managed, hyperphagia along with the resulting obesity and the associated negative behavior usually are the central difficulties for which families seek help. However, hyperphagia is only one of many behaviors in the food-related behavioral constellation, which also includes preoccupations surrounding food; food-seeking/foraging; sneaking, hiding, and hoarding; eating unusual food-related items (sticks of butter, used cooking grease, dog food, decaying or rotten food, food-flavored items such as shampoos, etc.); and for many, manipulative and occasionally illegal behaviors in an effort to obtain food.

Hyperphagia is not unique to PWS; several syndromes and conditions include some degree of hyperphagia (e.g., fragile X syndrome, Kleine-Levin syndrome, and "normal" obesity). Nonetheless clear differences distinguish the eating patterns associated with PWS from those of other conditions. Foremost among these is the age-related emergence, rapid escalation, and intensification of hyperphagia following several years of normal to poor eating and often failure-to-thrive.[30] Further distinguishing characteristics include duration of eating, amount of food eaten, and a delayed to absent deceleration of eating when satiety should be apparent. Further, the hyperphagia remains refractory to even the most recent anorexigenic medications.

Several studies have compared the duration of eating of those with PWS with that of control subjects (either obese of non-syndromic origin, normal weight, or other genetically disordered individuals).[48,57,100] Common to all studies was full access to food by all subjects for a full hour. And in all studies those with PWS, except for the very youngest participants, continued to eat throughout the hour, while control subjects stopped eating within 15 to 20 minutes. For most nonaffected individuals, a deceleration of eating is observed prior to complete eating cessation. Many with PWS showed a slight tendency toward longer pauses between bites near the end of the study hour, however there was no indication that those with PWS would have totally ceased eating had food continued to be made available.

Noting the tendency toward slightly longer pauses near the end of the study hour, several follow-up studies used a similar methodology with the exception that when the participants with PWS appeared to be finished, they were asked to leave the room.[2,70] With this slight modification, all participants with PWS finished eating in 45 minutes. This was still substantially longer than control subjects, who, as before, completed eating in 15 to 20 minutes. In addition to an extended duration of eating compared with control participants, those with Prader-Willi syndrome consumed far greater quantities of food than nonaffected controls. Several authors reported that even when those with PWS appeared sated and stopped eating, the satiety was short-lived and food seeking resumed far more rapidly than in an unaffected population.[57]

The failure of appetite control is also reflected in studies designed to look at food choices of persons with Prader-Willi syndrome, particularly those that look at a choice between a larger quantity of food versus a smaller quantity but more desirable food.[13,51,85] A recent study offered a choice between immediate access to a small amount of a previously determined favorite food of the participant versus a larger amount of a less-preferred food following a brief time delay.[64] Seven persons with PWS and six matched obese persons participated. In all instances, those with PWS opted for the delay followed by the larger quantity, while the control group was equally likely to select the smaller amount of more-desired food as to delay for quantity. A similarly designed, larger follow-up study offered an immediate but small amount of a desired food versus a delay followed by a larger amount of the same food. Similar results were obtained.

That food seeking is prevalent among those with PWS is not surprising, given these results, and indeed at least 50% of those affected actively seek food[31] including foraging for food and food stealing.[5,62] Food seeking typically occurs covertly. Clinical data and anecdotal reports are rife with stories of food-seeking incidents by affected individuals and the impact of these behaviors on obesity management, family stress, and the ultimate adjustment of both the affected individual and his/her family[1,17,43]. For some, food seeking is opportunistic; that is, if food is accessible and usual caretakers are absent or otherwise preoccupied (e.g., parent talking with a friend), the affected individual seizes the opportunity, often hiding the prize (usually in clothes) for later consumption. Others are more active in *making* opportunities, such as "borrowing" money from a parent's wallet for later food purchase, or asking to use the school restroom and while en route "checking out the desk" of a teacher known to keep candy in a desk drawer. At the most severe end, many look for every opportunity to escape the confines of a caregiver so they can seek food, be it from neighbors, local groceries, or any other source of abundant food.

In addition to extraordinary food-seeking and eating behaviors, even ordinary conversation reflects a constant preoccupation with food. Clinical descriptions indicate a constant need to know when the next food will be available, what that food might be, how much will be served, how it will be cooked, and an assurance that the information provided is certain. In the larger environment, food or restaurant advertisements, newspaper and magazine pictures and recipes, and other visual reminders of food frequently become the subject of discussion and "collections." And jigsaw puzzles, a favorite pastime of many persons with PWS, become even more enjoyable when the subject of the puzzle is food related.

Etiology of the Abnormal Food-Related Constellation

The etiology of hyperphagia in PWS has long been attributed to a hypothalamically mediated failure of satiety control,[21,57,70] a view at least partially supported by post-mortem documentation of specific hypothalamic abnormalities.[81] However, other emerging possibilities suggest a far more complex etiology than previously considered,[11,56] including a complete theoretical reorientation that views PWS as a *starvation syndrome* rather than an obesity syndrome. In this view, the obesity of PWS is seen as resulting from a physiologic signaling defect indicating that the body is in a constant state of starvation, similar to that of malnourished youngsters.[59,68] While the implications of this reorientation compared with the view of PWS as an obesity syndrome are beyond the scope of this chapter, such a possibility suggests a basis for explaining the failure of the food-related behaviors to respond to anorexigenic or obesity medications.

In the absence of adjunctive medications, treatment remains primarily behavioral and preventive. Strategies for preventing and coping with this behavior vary according to individual characteristics and family needs.[52] A more complete discussion of these strategies is found elsewhere in this volume (see Chapters 12 and 18).

Cognitive Characteristics

Level of Intellectual Functioning

Decreased intellectual functioning, presumably in the form of mental retardation, was among the four original defining characteristics of PWS.[32,71,99] Those studies[12,22,33,39,55,63,99] in which individual test results or ranges have been reported document a wide range of intellectual abilities, with overall IQ scores ranging from 12 to 100. Curfs's and Fryns's[23] oft-quoted summary of 57 studies including 575 individuals of unspecified ages from a number of countries yielded the following distribution of IQ scores: Normal 4.9%, Borderline 27.8%, Mild Mental Retardation 34.4%, Moderate Mental Retardation 27.3%, Severe to Profound Retardation 5.6%. However, methodologic concerns render these figures questionable. First, the included studies varied from case reports to large surveys. Few of the study populations were genetically confirmed, raising the likelihood that the studies included those who did not have PWS. Further, few authors reported the basis for determining mental retardation, a fairly important issue for the time period, considering IQ thresholds for mental retardation were substantially higher than present criteria. (Prior to 1973, Full Scale IQ scores of less than 85 were considered in the mental retardation range. A 1973 revision of criteria lowered thresholds for mental retardation to Full Scale IQ scores of less than 70.) Further, many studies failed to include measures of adaptive behavior, relying solely on IQ scores to determine the presence of mental retardation, thus failing dual criteria requirements.

In addition to uncertainties regarding the criteria for determining the level of cognitive functioning, there are equally important methodological concerns regarding IQ measurement and the validity of the scores included in the Curfs and Fryns survey.[23] Recall that these percentages within IQ ranges were calculated by averaging across a number of studies. Both the instruments used and the testing methods varied widely across studies. Studies that use a broader range IQ test (e.g., the Wechsler series) obtain different IQ results than those using a more narrowly focused, verbally based instrument (e.g., earlier versions of the Stanford-Binet) that emphasizes verbal abilities more heavily than non-verbal abilities. Secondly, there was no standardization across testing methods nor among those administering the tests. Thus some studies administered all the required subtests while others administered a subset and prorated overall scores from that subset. In addition, some studies required completion of all test protocols at a single sitting, while others spread the testing over several days in order to minimize interference from short attention spans and fatigue. Thus, differences in scores may reflect differences in tests and administration rather than real differences in intellectual abilities. Further, there was a wide age range included in the studies. For example, in a study including those from age 4 years to age 44 years, reported IQ scores may include many who are age 4 and whose test results indicate IQ scores obtained at age 4. This same study may also include three persons who are age 20. However, the included IQ scores were measured at different ages for each individual, such as age 12 for one, 16 for another, and 20 for the third. Finally, there is no accounting for the impact of educational differences on IQ scores among the studies.

A valid determination of cognitive ranges requires: (1) genetic confirmation of the diagnosis; (2) directly and validly administered, age-appropriate standardized testing; and (3) measures of adaptive behavior. While several recent studies included genetic confirmation of the study population, few included direct IQ assessment and determination of adaptive status, depending instead on previous school-based or developmental clinic testing.[43] Others utilized prorated scores from selected subtests or combinations of these methods[1,39,50,78] Five recent studies meet the majority of these criteria. They are summarized in Table 8.1. The Israeli data[54] are remarkable for the large percentage of children testing in the normal range. The authors reported that this sample of 18 represents 86% of the total number of individuals with Prader-Willi syndrome known in Israel. Clearly this population has a distribution of IQ scores that is quite different from the remaining four studies, so it will be considered separately. Averaging across the four remaining studies, all with approximately the same number of participants, 25.5 % have IQ scores above 70, 43% score in the range of Mild Mental Retardation, 27.7% in the region of Moderate Mental Retardation, and 3.8% in the Severe to Profound range. When compared with Curfs's and Fryns's 1992 estimates,[23] this combined sample of 203 subjects yields substantially fewer who score in the Borderline to Normal range (32.7% versus 25.5%) and a substantially greater portion scoring in the category of Mild Mental Retardation (see Table 8.2). Recall, however, that these are cognitive scores only. Although most authors administered an adaptive behavior measure, most did not report these scores. If we assume that many with IQ scores in the 70–75 range would be reclassified as mildly mentally retarded on the basis of their adaptive behavior profiles, then these figures are only approximate, with the true percentages in the borderline range expected to be

Table 8.1. Recent Studies of IQ in Individuals with Prader-Willi Syndrome

Study	N	Avg Age	Percent of Subjects Testing in Each IQ Range			Mild MR	Mod. MR	Severe/ Profound
			Normal		Borderline			
Einfeld et al.[43]	46	17.7[a]		21.6*		64.9	13.5	0.0
Gross-Tsur et al.[54]	18	14.3[b]	17.0	(73.0)*	56.0	27.0	0.0	0.0
Descheemaeker et al.[28]	55	14.1[c]	9.0	(25.4)*	16.4	27.3	40.0	7.3
Whittington et al.[98]	55	21.0[d]	5.5	(31.0)*	25.5	41.8	27.2	0.0
			IQ ≥ 70			60–69	50–59	≤49
Roof et al.[72]	47	23.2[e]	24.0*			38.0	30.0	8.0

* Subjects testing at Normal or Borderline; the numbers in parentheses are sums of the Normal and Borderline subjects in those studies, provided for comparison purposes with the Einfeld et al. and Roof et al. data.
[a] Age range: NA. Only half of subjects genetically confirmed; most IQs from records. (S.L. Einfeld, personal communication, 2004)
[b] Age range: 8.3–23.5 yrs. Did not give a measure of adaptive functioning.
[c] Age range: 1–49 yrs. (A. Vogels, personal communication, 2004)
[d] Age range: 5–46 yrs.
[e] Age range: 10–55 yrs.

Table 8.2. Estimates of IQ in Prader-Willi Syndrome Based on Multiple Studies

Survey	N	Percent of Subjects Testing in Each Range				
		Normal	Borderline	Mild MR	Mod. MR	Severe/Profound
Curfs and Fryns[23a]	575		32.7	34.4	27.3	5.6
Author[b]	203		25.5	43.0	27.7	3.8

[a] Summary of 57 studies; subjects' ages unspecified.

[b] Averages of data from four recent studies shown in Table 8.1, excluding data from Gross-Tsur et al.[54]

somewhat lower while those in the mild mentally retarded range would be somewhat higher.

At least one group of authors[98] suggests that (with the exception of the Israeli group) the distribution of scores approximates the normal curve usually associated with IQ scores, but with a mean shifted downward 30–40 points. They attribute this to a "direct and specific effect on brain development because of the absence of expression of an imprinted gene that is part of the PWS genotype."[98] This explanation, however, cannot account for the very different distribution underlying the Israeli data. To the extent that the Israeli scores reflect real differences in cognitive functioning, the source of these differences must be investigated.

Stability of Cognitive Functioning

In addition to the levels of cognitive ability, several additional aspects of cognitive functioning have been investigated. Early cross-sectional reports suggested that IQ scores in PWS decline with age[22,32,76] Subsequent studies failed to support a decline in IQ scores over time,[39,53] leading to the conclusion that, in contrast to other genetic disorders (e.g., fragile X and Down syndromes) the cognitive trajectories of those with PWS appear to be stable over time. However, the issue is again raised when examining the Israeli and Belgian data (data provided by A. Vogels; personal communication, 2004). In the Israeli data, significant age-related differences are apparent among those with a deletion. Subjects 12 years and under had significantly higher Full Scale IQ scores (78 versus 67, $p < 0.10$) and Verbal IQ Scores (86 versus 66, $p < 0.05$) than those over 12. No age-related differences were found in Performance IQ or Child Behavior Checklist Scores. The UPD sample size was insufficient to examine this issue. Similarly, dividing the Belgian data into those over (IQ = 57.2) and under age 12 (IQ = 63.8), a trend toward a significant reduction ($p = 0.09$) in IQ scores is noted for the older group. The English data (J.E. Whittington, personal communication, 2004), however, do not follow this trend. Full clarification of this issue requires longitudinal studies of several cohorts of affected individuals.

Academic Underachievement

It is generally accepted that across the cognitive spectrum, academic performance is substantially below that predicted by overall IQ scores, although exceptions are noted.[9,39,72,90,97] Part of this discrepancy may be a computational artifact associated with the calculation of Full Scale, Verbal, and Performance IQ. When test protocols are closely examined, most affected individuals demonstrate wide individual variability among the subtest scores comprising an overall IQ (subtest scatter). This uneven pattern of cognitive abilities is similar to that seen in children with learning disabilities.[26,54] When significant subtest variability exists, overall IQ scores are an unreliable and frequently inaccurate predictor of cognitive abilities, even among a population unaffected with PWS.

The remainder of the discrepancy is most likely attributable to specific areas of learning disability as indicated by the subtest scatter. The three areas most impacted are variously cited as reading, spelling, and arithmetic. The impact of these learning disabilities on academic underachievement varies by academic domain and by genetic subgroup. Roof et al.[72] compared the intellectual features and academic achievement of 24 persons with a deletion subtype and 14 persons with UPD. On average, those with UPD had significantly higher Verbal IQ scores than those with a deletion. Additionally, Verbal IQ scores exceeded Performance IQ scores for those with UPD, while those with a deletion evidenced the opposite pattern. When Verbal and Performance scores are averaged to yield an overall IQ, the IQ scores are not significantly different between the two groups. Thus, it appears that those with UPD may have relatively greater verbal skills, while those with a deletion evidence relatively greater visual-perceptual-spatial abilities. Since each of these relative strengths would have a differential impact on academic outcomes, there was no overall difference in performance between the subtypes. By contrast, two recent studies found that, on average, those with a deletion do substantially poorer on reading and spelling than do those with UPD (many of whom reach expected achievement levels in these two areas). There is little difference between the two groups on math skills, which are uniformly depressed.[9,97] In addition, among those with a deletion, the size of the deletion differentially impacts academic outcomes. Those with a type I deletion lose an additional 500 kb of genetic material in addition to what is missing in type II deletions.[9] Measures of academic achievement are strikingly poorer for those with a type I deletion than for those with a type II. To the extent that these differences exist, they may further reflect differences in brain development due to varying amounts of material on the paternally contributed member of the chromosome 15 pair.

Specific Deficits in Learning, Memory, and Cognitive Processing

A number of investigators have examined the relationship between intellectual deficits, academic underachievement, and specific cognitive processing deficits. Findings suggest deficits in visual and auditory attention and short-term memory, whereas long-term memory appears less impaired.

There are few studies on learning and memory in PWS. Warren and Hunt[90] found children with PWS performed less well on a picture recognition task than children with mental retardation of unknown etiology matched for chronological age and IQ. The two groups of children performed similarly on a task meant to measure access to long-term memory. The authors concluded children with PWS have a deficit in short-term visual memory, but not in long-term visual storage. The authors also compared cognitive capabilities of adults with PWS and controls matched on mental age and IQ. PWS participants with difficulty in short-term memory processing lost more information that was learned over time compared with controls. A recent study by Stauder et al.[75] extends these earlier findings, documenting that the auditory modality in PWS is even more affected than the visual and that short-term memory is impaired. As with other areas of cognitive processing, genetic subtype exerts an influence. The recent work of Joseph at al.[65] found that visual memory is a specific strength among individuals with maternal disomy. Specifically, she found that the rate of short-term memory decay among individuals with maternal disomy was considerably slower than that of either PWS individuals with deletions or matched controls.

Visual perception, organization, and puzzle-solving skills have been reported as relative strengths in some people with PWS. Taylor and Caldwell[84] reported Wechsler Adult Intelligence Scale (WAIS) subscores for adults with PWS and obese control participants matched for overall IQ. The highest subtest scores for the participants with PWS were on picture completion, object assembly, and block design. Similarly, block design on the Wechsler Intelligence Scale for Children (WISC) emerged as a strength for 9 of 26 children with PWS,[26] suggesting that some individuals with PWS have the ability to recognize and evaluate figural relations greater than would be expected based on other aspects of cognitive functioning.

To the extent that there are fundamental visual processing differences between genetic subtypes, these differences would be reflected in more basic cognitive-perceptual processes. Fox et al.[49] looked at subtype differences in the discrimination of familiar shapes (square, rectangle) generated through the motion of random elements. While not in normal ranges, the deletion group performed as well as others with non-specific intellectual disability, but the UPD group was significantly worse. However, in a follow-up study using photographs of familiar non-food *and* food-related objects, the UPD group demonstrated better short-term visual memory than the deletion group. Thus the question of subtype differences in basic cognitive processes and intellectual abilities remains open.

Additional cognitive processing differences are suggested in the work by Dykens et al.[39] Using the Kaufman Assessment Battery for children, the authors report significant sequential processing deficits with relatively stronger simultaneous processing abilities in a group of adults with PWS of mixed genetic origin. This must be considered a tentative finding as the population tested was outside the age ranges for which the test was standardized and normed. Reports of superior puzzle-solving ability in PWS individuals would be consistent with this hypothesis.[35,60] Additional studies of the basic cognitive processes (visual and auditory perception, visual and auditory processing, and memory) and learning styles are critical for designing more effective educational and behavioral programs.

Language Processing Deficits

The speech and language characteristics of those with PWS are dealt with at length elsewhere in this volume (see Chapter 9). This section provides a brief summary of these findings as one aspect of the neuro-developmental and neurocognitive components of PWS. A number of authors document delays in spoken language development (e.g., Kleppe[66]). While these delays may, in part, reflect poor muscle tone and reduced breathing capacity leading to reduced oral motor skills, it also appears that deficits in understanding and using language impact both spoken and written language. Language development typically follows a progression from hearing and understanding (receptive language) to speaking (expressive language) single words, two-word phrases, two-word sentences, and eventually interactive conversation. A number of speech-related difficulties have been found: (1) both delays and deficiencies in receptive language including difficulty with auditory discrimination (later, phonics or sounding out words is usually impaired), (2) vocabulary deficits, (3) sub-optimal understanding of sentences,[82] (4) limited ability to talk in sentences, and (5) poor understanding and use of language in a communicative context (pragmatic skills).

Higher-Order Processing Deficits

While a number of authors allude to higher-order processing deficits, little direct research has been done in this area. However, indirect evidence strongly indicates marked difficulties in this area. Higher-order cognitive processing encompasses such abilities as abstract thinking, executive functions, and metacognitive abilities. A number of authors describe the concrete thinking style associated with PWS across cognitive capacities. Thus, many affected individuals will show an inability to change perspectives even when proven wrong[79,82] and an inability to generalize from one situation to another or to see the commonalities across situations. In addition, many affected individuals are co-morbid for Attention-Deficit Disorder (ADD), demonstrating both attentional deficits and impulsivity in multiple settings.[54,95] Metacognitive abilities are defined as "thinking about thinking" or the use of executive processes in overseeing and regulating cognitive processes. Both the concreteness and rigidity of thinking, combined with attentional deficits and the judgment failures of impulsivity give evidence to a marked lack of metacognitive processes. Thus, both academic and social functioning are impacted by an inability to mobilize executive functions and use memory, visuomotor skills, language, and judgment in concert to solve multifaceted problems.

Behavior and Personality

Not illogically, both early and recent work in this area has been heavily focused on understanding the developmental trajectory, range, and severity of problem behaviors frequently associated with PWS. There has been relatively less focus on the positive behavior characteristics and personality strengths. In part this seemingly skewed focus reflects a continuing critical need for effective intervention programs. It may also reflect a real characteristic of the behavioral picture associated with PWS, i.e., a tendency toward a "negative" personality profile with a preponderance of difficult behaviors. Multiple studies using multiple behavior measures clearly document that, separate from the food-related behavioral difficulties, affected individuals are more prone to behavioral disturbances including age-inappropriate temper tantrums, stubbornness, skin picking, impulsivity, and ritualistic and repetitive behaviors[14,18,38,41,55,80,95] Further, the overall rate, severity, and chronicity of these disturbances are frequently more intense than those associated with comparable genetic or cognitive impairments[17,25,40] or other obese groups.[1] However, this difficult behavior picture is largely absent for the first few years of life. Indeed, most authors agree that, on the whole, infants and young toddlers with PWS are affectionate, placid, generally cheerful, largely compliant, and usually cooperative. Further, most authors agree that as the hyperphagia emerges, a separate and distinctly negative behavioral shift is also observed.[4] Unfortunately, the historic reliance on hyperphagia for diagnostic confirmation delayed systematic observation of these early behavioral characteristics until genetic technology supported early diagnosis. Thus, until recently, there was little available data regarding the behavioral and emotional characteristics of the first 2 years of life beyond that found in routine medical records. Therefore the question to be posed is: Is this shift simply a normal developmental process (i.e., the "terrible twos"), or is this an age-related expression of Prader-Willi syndrome—or both?

Behavior Development Across Time

At least three recent studies have looked at the nature of behavior changes in toddlers with PWS.[1,29] Akefeldt and Gillberg[1] examined three groups of affected individuals seeking to determine (1) which behavior and personality characteristics in PWS are primarily linked to the syndrome and not to cognitive levels or weight and (2) if behavior problems are age-related. The youngest group ranged in age from 8 months to 43 months; a second group ranged from 7 to 36 years of age and had received no medication treatment of any sort (either psychotropics or growth hormone); while a third group ranged in age from 4 years 2 months to 36 years 3 months and had received some kind of medication treatment. An age- and BMI-matched comparison group, approximately half of whom were also mentally retarded, served as controls. Behavior was measured with (1) the Swedish translation of the Rutter scales, a parent report instrument of behavior problems in the home setting; (2) the Asperger and High Functioning Autism

Screening Questionnaire; and (3) the Eating Attitudes Test. All but the youngest group were also administered the Birleson Depression Questionnaire. In addition, the youngest group had a mean Rutter score of 2.8 while the two older groups had mean scores of 11.6 and 12.8, respectively. Within the youngest group, parents reported an accelerating interest in food and an increased need for routine beginning around age 2 and a subsequent increased tendency toward stubbornness emerging around age 3. While some of these changes may reflect normal developmental processes (e.g., the "terrible twos"), unlike typically developing youngsters, the data indicate that those with PWS evidence no remission from this stage. Instead the data point to an evolving, increasingly difficult behavioral picture that intensifies with age.

Individuals with PWS aged 7 years and older showed a specific behavior pattern and characteristic personality traits that clearly differentiated these youngsters from the controls. These were skin picking; fussy, overly particular behavior; insistence on certain routines; aggressive behavior; very changeable mood; and preoccupation with food. Many of these behaviors were already present in those younger than age 5. A similar study of 58 individuals aged 3 to 29 years yielded essentially the same findings.[77]

Dimitropoulos et al.[29] sampled 105 children ranging in age from 2 to 6 years. These children were compared with an age-matched group of children with Down syndrome and a separate age-matched group of typically developing children. All groups were compared on the Early Child Developmental (age 2) or the Preschool Child Developmental (age 3) Inventories, a tantrum behavior survey, and a Compulsive Behavior Checklist for Clients with Mental Retardation. The children with PWS were clearly distinguished from both the youngsters with Down syndrome and the typically developing children. All groups demonstrated an emergence and escalation of both food-related and other tantrums; a shorter tolerance for frustration combined with an overreaction to frustration; repetitive and ritualistic behavior; and other behavior problems including increasing oppositional tendencies, a lessened ability to "go with the flow," and increasing stubbornness and "rigidity." However, the two comparison groups exhibited such behaviors *only transiently*; that is, the problems appeared and then subsided. The development of such behaviors in those with PWS was not transient, instead seeming to escalate with increasing age. Additionally, the onset of these behaviors in the PWS group was independent of intellectual, language, and motor abilities. For some with PWS, skin picking was already beginning to emerge. Taken together, these studies strongly indicate that the early emergence of unremitting negative behaviors is more likely a syndrome-driven phenomenon rather than an age-related, normal developmental process. However, since these scores are cross-sectional rather than longitudinal (i.e., following the same group over time), further investigation is warranted.

Other authors have also looked at the second shift that was previously indicated as occurring somewhere around 7 to 8 years of age. These changes include an escalation of the earlier-described behavioral pattern along with a significantly increased rigidity of behavior,[1,24] including the tendency to "get stuck" on a thought or question; increasing concerns for real and imagined worries; less responsiveness to caregivers' efforts to elicit flexibility, to redirection, or to "calming down until reasonable" alternatives; a tendency to "fly off the handle"; and a clear falling behind age-matched peers in social skills. At this age point, both skin picking and daytime sleepiness stand out as major behaviors that distinguish this group of children with PWS both from younger children with the syndrome and from other age-matched children with mental retardation. At the level of personality traits, parents of children with PWS report their children as less agreeable, having lower emotional stability and openness, and showing increased irritability compared with age-matched, typically developing peers.[86]

Emotional lability; temper tantrums; skin picking; repetitive, ritualistic, and compulsivelike behaviors; and hoarding become particularly prevalent in adolescents with PWS, distinguishing this age group from both younger children and older individuals with PWS, as well as from typical adolescents.[14,38,55,91] Furthermore, these difficulties persist into early adulthood.[25,43,53,95] Although some behavior modulation is often seen in later ages, nonetheless problematic behaviors still exceed those seen in other comparison groups.[17,43]

Unlike many other genetic syndromes, the behavior differences among those with PWS appear to be largely independent of gender, cognitive level, or weight, although an occasional study has reported a sex- or weight-specific finding.[24,36,95] By contrast, *genetic subtype* may differentially influence the behavioral picture. An early study by Wenger et al.[92] indicated no difference between those with a deletion and those without; however this study was conducted prior to routine genetic confirmation, particularly of those with uniparental disomy, and may have included individuals who did not have PWS. Dykens et al.[37] compared the problematic behavior of 23 genetically confirmed individuals with a deletion and that of 23 age- and gender-matched PWS persons with genetically confirmed UPD. Their findings suggested that, compared with those with UPD, individuals with a deletion had both a greater quantity of and more severe problematic behaviors. Specifically, those with a deletion had more problematic eating behaviors, were more underactive, were more withdrawn, sulked more, more often skin-picked and bit their nails, and engaged in more hoarding behavior. Those with a deletion also obtained higher severity scores on the Yale-Brown Obsessive Compulsive Scale (Y-BOCS) than did the UPD group. Butler et al.[9] compared 12 individuals with type I deletions to 14 individuals with type II deletions and 21 with UPD. In general, those with type I deletions had poorer behavior and psychological functioning coupled with reduced independent behaviors than those with either type II deletions or UPD. In most, but not all, instances those with a type II deletion scored midway between those with a type I deletion and those with UPD. However, studies by Gross-Tsur[54] and Steinhausen et al.[77] found no behavioral differences between the typical deletion and UPD genetic subtypes. The behavioral impact of genetic subtype appears to be extremely complex, requiring further research to fully understand. Indeed, genetic subtype may have a greater influence on the later development of psychiatric disorders, a subject to which we will later return.

Vogels et al.[89] assessed the *personality* dimensions of extroversion, agreeableness, consciousness, emotional stability, and openness to experience. In general, those with PWS scored more negatively on all dimensions than age-matched controls. This was particularly true for the subdimension of benevolence. A further difference was found between those with the isodisomy form of UPD (two identical maternal chromosome 15s) and those with imprinting center defects. The children with isodisomy UPD were significantly more benevolent and conscious (or more empathic and considerate) of others than were children with imprinting center defects. This is one of the few studies to assess personality on standard personality measures.

Adaptive Behavior

There has been very little formal assessment of adaptive behavior functioning in persons with PWS. Taylor[83] reported Adaptive Behavior Scale (ABS) data from an unpublished study conducted by Taylor and Caldwell.[84] ABS scores of adults with PWS were compared with those of a group of intellectually similar, obese individuals without the syndrome. The only significant difference between the groups on Part I of the ABS was in the physical development category, where the participants with PWS had scores that were 34% below those of the control group. Dykens et al.[38] attempted to establish the adaptive behavior profile of adolescents and adults using the Vineland Adaptive Behavior Scales. Daily living skills proved an adaptive strength for the group as a whole, while socialization, particularly coping skills, emerged as a relative weakness. Interestingly, the authors reported that daily living skills become more of a strength with increasing age. Similar results were reported by Holland et al.[58] Thompson and Butler (unpublished data) also found that PWS subjects and IQ-matched controls differed substantially in their degree of independent community living skills. The subjects with PWS appeared significantly less competent than controls. While this may indicate biologically based differences in cognitive ability, it more likely reflects the far more restricted lives led by most individuals with PWS (and therefore more limited opportunities to develop skills) due to the concerns of caregivers about access to food in uncontrolled settings.

Hoarding, Repeating, Ordering, Need for Sameness— Obsessive-Compulsive, Developmental Arrest, or Autistic Spectrum Disorder

Separate from food-related behaviors and fixations, individuals with PWS exhibit other behaviors that are repetitive, ritualistic, and appear both driven and compulsive. For many, these behaviors are sufficiently problematic and disruptive that management and adjustment are negatively impacted throughout the life span. Several authors studying these behaviors assert a genetic tie between PWS and Obsessive-

Compulsive Disorder (OCD)[41,74] and encourage treatment with medications for obsessive-compulsive disorders.[34,42] Other investigators, however, view these same behaviors quite differently, remaining skeptical of a genetically based PWS-OCD link.[18,58,94]

Arguing that these behaviors constitute "full-blown Obsessive-Compulsive Disorder (OCD)," Dykens et al.[41] compared the scores of 91 older children and adults with PWS and an age- and gender-matched (cognitively normal) comparison group who had previously been diagnosed with OCD. The PWS group yielded significantly elevated scores on specific subscales of the Yale-Brown Obsessive Compulsive Scale. These included hoarding, ordering and arranging objects according to a certain set of rules, and the repetitive asking or telling subscales. The scores of the PWS group were comparable both in number and severity to the comparison group. A subsequent study by the same authors found that these characteristics were more severe and problematic in those with a deletion than in those with UPD,[37] leading the authors to conclude that the pathogenesis of PWS appears to predispose many individuals to obsessive-compulsive behavior, if not full-blown OCD.[42]

By contrast, Feurer et al.[47] used the Compulsive Behavior Checklist (CBC) to study 53 PWS persons with a deletion and 12 with UPD between the ages of 4 and 41 years. The CBC was designed to gather information about compulsive behaviors specifically in persons with intellectual disability. Their findings indicate that, separate from skin picking, the compulsive-type behaviors observed in those with PWS are a single behavioral dimension that is quite different in character from classic OCD, and they question the usefulness of applying such diagnostic terms to the PWS population. Common to both sets of studies was the use of volunteer subjects. It has often been found that people volunteer for studies because they have the disorder being studied, so the study population may contain a more seriously affected group than would be found in the general population. Study results that are based on such a sample may wrongly indicate an association when none exists.

In an attempt to avoid the biases associated with a volunteer study population, a recent study attempted to assess *all* persons with PWS residing in a circumscribed geographic region of the United Kingdom.[18] Theoretically, by sampling from the whole population, certain types of bias found in volunteer subject populations are eliminated. Ninety-three persons with PWS ranging in age from 5 to >31 years were compared with 68 intellectually impaired persons of other etiologies on the Developmental Behavior Checklist, the Aberrant Behavior Checklist, and the Vineland Adaptive Behavior Scale. The authors report a specific lack of obsessive symptoms but did observe a high prevalence of ritualistic behaviors in the PWS population. Although ritualistic, these behaviors are quite different in character from those typically described as obsessive-compulsive behaviors or those associated with classic obsessive-compulsive disorder.

Continuing to characterize the behaviors associated with PWS as a variant of obsessive-compulsive disorder is seen by many as counterproductive. Current criteria for OCD include the presence of obsessions, compulsions, and psychological distress in the affected individual, based on the recognition that such behaviors are (1) out of the range of normal and (2) negatively impacting the affected person's ability to function. While studies clearly document repetitive, ritualistic, and compulsivelike behaviors as salient characteristics of PWS, clear evidence of classic compulsions and obsessional thinking is lacking. This latter argument is frequently countered with an argument that persons with PWS have difficulty assessing and expressing the presence of the internalizing components of obsessions. A more important argument, however, is the clear indication that persons with PWS do not view these behaviors as out of the range of normal, nor do they appear to suffer psychological distress as a result; rather, distress most frequently occurs when the person is prevented from engaging in a favored behavior. Indeed it is more often the caregiver who is bothered by these behaviors than it is the affected individual.[58] Recall that these behaviors appear during the same time frame and are similar in character to those found in typically developing children of the same age. Further, recall that children with PWS are distinguished from the typically developing children in that the latter develop these behaviors only transiently, while the children with PWS continue to display such behaviors. Thus, Clarke et al.[18] and Holland et al.[58] asserted that the etiology of these behaviors is a "specific pattern of atypical and arrested brain development such that the characteristic rituals and compulsions of early childhood continue and only resolve if development goes beyond that particular developmental phase."[58]

Yet another point of view suggests that the pattern and developmental trajectory of these behaviors is most compatible with Autistic Spectrum Disorders. Four criteria define these disorders: (1) an absence, delay, or impairments in both verbal and nonverbal language; (2) qualitative impairments in reciprocal social interaction; (3) a restricted range of interests and stereotyped patterns of behaviors and activities; and (4) a range of cognitive deficits.

At the most severe end of the spectrum are those with frank autism; however a milder variant, Asperger syndrome, has been described. Two studies directly address the presence of Autistic Spectrum Disorders in Prader-Willi syndrome. In a controlled study investigating the behavior and personality characteristics in a Swedish population of persons with PWS, Akefeldt and Gillberg[1] administered both verbal and nonverbal IQ tests, problematic behavior questionnaires, and the Asperger Syndrome and High Functioning Autism Screening Questionnaire to 44 persons with PWS ranging in age from 8 months to 36 years. The latter was used as a measure of social skills. Scores ranged from 0 to 54; 19 and above is considered to indicate the presence of Autistic Spectrum Disorders. The group of children aged 8 months to 3 years 8 months had a mean score of 3.4; however those whose ages were 7 years and up had a mean score of 19.1, indicating the presence of Autistic Spectrum Disorders. Since only mean scores were given, we do not know what percentage of the groups scored in the Autistic Spectrum range. Both groups scored higher on this measure than a control group.

In a study designed to look at Autistic Spectrum Disorders as the unifying framework for understanding the complex behavioral picture associated with PWS, Whitman[94] had parents complete the Autism Behavior Rating Scale, the Gilliam Autism Rating Scale, the Australian Asperger Scale, and the Social Skills Rating Scale, and directly assessed 18 subjects with PWS using age-appropriate IQ tests and a battery of language tests to assess language development, language processing, and pragmatic language. The results indicated that, under the age of 7, all children exhibited a number of autistic features (hand flapping, gaze avoidance, some tactile defensiveness, and abnormal use of language), and 30% qualified for a diagnosis of autism; by the age of 7, 85% qualified for a diagnosis of Asperger syndrome or "high functioning autism."

Deficits in social reciprocity have recently been asserted as the cardinal feature of Autistic Spectrum Disorders.[20] At the most fundamental level, successful engaging in reciprocal social interactions depends on two precursor abilities: social perception and social cognition, both of which appear to be impaired in those with PWS. Pictures representing the four emotions "scared, happy, angry, and sad" were presented to a group of children with PWS, and two additional groups of similar cognitive ability, one with a known genetic syndrome (Williams syndrome) and a group of children with cognitive impairments of unknown origin. With one exception, the children with PWS were less able to appropriately identify the emotion than were those in the other two groups. In the final instance, both the children with PWS and those with nonspecific mental retardation performed equally poorly and less well than those with Williams syndrome.

Social cognition was identified as an area of cognitive weakness in a study of 11 adolescents, aged 10.1 to 17.1 years, with PWS. Participants were asked to interpret the intention of characters in stories involving lies, jokes, and broken promises. Few of them were able to identify lies or jokes or to differentiate between a promise broken intentionally or unintentionally, which points to difficulties in interpreting non-literal language and social situations. Similarly, despite arguing that these behaviors signal a developmental arrest, both Clarke et al.[18] and Holland et al.[58] reported that these behaviors segregate with and correlate most strongly with autistic symptoms.

Recent genetic findings strongly support an autism-PWS association while suggesting that the PWS-OCD association may be exaggerated. Using ordered-subset analysis (OSD), Dementieva et al.[27] identified a homogeneous subgroup of families with Autistic Spectrum Disorders with a primary symptom complex described as "insistence on sameness." This complex includes three specific behavioral traits: (1) difficulties with minor changes in personal routine or environment, (2) resistance to trivial changes in the environment, and (3) compulsions/rituals. A second group was identified as those whose symptoms were primarily repetitive sensory and motor behaviors and interests. DNA studies of the "insistence on sameness" mapped the chromosomal abnormalities to chromosome 15q11–q13. A similar methodology investigated a group of "normal" individuals with diagnosed obsessive-compulsive disorder. These DNA studies mapped OCD to the X chromosome at the FRAXE site. Additional behavioral and genetic studies are needed to verify the PWS-Autistic Spectrum relationship; nonetheless Autistic Spectrum Disorders as a unifying framework for describing and understanding the behavioral complex associated with PWS appears promising.

Psychotic Symptoms, Psychotic Episodes, and Psychiatric Disorders

The presence of psychiatric symptoms and multiple psychiatric disorders has long been associated with PWS[8,55,67] with both an incidence rate exceeding that found in a general population of those with mental retardation and with a predominance of those disorders falling in the psychotic disorder spectrum.[16,46] And, while research indeed suggests a slightly increased rate of psychotic illness in adults with PWS, the nature of this illness appears to be substantially different in character and duration than that found in a non-PWS affected population.[19] Indeed, one group of authors suggests that rather than traditional DSM-IV (*Diagnostic and Statistical Manual of Mental Disorders*, Fourth Edition) nomenclature, the term "PWS Psychiatric syndrome" is a more appropriate nomenclature for the thought disturbances associated with PWS.[88]

Bartolucci and Younger[3] describe three distinct neuropsychiatric symptom complexes associated with PWS. The first complex includes exaggerations of behaviors commonly associated with the syndrome, including oppositional tendencies, explosive episodes, and occasional antisocial tendencies. Because of the biological basis, environmental reactivity, and fluctuating nature of this complex, the authors term this complex a Trait Fluctuation (TF) disorder. While most affected individuals will normally exhibit fluctuations or ups and downs in the severity of these traits, for some this fluctuation has predictable cycles of occurrence and severity that exceeds normal ups and downs, and they may qualify as having a cyclical psychiatric disorder.

The authors define a second complex termed Lethargic-Refusal States, characterized by a withdrawing from usual activities and interactions, a refusal to eat or drink, a complete inattention to hygiene and self-care often accompanied by a nonorganic incontinence, and remaining totally bedridden. Not infrequently, there are disordered thoughts particularly around the possibility of food being purposely poisoned. These episodes are characterized by sudden onset, with a self-limiting course of several weeks and frequently spontaneous remission.

Finally, the authors describe Florid Psychotic States consisting of auditory or visual hallucinations, mood disturbances, and delusions accompanied by fear or occasionally abnormal elation. Subsequent investigations of those with symptoms in this latter group have led to the relabeling of this complex as atypical Cycloid Psychoses[87] and have raised speculation that those with a UPD subtype may be genetically more vulnerable to developing this disorder. Vogels et al.[89] sampled an older PWS-affected group with known psychiatric problems. The group was divided into those with psychoses and those with mood disorders. The authors then conducted a retrospective review of early childhood behavior patterns and found that those with psychoses and either UPD or an imprinting center defect had an early childhood behavior pattern that was "active and extroverted." Those with mood disorders, all of whom had the deletion form of PWS, had a childhood behavior pattern of "passive and introverted." The authors note that these two behavior patterns are the extremes of a continuum. Many with PWS fall somewhere on a line between these two extremes. A subsequent population-based study in the U.K. found five young adults with PWS and psychoses. Since these young adults all were of the UPD genetic subtype, the assertion that UPD inevitably leads to psychoses has emerged.[6] What is unspecified in these studies, however, is how many of each genetic subtype were *not* affected with psychiatric difficulties. And, were there other commonalities besides genetic subtype among those who ultimately became ill?

To pursue psychoses as an inevitable outcome of a particular molecular subtype seems counterproductive. Indeed in the same article,[87] Verhoeven et al. also indicated that a subset of affected individuals with the deletion subtype appeared to be more vulnerable to serious mood disorders, among them bipolar disorder. While psychotic disorders are dramatic in presentation and call for immediate intervention, acute and recurring mood disorders are far more prevalent in this population[28,96] and may account for the majority of those described as the TF subgroup by Bartolucci and Younger. Clinical observations suggest that independent of a genetic vulnerability, the previously described cognitive rigidity, combined with difficulty labeling and expressing feelings and poor coping skills, may predispose those with PWS to an increased incidence of mood disorders, particularly when a major loss is involved. While genetic subtype may be associated with specific expressions of psychiatric difficulty, it seems much more fruitful to understand the nature of environmental stressors that catapult a small subset of individuals with PWS into these more serious psychiatric disturbances. Research regarding this association is ongoing.

References

1. Akefeldt A, Gillberg C. Behavior and personality characteristics of children and young adults with Prader-Willi syndrome: a controlled study. *Journal of the American Academy of Child and Adolescent Psychiatry.* 1999; 38(6):761–769.
2. Barkeling B, Rossner S, Sjoberg A. Methodological studies on single meal food intake characteristics in normal weight and obese men and women. *International Journal of Obesity.* 1995;19:284–290.
3. Bartolucci G, Younger J. Tentative classification of neuropsychiatric disturbances in Prader-Willi syndrome. *Journal of Intellectual Disability Research.* 1994;38:621–629.
4. Beange H, Caradus B. The Prader-Willi syndrome. *Australian Journal of Mental Retardation.* 1974;3:9–11.

5. Bistrian BR, Blackburn GL, Stanbury JB. Metabolic aspects of a protein-sparing modified fast in the dietary management of Prader-Willi obesity. *New England Journal of Medicine.* 1977;296(14):774–779.

6. Boer H, Holland AJ, Whittington JE, Butler JV, Webb T, Clarke DJ. Psychotic illness in people with Prader-Willi syndrome due to chromosome 15 maternal uniparental disomy. *Lancet.* 2002;359:135–136.

7. Branson C. Speech and language characteristics of children with Prader-Willi syndrome. In: Holm VA, Sulzbacher S, Pipes PL, eds. *The Prader-Willi Syndrome.* Baltimore, MD: University Park Press; 1981:179–183.

8. Bray GA, Dahms W, Swerfdloff RTS, Fiser RH, Atkinson RL, Carrel RE. The Prader-Willi syndrome: a study of 40 patients and a review of the literature. *Medicine.* 1983;62(2):59–80.

9. Butler MG, Bittel DC, Kibiryeva N, Talebizadeh Z, Thompson T. Behavioral differences among subjects with Prader-Willi syndrome and type I or type II deletion and maternal disomy. *Pediatrics.* 2004;113(3): 565–573.

10. Butler MG, Carlson MG, Schmidt DE, Feurer ID, Thompson T. Plasma cholecystokinin levels in Prader-Willi and obese subjects. *American Journal of Medical Genetics.* 2000;95:67–70.

11. Butler MG, Dasouki M, Bittel D, Hunter S, Naini A, DiMauro S. Coenzyme Q10 levels in Prader-Willi syndrome: comparison with obese and non-obese subjects. *American Journal of Medical Genetics.* 2003; 119A(2):168–171.

12. Butler MG, Meaney FG, Palmer CG. Clinical and cytogenetic survey of 39 individuals with Prader-Labhart-Willi syndrome. *American Journal of Medical Genetics.* 1986;23:793–809.

13. Caldwell ML, Taylor RL. A clinical note on the food pereference of individuals with Prader-Willi syndrome: the need for empirical research. *Journal of Mental Deficiency Research.* 1983;30:347–354.

14. Cassidy S. Prader-Willi syndrome. *Current Problems in Pediatrics.* 1984;14: 1–55.

15. Clarke D. Psychopharmacology of severe self-injury associated with learning disabilities. *British Journal of Psychiatry.* 1998;172(b):385–381.

16. Clarke DJ, Boer H. Problem behaviors associated with deletion Prader-Willi, Smith-Magenis and cri du chat syndromes. *American Journal of Mental Retardation.* 1998;103:264–271.

17. Clarke DJ, Boer H, Chung MC, Sturmey P, Webb T. Maladaptive behavior in Prader-Willi syndrome in adult life. *Journal of Intellectual Disability Research.* 1996;40:159–165.

18. Clarke DJ, Boer H, Whittington J, Holland A, Butler J, Webb T. Prader-Willi syndrome, compulsive and ritualistic behaviors: the first population-based survey. *British Journal of Psychiatry.* 2002;189:358–362.

19. Clarke DJ, Webb T, Bachman-Clarke J. Prader-Willi syndrome and psychotic symptoms: report of a further case. *The Journal of Psychological Medicine.* 1995;12:27–29.

20. Constantino JN, Przybeck T, Friesen D, Todd R. Reciprocal social behavior in children with and without pervasive developmental disorders. *Journal of Developmental and Behavioral Pediatrics.* 2000;21:2–11.

21. Couper RTL, Raymond J, Couper JJ, Coulthard K, Haan E. Double blind placebo controlled trial of intranasal cholecystokinin octapeptide as a satiety agent in Prader Willi syndrome. *Journal of Paediatric and Child Health.* 1997;33:A7.

22. Crnic KA, Sulzbacher SJ, Snow J, Holm VA. Preventing mental retardation associated with gross obesity in the Prader-Willi syndrome. *Pediatrics.* 1980;66:787–789.

23. Curfs LMG, Fryns JP. Prader-Willi syndrome: a review with special attention to the cognitive and behavioral profile. *Birth Defects.* 1992;28(1):99–104.

24. Curfs LMG, Hoondert B, van Lieshout CFM, Fryns JP. Personality profiles of youngsters with Prader-Willi syndrome and youngsters attending regular schools. *Journal of Intellectual Disability Research.* 1995;39(3):241–248.

25. Curfs LMG, Verhulst FC, Fryns JP. Behavioral and emotional problems in youngsters with Prader-Willi syndrome. *Genetic Counseling.* 1991;2(1):33–41.

26. Curfs LMG, Wiegers A, Sommers JRM, Borghgraef M, Fryns JP. Strengths and weaknesses in the cognitive profile of youngsters with Prader-Willi syndrome. *Clinical Genetics.* 1991;40:430–434.

27. Dementieva Y, Shou Y, Cuccaro ML, et al. Linkage of autistic disorder to chromosome 15Q11–Q13 using phenotypic subtypes. *American Journal of Human Genetics.* 2003;72:529–538.

28. Descheemaeker MJ, Vogels A, Govers V, et al. Prader-Willi syndrome: new insights in the behavioural and psychiatric spectrum. *Journal of Intellectual Disability Research.* 2002;46(1):41–50.

29. Dimitropoulos A, Feurer ID, Butler MG, Thompson T. Emergence of compulsive behavior and tantrums in children with Prader-Willi syndrome. *American Journal of Mental Retardation.* 2001;106(1):39–51.

30. Dimitropoulos A, Feurer ID, Roof E, et al. Appetitive behavior, compulsivity and neurochemistry in Prader-Willi syndrome. *Mental Retardation and Developmental Disability Research Reviews.* 2000;6:125–130.

31. Donaldson MDC, Chu CE, Cooke A, Wilson A, Greene SA, Stephenson JBP. The Prader-Willi syndrome. *Archives of Disease in Childhood.* 1994;70:58–63.

32. Dunn HG. The Prader-Labhart-Willi syndrome: review of the literature and report of nine cases. *Acta Paediatrica Scandinavica Supplement.* 1968;186:3–38.

33. Dunn HG, Tze WJ, Alisharan RM, Schulzer M. Clinical experience with 23 cases of Prader-Willi syndrome. In: Holm VA, Sulzbacher SJ, Pipes P, eds. *Prader-Willi Syndrome.* Baltimore, MD: University Park Press; 1981:69–88.

34. Dykens E. Annotation: psychopathology in children with intellectual disability. *Journal of Child Psychology and Psychiatry.* 2000;41(4):407–417.

35. Dykens EM. Are jigsaw puzzle skills "spared" in persons with Prader-Willi syndrome? *Journal of the American Academy of Child and Adolescent Psychiatry.* 2002;43:343–352.

36. Dykens E, Cassidy S. Correlates of maladaptive behavior in children and adults with Prader-Willi syndrome. *American Journal of Medical Genetics.* 1995;60:546–549.

37. Dykens E, Cassidy S, King B. Maladaptive behavior differences in Prader-Willi syndrome due to paternal deletion versus maternal uniparental disomy. *American Journal of Mental Retardation.* 1999;104(1):67–77.

38. Dykens E, Hodapp R, Walsh K, Nash L. Adaptive and maladaptive behavior in Prader-Willi syndrome. *Journal of the American Academy of Child and Adolescent Psychiatry.* 1992;31(6):1131–1136.

39. Dykens E, Hodapp R, Walsh K, Nash L. Profiles, correlates, and trajectories of intelligence in Prader-Willi syndrome. *Journal of the American Academy of Child and Adolescent Psychiatry.* 1992;31(6):1125–1130.

40. Dykens EM, Kasari C. Maladaptive behavior in children with Prader-Willi syndrome, Down syndrome, and nonspecific mental retardation. *American Journal of Mental Retardation.* 1997;102:228–237.

41. Dykens E, Leckman J, Cassidy SB. Obsessions and compulsions in Prader-Willi syndrome. *Journal of Child Psychology and Psychiatry.* 1996;37: 995–1002.

42. Dykens E, Shah B. Psychiatric disorders in Prader-Willi syndrome: epidemiology and management. *CNS Drugs.* 2003;17(3):167–178.

43. Einfeld SL, Smith A, Durvasula S, Florio T, Tonge BJ. Behavior and emotional disturbances in Prader-Willi syndrome. *American Journal of Medical Genetics.* 1999;82:123–127.

44. Einfeld SL, Tonge BJ. *Manual for the Developmental Behavior Checklist (Primary Carer Version).* Melbourne, Australia: Monash University Centre for Developmental Psychiatry; 1994.

45. Einfeld S, Tonge B. The Developmental Behavior Checklist: the development and validation of an instrument to assess behavioral and emotional disturbance in children and adolescents with mental retardation. *Journal of Autism and Developmental Disorders.* 1995;25:81–104.

46. Ewald H, Mors O, Flint T, Kruse TA. Linkage analysis between manic depressive illness and the region on chromosome 15q involved in Prader-Willi syndrome, including two GABAA receptor subtype genes. *Human Heredity.* 1994;44(5):287–294.

47. Feurer ID, Dimitropoulous A, Stone WL, Roof E, Butler MG, Thompson T. The latent variable structure of the Compulsive Behaviour Checklist in people with Prader-Willi syndrome. *Journal of Intellectual Disability Research.* 1998;42(6):472–480.

48. Fieldstone A, Zipf WB, Sarter MF, Berntson CB. Food intake in Prader-Willi syndrome and controls with obesity after administration of benzodiazepine receptor agonist. *Obesity Research.* 1998;6:29–33.

49. Fox R, Yang G, Feurer ID, Butler MG, Thompson T. Kinetic form discrimination in Prader-Willi syndrome. *Journal of Intellectual Disability Research.* 2001;45(4):317–325.

50. Gilmour J, Skuse D, Pembrey M. Hyperphagic short stature and Prader-Willi syndrome: a comparison of behavioral phenotypes, genotypes and indices of stress. *British Journal of Psychiatry.* 2001;179:129–137.

51. Glover D, Maltzman I, Williams C. Food preferences among individuals with and without Prader-Willi syndrome. *American Journal of Mental Retardation.* 1996;101:195–205.

52. Goldberg D, Garrett C, Van Riper C, Warzak W. Coping with Prader-Willi syndrome. *Journal of the American Dietetic Association.* 2002;102(4):537–542.

53. Greenswag LR. Adults with Prader-Willi syndrome: a survey of 232 Cases. *Developmental Medicine and Child Neurology.* 1987;29:145–152.

54. Gross-Tsur B, Landau Y, Benarroch F, Wertman-Elad R, Shalev R. Cognition, attention, and behavior in Prader-Willi syndrome. *Journal of Child Neurology.* 2001;16(4):288–290.

55. Hall BD, Smith D. Prader-Willi syndrome: a resume of 32 cases including an instance of affected first cousins one of whom is of normal stature and intelligence. *Journal of Pediatrics.* 1972;81(2):286–293.

56. Haqq AM, Stadler DD, Rosenfeld RG, et al. Circulating ghrelin levels are suppressed by meals and octreotide therapy in children with Prader-Willi syndrome. *Journal of Clinical Endocrinology and Metabolism.* 2003;88(8): 3573–3576.

57. Holland AJ, Treasure J, Coskeran P, Dallow J, Milton N, Hillhouse E. Measurement of excessive appetite and metabolic changes in Prader-Willi syndrome. *International Journal of Obesity.* 1993;17:527–532.

58. Holland AJ, Whittington J, Butler J, Webb T, Boer H, Clarke D. Behavioural phenotypes associated with specific genetic disorders: evidence from a

population based study of people with Prader-Willi syndrome. *Psychological Medicine.* 2003;33:141–153.

59. Holland A, Whittington J, Hinton E. The paradox of Prader-Willi syndrome: a genetic model of starvation. *Lancet.* 2003;362(9388):989–991.

60. Holm V. The diagnosis of Prader-Willi syndrome. In: Holm V, Sulzbacher S, Pipes P, eds. *Prader-Willi Syndrome.* Baltimore, MD: University Park Press; 1981:27–44.

61. Holm V, Cassidy S, Butler MG, et al. Prader-Willi syndrome: consensus diagnostic criteria. *Pediatrics.* 1993;91:398–402.

62. Holm V, Pipes P. Food and children with Prader-Willi syndrome. *American Journal of Diseases of Children.* 1976;130:1063–1067.

63. Jancar J. Prader-Willi syndrome (hypotonia, obesity, hypogonadism, growth and mental retardation). *Journal of Mental Deficiency Research.* 1971;15:20.

64. Joseph B, Egli M, Koppekin A, Thompson T. Food choice in people with Prader-Willi syndrome: quantity and relative preference. *American Journal of Mental Retardation.* 2002;107(2):128–135.

65. Joseph B, Egli M, Sutcliffe J, Thompson T. Possible dosage effect of maternally expressed genes on visual recognition memory in Prader-Willi syndrome. *American Journal of Medical Genetics.* 2001;105:71–75.

66. Kleppe SA, Katayama KM, Shipley KG, Foushee DR. The speech and language characteristics of children with Prader-Willi Syndrome. *Journal of Speech and Hearing Disorders.* 1990;55:300–309.

67. Kollrack HW, Wolff D. [Paranoid-hallucinatory psychosis in the Prader-Labhart-Willi-Fanconi syndrome]. *Acta Paedopsychiatrica.* 1966;33:309–314.

68. Lee, P. Model for a peripheral signaling defect in Prader-Willi syndrome. In: Eiholzer U, l'Allemand D, Zipf W, eds. *Prader-Willi Syndrome as a Model for Obesity.* Basel, Switzerland: Karger; 2003:70–81.

69. Lewis B, Freebairn L, Heeger S, Cassidy S. Speech and language skills of individuals with Prader-Willi syndrome. *American Journal of Speech Language Pathology.* 2002;11:285–294.

70. Lindgren AC, Barkeling B, Hagg A, Ritzen M, Marcus C, Rossner S. Eating behavior in Prader-Willi syndrome, normal weight, and obese control groups. *Journal of Pediatrics.* 2000;137(1):50–55.

71. Prader A, Labhart A, Willi H. Ein syndrom von adipositas, kleinwuchs, kryptorchismus und oligophrenie nach myatonieartigem zustand im neugeborenalter. *Schweizerische Medizinische Wochenschrift.* 1956;86:1260–1261.

72. Roof E, Stone W, MacLean W, Feurer ID, Thompson T, Butler MG. Intellectual characteristics of Prader-Willi syndrome: comparison of genetic subtypes. *Journal of Intellectual Disability Research.* 2000;44 (Pt 1):25–30.

73. Smith D. *Recognizable Patterns of Human Malformation: Genetic,Embryologic and Clinical Aspects.* 2nd ed. Philadelphia, PA: W.B. Saunders; 1976.

74. State M, Dykens E, Rosner B, Martin A, King B. Obsessive-compulsive symptoms in Prader-Willi and "Prader-Willi-like" patients. *Journal of the American Academy of Child and Adolescent Psychiatry.* 1999;38(3):329–334.

75. Stauder JE, Brinkman MJ, Curfs LMG. Multi-modal P3 deflation of event-related brain activity in Prader-Willi syndrome. *Neuroscience Letters.* 2002;327(2):99–102.

76. Stein DJ, Hutt CS, Spitz JL, Hollander E. Compulsive picking and obsessive-compulsive disorder. *Psychosomatics.* 1993;34(2):177–181.

77. Steinhausen HC, Eiholzer U, Hauffa B, Malin Z. Behavioural and emotional disturbances in people with Prader-Willi syndrome. *Journal of Intellectual Disability Research.* 2004;48(1):47–52.

78. Steinhausen HC, Gontard A, Spohr HL, et al. Behavioral phenotypes in four mental retardation syndromes: fetal alcohol syndrome, Prader-Willi syndrome, fragile X syndrome, and tuberosis sclerosis. *American Journal of Medical Genetics.* 2002;111:381–387.

79. Sullivan K, Tager-Flusberg H. Higher-order mental state understanding in adolescents with Prader-Willi syndrome. *The Endocrinologist.* 2000; 10(4)Suppl 1:38S–40S.

80. Sulzbacher SJ, Crnic K, Snow J. Behavioral and cognitive disabilities in Prader-Willi syndrome. In: Sulzbacher S, Holm V, Pipes PL, eds. *Prader-Willi Syndrome.* Baltimore, MD: University Park Press; 1981.

81. Swaab DF. Alterations in the hypothalamic paraventricular nucleus and its oxytocin neurons (putative satiety cells) in Prader-Willi syndrome: a study of five cases. *Journal of Clinical Endocrinology and Metabolism.* 1995;80:573–379.

82. Tager-Flusberg H, Sullivan K. A componential view of theory of mind: evidence from Williams syndrome. *Cognition.* 2000;76:59–89.

83. Taylor RL. Cognitive and behavioral characteristics. In: Caldwell ML, Taylor R, eds. *Prader-Willi Syndrome: Selected Research and Management Issues.* New York, NY: Springer-Verlag; 1988:29–42.

84. Taylor R, Caldwell ML. Psychometric performances of handicapped obese individuals with and without Prader-Willi syndrome. Paper presented at the Meeting of the American Association on Mental Deficiency, Dallas, TX; November 1983.

85. Taylor RL, Caldwell ML. Type and strength of food preferences of individuals with Prader-Willi syndrome. *American Journal of Mental Deficiency.* 1985;29:109–112.

86. van Lieshout CFM, De Meyer R, Curfs L, Fryns JP. Family contexts, parental behaviour and personality profiles of children and adolescents with Prader-Willi, fragile-X, or Williams syndrome. *Journal of Child Psychology and Psychiatry.* 1998;39(5):699–710.

87. Verhoeven WMA, Curfs L, Tuinier S. Prader-Willi syndrome and cycloid psychoses. *Journal of Intellectual Disability Research.* 1998;42(6):455–462.

88. Verhoeven WMA, Tunier S, Curfs LMG. Prader-Willi psychiatric syndrome and velo-cardio-facial psychiatric syndrome. *Genetic Counseling.* 2000;11(3):205–213.

89. Vogels A, Descheemaeker MJ, Govers V, Fryns JP. Intravert and extravert behaviors: two different behavioural types in children with the Prader-Willi syndrome. Proceedings, Fifth International and Ninth Australasian Prader-Willi Syndrome Scientific and General Conference, Christchurch, NZ; April 2004:72.

90. Warren HL, Hunt E. Cognitive processing in children with Prader-Willi syndrome. In: Holm VA, Sulzbacher SJ, Pipes PL, eds. *The Prader-Willi Syndrome.* Baltimore, MD: University Park Press; 1981:161–178.

91. Waters J, Clarke DJ, Corbett JA. Educational and occupational outcome in Prader-Willi syndrome. *Child Care, Health and Development.* 1990;16: 271–282.

92. Wenger SL, Hanchett JM, Steel MW, Maier BV, Golden WL. Clinical comparison of 598 Prader-Willi patients with and without the 15(q12) deletion. *American Journal of Medical Genetics.* 1987;28:881–887.

93. Whitman BY. Understanding and managing the behavioral and psychological components of Prader-Willi syndrome. *Prader-Willi Perspectives.* 1995;3(3):3–11.

94. Whitman BY. Prader-Willi syndrome and autistic disorder. Paper presented at the Fourth International Prader-Willi Syndrome Association Conference, Minneapolis, MN; June 2001.

95. Whitman B, Accardo P. Emotional symptoms in Prader-Willi syndrome adolescents. *American Journal of Medical Genetics.* 1987;28(12): 897–905.
96. Whitman B, Accardo P. Prader-Willi syndrome as a three stage disorder. *Journal of the Royal Society of Medicine.* 1989;82(7):448. (letter)
97. Whittington JE, Holland A, Webb T, Butler J, Clarke D, Boer H. Academic underachievement by people with Prader-Willi syndrome. *Journal of Intellectual Disability Research.* 2004;48(2):188–200.
98. Whittington JE, Holland A, Webb T, Butler J, Clarke D, Boer H. Cognitive abilities and genotype in a population-based sample of people with Prader-Willi syndrome. *Journal of Intellectual Disability Research.* 2004;48(2): 172–187.
99. Zellweger H, Schneider G, Johannsson H. Syndrome of hyptonia-hypomentia-hypogonadism-obesity (HHHO) or Prader-Willi syndrome. *American Journal of Diseases of Children.* 1968;115:558–598.
100. Zipf WB, Berntson GG. Characteristics of abnormal food-intake patterns in children with Prader-Willi syndrome and study of effects of nalozone. *American Journal of Clinical Nutrition.* 1987;46(2):277–281.

9

Speech and Language Disorders Associated with Prader-Willi Syndrome

Barbara A. Lewis

The speech and language skills of individuals with Prader-Willi syndrome (PWS) differ greatly in the severity and type of deficits that they present, ranging from individuals who are nonverbal to those who acquire normal speech and language skills by adulthood. Due to the low incidence of the disorder—1 in 10,000 to 15,000 individuals[4,5]—professionals such as speech-language pathologists, physical therapists, and occupational therapists may encounter only a few individuals with Prader-Willi syndrome in their practice. An understanding of the characteristics of Prader-Willi syndrome that may impact speech and language abilities will allow the professional to evaluate each individual for potential contributing factors to communication deficits and to plan appropriate intervention strategies. The goals of this chapter are to summarize the existing literature on the speech and language skills of those with PWS, to describe features of PWS that may contribute to speech and language disorders, to chart a developmental course of the speech and language skills, and to suggest possible intervention strategies.

Speech and Language Characteristics

The speech and language skills of individuals with Prader-Willi syndrome are reported to be below expectations based on intellectual levels.[3,11,14,18] Although great variability exists in the speech and language skills of individuals with Prader-Willi syndrome, several common features have been noted. These include poor speech-sound development, reduced oral motor skills, and language deficits. Speech is often characterized by imprecise articulation, hypernasality, flat intonation patterns, an abnormal pitch, and a harsh voice quality. Prosody, or the melody of speech, may also be disrupted. A slow rate of speech, typical of a flaccid dysarthria, may be observed. In addition to these speech difficulties, the individual with Prader-Willi syndrome may have language problems. Language problems include deficits in vocabulary, grammar, morphology, narrative abilities, and pragmatics. Table 9.1 illustrates the clinical features of PWS and their potential

Table 9.1. Clinical Characteristics of Prader-Willi Syndrome and Potential Impact on Speech and Language

Clinical Feature	Impact on Speech/Language
Mouth	
• narrow overjet • narrow palatal arch • micrognathia →	reduced articulatory skills
Larynx	
• altered growth due to endocrine dysfunction →	pitch variations
Dentition	
• decay due to reduced saliva output and enamel hypoplasia →	reduced articulatory skills
Hypotonia	
• poor velopharyngeal movement → • stretching of muscles of larynx → • slow movement of articulators →	hypernasality or hyponasality vocal pitch and quality variations imprecise articulation and slow rate of speech
Cognitive	
• mental retardation • sequencing problems →	delayed receptive/expressive language poor narrative skills
Behavioral Disturbances	
• temper tantrums • stubbornness → • manipulative behavior • depression • emotional lability • compulsive behavior • skin picking • difficulty relating to peers • poor social relationships • difficulty detecting social cues • argumentative	poor pragmatic skills

Source: B. A. Lewis, L. Freebairn, S. Heeger, and S. B. Cassidy, "Speech and Language Skills of Individuals with Prader-Willi Syndrome," *American Journal of Speech-Language Pathology*, 2002;11:285–294.[16] Copyright by the American Speech-Language-Hearing Association. Reprinted with permission.

impact on speech and language abilities. The following section will summarize findings of each speech and language domain.

Speech-Sound Development

A distinguishing feature of PWS is poor speech-sound development.[10] Speech-sound disorders include both errors of articulation or phonetic structure (errors due to poor motor abilities associated with the production of speech-sounds) and phonological errors (errors in applying linguistic rules to combine sounds to form words). Individuals with PWS may exhibit deficits in either articulation or phonology or both. Several factors may account for the poor speech-sound development of individuals with PWS, including oral structure abnormalities, abnormal saliva,[12,23] hypotonia, poor phonological skills, and cognitive deficits. Reduced breath support for speech may also be noted. Oral structures of the mouth and jaw that may impact articulation skills include a narrow overjet, a narrow palatal arch, and micrognathia. However, it is more likely that poor oral motor skills, especially reduced tongue elevation for speech and slower alternating movements of the articulators, account for poor speech-sound skills in PWS.[1,3,14] Speech characteristics are often similar to those reported for flaccid dysarthria.[14] Speech-sound errors that have been noted in individuals with PWS include sound distortions and omissions, vowel errors, simplification of consonant blends, and difficulty sequencing syllables. Phonemes that are motorically complex, such as /s/, /r/, /sh/, and blends are usually the most difficult.[14] As with most speech-sound disorders, single word utterances are often more intelligible than conversational speech, and the phonetic environment of the target sound can greatly influence speech intelligibility. For example, an individual might have difficulty producing the "r" sound in the word *crab* when it's preceded by the "qu" sound in the phrase "quiet crab's claws."

Other authors[10] have postulated that poor speech-sound development is the result of poor phonology skills, a component of a more general language deficit in PWS. Downey and Knutson[8] report that most individuals with PWS present with delayed speech-sound development characterized by phonological patterns typical of younger, normally developing children. However, some individuals demonstrate atypical patterns such as a phonological disorder or an apraxia of speech.[18]

Apraxia of speech is a severe speech-sound disorder that includes impairments in syllable sequencing, prosody, and speech-sound characteristics. Although the etiology of apraxia of speech is not well understood, it is presumed to result from impairment in the motor programming aspects of speech-sound production.[20] Rare cases of apraxia of speech have been reported in individuals with PWS. Children with apraxia of speech often do not develop intelligible speech until well into school age. Augmentative/alternative communication intervention (AAC—e.g., sign language and communication boards) may be employed to eliminate some of the frustration that the individual experiences in communication. AAC systems allow the individual to build vocabulary and pragmatic language skills while oral speech skills are developing.

Voice Characteristics

The voice of the individual with PWS may differ in pitch, quality, intensity, and resonance from that expected for his/her age and gender. Voice characteristics reported for individuals with PWS include a high-pitched voice, harsh/hoarse voice quality, inadequate vocal intensity, and hypernasality.[1,8] Hypotonia and altered growth of the larynx may result in a pitch that is too high or low. Nasal resonance may be disrupted by sluggish velopharyngeal movement and/or inadequate velopharyngeal closure potentially due to hypotonia. Poor velopharyngeal functioning may result in hypernasality, nasal emission, nasal snorting, weak plosive consonant sounds, and unusual manner of sound productions. Although hypernasality is most frequently reported in individuals with PWS, hyponasality has also been noted.[16] Growth hormone therapy, sometimes utilized with PWS individuals, may also affect voice characteristics.[15] Surgical procedures such as those employed for children with cleft palates (e.g., a pharyngeal flap) may reduce hypernasality and improve speech intelligibility. It should be noted, however, that often speech-sound errors persist even though hypernasality has been reduced.

Fluency

Fluency disorders (stuttering and cluttering) are not frequently observed in individuals with PWS.[8] However, to date, a systematic study of the fluency and prosody characteristics of children with PWS has not been reported. Clinical observations of conversational speech suggest that interjections, revisions, and word repetitions may be related to cognitive and language deficits.[14] A slow rate due to poor oral motor skills and a monotone may disrupt the flow and melody of speech.

Language Skills

Individuals with PWS frequently demonstrate poor receptive and expressive language skills, with expressive language more impaired than receptive skills.[3,14,18] Analysis of conversational speech samples indicates that individuals with PWS employ a shorter mean length of utterance (MLU) than their peers. Several authors have described patterns of cognitive strengths and weaknesses frequently observed in PWS that might impact language abilities.[9] Specific deficits have been reported in auditory short-term memory,[9] linear or temporal order processing, and auditory verbal processing skills.[7] Poor speech-sound development may also affect language skills. For example, the acquisition of grammatical markers (morphology) may be delayed both because the child cannot produce the /s/ phoneme to form the plural and because the child does not understand the concept of plurals.[8]

Few studies have examined narrative skills of individuals with PWS. Narratives are accounts of events either real or imaginary. Narratives include storytelling, scripts, schema, and episodic memory. Narratives contain a chronological sequence of events and causal relationships. For an individual to use narratives successfully, he/she must form a topic-centered story, use specific vocabulary, sequence events within the story, describe relationships between people and events, and use correct story grammar. Story grammar includes a setting, a beginning, reaction, goal, attempt, outcome, and ending. Narrative skills are essential to social development as they promote conversational skills. In addition, good narrative skills are necessary for academic success as they promote reading and writing, develop organizational skills, and build linguistic abilities.

Clinical observation suggests that both children and adults with PWS have great difficulty with story retelling tasks. One study[17] examined the narrative abilities of 19 individuals with PWS, ages 3 to 30 years. A simple story entitled "The Fox and Bear" was read, and the individual was asked to retell the story. Participants showed deficits in recalling grammar elements and content items and had difficulty answering both factual and inferential questions about the story. While poor language skills may account for some difficulty with narratives, deficits in other cognitive skills such as temporal sequencing abilities, auditory short-term memory, and poor auditory processing skills may also contribute.[7,9] While narrative skills appear to develop into adulthood, the narrative abilities of the individual with PWS lag behind other language skills. Poor narrative skills may contribute to deficient conversational skills in adolescents and adults with PWS and thus impact social and job-related communication skills.

Pragmatic deficits, including problems with maintaining a topic, judging appropriate proximity to the conversational partner, and turn taking, have also been observed.[8] Pragmatic skills may be influenced by a number of behavioral disturbances frequently noted in individuals with PWS.[22] For example, temper tantrums, compulsive behavior, and skin picking may interfere with peer relationships. Children with PWS have difficulty with social relationships. Sullivan and Tager-Flusberg[21] demonstrated that children with PWS were less likely than IQ-matched children with Williams syndrome to show appropriate empathetic responses. A study of 30 adults with PWS reported perseverative speech.[6] Such pragmatic deficits may impede progress in therapy. Pragmatic language skills may vary by activity, routine, and environment. In adulthood, poor pragmatic language skills may create difficulties in the workplace and with interpersonal relationships.

Written Language Skills

Surprisingly, despite oral language deficits, individuals with PWS show relative strengths in written language skills. Strengths associated with PWS include vocabulary knowledge and reading decoding (i.e., sounding out words).[9] However, some individuals may present with poor reading comprehension skills possibly due to language deficits as described above. Visual spatial skills that have been reported as a relative strength[9] may also contribute to good reading decoding ability. However, variability has been noted in these patterns of strengths and weaknesses. Curfs, Wiegers, Sommers, Borghgraef, and Fryns[7] reported that 10 of 26 subjects with PWS had performance IQs at least 15 points higher than their verbal IQs, 3 had verbal IQs at least 15 points higher than their performance IQs, and 13 subjects did not show any discrepancy. In summary, individuals with PWS present with speech, language, and cognitive deficits that impede their communication skills.

Developmental Course of Speech and Language Skills in PWS

The speech-language pathologist may become involved with the child with PWS soon after birth. In infancy, children with PWS present with a weak cry and early feeding difficulties most likely due to hypotonia. Reduced babbling and signs of early language delay are often observed.

All children begin acquiring word understanding essentially from birth (receptive language). Among typically developing children, expressive language follows soon after with cooing at 3 months, babbling at 6 months, and consonants in the form of "dada" and "mama" at around 8 months. For most, by 10 months of age, "dada" and "mama" are used discriminately, and there is evidence that the child understands the meaning of the word "no"; by 12 months of age, most children have acquired at least two words in addition to "dada" and "mama."

By contrast, children with PWS are 18 months of age before they begin to verbally evidence a vocabulary, combine words, and develop early syntax. A substantial number of affected children are much later in acquiring speech; some may be as late as 6 years of age. Oral motor skills remain poor and children exhibit many speech-sound errors that result in unintelligible speech. Pragmatic difficulties may be noted due to poor social skills and the emergence of behavioral disturbances. If the child's speech is highly unintelligible, AAC may be considered, including sign language or communication boards. AAC is usually transitional until oral speech abilities improve, and it alleviates some of the frustration that the child and caretaker may experience.

At school age (6 to 12 years), children with PWS are usually enrolled in speech and language therapy through the school. Articulation errors remain, with children less intelligible in connected speech than in single words. Receptive and expressive language skills lag behind those of their peers. Voice problems, as described above, may be observed as the child produces longer utterances. As noted previously, reading decoding emerges as a relative strength and children often become fluent readers. However, language difficulties may result in poor reading comprehension.

In adolescence and adulthood, the individual with PWS may continue to demonstrate communication difficulties including residual articulation errors, vocabulary deficits, poor conversational and pragmatic skills, and inappropriate pitch. Some individuals with PWS do achieve normal articulation skills. The individual with PWS may exhibit behavioral traits that are disruptive to good communication skills such as inappropriate laughter. Continued work on conversational speech is essential to adjustment and success in the workplace. Emphasis should be placed on functional language skills and life-skills training.

Summary of a Clinical Research Study

Previous research on the speech and language skills of individuals with PWS has been based on a small number of individuals and has not examined the developmental course of the speech and language disorder. Many studies have not distinguished between speech-sound errors due to poor oral motor skills and structural deviations and errors due to phonological deficits. Further, studies have not attempted to associate speech and language characteristics to a particular chromosome 15 abnormality (i.e., paternal deletion, uniparental disomy, or a transloca-tion). A study of the speech and language abilities of a relatively large cohort of individuals with PWS representing three age groups (infant/preschool, school-age, and adolescent/adult) was undertaken. The details of this study are reported elsewhere (see Lewis, Freebairn, Heeger, and Cassidy[16]). The findings of this study are summarized below to illustrate the variability and range of speech and language skills found in individuals with PWS.

Participants

The participants were 32 individuals (16 males and 16 females), ages 6 months to 42 years. All met the diagnostic criteria for PWS,[13] and diagnoses were confirmed by chromosomal analysis.

Measures

Measures were selected to assess oral motor skills, articulation and phonology, receptive and expressive language skills, prosody/voice characteristics, reading, and narrative abilities. Assessments included standardized tests as well as a spontaneous speech sample analysis. Standardized tests varied according to the age and intellectual abilities of the individual.

A rating scale was adopted to summarize data across various age and skill levels. The speech and language skills of each subject were rated independently by two licensed and certified speech-language pathologists. Receptive and expressive language, articulation, oral motor skills, fluency, narrative ability, and reading skills were scored as normal, mildly impaired, moderately impaired, or severely impaired. Pitch was rated as normal, high, or low for a participant's age and gender, based on criteria proposed by Boone and McFarlane.[2] Voice quality was rated as normal, soft, harsh, hoarse, or strained. Resonance characteristics of speech were rated as normal, hypernasal, or hyponasal. The rate of speech was classified as slow, normal, or fast. A monotone quality to the spontaneous speech sample was also noted.

Results and Conclusions

All participants reported a history of communication difficulties and/or were enrolled in speech and language therapy. The majority of children (83%) received speech therapy prior to the age of 3 years, all had received therapy during the school-age years, and two (20%) con-tinued to receive speech therapy into adolescence and adulthood. This suggests that therapy needs for the individual with PWS are identified early and required for most individuals across the life span.

Oral motor deficits and associated speech-sound disorders are preva-lent in the PWS population, with 90.6% demonstrating mild to severe oral motor deficits including poor tongue mobility, shortness of palate, and incoordination of the articulators (see Table 9.2). Mild to severe articulation impairment was observed in 92% of the participants, with younger subjects more severely impaired. A variety of sound

Table 9.2. Speech Characteristics and Deficits in Individuals with Prader-Willi Syndrome

Speech Characteristic	Ratings of Speech Characteristics Percentage of Subjects (N)			
	Normal	Mild	Moderate	Severe
Oral Motor Skills* N = 29	9.4%	31.3%	31.3%	18.8%
Articulation Skills N = 25	8%	12%	36%	44%

* 9.2% of participants could not be tested but did have oral motor difficulties.

substitutions and distortions as well as phonological processing errors were noted. This supports previous suggestions that the speech-sound errors observed in PWS are the result of both poor oral motor skills and concomitant language deficits.

As predicted, receptive and expressive language deficits based on age normative data were observed in the majority of individuals, with 90.5% presenting with receptive language delays and 91.7% presenting with expressive language delays. In addition, vocabulary, pragmatic, and narrative deficits were observed. However, most of the participants who were school-age or older were able to read fluently (83.3%). Reading comprehension was not assessed. Future studies should examine reading comprehension abilities relative to reading decoding skills.

Ratings of pitch and nasality revealed great variability. Thirty-five percent of the participants presented with a high pitch, 30% with a low pitch, and 35% with a pitch appropriate for their age and gender (see Table 9.3). While hypernasality was frequently observed (70.6%), hyponasality was also noted (17.6%). It is not known whether or not the same factors that contributed to the hypernasality (hypotonia and oral structure) also contributed to the hyponasality that was observed. Further research is needed in this area.

Comparisons between age-matched individuals with uniparental disomy (UPD) and individuals with chromosome 15q deletions were inconclusive. On two of the speech-language measures UPD subjects received better ratings than did the deletion subjects; on two other measures the deletion subjects received better ratings than did the UPD

Table 9.3. Voice Characteristics in Individuals with Prader-Willi Syndrome

Voice Characteristic	Ratings of Voice Characteristics Percentage of Subjects (N)		
Pitch N = 20	Normal 35%	Low 30%	High 35%
Resonance N = 17	Normal 11.8%	Hyponasal 17.6%	Hypernasal 70.6%

subjects; and in one case the ratings for the UPD subjects were comparable. Further research is needed to draw associations between the type of chromosome abnormality and the clinical presentation.

Therapeutic Implications

The individual with PWS will require the services of a speech-language pathologist from infancy through adulthood. A team approach that includes an occupational therapist, physical therapist, dietitian, psychologist, physician, speech-language pathologist, genetic counselor, social worker, and educational specialist provides the optimal management strategy for the child or adult with PWS. Early intervention begins in infancy with a focus on improving oral motor skills for feeding. Continued monitoring of speech and language skills is important as the degree of hypotonia changes over time.[18] Assessment includes standardized and nonstandardized measures to assess oral structures and functions, speech-sounds, and receptive/expressive language skills. Later on, with the development of conversational speech, voice, fluency, and resonance characteristics may be assessed.

It is essential that the speech-language pathologist be aware of the unique characteristics of the syndrome that may impact on speech-language development. For example, in some children drooling is a sign of poor oral motor control. However, children with PWS seldom drool due to reduced saliva output. The speech-language pathologist may incorrectly assume that oral motor skills are intact since drooling is not observed. Further, reduced saliva output may cause dental decay, thus impairing articulation. Table 9.1 summarizes some of the characteristics of PWS that may contribute to speech and language impairment.

In addition to articulation and language therapy, intervention should emphasize social skills and the pragmatic use of language.[8] As shown in Table 9.1, many of the behavioral characteristics associated with PWS impede good pragmatic language ability. Early and ongoing training of social skills will assist the individual with PWS in maintaining appropriate social and interpersonal interactions.

Caretakers and professionals should also be aware of the wide range of communication deficits that are associated with PWS. Therapy should be tailored to address the specific speech and language deficits observed, rather than employing a cookbook approach. Therapy should include an emphasis on the development of oral motor skills. Imitation of movements of the tongue, lips, jaws, and palate may be incorporated into games (see Orr[19] for oral motor games for children). Oral motor skills may be trained in both speech-sound and nonspeech activities. Isolated movements may be mastered first, followed by sequential and motorically complex movements of the articulators.

Future Directions

The speech and language skills of individuals with PWS have not been as well described as those of syndromes with a higher prevalence, such as Down syndrome. While individual case studies have been useful in outlining some of the characteristics of the speech and language disorders associated with PWS, such case studies have failed to describe the great variability of these skills found in PWS. Larger cohort studies are needed to understand the range of speech and language skills that individuals with PWS present. Specific cognitive strengths and weaknesses associated with PWS may impact the speech and language skills of an individual with PWS in a unique way. Comparison groups of children with similar IQ ranges may be employed to highlight the distinctive aspects of PWS. Therapists and other professionals should be acquainted with the features of PWS that may potentially influence communication development. Therapy programs designed for children without PWS may not be appropriate for the child with PWS.

As new medical treatments are employed with the PWS population, such as growth hormone treatment, continued research is needed to determine its impact on speech and language. Therapy strategies may be modified to augment these medical interventions.

Glossary

apraxia (of speech)—nonlinguistic sensorimotor disorder of articulation, characterized by impaired capacity to program position of speech musculature and sequence of muscle movements for the volitional production of phonemes.

articulators—the teeth, lips, and tongue, as they are involved in the production of meaningful sounds.

cluttering—speech characterized by overuse of fillers, rapid rate, and word and phrase repetitions. Unlike stuttering, the individual is usually unaware of the difficulty.

flaccid dysarthria—faulty speech production due to motor difficulties resulting from hypotonic (decreased) muscle tone, characterized by imprecise consonants and irregular articulation. Respiration, voice, fluency, and prosody (melody of speech) may be hindered as well. Both volitional and automatic actions, including chewing, swallowing, and other oral motor movements, may also be impaired. Anarthria, or the inability to articulate at all, is the result of severe neuromuscular involvement.

interjections—the insertion of extra sounds or words that do not add to or modify the meaning of the sentence, such as "you know" or "like."

larynx—the upper part of the trachea (windpipe); contains the vocal cords.

micrognathia—a small jaw.

morphology—the form and internal structure of words; the transformation of words in such ways as tense and number.

nasal emissions—airflow directed via the nasal cavity that passes out the nose rather than the more normal route, through the oral cavity.

nasal snorting—airflow directed into the nasal cavity producing a snorting sound.

pharyngeal flap—a surgical procedure designed to correct velopharyngeal insufficiency.

phoneme—the smallest unit of sound in any particular language; the English language designates approximately 44 different phonemes.

plosive consonants—p, b, t, d, k, and g.

velopharyngeal—of or relating to the structures of the soft palate and the pharynx.

References

1. Akefeldt A, Akefeldt B, Gillberg C. Voice, speech and language characteristics of children with Prader-Willi syndrome. *Journal of Intellectual Disability Research.* 1997;41(4):302–311.
2. Boone DR, McFarlane SC. *The Voice and Voice Therapy.* 4th ed. Englewood Cliffs, NJ: Prentice Hall; 1998.
3. Branson C. Speech and language characteristics of children with Prader-Willi syndrome. In: Holm VA, Sulzbacher S, Pipes PL, eds. *The Prader-Willi Syndrome.* Baltimore, MD: University Park Press; 1981:179–183.
4. Burd L, Veseley B, Martsolf J, Kerbeshian J. Prevalence study of Prader-Willi syndrome in North Dakota. *American Journal of Medical Genetics.* 1990;37:97–99.
5. Cassidy SB. Prader-Willi syndrome. *Journal of Medical Genetics.* 1997;34:917–923.
6. Clarke DJ, Boer H, Chung MC, Sturmey P, Webb T. Maladaptive behavior in Prader-Willi syndrome in adult life. *Journal of Intellectual Disability Research.* 1996;40(2):159–165.
7. Curfs LM, Wiegers AM, Sommers JR, Borghgraef M, Fryns JP. Strengths and weaknesses in the cognitive profile of youngsters with Prader-Willi syndrome. *Clinical Genetics.* 1991;40(6):430–434.
8. Downey DA, Knutson CL. Speech and language issues. In: Greenswag LR, Alexander RC, eds. *Management of Prader-Willi Syndrome.* 2nd ed. New York, NY: Springer-Verlag; 1995:142–155.
9. Dykens EM, Hodapp RM, Walsh K, Nash LJ. Profiles, correlates, and trajectories of intelligence in Prader-Willi syndrome. *Journal of the American Academy of Child and Adolescent Psychiatry.* 1992;31(6):1125–1130.
10. Dyson AT, Lombardino LJ. Phonological abilities of a preschool child with Prader-Willi syndrome. *Journal of Speech and Hearing Disorders.* 1989;54:44–48.
11. Edmonston NK. Management of speech and language impairment in a case of Prader-Willi syndrome. *Language, Speech, and Hearing Services in the Schools.* 1982;13:241–245.
12. Hart S, Poshva C. Salivary abnormalities in Prader-Willi syndrome. PWSA (USA) 16th Annual Scientific Conference. Atlanta, GA; 1994.
13. Holm VA, Cassidy SB, Butler MG et al. Prader-Willi syndrome: consensus diagnostic criteria. *Pediatrics.* 1993;91(2):398–402.
14. Kleppe SA, Katayama KM, Shipley KG, Foushee DR. The speech and language characteristics of children with Prader-Willi syndrome. *Journal of Speech and Hearing Disorders.* 1990;55:300–309.

15. Lee PDK, Allen DB, Angulo MA et al. Consensus statement—Prader-Willi syndrome: growth hormone GH/insulin-like growth factor axis deficiency and GH treatment. *The Endocrinologist.* 2000;10(4)s:71S–74S.

16. Lewis BA, Freebairn LA, Heeger S, Cassidy SB. Speech and language skills of individuals with Prader-Willi syndrome. *American Journal of Speech and Language Pathology.* 2002;11:285–294.

17. Lewis BA, Freebairn LA, Sieg FL, Cassidy SB. Narrative skills of individuals with Prader Willi syndrome. PWSA (USA) 15th Annual Scientific Conference. Pittsburgh, PA; 2000.

18. Munson-Davis JA. Speech and language development in Prader-Willi syndrome. In: Greenswag LR, Alexander RC, eds. *Management of Prader-Willi Syndrome.* New York, NY: Springer-Verlag; 1988:124–133.

19. Orr C. *Mouth Madness, Oral Motor Activities for Children.* San Antonio, TX: The Psychological Corporation; 1998.

20. Shriberg LD, Aram DM, Kwiatkowski J. Developmental apraxia of speech: I descriptive perspectives. *Journal of Speech Language and Hearing Research.* 1997;40:273–285.

21. Sullivan K, Tager-Flusberg H. Empathetic responses of young children with Prader-Willi syndrome to the distress of others. PWSA (USA) 13th Annual Scientific Conference. Columbus, OH; 1998.

22. Thompson T, Butler MG, MacLean WE, Joseph B. Prader-Willi syndrome: genetics and behavior. *Peabody Journal of Education.* 1996;71(4):187–212.

23. Young W. Syndromes with salivary dysfunction predispose to tooth wear: case reports of congenital dysfunction of major salivary glands, Prader-Willi, congenital rubella, and Sjogren's syndromes. *Oral Surgery, Oral Medicine, Oral Pathology, Oral Radiology and Endodontics.* 2001;92(1):38–48.

10

Motor and Developmental Interventions

Toni Goelz

What is physical therapy, and why is it so important that physical therapy be included as a component of the multidisciplinary approach to managing the care of the individual with Prader-Willi syndrome? Physical therapy, as a profession, assumes a role in the "diagnosis and treatment of movement dysfunctions and the enhancement of physical health and functional abilities."[2] From birth through adulthood, persons affected with Prader-Willi syndrome (PWS) are subject to postural, movement, and developmental dysfunctions. The low muscle tone and absence of normal primitive reflexes in the neonate (newborn) with Prader-Willi syndrome prevent typical movement and postures. Thus, during the child's early development, even the most fundamental milestones are delayed (Table 10.1). The preschool years can herald the onset of obesity, which further impacts movement and activity. The school years often add challenges from learning and behavioral deficits that can be further complicated by motor deficits. By adolescence, the cumulative impact of gravity and poor motor, postural, and muscular development significantly challenge both the spine and most joints. If not previously encountered, in adulthood, sleep is frequently disrupted by respiratory problems and decreased oxygen saturation. Further, osteoporosis substantially raises the risk for fractures in adults with PWS. Since physical therapy intervention at each of these life stages can prevent and remediate obstacles to function and independence, optimum and comprehensive care of infants, children, and adults with Prader-Willi syndrome dictates physical therapy as one of the disciplines providing care. Of equal or even greater importance, the physical therapist is another team member who can support and educate parents and caregivers navigating the challenges and joys of the child with Prader-Willi syndrome from infancy through adulthood.

All infants with Prader-Willi syndrome should be evaluated and followed for intervention by physical, occupational, and speech therapists. With identification of the risk for delay, intervention can begin. All of the states in the union have developmental intervention programs, which serve children until the age of 3. During the birth

Table 10.1. Developmental Milestones

		Typically Developing Child	Child with Prader-Willi Syndrome
Motor	Independent sitting	6–8 mos.	11–12 mos.
	Crawling	8–12 mos.	15–18 mos.
	Walking	9–18 mos.	24–27 mos.
Language	First words	10–12 mos.	18–72 mos.

Source: Butler, Hanchett, and Thompson (see Chapter 1, this volume) and Lewis et al., "Speech and language skills of individuals with Prader-Willi syndrome," *American Journal of Speech and Language Pathology.* 2002;11:285–294.

to 3-year-old period, there can be much overlap between the motor interventions provided by a physical therapist and those provided by an occupational therapist. The speech therapist, who may also be enlisted to work on feeding with the neonate, may stay on the case to assist with speech and language skills at the 12-to-24-month level. At the age of 3, the children are transitioned to early childhood programs, which continue the goals of developmental intervention for those children still in need.

This chapter provides an in-depth examination of the neuromuscular and concomitant developmental concerns resulting from PWS across the life span and highlights the role of physical therapy interventions for these concerns.

Birth to 3 Months

As with many congenital disorders, evidence of Prader-Willi syndrome typically presents *in utero*. Fetal movement that is both limited and of low velocity frequently is reported by the second trimester. Both decreased fetal movements during pregnancy and depressed Apgar scores at birth result from the low muscle tone, or hypotonia, which is the most classic presenting characteristic of the neonate with Prader-Willi syndrome. The hypotonia may also account for a decreased state of arousal and poor respiratory responses in the neonatal period. Further, the newborn with PWS may be remarkably inactive, with none of the unorganized responses to auditory and visual stimuli or the expected reflexive, random movements. Primitive reflexes—including the sucking reflex—may be decreased or absent. Newborns with PWS usually cry weakly or not at all, reflecting weak respiratory and oral motor musculature, as well as energy conservation.

Inadequate oral motor control results from low tone through the face and mouth, combined with a poor sucking reflex and easy fatigability. It is this inadequate oral motor control that is the etiology (cause) of the early feeding difficulties prevalent in newborns with PWS. Perhaps no frustration is as profound as the inability of the infant to obtain, or the parent to provide nourishment to their child during this time when brain growth and development demand adequate nutrition. No

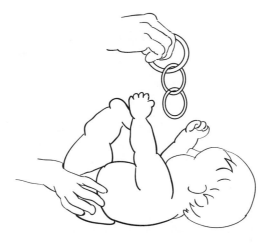

Figure 10.1. Midline skills.

caretaker should have to struggle with this dilemma without the support and assistance of professionals skilled in techniques for remediating feeding difficulties. In addition to physical therapists, occupational therapists and speech therapists can provide assistance in special handling and feeding techniques and have access to an arsenal of feeders, nipples, and other gadgets to facilitate oral motor skills and success with nourishing.

For an infant with normal muscle tone, the first 3 months of life are characterized by early acquisition of control over reflexive movements and strengthening in gravity-eliminated postures—those postures in which the body does not counter the full force of gravity, such as side-lying. By contrast, for the child with PWS, marked delays in achieving the midline control and early antigravity positioning normally expected by 3 months of age result from the hypotonia. Neuromotor intervention during these first 3 months focuses on achieving midline awareness (e.g., hand to mouth), midline posture and skills (holding head in the middle of the body), and on teaching stabilization techniques by way of weight bearing through the upper extremities and trunk.

Cognitive and perceptual growth depend on an infant's ability to attend to stimuli by body orientation—typically orientation to the midline—and on an infant's ability to be successful with early motor feedback by batting at toys, by shifting eyes and head position, and by smiling and babbling. By 3 months of age, the typically developing infant begins to recognize the potential for intervention between his body and what he sees, hears, and feels. Hand-eye awareness becomes evident as a baby brings arms towards the midline to bat at objects he sees (see Figure 10.1). Activation of arms and legs in response to auditory stimuli signals cognizance of voices and noisy toys. A baby's early babbling and oral motor responses to a cooing admirer represent first successes in the area of speech and language. For the 3-month-old with PWS, low tone, an inability to orient to the midline, and an inability to stabilize posture—even with gravity eliminated—prevent those vital early perceptual and interactive experiences. Therefore, effective motor interventions must incorporate the needs of the perceptual modalities vision, hearing, touch, and taste.

Since early motor interventions that may be critical to long-term outcomes depend on appropriate recognition of need, it follows that the most important variable for a child at risk for developmental delay—regardless of the etiology of the delay—is the early identification of the potential for delay and the immediate onset of intervention. Standardized testing of baseline skills provides the basis for objective measurement of progress. The Gesell[12] and Peabody Developmental Motor Scales II[10] are two tools that can be used to assess babies as young as the neonatal period.

3 to 6 Months

For infants benefiting from normal muscle tone, the 3rd through 6th months of development herald the first motor successes against gravity and the first experiences with mobility. Typically developing children learn stability through the shoulder and pelvic girdles by the 3rd month, providing a foundation for building the strength that will control the body that has to compete with gravity. During the second 3-month period, most babies learn to roll from back to tummy, to sit with as little as just guarding assistance, to push up from lying on the tummy to peer over the top of the crib's bumper pads, and to marine crawl. Gross motor accomplishments are accompanied by rapidly emerging fine motor skills. By 6 months of normal development, most babies successfully extend their reach into all planes, grasp toys with the thumb side of the hands, and orally explore all items grasped.

Most 3-month-old infants with Prader-Willi syndrome haven't yet developed stabilization techniques, therefore they should not be expected to have either the head and neck control (Figure 10.2) or the shoulder and abdominal strength to roll, nor will they be able to assist with supported sitting. Most will remain very dependent from

Figure 10.2. Head lag.

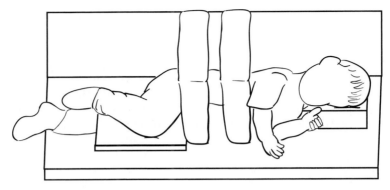

Figure 10.3. Side lyer.

the tummy position and may not attempt raising the head and neck against gravity. Weakness and instability at the shoulder girdles interfere with midline play. Expect a 3-to-6-month-old baby with Prader-Willi syndrome to grasp a toy directly placed in hand but not to bring the toy to the midline for simultaneous visual, oral, and tactile explorations.

Positioning devices at this age can benefit both babies and caregivers. Molded seats and side-lying devices (Figure 10.3) can be very valuable for positioning the infant to midline and to begin upright positioning against gravity.

Every opportunity to handle a low-tone baby during this crucial period of development is an intervention and therapy opportunity. Involvement of physical and/or occupational therapists should continue and may even need to intensify during the 3rd through 6th months. If scarcity of therapists prevents direct delivery of regular services, the therapy staff can be relied upon to educate caregivers on handling and techniques to facilitate achieving developmental milestones. The parents should be able to rely on the therapy staff for guidance and information as they navigate the challenges of nurturing a special needs child.

6 to 9 Months

From 6 to 9 months a typically developing baby continuously strengthens the abdominal, shoulder, and hip musculature. Rolling matures into a controlled and segmented process (Figure 10.4). Sitting is now independent and confident. Mobility is accomplished by efficient marine crawling and by scooting and pivoting in sitting. All-fours rocking lays the groundwork for mobility from all-fours. Maturation to the thumb side of the hand and forearm enables a grasp involving the thumb and index finger. Rotation of the forearm provides a mature communication between the right and left upper extremities. Normal strength and tone through the oral motor structures enable proficiency with spoon-feeding and nutrition through finger-feeding. Babbling and imitation of intonations and sounds are normal expectations and important precursors to spoken words.

The 6-to-9-month-old baby with PWS has probably developed supine to side-lying rolling skills. Rolling is likely to be initiated with a hyperextension of the head and neck and executed without segmentation. While the pattern of movement for rolling may not be optimum, this accomplishment is major and is often reflected in the child's obvious joy at this very early mobility.

Early success in the prone position often follows success with rolling. However, because of the hypotonia, the subsequent difficulty stabilizing the shoulder girdle for weight bearing, and the inherent weakness of the trunk and shoulder musculature, the baby with PWS struggles in the tummy position. Expect the baby to be intolerant when positioned prone. Expect the child to require assistance supporting weight on the forearms and lifting the head to and beyond horizontal. Positioning a baby in prone with a small roll at the chest assists with prone extension and makes the prone-on-elbows position a less daunting task for a child who might otherwise feel defeated in the tummy position. Though lacking the confident elongation and stabilization of normal development, the baby will eventually learn to position against gravity by stacking the head on the shoulders and weight bearing on elbows held close to the chest.

In the absence of optimum strength through the antigravity muscles, the child with Prader-Willi syndrome is unlikely to sit independently before 12 months of age. Frequently positioning the 6-to-9-month child with PWS in supported sitting facilitates the development of a sitting posture and early balance awareness. The caretaker's positioning hands should be held proximally (high on the body) as needed—even to the shoulders and neck (Figure 10.5). As the baby gives feedback that he can manage the postural challenges, the hands-on support should be moved distally (to a point lower on the body) to appropriately challenge postural skills.

Weakness against gravity is also reflected in a lack of progression through fine motor skills. Even at 9 months of age, most children

Figure 10.4. Segmented rolling.

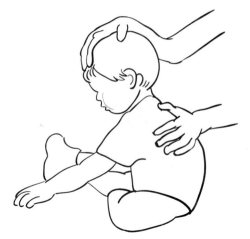

Figure 10.5. Head holding.

with PWS have insufficiently developed shoulder and upper body strength to support arms against gravity and are unlikely to have the strength to extend the wrists against gravity. Without antigravity wrist extension, reach and prehension patterns are unable to progress.

Many parents of 6-to-9-month-old babies with PWS may be unaware that feeding problems still exist. However, in all likelihood, the problematic mechanics of feeding persist. The parents and the baby have simply learned to compensate and to work with the challenges. Feeding skills may be reevaluated at this juncture to insure that nutritional needs are still being adequately met. Because of continuing oral motor concerns, the 9-month-old child with PWS may remain remarkably quiet and may only infrequently utter immature verbalizations.

9 to 12 Months

When antigravity strength has evolved, and a normally developing child has laid the groundwork for normal movement patterns and postures, the 9th through 12th months signal boundless opportunities to get up and go. By 10 months of age, the child is transitioning into and out of sitting, scooting and pivoting in sitting, and accomplishing mobility in the all-fours position (Figure 10.6). At 11 to 12 months, the child begins pulling to standing, cruising along furniture, practicing independent standing balance, and preparing the foundation for independent ambulation. In the fine motor realm, a neat superior pincer (thumb and index finger grasp) and a controlled voluntary release enable play with blocks, puzzles, books, and crayons. These fine motor skills facilitate higher learning and manipulation of environment. Oral motor maturation at this time is signaled by independence with finger-feeding and drinking from a cup, accepting spoon-fed table foods, and independent attempts to spoon-feed. Single words said spontaneously and with meaning emerge at 12 months.

Figure 10.6. Crawling.

On average, the baby with PWS will, if placed, sit independently by 12 months of age. Because of strength and balance deficits, the child's early sitting is likely to be accomplished with supplemental gross stabilization techniques. Further, instead of an upright posture with elongation through the neck and trunk, the child with PWS is likely to persist with a forward flexed (slumped) trunk posture, a posteriorly tilted pelvis, and shortened neck with head-stacked posture (Figure 10.7). This child will require the stabilization of arm propping much longer than a normotonic child will.

By 12 months of age, the child with PWS is typically independent with rolling. Expect that, even at 12 months, the rolling skills of a child with PWS will still be accomplished with neck and upper-trunk hyper-extension and with little or no segmentation.

During age 9 to 12 months the child with PWS may have early success on all-fours, and even with all-fours mobility. Because of the

Figure 10.7. Foward flexed sitting.

Figure 10.8. Locked knees.

low tone and accompanying postural weaknesses, the child will struggle to accomplish simultaneous weight bearing on hands and knees. Some stabilize themselves by using their head to help support the weight. Hips are likely held in a widely abducted—almost frog leg—posture. If a child with PWS is accomplishing mobility from the prone or all-fours position by 9 to 12 months, mobility is not likely to feature the segmentation and reciprocating quality of normal development. A bunny-hopping or inch-worming quality is the more likely expectation.

By 12 months of age, if the child with PWS is placed in a standing position, the standing will likely be accomplished with the lower extremities stabilized by locked knees (Figure 10.8), with a wide base of support, and with heavy reliance on the upper body leaning on a stationary support. The ability to independently pull to standing is often delayed well beyond the time that the child has gained confidence with supported standing.

The progression of fine motor skills for this child is very contingent upon mastery of balance and on strengthening the shoulders and upper body. The propping required in early sitting precedes the wrist extension required for fine motor maturation. However, as long as the child relies on his arms for propping, his opportunities to use his arms and hands to reach, to grasp, and to manipulate the environment remain limited.

At 12 months of age, the baby with PWS is still expected to be very quiet, with limited babbling and without the single-word utterances of a normal 1-year-old.

Frequent weight checks to insure that the 1-year-old with PWS is still gaining weight and maintaining his growth curve remain imperative.

Twelve months is an appropriate age to initiate standardized testing if testing has not already begun. Most standardized testing—such as the Peabody Developmental Motor Scales II—can be repeated at 6-month intervals.

12 to 18 Months

Twelve to 18 months is the time in normal development when toddlers align their upright posture and refine their gait. Two to 3 months after a child begins ambulating without assistance, he will begin to demonstrate a heel strike at initial contact and an arm swing. Expect the normally developing toddler to stand with much anterior pelvic tilt of pelvis—giving the child the typical exaggeration of the curve of the low back. When the toddler begins climbing over obstacles on the floor and climbing up stairs or onto furniture, he begins strengthening the muscles of the pelvic girdle, which will ultimately tilt the pelvis posteriorly and lead to the posture associated with appropriate spinal curves.

At 12 to 18 months of age, the child with PWS will likely continue demonstrating gross and fine motor delays of 50% or more. In all likelihood, the child will sit independently. However, the level of independence and the quality of that sitting independence will vary greatly. Some children may be independent by propping if placed in sitting. Other affected children may be transitioning to sitting, maintaining sitting without upper extremity support, and demonstrating an ability to pivot and scoot for sitting mobility.

When the child with PWS begins transitioning to sitting, it will likely be by transitioning without segmentation and by pushing back through the hips, often into a reverse tailor, or W-sitting position (Figure 10.9). When possible, it is best to avoid the reverse tailor sitting position, because this position enables sitting without abdominal stabilization and interferes with the development of gluteal strength and the normal pelvic position. Delays or absence of normal hip external rotation and delays in maturation of gait patterns are often attributed to reliance on reverse tailor sitting—even in the normal population.

Figure 10.9. W sitting.

Mobility in the prone or all-fours position is a realistic expectation for most Prader-Willi children of 18 months of age. Quality of prone mobility may vary from marine or tummy crawling accomplished with forearms and inner thighs to all-fours creeping. The influences of hypotonia and the associated weakness will likely remain evident in the upper extremities, which may still be locked at elbows and elevated at the shoulders. Hypotonia will be evident in the lower extremities by a broadened base of support and by lack of reciprocated movement.

With increasing independence against gravity, balance awareness evolves. With improved balance, the child is able to release upper extremities from their need to prop the torso. With the upper extremities freed of propping responsibilities, the child with PWS has increased opportunities to practice fine motor manipulations. Fine motor progression at this point in maturation is most profoundly challenged by the low muscle tone of the forearms, wrists, and digits.

During the child's 2nd year, the feeding difficulties subside and weight gain stabilizes. Continued monitoring of the child's growth to insure that the growth curve is maintained is still essential.

Therapy goals for a 12-to-18-month-old child with PWS emphasize balance and strengthening of the shoulder and pelvic girdles. Sit skis can be used to discourage reverse tailor sitting and to facilitate abdominal function. Slanted stools, gym balls, and simple benches can be used to challenge sitting balance against gravity. Opportunities to short sit on benches and small chairs offer alternatives when reverse tailor sitting is difficult to discourage. The prone position also offers important options for strengthening shoulder and pelvic musculature. Whether used to give the child a ride, or propelled by the child's own power, scooter boards are important tools to encourage effort and enjoyment from the prone position.

Speech and occupational therapists should continue monitoring and strengthening oral motor musculature and skills to promote achievement of feeding and speech goals.

18 to 24 Months

By 2 years of age, a child with normal motor development can skillfully jump up and down and forward. Most will ascend and descend stairs from standing—usually by placing both feet on each step. A previously wide-based gait now features a narrow base of support, and heel strike at initial contact is reliable at 2 years.

A child with Prader-Willi syndrome will likely begin to ambulate independently at about 24 months of age. Gait patterns are immature and typical of an early toddler gait—characterized by a wide base of support, small steps, and anteriorly tilted pelvic posture (Figure 10.10). In those with normal motor development, body structure and movement patterns change as strength increases and as the child experiences gait and movement through space. By contrast, low muscle tone and subsequent insufficient extensor muscle strength prevent the child with PWS from achieving normal maturational progression of posture and

Figure 10.10. Wide based gait.

gait. In addition, because many children with PWS experience slow growth during infancy and early childhood, linear growth deficits may be another factor limiting skeletal remodeling, which also improves the efficiency of locomotion.

The physical therapist working with affected toddlers is likely working on goals to facilitate independence with standing and gait. Interventions may include adjunctive orthotics for feet and ankles. The orthotics may be as simple as UCBLs (University of California Berkeley Laboratory shoe inserts), which address pronation by stabilizing the heel. The orthotic of choice might be an SMO (supramalleolar orthotic), which offers a greater degree of frontal plane support. For children whose hypotonic lower leg muscles require ankle support in the sagittal plane, an ankle foot orthosis (AFO) can prevent excessive ankle extension.

24 to 60 Months

The typically developing child's rapid rate of growth between the ages of 2 and 5 years moves the child's center of gravity closer to the lower extremities.[14] This child demonstrates locomotion skills reflecting muscle tone, strength, and length that insures balance and equilibrium. Thus, by age 3, the normally developing child can ascend and descend stairs with alternating feet and without need for handrail assistance. By age 4, most can hop on one foot and can tandem walk on a line on the floor. By age 5, most can skip, can balance on one foot for 10 seconds, and can perform sit-ups.

During early childhood, the gross and fine motor delays of the child with PWS become less glaring. By age 2, some children with PWS have enough muscle strength and control to achieve ambulation and mobility milestones. Though the child may be walking, climbing, and transitioning, these skills lack quality, maturity, and refinement. Assessment on a standardized test, such as the Peabody Developmental Motor Scales II, will define the delays more specifically.

Problems compound for the child with PWS during the early childhood period, as this is the time associated with rapid weight gain.[3] Weight gain for a child with PWS typically does not represent muscle and length. For the child with PWS, the growth of body fat exceeds the growth rate of bone and muscle.[3] The challenges of managing posture, equilibrium, and mobility increase when weight gain further compromises limited strength, lack of muscle tone, and short stature.

The physical therapist working with the preschool child with Prader-Willi syndrome should define goals to strengthen the proximal and core musculature and to challenge endurance. During this period of rapid weight gain, aerobic activities not only address muscle tone but also boost metabolic rate and increase energy expenditure. Scooter-board play, games of wheelbarrow, crab walking, climbing skills (on simple playground equipment or on more sophisticated rock walls), riding toys, swimming, and water play are all gross motor activities that strengthen muscles and challenge aerobic capacity. Horseback riding, or hippotherapy, can be a very beneficial activity for children as young as preschool age. Therapists use horses as therapeutic aids to address muscle strengthening, balance and equilibrium, sensorimotor, conditioning, postural, and even speech goals.[1]

The occupational therapist should be evaluating and remediating deficiencies in activities ranging from grapho-motor skills, to perceptual skills, to dressing and activities of daily living. The occupational therapist is frequently the team member who assumes responsibility for evaluating the child with Prader-Willi syndrome for sensory integrative dysfunction. Sensory integration refers to the brain's ability to organize the information received from all the body's sensory modalities—vision, hearing, tactile, taste, position in space, pull of gravity, and movement.[4] When development is typical, children receive and organize sensory input to produce well coordinated movement, behavior, and self-image. For the child with Prader-Willi syndrome, poor muscle tone and short stature are just two of the many factors that impact normal sensory integration. The potential value of a sensory integrative evaluation and sensory integrative therapy for the affected child should always be considered during this early childhood period.

Similarly, at this point, speech therapists might be assisting the Prader-Willi child with articulation deficits as well as with language content.

From Preschool Through Adolescence

In those developing typically, the center of gravity continues to lower as body length progresses to adult height.[14] Muscle bulk and strength guide the body's postural development and movement patterns. Normal cognitive skills and motor maturation accommodate a person's inherent need to move and to use the body in recreation and play.

For the child with PWS, the phenotypic body shape that became evident in childhood remains throughout adolescence and adulthood. Without growth hormone intervention, the lack of a pre-pubertal growth spurt and the failure to fully develop secondary sex characteristics (e.g., waist and hip body form) serve to amplify the already present impact of hypotonia and insufficient lean muscle mass. Thus, in comparison with peers, the child with PWS will be short, particularly when compared with grown family members, and will have small hands and feet. Lacking growth hormone intervention, the average height of adults with Prader-Willi syndrome is below the 5th percentile.[6]

Further, by adolescence, continuing rapid growth of body fat may result in total body fat of 40% to 50%. The excessive body fat is most likely to accumulate at the body's midsection and thighs.[8,9] The combination of low tone, postural muscle weakness, high center of gravity, and excessive body fat insure that posture will fall outside of normal plumb line alignment, preventing normal gait and movement. Inactivity is a logical result of the affected teenager's body shape and physical status. Cognitive and oppositional behavioral tendencies are additional factors impacting these teens' and adults' potential for improving body size and shape.

Patello-Femoral Syndrome

Patello-femoral syndrome is a common cause of knee pain that is grossly underreported in all teens. Even when musculoskeletal development and lean body mass are within normal limits, muscle and soft tissue imbalances at the hip and knee are a frequent cause of patello-femoral knee pain. The hypotonia and typical body shape render the teen and young adult with PWS even more susceptible to this very common cause of knee pain. Patello-femoral syndrome results when the patella, or kneecap, does not track accurately through the patellar groove during flexion and extension movements. The faulty tracking occurs as a result of a variety of musculoskeletal factors.

Of the four quadriceps muscles, the lateral quadriceps (vastus lateralis) is the strongest and most likely to pull the kneecap laterally (to the side), away from the appropriate pull path. The medial quadriceps (the vastus medialis oblique), which should activate to pull the kneecap medially (to the middle), is the smallest and least powerful of the quadriceps. Most commonly in those with PWS, the exaggerated anterior pelvic tilt keeps the femurs internally rotated, further preventing the activation of the vastus medialis oblique.

The gluteus medius muscles located at the lateral aspect of the hips are important pelvic stabilizers. Even in the normotonic population, the gluteus medius musculature is frequently too weak to prevent dropping of the pelvis, further encouraging patellar malalignment. This weakness, as well as hyperextension of the knees (genu recurvatum), is a common postural characteristic among those with PWS. Genu recurvatum contributes to patello-femoral dysfunction because the knee rests in extension without benefit of any quad muscle input.

Furthermore, knee hyperextension, or recurvatum, occurs with the femur fully internally rotated.

Patello-femoral pain occurs when bending or straightening the knee. Those troubled with patello-femoral syndrome will experience pain when ascending and descending stairs, and will frequently describe knee pain upon rising after extended periods of sitting. The most dramatic sequelae (long-term outcome) of patello-femoral dysfunction is subluxation, or dislocation, of the kneecap. Patella dislocation typically has a sudden onset, often with no identifiable precipitating event. If patellar relocation (return to normal place or position) does not happen spontaneously, it can easily be put back in place in the emergency department. Immobilization in a splint or cast offers rest and an opportunity for swelling to subside.

Preventing repeat patellar subluxations (dislocations) is of paramount importance. While surgical stabilization is one treatment option, a strengthening program for the medial quadriceps and gluteus medius musculature, combined with a stretching program for the iliotibial band of the lateral thigh and hamstring muscles and a patello-femoral taping program[13] offers a conservative approach that can successfully rehabilitate the knee and prevent the need for surgical intervention. Patello-femoral reeducation has a high rate of success in the normotonic population and offers a reasonably conservative option for members of the Prader-Willi population who are cognitively able to understand and execute the exercises, and who have the support persons to insure compliance with the program.

Scoliosis

In the normotonic population, idiopathic (specific cause unknown) scoliosis occurs in 2% to 3% of children aged 7 to 16 years.[7] Those with Prader-Willi syndrome are more at risk for neuromuscular scoliosis, presumably as a result of low muscle tone. Approximately 62% to 68% of the Prader-Willi population have scoliosis with a structural change of at least 10 degrees.[11] The high prevalence of neuromuscular scoliosis among children and teens with PWS necessitates careful monitoring by radiographic studies, especially during periods of rapid growth. Progressive curves (curves that have increased by 5 or more degrees on two consecutive examinations) should be evaluated radiographically every 4 to 6 months. The treatment of neuromuscular scoliosis ranges from careful monitoring to bracing to surgical stabilization.

Inactivity and Obesity

The school-age child with Prader-Willi syndrome will most likely be active and playful with school peers. However, the child most likely will be unable to compete or even maintain pace with peers. Standardized testing reveals objective data for better understanding the motor difficulties encountered by school-age children with PWS. The Bruininks-Oseretsky Test of Motor Proficiency is a standardized assessment for children aged 4 to 14 years[5] that defines specific areas of weakness: muscle strength, speed, balance, coordination, and bilateral skills.

The genetically driven hypotonia, weak musculature, and higher fat-to-lean body mass ratio lead to motor delays and frank inability to acquire the needed skills for developing strong and lean bodies. Rarely do organized and intramural sports make a place on the team for a child with Prader-Willi syndrome.

The physical therapist working with the school-age child with Prader-Willi syndrome must address these issues of inactivity and impending or frank obesity. While the children are still in school, adaptive physical education programs can be invaluable for maintaining an activity level and increasing strength, muscle tone, and aerobic capacity. Modified track, swimming and water play, tricycles and bicycles—adapted as necessary, and all forms of gym and playground play can be fun, therapeutic, and safe.

Special Olympics programs also are a wonderful option for families of children with Prader-Willi syndrome. Special Olympics programs can provide the means and motivation for dealing with weight, weakness, and inactivity through organized and supportive sports and play. Beyond the physical benefits, Special Olympics offers affected children the joy of movement and play, the thrill of competition, and the comradeship of a team, which might otherwise be unavailable for this population.

The Adult with Prader-Willi Syndrome

The U.S. Food and Drug Administration (FDA) approved growth hormone treatment for children with Prader-Willi syndrome in the year 2000. That milestone may change the future for Prader-Willi children who will reach adulthood in the next decade or two. For the present, adults with Prader-Willi syndrome face challenges that follow them from childhood and other challenges that emerge in adulthood.

The postural challenges that first emerged in childhood often become more problematic in adulthood. The persistent ramifications of hypotonia, small stature, a high center of gravity, increased body weight, decreased opportunities for physical activities in the post-school-age population, and gravity all conspire against upright, plumb-line postural alignment. Simple techniques can curb the forward flexed posture that so often typifies the adult with PWS.

Time spent each day in the prone or tummy-lying position with weight propped on forearms strengthens the back extensors and shoulders and stretches the flexors of the trunk and hips. Upper extremity and upper body strengthening programs can modify the stereotypic head-forward and shoulders-rounded posture. Such upper body strengthening and stretching programs might include swimming, wand exercises for the shoulders, pulley exercises, range of motion exercises done with resistive bands, or even cuff weights starting at 1/2 pound and increased in 1/2-pound increments.

Adults with PWS are also at increased risk for osteoporosis and related bone fractures (see Chapter 5). Physical therapy and exercise programming should be carefully coordinated with medical monitoring and treatment for affected individuals. Walking, running, horseback riding, and jumping may not be optimal activities for a person with osteoporosis, since these activities may further increase the risk for fracture in the lumbar spine and hips (the major weight-bearing parts of the skeleton). Nonweight-bearing exercises, such as swimming and water aerobics, may be preferable. However, if a person is being medically treated for osteoporosis (e.g., with estrogens or bisphosphonate) and there is a documented improvement in bone density, then weight-bearing exercises can be judiciously reintroduced into the exercise regimen after consultation with the treating physician.

In the absence of osteoporosis, a walking program is an excellent weight-bearing activity for adults with PWS and very adequately stresses the long bones, facilitating calcium fill. It also provides a manageable aerobic exercise program. Walking is easily executed by people of all skill levels and requires no special equipment other than a pair of supportive shoes. When walking outdoors is impossible, the use of a treadmill can be considered. However, balance issues make this difficult for many with the syndrome, so this should not be attempted without proper supervision and attention to safety.

The need for aerobic activity does not decrease as the person with PWS reaches adulthood. The adult with PWS is predisposed by obesity to sleep apnea and hypoventilation syndromes. Aerobic activity offers the best defense against obesity and the best means of strengthening lung capacity.

Adults typically have fewer opportunities than children for organized physical activities, and this is particularly true for adults with physical challenges. As previously indicated, Special Olympics programs are invaluable in fostering motivation for physical activity and providing opportunities for organized, varied motor challenges.

Across the nation in large urban areas as well as in small rural locales, communities are building centers for fitness and education. These community centers are rising to the challenges of the nation's increased need for physical activity. These centers offer opportunities for all ages and skill levels to enjoy water play, court sports, and organized group exercise. Many of these wellness and fitness centers provide opportunities for those with special needs to utilize their facility.

Conclusion

Since 1956, when Swiss physicians Prader, Labhart, and Willi first described Prader-Willi syndrome, much has improved for affected children and adults. An increasing knowledge base and exciting treatment options, such as growth hormone, offer increased optimism for quality of life. At all stages of development, and throughout the life span of the person with Prader-Willi syndrome, motor intervention and activity modalities provide the tools for dealing with the challenges of hypotonia, developmental delays, balance deficits, orthopedic anomalies, and obesity. From the infant, who has perhaps not yet even been diagnosed, to the adult seeking meaningful work through vocational training, every individual with Prader-Willi syndrome should have care that includes physical, occupational, and speech therapists as essential team members.

References

1. Agarwal JM. Horseback riding for all ages. *The Gathered View, Newsletter of the Prader-Willi Syndrome Association (USA).* 2003;28(3):8–9.
2. American Physical Therapy Association. *APTA Mission Statement.* Alexandria, VA: American Physical Therapy Association; 2003.
3. Angulo MA. *Prader-Willi Syndrome: A Guide for Families and Professionals.* Sarasota, FL: Prader-Willi Syndrome Association (USA); 2001:4, 9.
4. Ayres AJ. *Sensory Integration and The Child.* Los Angeles, CA: Western Psychological Services; 1979.
5. Bruininks RH. *Bruininks-Oseretsky Test of Motor Proficiency.* Circle Pines, MN: American Guidance Service, Inc.; 1978.
6. Butler MG, Meaney FJ. An anthropometric study of 38 individuals with Prader-Labhart-Willi syndrome. *American Journal of Medical Genetics.* 1987; 26:445–455.
7. Campbell SK. *Physical Therapy for Children.* Philadelphia, PA: W.B. Saunders Company; 1994.
8. Carrel A, Myers S, Whitman BY, Allen DB. Growth hormone improves body composition, fat utilization, physical strength and agility, and growth in Prader-Willi syndrome: a controlled study. *Journal of Pediatrics.* 1999; 134(2):215–221.
9. Cassidy S. Prader-Willi syndrome. *Current Problems in Pediatrics.* 1984;14: 1–55.
10. Folio MR, Fewell RR. *Peabody Developmental Motor Scales.* 2nd ed. Austin, TX: Pro-Ed; 2000.
11. Holm VA, Laurnen EL. Prader-Willi syndrome and scoliosis. *Developmental Medicine and Child Neurology.* 1981;23:192–201.
12. Knobloch H, Stevens F, Malone A. *Manual of Developmental Diagnosis: The Administration and Interpretation of the Revised Gesell and Armtruda Developmental and Neurological Examination.* Hagerstown, MD: Harper & Row; 1980.
13. McConnell J. *The Advanced McConnell Patellofemoral Treatment Plan* [course manual, June 1995]. Sammanish, WA: The McConnell Institute; 1995.
14. Wenger DR, Rang M. *The Art and Practice of Children's Orthopaedics.* New York, NY: Raven Press, Ltd.; 1993.

11

Educational Considerations for Children with Prader-Willi Syndrome

Naomi Chedd, Karen Levine, and Robert H. Wharton[†]

Children with Prader-Willi syndrome (PWS) present unique characteristics, needs, and challenges to traditional educational environments. Providing an optimal educational experience requires that parents, teachers, and associated service providers be familiar with the issues and educational options for affected children and their families. This chapter provides an overview of the many and varied considerations involved in providing educational opportunities to children with PWS from infancy through adolescence. It delineates specific considerations associated with age and developmental milestones and offers practical suggestions for maximizing the educational experience of students with PWS. In addition, it provides information on current United States (U.S.) federal legislation affecting children with disabilities, including the No Child Left Behind Act of 2001, the nation's latest general education law.

U.S. Education Legislation

U.S. federal law requires that all children with a disability have a written plan that describes each area of educational need and specifies how these educational needs will be met. As of this writing, the current special education law articulating these requirements is the Individuals with Disabilities Education Improvement Act of 2004 (Public Law 108–446), known as IDEA or IDEA 2004. Although the law became effective in July 2005, amending a previous version of IDEA, new regulations have not been finalized (expected in early 2006).

For children under age 3, this plan is accomplished through an Individualized Family Service Plan (IFSP), described in Part C of IDEA. The law covering this age group calls for statewide, comprehensive, interagency service programs for all infants and toddlers with disabilities and their families. It also requires that the IFSP outline services that will be provided to meet the infant's or toddler's developmental needs

[†] Deceased.

in one or more of the following areas: physical development, cognitive development, speech and language development, psychosocial development, and self-help skills.

In contrast to the original 1986 provisions for early intervention services (specified in P.L. 99–457), the current version of the law covering infants and toddlers requires that services for children under 3 be provided in "natural environments," which may include parks, public libraries, and other community settings as well as homes and day care facilities. If intervention services are provided in other settings, such as a medical center, the IFSP must offer justification for providing these services in a "more restrictive" environment.

Part B of IDEA focuses on children between ages 3 and 21 and requires that school systems provide a plan for identification, evaluation, and provision of special education and related services. Once a child is identified, evaluated, and found in need of special education services, an Individualized Education Plan (IEP) must be implemented. The IEP is an educational blueprint of all aspects of that child's special education needs and the resulting services to be provided. It is, in essence, a contract between a school and a parent regarding the type and extent of educational and related services to be employed for educating the child. A main difference between Parts B and C of the IDEA law relates to the role of the family. Whereas Part C focuses on supporting the family in their capacity to provide a nurturing environment for infants and toddlers and enhancing coordination of services required by both the child and family, Part B focuses on the role of public schools, guaranteeing that children with disabilities receive a free and appropriate public education (FAPE) in the least restrictive environment (LRE). IDEA, through the amendments passed in 1997, also mandates that all students have access to the general curriculum, rather than being educated with significantly different content and programming.

In addition, differences and changes in the law reflect the needs of the children and parents at different points along the child's developmental continuum and an increased awareness of the importance of the family as part of the team. The 1997 changes shifted the focus somewhat, from *access to education* for children with disabilities to *improving and demonstrating educational results*, and the 2004 IDEA amendments continue that emphasis, requiring higher standards for qualification of special education teachers.

Another recently enacted federal law that has considerable implications for children with special needs is the No Child Left Behind Act of 2001 (P.L. 107–110), known as "No Child Left Behind" or NCLB, enacted in 2002. According to the U.S. Department of Education, NCLB was designed to work with IDEA and is based on the assumption that all children, including children with disabilities, can meet high educational standards. Like IDEA, it states that students with disabilities must have access to the general curriculum. But it also requires that teachers in all settings—a general education classroom, integrated setting, or substantially separate setting, such as a resource room—need to be skilled in teaching all students. It differs from IDEA in that it makes schools accountable for student performance.

It is helpful, indeed necessary, for all families engaged in educational planning for their children with PWS to familiarize themselves with the latest provisions of the federal education laws. Several excellent sources of information on these laws can be found on the Internet, including the U.S. Department of Education (www.ed.gov), the National Dissemination Center for Children with Disabilities (www.nichcy.org), and Wrightslaw (www.wrightslaw.com). In addition, getting information about individual state laws and local practices will better enable parents to secure the most comprehensive and appropriate education possible for their children.

Qualifying for Special Services

To be eligible for services under IDEA, a child must have an identifiable condition that interferes with educational progress and school performance to the extent that special education services are required. The law requires that children with a suspected disability be evaluated by an interdisciplinary team, following which an expanded team that also includes the parents designs the child's IEP based on the results of the evaluation; that children be educated in the least restrictive environment; that services be provided when deemed necessary by the IEP team; and that child and parent rights to "due process" be protected. That is, if parents disagree with team recommendations for services or placement, they have a right to a hearing in a court of law.

Program Models

There are variations in classroom options for preschool and school-age children with PWS. Which option is best depends on both the needs of the child and the support services school systems are able to provide in regular and specialized settings. Because children's needs change over time, planning and placement require periodic review and often modification. Any model must incorporate understanding and support for educational, emotional, and social development while normalizing all experiences as much as possible and optimizing quality of life.

Although terms may be defined and interpreted slightly differently, most schools recognize several program models. An "inclusion model" involves placing children with disabilities such as PWS in regular education settings while providing individualized accommodations and support as needed. Where its interpretation is flexible and a broad array of educational and behavioral supports is available, it can be very effective. One advantage is the implication that *all* children are included and valued regardless of ability or capacity to accept ongoing educational challenges. The classroom environment is altered to accommodate all children's needs. Another advantage is that children have an opportunity to develop friendships among classmates with whom they might not otherwise interact.

For children with PWS this model *must* incorporate implementation of support services when difficulties arise and, preferably, be in place before difficulties arise. For example, psychosocial/behavioral consultation may be beneficial in situations such as managing food issues, managing frustration around transitions, and facilitating development of friendships.

Computer instruction and use of a computer in class may be helpful accommodations as well. Other services (speech, physical, and occupational therapy) can either be incorporated into the classroom or take place outside class. It is often necessary to have a classroom aide or paraprofessional to assist with social, educational, and behavioral matters. Depending on the needs of the child, either a one-to-one aide or an aide who works with several children can be effective. However, in both cases, it is necessary for the aide to get appropriate, PWS-specific training in facilitating social interactions between the child with PWS and peers, as well as being able to modify work as needed and provide some behavioral interventions.

Social facilitation can occur by having the child with PWS sit next to classmates during seated group activities, by prompting conversations about topics familiar to the child with PWS, and getting to know all classmates well enough to earn their trust and gain their cooperation. When the inclusion model is not flexible, there are insufficient supports, or appropriate staff training isn't provided, children with PWS may be misunderstood, isolated from their peers, and have fewer opportunities for academic, behavioral, and social development and success.

When an inclusion model is not indicated for the child, other program formats may be pursued. For children with more significant learning and/or behavioral problems who are placed in schools that have large classes and few supports, a smaller, specialized class for most or all academic work may be beneficial. However, classrooms for children with behavioral disturbances (often termed BD classrooms) are not usually appropriate for children with PWS because their behaviors, as well as the underlying causes, tend to be very different from those of children typically placed in such an environment. When classroom teachers need to develop specific management strategies, consultation with a behavioral psychologist may be beneficial. It would be particularly helpful to work closely with a behavioral psychologist who has had experience in working with children with PWS.

Regardless of the classroom model, benefits to social-emotional development accrue from integrated experiences. Integration into a class of typically developing students can be successful for structured activities such as hands-on science or art projects and story times but poses extra problems at lunch and snack times. The concept of reverse integration also works in some instances. This involves having a child, or several children, without special needs join the special classroom to participate with the child with PWS in activities facilitated by the teacher. The child or children can subsequently "host" the child with PWS in the regular classroom.

Some children, especially older children who are experiencing significant behavioral challenges, may benefit at some point from placement in a residential school where predictable structure, a high staff-student ratio, and limited access to food create an atmosphere that greatly reduces overall stress and stress-induced behaviors. Clearly, even in this type of program, the family remains a vital emotional presence and support; close communication and cooperation between the residential program and the home are essential. Sometimes the residential model can provide "crisis intervention" until the circumstances surrounding initial placement stabilize. After a period of time, the child's success should be reevaluated and the child may either return home and attend a public or other day school or remain in the residential program. Ongoing evaluation of the child's needs is essential.

No matter which program is selected by the team, teachers and all other school personnel working with the child *must* be taught about food, behavior, and social issues. Such education provides information on which to base teaching methods as well as management strategies for restricting assess to food, thereby reducing frustration for the child and staff. Although some behaviors may appear voluntary, such as being extremely stubborn or perseverative, it is essential to help the school/community understand that children with PWS may, at times, display behaviors over which they have little or no control. It is equally important to help school staff recognize the particular strengths of the individual with PWS and gear teaching/learning strategies toward them.

Medical Issues That Affect the Educational Process

Educators need to be aware of several medical and psychological features associated with PWS, particularly an altered level of arousal, the pronounced appetite disturbance previously referenced, and diminished muscle tone and motor planning skills. These features impact on both classroom performance and perceptions of children by teachers and classmates.

Children with PWS frequently demonstrate a diminished sense of arousal. Whereas some children with other medical conditions can be excessively active, the majority of children with PWS are underactive. This characteristic manifests itself as a general lack of dramatic affect, decreased initiation of activities, and a frequent lack of enthusiasm. For example, some children tend to fall asleep in sedentary situations, during a long lecture or silent reading period. Others demonstrate only minimal interest in classroom activities. This feature, sometimes misinterpreted as emotionally based lethargy, attention deficit, or an inability to participate, actually is part of an altered level of arousal mediated by the central nervous system and is unrelated to intelligence.

Most disabling is the inability of children with PWS to control the drive to eat. Unlike arousal, however, this appetite disorder is one of excess rather than underactivity. This characteristic, while mild in some children and profound in others, is moderated by brain chemicals and beyond internal controls. The lack of capacity for control is not related to cognitive ability, disruptions in the home environment, or a need for comfort or emotional or physical nourishment. Children with PWS can no more control their appetite than they can control a sneeze. It is imperative that the child's environment be modified; trying to change a child's desire for food or food-seeking behavior is futile without limiting access.

Another physical feature associated with this syndrome is low muscle tone and a diminished ability to engage successfully in tasks requiring substantial motor planning skills. Children tend to appear quite weak, and even simple motor tasks such as dressing can be challenging regardless of cognitive level or weight. Moving about the classroom, changing rooms—especially where stairs are required, and engaging in exercise routines or athletic events may be arduous and trying. The inability to match the efforts of nonaffected peers in these areas must be understood on the basis of underlying problems of muscle strength and motor control.

Educational Issues Across Developmental Stages

Infancy and Toddlerhood (Birth to 3 Years)

Because infancy and toddlerhood are periods of rapid development and learning, they are crucial times for children with PWS. Infants with PWS, like other infants, develop strong attachments to parents, siblings, and other caregivers. Separation anxiety, happiness at greeting familiar people, and the particular smiles and eye contact for parents are all part of the special relationships that form between infant and family. Aside from being a delightful quality, this strong social orientation is a vital strength upon which to build a successful learning and intervention program, right from the start. In the first few months, however, sleepiness and nonresponsiveness make nurturing and early attachment to caregivers a challenge. Low muscle tone and generalized weakness cause delay in reaching gross motor milestones such as sitting, crawling, and walking. There is also lag in development of fine motor skills.

Because of developmental delays, infants can benefit from interventions beginning as early as 1 month of age. Feeding and oral motor skills, receptive and expressive communication achievements, and both gross and fine motor development should be assessed and addressed throughout infancy. Programming should focus on physical stimulation, communication, and socialization. Regular "baby" play activities are important. Singing, nursery rhymes, books, mirror play, bubbles, pictures, rough-and-tumble romping, and cuddling are vital developmental activities. However, adjustments are required to compensate for motor and speech limitations, i.e., providing more physical support, exaggerating affect and facial expression, and reinforcing any and all vocalizations and other attempts to communicate. In addition to directly assisting with motor development, physical and occupational therapists should be consulted regarding strategies to help the child compensate for limitations (see Chapter 10).

During these first years, due to output delays in the motor and speech systems, cognitive processes tend to develop ahead of physical skills. That is, the capacity for thinking and understanding outdistances verbal expression. Expressive language (speech) generally lags behind receptive language (word comprehension). Receptive language tends to correlate highly with cognitive development. The cause of this initial speech delay is poorly understood but is thought to be due, in part, to oral motor difficulty (see Chapter 9). The delay frequently causes frustration for both the infant/toddler and the family. Further, limited expressive skills and motor skills make developmental assessment difficult and can result in underestimation of cognitive abilities.

As soon as difficulties are noticed or diagnosis is suspected, early intervention services should be ordered by the pediatrician. In some states these services are administered through the Department of Public Health, while in other states these services are administered through the Department of Education. Some early intervention programs offer speech, physical therapy, occupational therapy, and play groups, while in other locales these therapies must be funded through other sources. Whatever source is utilized, these specialized interventions should be taught to caregivers and incorporated into daily and weekly routines that encourage optimal adaptation over time. In addition to these special services, infants and toddlers can benefit from experiencing regular day care and playgroups that provide rich opportunities for enhancing social communication and play skills and offer families normalizing experiences.

Preschool (3 to 5 years)

Toddlers, with or without PWS, begin to declare themselves as individuals as they progress through the developmental milestones of walking and talking. Individual personalities emerge as youngsters learn to express themselves, make demands, and struggle to gain control of their environment. Interests, needs, and preferences may be expressed with great conviction, and adult directions may be met with opposition. During this era children *know* what they want and *demand* immediate fulfillment. When gratification is not immediate or goals of others interfere, frustration can evolve into a full-blown tantrum, especially if verbal expression is limited. Who hasn't heard of the "terrible twos"? For children with PWS, this phase may begin a little later and last a little longer.

Although preschool children with PWS exhibit delayed motor and speech abilities, they still crave independence. Delays in achieving these milestones can contribute to resistance to physical activities, frustration due to their inability to communicate, and tantrums. Difficulties adapting to change, controlling frustration, or becoming calm once upset may occur; tantrums may also be more frequent and intense.

Implementing positive behavioral strategies, speech therapy, and physical therapy, and creating a predictable environment are essential.

At the age of 3 years, children become eligible for comprehensive school programs and enter the public education system through the process of evaluation by an educational team. Cognitive, communicative, social, and behavioral skills are assessed in order to guide educational approaches, goals, and placements. In addition to including complete transitional summaries and recommendations from a child's early intervention program, supplying the school team with educational information about PWS can be beneficial. A number of such publications are available through the national and international Prader-Willi syndrome organizations (see Appendices E and F).

Part of the process for determining educational needs and developing the IEP involves standardized intelligence tests. Interpreting test results requires an awareness of the broad spectrum and characteristic patterns of learning difficulties in children with PWS. Some function in the average or borderline range of intellectual abilities, but with accompanying learning disabilities; others are in the mild-to-moderate range of mental retardation; and a small percentage are severely impaired. The numbers of individuals in these categories are not known precisely, and figures cited vary across studies.[1,2,7,8,9] However, taken together, the literature suggests that at least half of the children with PWS test in the average-to-borderline range. Some studies suggest that there is no correlation between weight and intelligence and that intelligence remains stable throughout development.[4] Regardless of IQ scores, children with PWS show substantial scatter in ability levels across domains. Moreover, since IQ scores are derived from averaging performances across skill areas, and because children with PWS usually have motor and speech delays, scores reported during preschool years may underestimate cognitive levels.

Measures of receptive vocabulary such as the Peabody Picture Vocabulary Test III[3] are better predictors of cognitive development at this age. This test, which requires minimal motor and speech skills, has good correlation with overall intelligence in the general population but is not valid for children who have substantial attention difficulties. Intelligence test scores may become more meaningful for older school-age children with PWS, whose speech and motor skills typically catch up with their other areas of development and for whom there is less discrepancy in performance level across domains. However, when older children continue to show substantial scatter in skill levels across domains, the validity of the IQ scores remains questionable.

While IQ scores for preschoolers are not particularly useful for predictive purposes, the testing results can be quite helpful, indicating learning strengths and styles. The specific test chosen is less important than how test data are interpreted. Reports should discuss performance levels across areas of strengths and weaknesses, with particular attention to learning style and functional supports needed and should be based on assessment by a psychologist or educator experienced in evaluating children with developmental disabilities.

Children with PWS who need specific educational and behavioral intervention may benefit substantially from a specialized preschool experience, such as an integrated, special needs, language-based classroom. Others with fewer needs may benefit from enrollment in a regular preschool where, although less individualized attention is provided, there are substantial advantages gained from socialization, language, play, and group experiences. A common successful approach is enrollment in a combination of special and regular programming, or in an integrated program that includes some children without special needs. Regardless of ability, most children with the syndrome are able to handle two programs as long as approaches and daily activities remain consistent.

Preschool IEPs should include a regular, preacademic curriculum in a developmentally appropriate context. Most children will continue to benefit from continuation of earlier assistance in speech and physical/occupational therapies (see Chapters 9 and 10). These services are most effective when therapists spend part of their time in individual and/or small group interventions and some time consulting with the classroom teacher about how to incorporate therapy goals into the regular curriculum and classroom activities.

In addition to preacademic learning, preschool IEPs and programming should address behavioral and socialization needs, since social development tends to present difficulties as children get older.[4] Early social skills training, including pragmatics (the practical use of language), is crucial and can help children develop understanding of some of the more subtle aspects of socialization. These cues for social skills are best taught through teacher facilitation of child-to-child interactions and social and dramatic play. Because social skills are so important for future success, both in and outside of school, and because children with PWS require intensive teaching and numerous repetitions, a social skills/pragmatics program should be a priority in every IEP.

Preschoolers sometimes display behaviors that can interfere with learning and acceptance in the classroom, particularly those that result in class disruption caused by schedule changes. These behaviors should be addressed through the IEP, and anticipation is the first line of defense against outbursts. A predictable daily routine is important because there is a tendency to express distress and excessive anger in response to apparently minor alterations in planned activities. A picture schedule for daily routines and a wall calendar for upcoming events can help, and providing information about changes ahead of time may prevent or reduce upsets. Bringing an anticipatory object such as a ball when going to physical therapy or a toy when going outside may ease transitions. Providing predictable "change-in-routine" signals, such as singing certain songs, also helps anticipation and adjustment. Once a child is upset, emotions may escalate rapidly and are difficult to control. When tantrums do occur, it is best to remove the child from the situation, remain nearby, ignore the behavior, and avoid scolding or even trying to reason. For children who show decreased interest and lack of arousal, interspersing motor activity with sedentary projects and seating the child near the teacher can help.

Perseveration is another behavior that challenges the patience of teachers and peers. It often takes the form of repetitive questioning and/or repeated engagement in singular play (such as tearing paper or drawing circles). When repetitive play seems to be soothing, it is beneficial to restrict the behavior to a certain place, perhaps a quiet area of the classroom, and to certain times that are predictably stressful, such as just before snack time, rather than attempting to eliminate it. Once a question has been answered, ignoring the redundancy and changing the subject are often sufficient for this age group.

Finally, at this age children not only observe that eating is unrestricted for their peers but also note that food may be accessible. Several strategies can help manage this situation. During school hours, edibles should be kept outside the classroom in cubbies or in high cabinets. Snacks should be served in child-specific portions in front of each child rather than in large serving bowls and allowing children to help themselves. Supervision is necessary whenever food is available. Children should not have to sit for long periods near others who are eating, nor should they be isolated from their peers. Teachers should be instructed not to use food as a reward for a child with PWS and, preferably, not in the classroom at all. Stickers, colorful markers, or small, safe toys are better choices. Food is an issue that requires teachers and families to function as a team, using structure and positive behavioral approaches that can be implemented at home and school. When teachers are aware of potential difficulties, the environment can be structured to minimize problems.

School-Age Children (6 to 12 Years)

Developmental goals of school-age children include concepts of mastery of tasks and pride in achievements. There is a desire to acquire knowledge, skills, and control of their environment. Providing many opportunities for success is vital to the development of healthy self-esteem during this period when awareness of differences emerges. Some children experience significant tantrums during this period in response to unexpected changes, frustrations, and limited access to food. Skin-picking behavior, common in children with PWS, may also surface during these years in response to frustration and anxiety. Social development continues, as do social challenges, especially in the development of friendships.

Parents can benefit from learning about this stage of educational growth with the assistance of a developmental psychologist, an education specialist, and/or pediatrician knowledgeable about PWS, school options, and opportunities. Psychologists play a major role in the development of IEPs that link parents and school systems. Management of food-related concerns should be included in planning, and recommendations should stress close supervision. During this age period, feeling successful at some physical activity is important in shaping lifelong attitudes toward exercise. Participation in both school-based activities as well as outside programs, such as Special Olympics, can be very rewarding, both personally and socially. Noncompetitive activities, such as swimming, walking, and group exercise classes can be good alternatives to team sports.

Interventions

Younger school-age children with PWS usually do well in regular classrooms when provided with extra services. Another approach involves regular morning kindergarten classes and a special language-based classroom in the afternoon that allows for individualized teaching. This format can be quite successful and may be continued through the elementary grades.

Meticulous attention must be given to the learning profiles of children with PWS as they begin their school experiences. Their learning profiles generally have characteristic strengths and weaknesses. Strengths, indicated in Table 11.1, are relative to their own abilities, not to peer performance. A particular strength in many children is long-term memory for information. This strength applies to academics as well as to events and names. While initially it may be more difficult to teach new material due to evidence of learning difficulties, it is worth the effort. Functional skills should be encouraged early in the curriculum using positive approaches, such as complementing the student frequently, showing high interest, and providing individualized attention. Classroom activities should move at a brisk pace with variety to maximize motivation.

Most children with PWS become skilled readers. Verbal information is best understood when presented in brief pieces and time is allowed to process it. Much can be learned through hands-on experiences. Absorbing visually presented information is also a common strength. Teachers should be aware that visual materials such as photos, illustrations, and videos are highly motivating, useful teaching aids for most children with PWS. Recent research has indicated that "simultaneous processing" is also a relative strength for individuals with PWS.[6] They are more adept at integrating and understanding information when it is presented as a whole and they understand the "big picture." In contrast, they have more difficulty with tasks involving sequential, or step-by-step, processing. It is important that teachers and parents take these strengths and weaknesses into account when designing educational programs or teaching any new skill.

Other learning difficulties often present in children with PWS fall into distinct areas (Table 11.2). One area of relative weakness is difficulty with short-term auditory memory,[2] which makes it a struggle to remember verbally presented information. Moreover, when a series of verbal directions or a list of steps or objects is presented, the demand

Table 11.1. Characteristic Learning Strengths

Long-term memory for information
Receptive language
Visually based learning through pictures, illustrations, videos
Hands-on experiences
Reading

Table 11.2. Characteristic Learning Weaknesses

Expressive language

Short-term auditory memory

Fine motor skills, related to strength, tone, and motor planning

Interpreting subtle social cues, learning subtle social norms

for understanding and response is compounded by limited expressive ability. It may be that the children have difficulty transferring auditory information from short- to long-term memory. However, when given the opportunity to do this through "rehearsal" and attaching meaning, information can be returned and recalled from long-term memory. The difficulty with being able to remember strings of verbally presented information can be misunderstood as disobedience because the child is unable to sequentially process "pieces" of the directions. It is not uncommon to find students very effective and productive for a given period of time. However, after a while, they appear to "lose it" and need to be retaught, a very frustrating situation for educators. Performance can improve through teaching rehearsal strategies, repeating directions, writing down procedures, using visuals, and modifying verbal instructions. A speech therapist can be helpful in working with the child as well as with the classroom staff to teach these strategies.

Fine motor skills and motor planning tasks, such as writing and drawing, also present as relative difficulties, although a few children are particularly good in this area. Most can improve over time and with practice. Use of a computer should be taught in the classroom beginning as early as kindergarten, and keyboarding can begin to be taught in third or fourth grade. Minimizing writing demands, providing alternative assignments (either color a picture of a train or find one in a magazine), as well as facilitating sufficient opportunities to practice new motor tasks can reduce frustration. Learning is more effective when the right answer can be checked from a multiple-choice format rather than tracing or writing out the answer.

Physical education and therapy are useful for developing strength, coordination, and balance as well as motor planning skills. Gross motor activities such as walking, swimming, and low-impact aerobics are good choices for children with PWS. Scheduling physical activities at predictable times several days each week can positively influence many aspects of life. Children should be encouraged to participate and keep personal records of their gross motor achievements for which they can earn rewards. Physical programming should enhance opportunities for socialization as well as the development of regular, healthy exercise patterns. However, comparisons or competition with others can be discouraging. The record-keeping system should be designed to celebrate personal progress and achievements with guaranteed successes.

While the behavioral and social challenges discussed in the previous section also apply to school-age children, perseveration and obsessive behaviors may intensify. Management should include answering the question once, ignoring repetitions, and changing the subject. Writing down the answer to the perseverative question on a card to which the child can refer can be helpful. A child who repeatedly asks, "Did I do a good job?" may proudly show an "answer card" that says, "I did a terrific job today." A child who continues to ask, "When do we go home?" can refer to a reminder card that says, "School ends at 2:45."

Home/school communication is essential. Sending a notebook back and forth across environments with information about activities, successes, strategies that work particularly well, and any special diet or behavioral issues can maximize consistency and successes for the child.

Adolescence

Adolescence is traditionally a challenging time for children, having to cope with increased pressures coming from all directions. For those with PWS, the growing awareness of the differences between themselves and their non-PWS peers occurs at the same time that being "just like" one's friends is so important. Individuals with PWS who observe changes in their peers (but none in themselves) are likely to demonstrate increased stress, which may create anger, resistance, and food-seeking behaviors that interfere with learning and adaptation.

Schools can benefit from regularly scheduled parent conferences and outside consultation from a professional familiar with PWS. For some teenagers specific behavioral programs can assist with modulating excessive behaviors (see Chapter 12), while others may require counseling or other forms of support. It should be remembered that out-of-control tantrums, excessive skin picking, and what may appear to be oppositional behavior reflect the inability to cope with stress and should be attended to promptly and comprehensively by school and outside specialists.

While challenges exist, there are positive aspects of adolescence. Many teenagers develop effective verbal skills and become active participants and contributors to school activities. This is a time during which vocational planning and work experiences should begin to be pursued (see Chapters 13 and 14). Being a good reader helps, and hobbies should be encouraged. While it is generally difficult for people with PWS to cultivate friendships independently, satisfying relationships are possible, especially if parents, teachers, and other supportive adults in the community are involved. Due to behavioral challenges—regardless of cognitive functioning—most are not yet ready for the full autonomy their same-age peers are achieving; most need continued supervision and protection. Both teachers and parents need to provide extra security in the school and home while seeking ways that adolescents with PWS can increase independence, participate in enjoyable activities, and improve their quality of life.

School programs should capitalize on strengths and teach coping skills for dealing with challenges. As with younger individuals, daily and weekly schedules should be as consistent as possible with minimal transitions. In addition to learning, behavioral, physical, and social considerations, prevocational teaching and planning become critical.

While academic work clearly continues to be a part of school programming, emphasis should be placed on practical uses of learning such as application of math skills to manage money, time, and use of public transportation. Skills necessary for community living should be part of any prework curriculum. Enhancement of social skills needs to continue, and educators should consider offering one of several excellent courses about sexuality that are designed specifically for persons with developmental disabilities.

Some adolescents may benefit from individual counseling in order to express feelings and work through frustrations associated with a desire for independence and conflicts surrounding the issue of "being different." Providing therapeutic counseling and support during a long walk has worked well for some adolescents with PWS, as this model also supports physical well-being while promoting good mental health. A support group of other individuals who face similar challenges is usually very helpful as well. More detailed discussion of educational issues for adolescents with PWS appears in a separate chapter (see Chapter 13).

Conclusion

Children and adolescents with PWS can achieve many successes in school and community-related activities. When educators understand each child as an individual with special strengths and needs, they then can determine how the attributes of PWS can contribute to successes and can assist families, school staff, and ultimately, the children themselves to cope with the challenges associated with the presence of this syndrome and maximize their functioning, independence, and happiness.

Editor's note: This chapter is adapted from "Educational Considerations," by Karen Levine, PhD, and Robert H. Wharton, MD, *Management of Prader-Willi: Syndrome,* 2nd Edition, Springer-Verlag, 1995.

References

1. Cassidy S. Prader-Willi syndrome. *Current Problems in Pediatrics.* 1984;14: 1–55.
2. Curfs LMG, Wiegers AM, Sommers JRM, Borghgraef M, Fryns JP. Strengths and weaknesses in the cognitive profile of youngsters with Prader-Willi syndrome. *Clinical Genetics.* 1991;40:430–434.
3. Dunn L, Dunn L. *Manual for the Peabody Picture Vocabulary Test.* 3rd ed. Circle Pines, MN: American Guidance Service; 1997.
4. Dykens E, Hodapp R, Walsh K, Nash L. Adaptive and maladaptive behavior in Prader-Willi syndrome. *Journal of the American Academy of Child and Adolescent Psychiatry.* 1992;31(6):1131–1136.
5. Dykens E, Hodapp R, Walsh K, Nash L. Profiles, correlates, and trajectories of intelligence in Prader-Willi syndrome. *Journal of the American Academy of Child and Adolescent Psychiatry.* 1992;31(6):1125–1130.

6. Fidler DH, Hodapp RM, Dykens EM. Behavioral phenotypes and special education: parent report of educational issues for children with Down syndrome, Prader-Willi syndrome, and Williams syndrome. *Journal of Special Education.* 2002;36(2):80–88.

7. Gabel S, Tarter RE, Gavaler J, Golden WL, Hegedus AM, Maier B. Neuropsychological capacity of Prader-Willi children: general and specific aspects of impairment. *Applied Research in Mental Retardation.* 1986;7:459–466.

8. Taylor RL. Cognitive and behavioral characteristics. In: Caldwell ML and Taylor RL, eds. *Prader-Willi Syndrome: Selected Research and Management Issues.* New York, NY: Springer-Verlag; 1988:29–42.

9. Warren HL, Hunt E. Cognitive processing in children with Prader-Willi syndrome. In: Holm VA, Sulzbacher SJ, Pipes P, eds. *The Prader-Willi Syndrome.* Baltimore, MD: University Park Press; 1981:161–178.

12

Tools for Psychological and Behavioral Management

Barbara Y. Whitman and Kevin Jackson

Prader-Willi syndrome (PWS) continues to unfold as one of the more complex genetic syndromes. Most medical texts now identify PWS as an obesity syndrome, with accompanying cognitive deficits and behavior difficulties. Certainly the pervasiveness, severity, and impact of the food-related behaviors underscore this description. However, this characterization fails to recognize the primary role of the accompanying behavioral issues in the adjustment and management of affected individuals.

Behaviorally, the young child with PWS is described as happy, personable, affectionate, cooperative, and compliant, despite the emerging hyperphagia.[7] In direct contrast, however, by the time persons with PWS reach chronological adolescence, unless obesity is life-threatening, behavioral difficulties—often severe—become the central issue, both for individuals with the syndrome and their families.[31] While not universal, many by adolescence are described as ego-centered, moody, stubborn, manipulative, and prone to temper tantrums and rages that can result in injury to themselves and others. Emotional instability may appear as depression, obsessions, and frank psychoses. In many instances, these behaviors remain resistant to traditional behavioral and psychopharmacological management and can result in family turmoil, failure of both community living and meaningful employment, and may require mental health consultation, alternative living arrangements, and possibly even institutional placement.

Despite the presence of these behavioral features and their impact on the adjustment of affected individuals, there is a paucity of behavior management research on this population. With rare exception,[12,28,61] behavioral research is descriptive in nature, directed toward determining the presence (or absence) of specific behaviors, skills, or abilities, and the differential impact of the causal genetic mechanism (chromosome 15 deletion versus uniparental disomy) on these characteristics.[8,10,24] While critical for defining "typical" characteristics of affected individuals, these studies provide limited guidance regarding the day-to-day functioning of a specific individual and tend to underestimate the impact of environmental factors affecting behavior. In addition,

Table 12.1. A Comparison of Normal Hypothalmic Functions with the Altered Physiologic Functioning Found in Prader-Willi Syndrome

Normal Hypothalamus	Prader-Willi Syndrome
Regulates appetite	Hyperphagia
Sensitivity to pain	Altered pain sensitivity (skin picking)
Regulates temperature	Temperature instability
Regulates day/night cycle	Altered sleep/wake cycles
Regulates emotions	Emotional excesses
Impacts memory	Short-term memory deficits
Regulates breathing	Central apnea

many studies, in order to have a sufficiently large sample, include an age range encompassing infants through middle to late adulthood, measured at a single point in time.[23,38] Such a methodology makes it difficult to assess age-related behavioral shifts.

The primacy of hypothalamic dysfunction has long been asserted.[33,34,39,59,70] The role of the hypothalamus, juxtaposed against characteristics of PWS, highlights the basis for that assertion (Table 12.1). However, the documented range of intractable behaviors and neurocognitive deficits, coupled with neuroimaging studies,[55,70] indicate that Prader-Willi syndrome: (1) is a *pervasive* neurodevelopmental syndrome, and (2) reflects a distributed central nervous system dysfunction that has yet to be fully described either anatomically or biochemically. From this perspective, the centrally driven, food-related behavioral constellation, albeit dramatic, is just one of the neurobehavioral abnormalities attendant to this disorder and, when appropriately addressed, perhaps the easiest of the behavioral abnormalities to manage. This chapter will detail typical behavioral characteristics associated with PWS, examine critical issues, and offer strategies for effective management. To frame this discussion, it is necessary to briefly review the known neurobehavioral characteristics of affected persons that impact management. A more extensive review can be found elsewhere in this volume (see Chapter 8).

Behavioral Overview

Food

Despite a broad-ranging "typical" behavioral profile, Prader-Willi syndrome remains most readily noted for, and recognized by, an evolving constellation of abnormal food-related behaviors. Neonates typically demonstrate feeding difficulties and an accompanying failure-to-thrive that is, in part, attributed to the characteristic hypotonia (severe) and weak suck. Feeding aids, including special nipples, feeders, and even surgical supplementation, are routinely employed. Feeding difficulties, generally resolved by the toddler period, give way to an evolving constellation of eating concerns including extreme and constant (over) eating, food foraging, food hoarding, preoccupation with food, manipulative and sometimes illicit behaviors directed at obtaining food, lessened liquid intake, and an accompanying aversion to activity even in the absence of obesity.

Part of the continuing (over)emphasis on the abnormal eating associated with PWS is historical. Early diagnosis for PWS is very recent. While the clinical constellation has been recognized since 1956, it was not until the late 1990s that testing provided the ability to genetically confirm the majority (99%) of those affected. Prior to that, diagnosis was often suspected, but delayed, until the emergence of the abnormal eating constellation and obesity. For those with a deletion, the average age at diagnosis was 6 years, and for those with uniparental disomy age 9.[5] Such diagnostic delays often led to an "intertwining" of the food-related behaviors with other behavioral components, so that the full range and impact of the behavioral profile as an issue, separate from the food-related behaviors, was not sufficiently appreciated. In fact, prior to the mid-1980s, uncontrolled obesity with the attendant health consequences was the central focus. Most affected individuals did not live beyond late adolescence or early adulthood. Fortunately, the advent of early diagnosis has allowed increased attention to weight management and the prevention of obesity for many (see Chapter 6). Additionally, the development of group homes specifically for those with PWS during the late 1980s and early 1990s has facilitated the development of effective weight reduction strategies, so that many affected individuals now live into middle and older age.[41] As weight and, concomitantly, health have become less of a central management focus, behavioral issues have become the central management focus.

Personality/Behavior and the Development of Behavior and Emotional Difficulties

Like their nonaffected peers, individuals with PWS demonstrate a range of personality traits, talents, interests, strengths, and weaknesses.[14–16] In addition, the genetic alteration seen in this population may be specifically associated with a number of frequently challenging behavior and personality traits.[29,38] These traits and characteristics (termed a behavioral phenotype), while common among the majority of affected individuals (Table 12.2), may differ in severity between individuals, and may differ in severity in a single individual at different times.

Further, empirical data support clinical observation and parents' reports of a sequence of distinct behavioral/personality "epochs."[21,65,66] As indicated, infants and young toddlers with PWS are usually described as cheerful, compliant, and cooperative. However, with the advent of emerging hyperphagia, several behavioral shifts occur, shifts that are independent of intellectual, language, or motor capacity.[19,20] This initial shift includes increasing oppositional tendencies, stubbornness, rigidity, increasing noncompliance, escalation in both food-related and other temper tantrums, shorter tolerance for frustration, overreaction to frustration, and a lessened ability to "go with the flow."

Table 12.2. Behavioral Comparisons of Those with Prader-Willi Syndrome and Cognitively Impaired Controls

	Percent Occurrence	
Behavior	Individuals with PWS	Cognitively Impaired Controls
Severe temper tantrums, e.g., stamps feet, slams doors	83	26
Gets obsessed with an idea or activity	72	43
Scratches or picks his/her skin	89	27
Eats nonfood items, e.g., dirt, grass, soap	17	14
Gorges on food; will do anything to get food (e.g., takes food out of garbage bins or steals food)	72	14
Stubborn, disobedient, or uncooperative	85	61
Cries easily for no reason or over small upsets	77	43
Shy	13	40
Underreacts to pain	65	36
Moves slowly, underactive, does little (e.g., only sits and watches others)	80	31
Smells, tastes, or licks objects	39	24
Tells lies	74	22
Sleeps too much	54	16
Steals	59	12
Upset and distressed over small changes in routine or environment	72	45
Mood changes rapidly for no apparent reason	63	44
Kicks, hits others	41	41
Abusive, swears at others	52	30
Impulsive, acts before thinking	59	52
Tense, anxious, worried	46	35
Thoughts are unconnected, different ideas are jumbled together, etc.	35	28

Source: Adapted from Einfeld et al., American Journal of Medical Genetics, 1999.[29]

Youngsters exhibit increased difficulty with transitions, even from one activity to another, and develop special routines and rituals that are difficult to avoid or circumvent. For many, attentional deficits are also noted.

A second shift is noted somewhere around 7 to 8 years of age with an escalation of previous patterns and a significantly increased rigidity of behavior,[1,14] including the tendency to "get stuck" on a thought or question, further evidence of attentional deficits impacting academics and behavior, increased worrying about both real and imagined problems, less response to redirection, and less benefit from "time out" and "calming" efforts. The need for invariance is noted along with excessive reactivity to changes in this structure and routine. A falling behind age-matched peers in social skills and the beginning of a chronic anxiety also become evident.

Finally, another shift is associated with the chronological onset of adolescence. The impact of chronic anxiety becomes clearly evident. Increased mood instability frequently is coupled with more outwardly directed expressions of frustration and anger. Immaturity, in the form of insistence on having one's way, regardless of the environment or others affected, often leads to social and relationship problems. For many, there is a continuing increase in reactivity to real or imagined hurts and injustices, and for some, frank emotional and thought disorders. These behavioral epochs occur in both males and females and appear to be independent of cognitive ability and BMI (body mass index).

Recent behavioral studies debate whether the constellation of repetitive speech, special routines, rituals, and skin picking are indicative of a specific psychiatric diagnosis,[26,57] whether the development of these behaviors and psychiatric difficulties is inevitable,[4] and whether an individual's causal genetic mechanism (i.e., deletion vs. uniparental disomy) implies increased risk for these behaviors.[38] Ultimately, these studies aim for a more parsimonious match between problem behaviors and intervention, including medication. To date, however, that yield has been sparse, with most research providing few guidelines for intervention in these behavioral symptoms. There are a number of reasons for this sparseness. First, while many people with the syndrome exhibit a broad range of behaviors that are problematic and unresponsive, in general these behaviors and "symptoms" do not meet criteria for a formal psychiatric diagnosis.[11] Moreover, while most affected individuals demonstrate a constant "baseline" of difficult behaviors that, with patience, are manageable by families without specific intervention, most do, at some point, acutely exhibit more severe behavior problems and outbursts. Often problematic behaviors are limited to a single environment (e.g., school only, home only, workshop only); when the problematic environment is addressed, the behaviors return to baseline. While many affected individuals may have episodes of extreme emotional disruption, and for some that disruption may become chronic, it remains unclear how much of this disruption (if any) is a direct function of Prader-Willi syndrome. It seems clear that affected individuals are more vulnerable to stresses that can lead to emotional disruption, but whether the disruption is differentially intensified due to the underlying genetic influence of Prader-Willi syndrome on emotional reactivity, or whether the disruption is simply a reaction to the concurrent pressures of having Prader-Willi syndrome in combination with environmental difficulties, remains unspecified. There is general evidence that an alteration of the genetic structure and the subsequent altering of neurobehavioral functioning may include an alteration in the way the affected individual perceives and responds to routine environments, normal stresses, and standard reward and punishment strategies. Systematic study of the proclivity toward emotional reactivity, specifically in persons with Prader-Willi syndrome, would be very helpful. An issue to be addressed later is the use of behavior change medication for this group, since a number of risks including an escalation of appetite may be incurred.

Cognition

While the characteristics of PWS include cognitive impairments, primarily mild mental retardation, data document a wide range of intelligence scores among affected individuals. When the findings of a number of studies are pooled, the following percentages emerge: Normal 4.9%, Borderline 27.8%, Mild Mental Retardation 34.4%, Moderate Mental Retardation 27.3%, and Severe to Profound Mental Retardation 5.6%.[13] However, tested IQ scores yield an inflated estimate of ultimate behavioral performance. Several additional factors further depress both academic performance and overall adjustment. First, even when tests indicate IQs in the normal range, cognitive abilities and learning capacities are unevenly developed. Difficulties in understanding and using language have an additional negative impact on cognitive functioning. Studies document deficits in vocabulary, receptive and expressive language, language comprehension, pragmatic language, poor discourse and conversational skills, and shortened length of utterances.[46] Problems in short-term memory as well as learning tasks that are sequential in nature present challenges that are often insurmountable.[25] Most affected individuals remain extremely concrete, rigid, and rule-governed in their thinking, lacking the capacity to develop abstract thinking and concepts. Collectively, these cognitive difficulties impact daily behavior and the overall ability to adapt.

How do these deficits impact day to day behavior? The term *mental retardation* indicates a slower-than-average "rate" of learning, a characteristic that is observed in persons with PWS across all levels of cognitive ability. In addition, it implies a slower-than-average rate of "processing" routine tasks, another phenomenon frequently observed in those with PWS—particularly when responding to precise verbal demands. Thus, if asked a question and rushed to provide an answer, there is often an observable increase in anxiety and frustration. If this happens several times in close succession, increased frustration may escalate to acting out.

Further, persons with PWS—though frequently highly verbal—often have difficulty explaining their actions. Even when behaviors are reasonable, affected individuals are frequently unable to articulate an explanation, escalating their frustration and raising the risk for aggression. Moreover, language-processing difficulties combined with slower processing may result in an individual's grasping only the initial elements of a conversation or set of instructions while additional aspects of the ongoing interaction are missed.

Sequential processing deficits[25] add further complications. Persons with sequential processing deficits have difficulty telling time and comprehending temporal order (days of the week, months of the year), have difficulty integrating elements into temporal or sequentially ordered groups, and have trouble following a schedule or sequence, such as difficulty remembering what to do and when to do it. Figuring out the order in which to tackle the components of a task is usually an ordeal for them.[64] An inability to figure out what needs to be done and the order in which it must be done—especially if the task is not part of the learned, daily, predictable routine—leads to a constant state of being overwhelmed and anxious, with an increased vulnerability to even minor stresses (e.g., being rushed to change from one task to another). Finally, an inability for abstract thinking is reflected in poor social relationships. Persons with PWS recognize only a limited number of emotions, perhaps as few as two to three emotions along a continuum from extremely happy to very sad. Limited recognition of their own as well as others' emotions, coupled with difficulty in understanding another's point of view,[58] predictably leads to problems getting along with others. Those who understand the world only in terms of "I want" or "I don't want" have difficulty with the social conventions and reciprocal behavior necessary for relationships. In sum, this combination of cognitive difficulties leads to difficulties learning and getting along with others. Since the management of the food-related behaviors is, in many ways, the easiest of the behavioral difficulties to manage, we turn to that issue first.

Managing the Food-Related Behavioral Constellation

Medication

One of the many complexities associated with Prader-Willi syndrome is the continuing unresponsiveness of the hyperphagia to conventional pharmacologic intervention. One stream of evidence suggests that the hyperphagia associated with PWS results from a central nervous system failure to signal satiety, rather than traditional hunger mechanisms. This implies that affected individuals are always hungry because they never feel full.[36,37] While some authors locate this failure in the hypothalamus, other authors postulate that dysregulation of one or more specific neurotransmitters causes the hyperphagia[17,30,48,59]; a variety of medications targeting these neurotransmitters demonstrate only spotty success. Currently there are two broad categories of approved pharmacologic interventions: (1) those that decrease food intake by reducing appetite or increasing satiety by their action on central nervous system neurotransmitters (appetite suppressants), and (2) a newer group of those that decrease nutrient absorption by their action in the gut.[69]

Historically (i.e., in the 1960s and 1970s), affected individuals were routinely prescribed appetite-suppressing amphetamines; however, their use was discontinued when they proved ineffective against the hyperphagia while often causing significant behavioral side effects.[67] A more recent group of appetite-regulating medications includes fenfluramine, phentermine, and sibutramine (Meridia). Selikowitz et al.[52] reported statistically significant weight loss in 8 of 15 patients participating in a doubly blinded treatment/placebo study of fenfluramine. A closer examination of the data indicates that 2 of those 8 also lost weight during the placebo portion of the study. Two other participants neither gained nor lost weight during the placebo period but gained weight during the medication trial. Subsequent safety concerns, however, excluded the use of both fenfluramine and phentermine. The side effects of sibutramine as well as those of absorption blockers such as orlistat, make them impractical for the extended use needed by those with PWS.

Based on the theory that the hyperphagia results from increased brain endorphins, Zlotkin et al.[71] tested the appetite-suppressing effects of naloxone in four adolescents with PWS, with no significant effect.

Thus until a safe and effective medication becomes available, management of the food-related aspects of the syndrome remains primarily behavioral and environmental. For nutritional management information and further discussion of appetite research, see Chapter 6.

Behavioral and Environmental Management

Experience of the past two decades has been instrumental in developing effective strategies for managing both weight and the accompanying food-related behavior constellation. Much of this understanding comes from two populations of young adults entering group living settings: (1) young adults whose weight has been well managed and who have never been obese, and (2) young adults who, prior to out-of-home placement, have been extremely obese. This experience underscores that, of the challenging behaviors associated with PWS, the food-related behavioral constellation is the easiest to manage.[38] Simply put, effective management requires the unyielding presence of four elements: (1) a physical environment structured so that food access is completely eliminated, (2) an appropriate dietary and exercise plan, (3) a procedure for ensuring that the affected person is always informed regarding the time and menu for the next meal or snack, and (4) elimination of all other avenues for obtaining food. Where such stringent management is new, the person with PWS may initially act out and attempt to dismantle all boundaries. For many, such behaviors have previously proven effective in undermining management attempts. However, with consistency and patience, this period of acting out usually subsides. Even when behavioral compliance has been accomplished, frequent conversations about food may continue. Many routinely ask what will be served at the next meal, ask when the next meal will be served, ask how the food will be prepared, ask "Is it time yet?", complain that there isn't enough food, and request additional portions. Nonetheless, with management that is both consistent and stringent, behavior eventually stabilizes and cooperation and compliance are less difficult to obtain. Parents and caregivers report that, over time, when the person with PWS realizes that excess food is no longer available, many appear to experience a marked reduction in anxiety and report feeling safer, knowing the limits are effective and inviolable.

Thus, while extremely simple in concept, total elimination of food access can be extraordinarily difficult to accomplish, because to do so requires eliminating food and money accessibility *in every environment*, including access to vending machines or even the remotest possibility of surreptitiously "running down the street to get a Coke." Food of one type or another is regularly found in virtually all human environments—classrooms, school buses, cars, offices, gymnasium dressing rooms, parking lots, playgrounds, and anywhere a trash can is present. Furthermore, many groups routinely fund-raise with some food product: band candy bars, Girl Scout cookies, and fruit cakes and pecans during the holidays are some that may be more difficult to avoid. To fully "food-proof" an environment, it may be necessary to lock cabinets, refrigerators, doors to the kitchen, and garbage cans and to alter traffic patterns for delivering groceries. And, while a family or caregiving setting may be able to make these alterations, equally important, but often more difficult to manage, are additional environments where the affected person may spend time, such as schools, workplaces, recreational programs, churches, and shopping malls. Individualized Education Plans (IEPs) and Individual Habilitation Plans (IHPs) should specify the needed environmental safeguards as part of the plan. However, it is rarely enough to merely specify that environments should be food-free, even when those managing the various setting are fully cooperative. Even the best-intentioned schools, churches, or workplaces will find it challenging and sometimes impossible to eliminate food. For this reason, affected individuals often require the assistance of an aide or increased supervision of some form to prevent food access. While vigilance and awareness is always necessary, there will inevitably be slips that should be handled matter-of-factly and subsequent dietary intake adjusted. Even in well-monitored environments it is essential to educate caregivers to detect pursuit of food and to deter the inevitable and often clever pursuit of food.

Using Food as a Reinforcer

In addition to social praise and attention, a variety of non-food items and privileges can effectively serve as reinforcers for most individuals. However, when several behavior change plans have proven ineffective, the possibility of using food reinforcers is inevitably raised. There currently are several points of view regarding this issue.

One point of view assumes that the nutritional management of PWS is analogous to, and as critical an issue as, the dietary management of diabetes. Therefore, appropriate management includes a restricted diet developed in consultation with a nutritionist or clinical dietitian and the primary care physician. The degree of restriction hinges on individual need; those needing to lose weight may require a more restricted caloric intake, while those needing to maintain weight or who are still in a growth phase may require a less restricted diet. A number of other variables impact caloric requirements, such as energy expended through exercise, use of supplemental growth hormone therapy, and the presence of diabetes. This point of view considers the use of food as a reinforcer "out of bounds." It assumes that such an approach is confusing and undermines the critical medical aspect of dietary requirements. Thus it follows as an absolute that any food that is *outside* the dietary and medical treatment plan cannot be used as a reinforcer. For example, a school setting that allows an "extra" Coke, candy bar, or ice cream to obtain compliance or alter behaviors would be considered both unsafe and inappropriate.

However, there are ways within this frame of reference to effectively use food as a reinforcer, when safely incorporated within the dietary and medical treatment guidelines. For example, individuals with Prader Willi syndrome rarely eat indiscriminately and have individual food preferences that remain constant.[40,42] Appropriately structured, these preferences can be employed as reinforcers. Thus allowing a preferred (and allowable) breakfast food choice (e.g., Cocoa Puffs) can be made contingent on timely completion of a prespecified behavior (e.g., being up, dressed, and ready for school on time). Similarly, special restaurant visits or trips to a store to purchase a food item may be used as a one-time reinforcer or on a regular schedule based on meeting certain specific behavioral goals over time (e.g., being up on time for school each day). In some instances, very low-calorie food items such as sugar-free candies, drinks, or gum may serve as strong reinforcers without significant impact on overall caloric consumption.

In the case of critically important behavior, a specific portion of daily calories may be allocated for use as reinforcement. For example, Keefer[42] examined the exercise behavior of seven individuals with Prader-Willi syndrome. None of the participants in this study exercised during baseline observations. Initially, most of the subjects engaged in some exercise with just verbal prompting. However, even with verbal encouragement, exercise participation diminished over a period of a few days for all but one individual. In phase two, the remaining six individuals were then provided with either extra calories (usually 100 calories) to be used at their discretion or specific extra food items based on completing exercise requirements. All of these participants resumed exercise and ultimately met the program goal of exercising continuously for 45 minutes, three times a week, and their hearts remained in a beneficial aerobic zone. One individual who weighed 500 pounds at the start of the study lost 180 pounds during the course of the project. Notably, no negative behaviors were observed associated with program participation. Predictably, there were problem behavior incidents from participants who exercised but whose efforts failed criteria for earning extra calories. There were no reports of obsessive verbalizing about the exercise or calories being earned.

A different point of view holds that, when used appropriately, and integrated into the overall dietary plan, food is an especially powerful reinforcer in this population. Thus, holders of this view assert that it would be a disservice to fail to take advantage of edibles as a reinforcer when either addressing problem behaviors previously resistant to change or when establishing critical behaviors such as exercise or cooperation with medical procedures or therapies. Since food is an especially powerful reinforcer, therefore, its use for reinforcement should be reserved for priority behaviors.

However, even those who embrace this point of view assert that *any* use of food as a reinforcer should be approached with caution, since the effects of any reinforcement procedure depend on establishing an appropriate link between the desired behavior and the reinforcer and reliably implementing the procedure. To avoid reinforcing inappropriate behavior, food should never be provided in the midst of a behavior incident as a means of stopping the behavior or gaining compliance. Similarly, food should never be provided in response to a threat (e.g., "Give me food or else."). Correct use of food as a reinforcer requires documentation that the individual can perform the desired behaviors. If appropriately designed, the reinforcer will be earned more often than not. The reinforcement plan should be discussed in advance with the participant at a time and place away from any behavior problem incident and in a private location free from interruption. When individuals participate in the development of their own behavior plan, there is greater compliance and an increased likelihood of success. To insure that the plan is clear to the participant, it is helpful for the individual to restate the plan, specifying the expected behavior and the consequences for meeting and not meeting expectations.[45] A written behavior contract can be an effective method for setting up such plans.

General Strategies and Approaches to Intervention

Successful behavior management requires at least three elements: (1) a consistent, supportive environment; (2) strategies for promoting positive behavior; and (3) tools for changing difficult behavior.

Environment

It has long been accepted that much of our behavior is influenced and controlled by the environment—both physical and social. It has also been established that many individuals with genetic disorders have altered and often skewed perceptions of the environment. That those with PWS are extremely sensitive and reactive to the environment also has long been clinically accepted and seems only logical when the impact of multiple cognitive/learning deficits on day-to-day functioning is considered. In addition to heightened environmental sensitivity, persons with PWS tend to overreact to and overemphasize the negative aspects (real or imagined) and underreact to the positive aspects of the environment. Thus the role of the surrounding environment in supporting or impeding positive behavior management and positive adjustment for those with PWS cannot be overemphasized.

Research documents that family environments (as well as school, vocational, and group home/supported living settings) can be characterized on a continuum ranging from chaotic to flexible, structured, and ultimately rigid.[35,49] Both flexible and structured environments are most effective when an individual's neurologic status allows reasonable levels of flexibility or an ability to "go with the flow." Both chaotic and extremely rigid environments are, at a minimum, unsupportive and can be totally unsuitable. As individual flexibility decreases, the need for more highly structured and predictable environments increases. The reduced capacity of those with PWS to be flexible, an inherent neurologically based reactivity, and an inability to employ sequential problem-solving strategies dictate that environmental structures that are routine, predictable, and consistent best minimize frustration and reactivity. This structure should include (1) those inviolate "family and house rules" regarding what is expected, acceptable and unacceptable behavior, along with specified consequences for misbehavior that are consistently administered; and (2) schedules that are, as much as possible, invariant in time and task routines. Consistency between parents or caregivers is critical since inconsistency invites frustration and acting out. In addition, the structure must anticipate and provide carefully designed opportunities for the individual to make choices. A simple example will illustrate. If Mom asks, "What do you want for breakfast?" in an open-ended fashion, it suggests (1) that all choices are valid and available and (2) that any choice will be provided. When the answer is "chocolate cake," the obvious response from Mother is, "You can't have chocolate cake for breakfast!" To the person with PWS, the offer implying an open-ended choice, followed by a denial of the selected choice, is confusing and *feels punitive*, raising the probability of a reactive tantrum. However, if Mom asks, "Do you want eggs or oatmeal for breakfast?" she is still offering a choice but one that is positively structured within the dietary plan, while at the same time providing the affected person with a choice and a sense of personal control. For both Mother and the person with PWS, the choice has been structured as a win-win outcome. In the event that the person with PWS answers, "I want (something different)," Mother has two options. She can respond, "That was not one of the choices; you need to pick either eggs or oatmeal," or she can respond, "OK, just for today I'll allow another choice." Whether Mother allows a different choice this one time in part depends on Mother's previous experience with this child and her assessment of whether granting this one exception is likely to lead to perseverative pleading for more exceptions. When as many positively structured choices as possible are a routine part of daily living, many areas of power struggle and frustration are avoided.

Strategies for Changing Behavior

Appropriately utilized, behavior analytic procedures can be successful in addressing behavior concerns in Prader-Willi syndrome.[3] Most published reports have focused on food-related behaviors,[2,50] but success has also been documented with other behaviors.[6,27] The success of behavioral interventions hinges on the ability of caregivers to provide the necessary intervention with consistency and integrity. In some cases a more technical and professionally designed behavior change program may be necessary to produce an acceptable behavior improvement. This section begins by describing general behavior analytic principles and change strategies, and then focuses on the application of these principles to those with PWS.

The behavioral approach is based on changing the consequences of behavior. The most successful behavioral approach includes a general strategy of emphasizing (or giving positive consequences for) positive social interactions and eliminating coercion.[45] Examples of coercion include questioning, arguing, threatening, using verbal or physical force, criticism, and taking away privileges. Research has shown that many problem behaviors are maintained by social attention and that positive attention can be a powerful consequence for strengthening appropriate behavior.[44] Coercion may stop a given instance of a problem behavior in the short run but, over time, makes things worse. Coercion motivates the desire to escape and to avoid the people and circumstances associated with the coercion. It can also directly trigger inappropriate emotional behavior and often motivates the individual to retaliate or get even with people and places associated with the coercion. These common effects of coercion may be heightened in individuals with Prader-Willi syndrome, who are known for their emotional lability. Coercion also provides social attention to problem behaviors, for many with Prader-Willi syndrome serving as an unintended, *positive* consequence, thereby directly strengthening these behaviors. It is well documented that, when used long-term, coercive social environments are highly predictive of antisocial behavior and other undesirable effects in children.[54,62] Coercion also provides a model of inappropriate behavior that may later be imitated.

The "praise and ignore" or "pivot" tool illustrates a positive change approach. Pivoting involves the use of attention as a reinforcer of appropriate behavior. The pivot tool involves (1) attending to (with praise or physical contact) appropriate behavior on a regular and consistent basis, while (2) ignoring harmless inappropriate behavior. When inappropriate behavior occurs, the strategy is to pivot attention away from the problem behavior to another task or the appropriate behavior of another individual, wait patiently for the targeted individual to engage in some appropriate behavior, then pivot attention and praise back to the individual. For example, if a child is asked to perform some task and responds with resistive arguing, the parent would immediately pivot attention to some other activity and wait until the child begins to do the task, then immediately praise the child while verbally acknowledging the appropriate behavior.

If a desired appropriate behavior is occurring too infrequently to provide sufficient opportunities for reinforcement, the "reinforced practice" tool can be used. This tool is especially appropriate for addressing tasks that the individual with Prader-Willi syndrome may have been consistently resistive to completing. The steps of this tool are: (1) decide what the individual is expected to do, (2) provide opportunities to learn and practice the behavior repeatedly with reinforcement, and (3) reinforce the behavior when it occurs in the real situation. For example, suppose a parent wants the child to carry his/her school materials into the bedroom after school each day instead of just dropping these items inside the door. The parent waits for a time when they both are calm and in a good mood. The parent then (1) introduces the topic in a positive way using actual school materials and (2) has the child pretend he/she is just arriving from school and (3) practice putting the materials in the bedroom. As soon as the behavior occurs, the parent (4) reinforces the behavior by, for example, delivering exaggerated praise, emphasizing that the child has done exactly what the parent expects. These steps should be repeatedly practiced, with reinforcement provided each time the sequence is appropriately performed. To be sure that the youngster fully understands what is being asked, the child should be asked to verbally restate what behavior is expected and when it is to occur. Following the practice session, the parent must closely observe the child the next time he/she returns from school; if the practiced behavior occurs, it should occasion enthusiastic praise.

Once the behavior has occurred a couple of times, it may not be necessary to provide praise at each occurrence of the behavior. If the behavior deteriorates unacceptably at a later time, the parent simply repeats the practice tool.

In addition to praise and attention, individual preferences usually offer a variety of other items or privileges that can successfully reinforce appropriate behavior.[45] Such tangible reinforcers are usually reserved for behaviors that are resistive to change or involve a greater amount of work (e.g., vacuuming several rooms).

Another powerful tool for making use of such reinforcers is the "behavior contract." A behavior contract is a written agreement that an individual will earn a desired item or privilege, based on the occurrence of a specifically defined behavior. Behavior contracts are usually used to address persistent problem behaviors, establish new responsibilities, or improve performance when the individual's responses have been inconsistent. For example, if an individual with Prader-Willi syndrome has trouble getting up and getting ready for school in the morning, and prompting this behavior often leads to tantrums, a contract may be helpful in addressing the morning routine. A contract must specify all tasks to be completed, completion times (e.g., fully dressed and in the kitchen by 7:15 a.m.), and other attitudinal or behavioral parameters (e.g., speaking with an inside voice rather than yelling, as in a tantrum). The contract should also specify a daily consequence for completing these tasks (e.g., sticker on a wall chart) and a long-term consequence (e.g., trip to the pet store) for consistent completion of these tasks (e.g., 4 out of 5 weekdays). Once the behavior is occurring consistently, the tangible reinforcer can be used to address another behavior and the original behavior can be maintained by the occasional praise and more natural consequences.

Dangerous behavior or behavior that is too disruptive to ignore can be addressed with the "stop-redirect-reinforce" technique. This tool requires (1) immediate termination of the behavior by interrupting it verbally and physically, (2) verbal redirection to an appropriate behavior (related, if possible), and (3) immediate provision of positive attention when the redirected (or any appropriate) behavior occurs.

For behaviors that have not responded well to other attempts at change, or when the process of stop-redirect-reinforce has been unsuccessful, "time-out" (i.e., time-out from positive reinforcers—not to be confused with the usual use of the term) may be used. Time-out is a specific detailed protocol that, when correctly employed, can be part of a very effective plan for eliminating serious problem behavior. Time-out should only be used with an individual that can be controlled physically by those expected to implement the time-out. Time-out will not be effective if improperly implemented, as often occurs. When first used, time-out often produces a burst of resistive or aggressive behavior. In some cases this may be so severe that the intervention cannot be used. It is recommended that a reliable reference[45] be consulted in developing a plan for use of time-out. Table 12.3 lists general behavioral analytic tools appropriate to specific categories of problem behaviors often demonstrated by persons with Prader-Willi syndrome. Table 12.4 details a number of specific behaviors associated with Prader-

Table 12.3. Problem Behavior Categories and Tools for Dealing with These Behaviors

Problem Behavior	Behavior to Increase or Reinforce	Behavioral Tool(s)
Annoying behavior that is physically harmless	Playing appropriately	Pivot (praise and ignore)
Arguing	Beginning, working on, or completing a task	
Minor skin picking	The absence of any problem behavior	
Not complying with a request (occasional)		
Not complying with a type of request (regularly)	Complying with the specific type of request targeted for change	• Pivot (praise and ignore) • Reinforced Practice • Behavior Contract
Not performing some assigned task or activity consistently	Performing the assigned task or activity	
Disruptive behavior that cannot be ignored	Specific appropriate behavior	• Stop-Redirect-Give Positive Consequences
Dangerous behavior, or a behavior that occurs in a situation that in the past has resulted in dangerous or disruptive behavior, that cannot be ignored	A positive behavior routine that is expected to occur	• Time-out (not for skin picking) • Reinforced Practice • Behavior Contract

Willi syndrome and specific behavioral analytic techniques that have proven successful for managing those behaviors at home and school.

The Use of Medication for Behavior Management

The use of psychotropic and neuroleptic medications (medications designed to alter behavior, mood, or thought by targeted action on brain chemistry) is increasingly advocated to (1) supplement behavior management strategies or (2) as a primary intervention in the case of severe emotional or cognitive disruption (e.g., depression, psychosis). The use of these medications in this population remains one of the more controversial management strategies, due in part to a paucity of research in this area. With little empirical data to guide clinical decisions, many clinicians have developed prescribing practices based on individual clinical experience. Widely differing, often contradictory experiences with similar medications and dosages result in vastly differing clinical opinions regarding medication use in general and regarding specific medications. An additional source of controversy is lodged in the assertion that medication is frequently used as a substitute for good behavior management, particularly for difficult and challenging behaviors, and particularly in community living settings. This assertion is not unfounded,[47] further complicating decisions regarding medication use in this population.

Given the importance of this issue, it may seem strange that there is so little research in this area. However, this lack of data begins to make

Table 12.4. Behavior Management Strategies for Home and School

Common Behavior Characteristics in Youngsters and Adults with PWS	Possible Management Strategies	
	Home	School
Rigid Thought Process and Inability to Be Behaviorally Flexible Youngsters and adults with Prader-Willi syndrome have a hard time being flexible and accommodating changes of schedule and routine. It is also common for people with PWS to receive and store information in a single order, so that they must go through the whole order to get any one piece. Thus, there is a strong need for routine, sameness, and consistency in the learning environment. Because of a problem in sequence processing, students are not always able to turn *what not to do* into *what to do*.	• Provide weekly and daily schedules in visual form. • Provide verbal anticipation of changes and allow for discussion. Do this in a safe area where person with PWS can share feelings. (The youngster needs time to adapt to this change.) • If there is a change, use visuals; put things in writing—lists, schedules (correct when changes are made). • Don't make promises you can't keep. • Break down tasks and activities into concise, orderly steps. For many youngsters it is helpful to provide pictures of the youngster/adult performing tasks and activities. • To resolve "stubborn issues" try using compromise. Both the parent/caregiver and the youngster/adult have to come up with a totally new solution. Not only is this a successful problem-solving strategy—it can also be a form of diversion. • Provide praise when being flexible.	• Provide weekly and daily schedules in form and, where needed, pictures of the youngster/adult performing tasks and activities. • Provide verbal anticipation of changes and allow for discussion. Do this in a safe area where they can share feelings. (The student needs time to adapt to this change.) • If there is a change, use visuals; put things in writing—lists, schedules (correct when changes are made). • Don't make promises you can't keep. • If able, communicate changes in personnel ahead of time—but not too far ahead. • Break down tasks and activities into concise, orderly steps. For many youngsters it is helpful to provide pictures of the youngster/adult performing tasks and activities. • To resolve "stubborn issues" try using compromise. Both the student and the educator have to come up with a totally new solution. Not only is this a successful problem-solving strategy—it can also be a form of diversion. • Provide praise when being flexible.
Perseverative or Obsessive Thinking This is the tendency to get "caught" on one thought, question, or issue—to the point where it overshadows all other thoughts and activities.	• Use reflection—have youngster/adult restate what you said. • Put in writing; use visuals. Carry a small notebook if needed. • Less is best—give one direction at a time and allow completion before the next direction.	• Use reflection—have student restate what you said. • Put in writing; use visuals. Carry a small notebook if needed. • Less is best—give less amount of work at one time rather than more. Add to the work as time allows.

This behavior can contribute to difficulty in transitioning from one topic/activity to another. Since affected individuals often have a great need to complete tasks, if they are stuck it can lead to loss of emotional control.

Tenuous Emotional Control or Short Frustration Tolerance

Any combination of life stressors can lead to emotional "dyscontrol." The result may be exhibited as challenging behaviors such as tantrums—yelling, swearing, aggression, destruction, self-injury.

During these episodes, the ability to reason is overridden by emotion.

- Avoid power struggles and ultimatums.
- Ignore (if possible).
- Don't give more information than is necessary, especially too far in advance.
- Use "strategic timing"—schedule an activity that the youngster/adult has difficulty ending for right before snack or lunch.
- Set limits: "I'll tell you two more times, then we move on to next topic. This is #1."

- Be aware of "person/environment overstimulation"—for instance large family gatherings, church settings, malls, grocery stores, etc. Have a previously practiced time-out procedure to allow calming down when getting overstimulated and give praise for using.
- Start the day off on the right foot by allowing time to go over the schedule for the day and work through any changes there may be. Putting the new schedule in writing often helps to decrease anxiety.
- At the start of the day, set daily behavioral expectations with the youngster. Limit to no more than 3. Communicate behaviors *you wish to see.* Make it a cooperative task that provides concrete behavior expectations. Put goals in writing. Avoid the word "Don't"... focus on the word "will" (e.g., "I *will* talk in a quiet voice"... instead of "Don't yell." "When I feel frustrated, I *will* tell Mr. Smith or another adult.")

- Avoid power struggles and ultimatums.
- Ignore (if possible).
- Don't give more information than is necessary, especially too far in advance.
- Use "strategic timing"—schedule the activity that the student has difficulty ending for right before snack or lunch.
- Set limits: "I'll tell you two more times, then we move on to next topic. This is #1."

- Be aware of "hallway overstimulation"—especially before the school day begins. Have student enter the building at a less popular entrance. If possible, have arrival time be 5–10 minutes after school starts. Dismiss early.
- Start the day off on the right foot by allowing time to go over the schedule for the day and work through any changes there may be. Putting the new schedule in writing often helps to decrease anxiety.
- At the start of the day set daily goals *with* the student. Limit to no more than 3. Communicate behaviors *you wish to see.* Make it a cooperative task that provides concrete behavior expectations. Put goals in writing. Avoid the word "Don't"... focus on the word "will" (e.g., "I *will* talk in a "quiet voice"... instead of "Don't yell." "When I feel frustrated, I *will* tell Mr. Smith or another adult.")

(Continued)

Table 12.4. Behavior Management Strategies for Home and School (*Continued*)

Common Behavior Characteristics in Youngsters and Adults with PWS	Possible Management Strategies	
	Home	School
Tenuous Emotional Control or Short Frustration Tolerance—continued	• Provide positive attention and praise when youngster/adult is maintaining control, especially in difficult situations. Celebrate success! • Encourage communication and acknowledging feelings. Words are important—Listen carefully! • Include the student in behavior plans. Having their input elicits cooperation and a sense of support. • Be a role model: "I always say 'darn' when I am angry. Let's try that for you . . . darn, darn, darn." Practice when the youngster/adult is *not* agitated or angry. • Depending on the youngster/adult and the situation, use humor. It is often effective. • Similarly to the process for overstimulation, anticipate build-up of frustrations and help him/her to remove self to "safe area." Create a key word or phrase that will alert the student that it is time to go. Practice using these words/ phrases when the student is calm. • Develop a plan and rehearse with the youngster/adult *what to do* if he/she feels angry or frustrated. Many students substitute a means of releasing this pent-up anger—long walks/exercise, ripping paper, tearing rags, popping packaging bubbles. • *Don't try reasoning when student is out of control. Limit discussion.*	• Provide positive attention and praise when student is maintaining control, especially in difficult situations. Celebrate success! • Encourage communication and acknowledging feelings. Words are important—Listen carefully! • Include the student in behavior plans. Having their input elicits cooperation and a sense of support. • Be a role model: "I always say 'darn' when I am angry. Let's try that for you . . . darn, darn, darn." Practice when the student is *not* agitated or angry. • Depending on the student and the situation, use humor. It is often effective. • Anticipate build-up of frustrations and help him/her to remove self to "safe area." Create a key word or phrase that will alert the student that it is time to go. Practice using these words/phrases when the student is calm. • Develop a plan and teach the student *what to do* if he/she feels angry or frustrated. Many students substitute a means of releasing this pent-up anger—long walks/exercise, ripping paper, tearing rags, popping packaging bubbles. • *Don't try reasoning when student is out of control. Limit discussion.* • Have a plan in place if student becomes more violent. Safety for all is a priority. Consistency in approach is imperative.

- Provide positive closure. Don't hold a grudge.
- If using consequences, they should be immediate and help the youngster/adult learn from the outburst—e.g., saying "I'm sorry," sending a note to say they are sorry.

Food-Related Behaviors and Dietary Restrictions

For people with PWS, the message of fullness never reaches the brain—they are always hungry. In addition to this craving for food, food is metabolized at such a slow rate that it causes extraordinary weight gain. Food must be monitored and the individual supervised at all times.

- Make sure access to food is restricted. This may require locked pantries, refrigerators, kitchen doors, and garbage cans.
- Educate and inform *all family members, neighbors, and friends* regarding the restricted dietary needs.
- Be aware of the many "tricks" the person may develop to get extra food beyond that dictated by his/her diet, and be alert for the many clever hiding places he/she may develop to store "illicit" food items.
- Supervise in kitchen, dining room, and in all food-related areas.
- Avoid providing the youngster or adult with money beyond that for a singular item for which he/she may be shopping (under supervision). Lock up all sources of money, including purses. Money buys food!
- Address any stealing or trading of food in private.
- Provide guidelines to school for treats or eating of extra food. Communication with all in school is very important.
- Follow calorie-controlled diet. If a special calorie diet is needed and served by the school, a prescription must be obtained from a health care provider and should be a part of the student's educational plan.

- Provide positive closure. Don't hold a grudge.
- If using consequences, they should be immediate and help the student learn from the outburst—e.g., saying "I'm sorry," sending a note to say they are sorry.
- Make sure lunch is placed with a bus driver and/or an assistant on the ride to school.
- Educate and inform *all people* working with this student, including bus drivers, custodians, secretaries, and volunteers.
- If the student states he/she has not had breakfast, call parents or caregiver before giving more food. (Oftentimes they say this to get more food.)
- Supervise in lunchroom and in all food-related areas–including vending machine areas. In some cases, student may need to eat in classroom (with peer/friend).
- Many require supervision in hallways or near unlocked lockers at all times.
- Avoid allowing the student to have money. Lock up all sources of money, including purses. Money buys food!
- Address any stealing or trading of food in private.
- Follow guidelines for treats or eating of extra food. Communication with home is very important.
- Follow calorie-controlled diet. If a special calorie diet is needed and served by the school, a prescription must be obtained from a health care provider and should be a part of the student's educational plan.

(Continued)

Table 12.4. Behavior Management Strategies for Home and School (*Continued*)

Common Behavior Characteristics in Youngsters and Adults with PWS	Possible Management Strategies	
	Home	School
Food-Related Behaviors and Dietary Restrictions—continued	• Don't delay snack or lunch; if this is necessary discuss ahead. • Limit availability and visibility of food. Be aware of candy dishes and other sources of food. • Praise situations where student does not take food when you see they could have. • Avoid using food as a reward or incentive. • Be aware of smells—there is nothing like the smell of popcorn to make a person with PWS agitated. • When going on a family outing, discuss all food-related issues *ahead of time*. Will you bring snack along or will it be purchased? If purchased, what will it be? Will the outing interfere with the time of a meal or snack? Provide supervision at large family get-togethers so that difficulties are avoided. • Obtain weekly weight if indicated. • Daily exercise should be a part of student's schedule.	• Don't delay snack or lunch; if this is necessary discuss ahead. • Limit availability and visibility of food. Be aware of candy dishes and other sources of food. • Praise situations where student does not take food when you see they could have. • Avoid using food as a reward or incentive. • Be aware of smells—there is nothing like the smell of popcorn to make a student with PWS agitated. • When going on a field trip or other outing, discuss all food-related issues *ahead of time*. Will you bring snack along or will it be purchased? If purchased, what will it be? Will the outing interfere with the time of a meal or snack? • Obtain weekly weight by school nurse if indicated. • Daily exercise should be a part of student's schedule.
Poor Stamina People with PWS tire more easily and may fall asleep during the day. Morning is typically their optimal performance time, when energy level is highest.	• Get person up and moving. Send on errand. Take a walk. • Schedule high energy, mobilizing activity after lunch. • Offer items/activities which stimulate large muscles and deep breathing—balloon blowing, party blowers. • Provide scheduled rest time or a quieter activity if needed.	• Get person up and moving. Send on errand. Take a walk. • Schedule high energy, mobilizing activity after lunch. • Offer items/activities which stimulate large muscles and deep breathing—balloon blowing, party blowers. • Provide scheduled rest time or a quieter activity if needed.

Scratching and Skin Picking

These two behaviors are often seen in individuals with PWS and may be worse during times of stress or boredom. Combined with a higher pain threshold, these behaviors can result in tissue damage if not controlled.

- Use diversion–provide activities to keep hands busy (coloring, computer time, play dough, hand-held games).
- Keep nails short. Apply lotion liberally—it keeps skin slippery. Skin that is soft and moisturized is more difficult to pick. Applying lotion can also be an effective diversion.
- Provide supervision. Reward and praise for not picking.
- Cover area with bandage or similar covering.
- Don't just tell him/her to stop picking—it won't work.
- Apply mosquito repellant before any walks or outside activity.

Difficulty with Peer and Other Social Interactions

While people with PWS want, need, and value friends, they often have difficulty with social interactions. It may also be difficult for them to be exposed to the unpredictability of others for long periods. The need for order often translates into fairness issues and comparing themselves to others, often resulting in anger.

- Many do better in small groups and at times alone.
- Pre-plan outings. Keep time short.
- Provided "supported" social outings, planned activities with a friend.
- Include child in planning activities that are of interest to him/her (board games, puzzles, computer games).
- Provide social skills classes that emphasize sharing, taking turns.

Source: Adapted with permission from "Information for School Staff: Supporting the Student Who Has Prader-Willi Syndrome," compiled by Barbara Dorn and available from the Prader-Willi Syndrome Association (USA).

sense when viewed both from a methodological framework and historically. Methodologically, the "gold standard" of drug studies is the two-group, treated versus control, procedure. This procedure is even more robust if both the patient and the clinician are "blind" as to who receives the medication and who receives the "placebo." However, to conduct such studies requires a medication with a strong indication of effectiveness and a sufficiently large population in which to conduct the study. To date, no medication has met the "strong indication" criterion in those with PWS. Secondly, the size of the patient population, even at the largest PWS centers, is quite small (80 to 120), so that obtaining a sufficiently large treatment and control group is difficult. Thirdly, not all affected individuals need medication; among those who do, not all need the same drug. Some may need a medication for anxiety while others need an antidepressant, further narrowing the size of the eligible research population. Finally, when a person is in sufficient behavioral crisis to require medication, the ethics of assigning them to a "placebo" control group is questionable. Some studies can compare two separate medications for the same disorder, such as two separate antidepressants, but again the overall size of the eligible population at any one center makes such studies extremely difficult.

There are also historical reasons for the lack of studies in this area. Following the original description of the syndrome, many affected individuals were treated both with standard appetite-suppressing medications and with "routine" mood and behavior-altering medications. Subsequently, the use of these medications was discontinued after early therapeutic trials with appetite-suppressing agents, thyroid supplements, diuretics, and psychotropic agents proved unsatisfactory.[7] Additionally, many psychotropic agents increase appetite, while others suppress cognitive capacities, so that, frequently, the use of medications *worsens* behaviors rather than helping. Indeed, following a small study in 1992, Tu et al. declared, "it is becoming increasingly clear that the use of neuroleptic and appetite suppressive medications in PWS is undesirable."[61] Subsequent efforts to find a reliable, appetite-suppressing medication have been, and remain, unsuccessful.[52]

Despite this gloomy report, there emerged several case reports suggesting positive behavioral effects with fluoxetine.[3,18,63] At the same time, a three-wave survey study of 106 persons with Prader-Willi syndrome, conducted between 1989 and 1995, and a separately sampled cohort in 1998 indicated that, at some point in time, most psychotropic medications had been employed for behavioral difficulties, either singly or in combination, with little success.[32,67,68] These studies indicated three medications that reliably modified behavior: haloperidol, thioridazine, and (with some variability) those medications termed selective serotonin reuptake inhibitors (SSRIs), of which fluoxetine was the prototype. Subsequent reports, however, indicate that SSRIs are neither universally effective, nor without side effects, including increased appetite.[43]

The last decade has seen the advent of several "new generation" medications. These have prompted renewed interest in studying medication impact in this population, as evidenced by several recent reports. Positive behavioral effects are reported for seven patients with PWS treated with risperidone.[22] Others, however, report serious side effects from this same medication.[51] Similarly, in two studies totaling 13 patients, 12 had mildly improved mood and some reduction in skin picking with topiramate, a new antiepileptic medication.[53,56] In the latter study, however, 4 of the 7 patients were also on other medications, primarily SSRIs. In addition, while several patients had a modest improvement in the amount and severity of skin picking, one displayed markedly worsened picking behaviors and initiated hair pulling, requiring additional medication for this behavior. Further reports suggest that the use of these medications must be approached with extreme caution, as a number of serious side effects have been reported (see Chapter 17).

Thus, it appears that the medication picture is not as bleak as that described by Tu[61] only a decade ago. A range of effective medications is currently available. However, as in the general population, some patients respond well to one medication and not to another, so trials of several medications may be necessary. And, while sparse, the available data do suggest some cautions to consider when prescribing for this population. These cautions are both physiologic and behavioral.

First, due to an altered body composition and metabolism, persons with Prader-Willi syndrome often have extremely idiosyncratic, frequently unpredictable, and often negative responses to medications. Additionally, many demonstrate much lower dose thresholds and tolerances and longer drug half-lives than those of a nonaffected population.[9] Further, symptom response may be evident much sooner than in other populations. It is not unusual to observe improvement with SSRI therapy in as little as 24 hours, particularly in the area of improved flexibility and reduced irritability. This rapidity of response has led to speculation that the effect may be nonspecific rather than as a classical antidepressant agent. Finally, side effects may be different, more intense, and longer lasting than in other populations.

An additional caution concerning the use of medication concerns the meaning of behavior. As has been described, many difficult behaviors are commonly present in the behavioral phenotype of persons with PWS, behaviors that often become exaggerated in frequency or severity with stress or physical illness. Many parents report feeling pressured to medicate their child when behaviors become exaggerated in response to stress, rather than altering the environment to manage the stress. For instance, the person unaffected by Prader-Willi syndrome facing the normal stress and grief that accompanies the loss of a parent experiences a long period of sadness often accompanied by irritability, depressed levels of performance, and withdrawing from social interactions. For most, time and "just getting through it" is the recommended treatment. Too frequently, a similar period of accommodation is denied the person with PWS under similar stress, as the change in behavior is ascribed to a deterioration of brain mechanisms associated with PWS rather than recognizing that the person with PWS may have a more intense and prolonged, yet normal, reaction to the stress of grief. Many parents report feeling that the rush to medication is for the convenience of the service provider (e.g., group home staff, teacher) rather than the needs of the person with PWS. Multiple experiences of this sort render many parents reluctant to consider medication, even when it is appropriate. Thus the use of these medications remains a tool that must be employed with great caution.

References

1. Akefeldt A, Gillberg C. Behavior and personality characteristics of children and young adults with Prader-Willi syndrome: a controlled study. *Journal of the American Academy of Child and Adolescent Psychiatry.* 1999; 38(6):761–769.
2. Altman K, Bondy A, Hirsch G. Behavioral treatment of obesity in patients with Prader-Willi syndrome. *Journal of Behavioral Medicine.* 1978;1(4): 403–412.
3. Benjamin E, Buot-Smith T. Naltrexone and fluoxetine in Prader-Willi syndrome. *Journal of the American Academy of Child and Adolescent Psychiatry.* 1993;32(4):870–873.
4. Boer H, Holland AJ, Whittington JE, Butler JV, Webb T, Clarke DJ. Psychotic illness in people with Prader-Willi syndrome due to chromosome 15 maternal uniparental disomy. *Lancet.* 2002;359:135–136.
5. Butler MG, Meaney F, Palmer CG. Clinical and cytogenetic survey of 39 individuals with Prader-Labhart-Willi syndrome. *American Journal of Medical Genetics.* 1986;23:793–809.
6. Carpenter S. Development of a young man with Prader-Willi syndrome and secondary functional encopresis. *Canadian Journal of Psychiatry.* 1989; 43:123–126.
7. Cassidy S. Prader-Willi syndrome. *Current Problems in Pediatrics.* 1984;14: 1–55.
8. Cassidy SB, Forsythe M, Heeger S, et al. Comparison of phenotype between patients with Prader-Willi syndrome due to deletion 15q and uniparental disomy 15. *American Journal of Medical Genetics.* 1997;68:443–440.
9. Clarke DJ. Self injurious and aggressive behaviors. In: O'Brien G, ed. *Behavioral Phenotypes in Clinical Practice.* London: MacKeith Press; 2002:16–30.
10. Clarke D, Boer H, Webb T, et al. Prader-Willi syndrome and psychotic symptoms: 1. Case descriptions and genetic studies. *Journal of Intellectual Disability Research.* 1998;42(6):440–450.
11. Clarke DJ, Boer H, Whittington J, Holland A, Butler J, Webb T. Prader-Willi syndrome, compulsive and ritualistic behaviors: the first population-based survey. *British Journal of Psychiatry.* 2002;189:358–362.
12. Coleman, D. *Enhancing Self-Management Skills in People with Prader-Willi Syndrome.* Scottsdale, AZ: The Devereaux School in New York, Red Hook; 1993.
13. Curfs L. Psychological profile and behavioral characteristics in the Prader-Willi syndrome. In: Cassidy SB, ed. *Prader-Willi Syndrome and Other Chromosome 15q Deletion Disorders.* New York, NY: Springer-Verlag; 1992:211–221.
14. Curfs LMG, Hoondert B, van Lieshout CFM, Fryns JP. Personality profiles of youngsters with Prader-Willi syndrome and youngsters attending regular schools. *Journal of Intellectual Disability Research.* 1995;39(3): 241–248.
15. Curfs LMG, Verhulst F, Fryns JP. Behavioral and emotional problems in youngsters with Prader-Willi syndrome. *Genetic Counseling.* 1991b;2(1): 33–41.

16. Curfs LMG, Wiegers AM, Sommers JRM, Borghgraef M, Fryns JP. Strengths and weaknesses in the cognitive profile of youngsters with Prader-Willi syndrome. *Clinical Genetics.* 1991a;40:430–434.

17. Dahir GA, Butler M. Is GABA-A receptor B subunit abnormality responsible for obesity in persons with Prader-Willi syndrome? *Dysmorphology and Clinical Genetics.* 1991;5:112–113.

18. Dech B, Budow L. The use of fluoxetine in an adolescent with Prader-Willi syndrome. *Journal of the American Academy of Child and Adolescent Psychiatry.* 1991;30:298–302.

19. Dimitropoulos A, Feurer ID, Butler MG, Thompson T. Emergence of compulsive behavior and tantrums in children with Prader-Willi syndrome. *American Journal of Mental Retardation.* 2001;106(1):39–51.

20. Dimitropoulos A, Feurer ID, Roof E, et al. Appetitive behavior, compulsivity, and neurochemistry in Prader-Willi syndrome. *Mental Retardation and Developmental Disabilities Research Reviews.* 2000;6(2):125–130.

21. Donaldson MD, Chu CE, Cooke A, Wilson A, Greene SA, Stephenson JB. The Prader-Willi syndrome. *Archives of Disease in Childhood.* 1994;70(1): 58–63.

22. Durst R, Rubin-Jabotinsky K, Raskin S, Katz G, Zislin J. Risperidone in treating behavioral disturbances of Prader-Willi syndrome. *Acta Psychiatrica Scandinavica.* 2000;102:461–465.

23. Dykens E, Cassidy S. Correlates of maladaptive behavior in children and adults with Prader-Willi syndrome. *American Journal of Medical Genetics.* 1995;60(6):546–549.

24. Dykens E, Cassidy S, King B. Maladaptive behavior differences in Prader-Willi syndrome due to paternal deletion versus maternal uniparental disomy. *American Journal of Mental Retardation.* 1999;104(1):67–77.

25. Dykens E, Hodapp R, Walsh K, Nash L. Profiles, correlates, and trajectories of intelligence in Prader-Willi syndrome. *Journal of the American Academy of Child and Adolescent Psychiatry.* 1992;31(6):1125–1130.

26. Dykens E, Leckman J, Cassidy SB. Obsessions and compulsions in Prader-Willi syndrome. *Journal of Child Psychology and Psychiatry.* 1996;37:995–1002.

27. Edmonston NK. Management of speech and language impairment in a case of Prader-Willi syndrome. *Language, Speech and Hearing Services in Schools.* 1982;13:241–246.

28. Eiholzer U, Nordmann Y, L'Allemand D, Schlumpf M, Schmid S, Kromeer-Hauschild K. Improving body composition and physical activity in Prader-Willi syndrome. *The Journal of Pediatrics.* 2003;142:73–78.

29. Einfeld S, Smith A, Durvasula S, Florio T, Tonge B. Behavior and emotional disturbance in Prader-Willi syndrome. *American Journal of Medical Genetics.* 1999;82(2):123–127.

30. Gabreels BA, Swaab DF, Seidah NG, van Duijnhoven HL, Martens GJ, van Leeuwen FW. Differential expression of the neuroendocrine polypeptide 7B2 in hypothalami of Prader-(Labhart)-Willi syndrome patients. *Brain Research.* 1994;657(1–2):281–293.

31. Greenswag LR. Adults with Prader-Willi syndrome: a survey of 232 Cases. *Developmental Medicine and Child Neurology.* 1987; 29:145–152.

32. Greenswag LR Whitman BY. Long-term follow-up of use of Prozac as a behavioral intervention in 57 persons with Prader-Willi syndrome. Proceedings, Second Prader-Willi Syndrome International Scientific Workshop and Conference, Oslo, Norway; June 1995.

33. Hanchett J. The brain and PWS. Paper presented at the Ontario PWSA annual Conference, Ontario, Canada; October 1993.

34. Hauffa B. *Prader-Willi Syndrome: A Clinical Overview.* Berkshire, UK: Colwood Healthworld; 2000.

35. Hoffman CH, Aultman D, Pipes P. A nutrition survey of and recommendations for individuals with Prader-Willi syndrome who live in group homes. *Journal of the American Dietetic Association.* 1992;92:823–833.

36. Holland AJ, Treasure J, Coskeran P, Dallow J, Milton N, Hillhouse E. Measurement of excessive appetite and metabolic changes in Prader-Willi syndrome. *International Journal of Obesity.* 1993;17:526–532.

37. Holland AJ, Treasure J, Coskeran P, Dallow J, Milton N, Hillhouse E. Characteristics of the eating disorder in Prader-Willi syndrome. *Journal of Intellectual Disability Research.* 1995;39:373–381.

38. Holland AJ, Whittington J, Butler J, Webb T, Boer H, Clarke D. Behavioural phenotypes associated with specific genetic disorders: evidence from a population based study of people with Prader-Willi syndrome. *Psychological Medicine.* 2003;33:141–153.

39. Inui A, Uemoto M, Takamiya S, Shibuya Y, Baba S, Kasuga M. A case of Prader-Willi syndrome with long-term mazindol treatment. *Archives of Internal Medicine.* 1997;157(4):464–465.

40. Joseph B, Egli M, Koppekin A, Thompson T. Food choice in people with Prader-Willi syndrome: quantity and relative preference. *American Journal of Mental Retardation.* 2002;107(2):128–135.

41. Kaufman H, Overton G, Leggott J, Clericuzio C. Prader-Willi syndrome: effects of group home placement on obese patients with diabetes. *Southern Medical Journal.* 1995;88(2):182–184.

42. Keefer NL. *Differential Reinforcement of Exercise by Individuals with Prader-Willi Syndrome.* Gainesville, FL: University of Florida; 2003.

43. Kohn Y, Weizman A, Apter A. Aggravation of food-related behavior in an adolescent with Prader-Willi syndrome treated with fluvoxamine and fluoxetine. *International Journal of Eating Disorders.* 2001;30(1):113–117.

44. Ladd GW. Peer relationships and social competence during early and middle childhood. *Annual Review of Psychology.* 1999;50:333–359.

45. Lathan GI. *The Power of Positive Parenting.* North Logan, UT: P & T Ink; 1998.

46. Lewis B, Freebairn L, Heeger S, Cassidy S. Speech and language skills of individuals with Prader-Willi syndrome. *American Journal of Speech-Language Pathology.* 2002;11:285–294.

47. Lewis MA, Lewis C, Leake B, King BH, Lindemann R. The quality of health care for adults with developmental disabilities. *Public Health Reports.* 2002;117(2):174–184.

48. Martin A, State M, Koenig K, et al. Prader-Willi syndrome. *American Journal of Psychiatry.* 1998;155(9):1265–1273.

49. Olson DH. Circumplex model of marital and family systems. *Journal of Family Therapy.* 2000;22:144–167.

50. Page TJ, Stanley AE, Richman GS, Deal RM, Iwata BA. Reduction of food theft and long-term maintenance of weight loss in a Prader-Willi adult. *Journal of Behavioral Therapy and Experimental Psychiatry.* 1983;14(3): 261–268.

51. Phan TG, Yu RY, Hersch MI. Hypothermia induced by risperidone and olanzapine in a patient with Prader-Willi syndrome. *Medical Journal of Australia.* 1998;169(4):230–231.

52. Selikowitz M, Sunman J, Prendergast A, Wright S. Fenfluramine in Prader-Willi syndrome. A double blind placebo controlled trial. *Archives of Disease in Childhood.* 1990;54:112–114.

53. Shapira NA, Lessig M, Lewis MH, Goodman WK, Driscoll DJ. Effects of topiramate in adults with Prader-Willi syndrome. *American Journal of Mental Retardation.* 2004;109(4):301–309.

54. Sidman M. *Coercion and Its Fallout*. Boston, Mass: Authors Cooperative; 2001.
55. Sieg K. Advances in understanding the neurobiology of Prader-Willi syndrome. *Prader-Willi Perspectives*. 1995;3(2):8–12.
56. Smathers S, Wilson J, Nigro M. Topirimate effectiveness in Prader-Willi syndrome. *Pediatric Neurology*. 2003;28(2):130–3.
57. State MW, Dykens EM. Genetics of childhood disorders: XV. Prader-Willi syndrome: genes, brain, and behavior. *Journal of the American Academy of Child and Adolescent Psychiatry*. 2000;39(6):797–800.
58. Sullivan K, Tager-Flusberg H. Empathetic responses of young children with Prader-Willi syndrome to the distress of others. Proceedings, 13th Annual Prader-Willi Syndrome Association (USA) Scientific Conference, Columbus, OH; July 1998.
59. Swaab DF. Alterations in the hypothalamic paraventricular nucleus and its oxytocin neurons (putative satiety cells) in Prader-Willi syndrome: a study of five cases. *Journal of Clinical Endocrinology and Metabolism*. 1995;80: 573–379.
60. Swillen A, Borghgraef M, Descheemaeker MJ, Plissart L, Rasenberg S, Curfs LMG. The effects of a curative self-supporting program on youngsters and adults with the Prader-Willi syndrome. Proceedings, 2nd Prader-Willi Syndrome International Scientific Workshop, Oslo, Norway; June 1995.
61. Tu J, Hartridge C, Izawa J. Psychopharmacogenetic aspects of Prader-Willi syndrome. *Journal of the American Academy of Child and Adolescent Psychiatry*. 1992;31(6):1137–1139.
62. Walker HM. *The Acting-Out Child: Coping with Classroom Disruption*. Longmont, CO: Sopris West; 1995.
63. Warnock JK, Kestenbaum T. Pharmacologic treatment of severe skin-picking behaviors in Prader-Willi syndrome. Two case reports. *Archives of Dermatology*. 1992;128(12):1623–1625.
64. Whitman BY. Understanding and managing the behavioral and psychological components of Prader-Willi syndrome. *Prader-Willi Perspectives*. 1995;3(3):3–11.
65. Whitman B, Accardo PJ. Emotional symptoms in Prader-Willi adolescents. *American Journal of Medical Genetics*. 1987;28:897–905.
66. Whitman B, Accardo PJ. Prader-Willi syndrome. *Journal of the Royal Society of Medicine*. 1989;82(7):448.
67. Whitman B, Greenswag LR. The use of psychotropic medications in persons with Prader-Willi syndrome. In: Cassidy SB, ed. *Prader-Willi Syndrome and Other Chromosome 15q Deletion Disorders*. New York, NY: Springer-Verlag; 1992:23–33.
68. Whitman BY, Greenswag LR, Boyt M. The use and impact of psychotropic medications for managing mood and behavior in persons with Prader-Willi syndrome. Proceedings, 3rd Prader-Willi Syndrome International Scientific Workshop and Conference, Lido Di Jesolo, Italy; May 1998.
69. Yanovski S, Yanovski J. Drug therapy: obesity. *The New England Journal of Medicine*. 2002;346(8):591–602.
70. Yoshii A, Krishnamoorthy K, Grand PE. Abnormal cortical development shown by 3D MRI in Prader-Willi syndrome. *Neurology*. 2002;59:644–645.
71. Zlotkin SH, Fettes I, Stallings VA. The effects of naltrexone, an oral beta-endorphin antagonist, in children with the Prader-Willi syndrome. *Journal of Clinical Endocrinology and Metabolism*. 1986;63(5):1229–1232.

13

Educational and Social Issues for Adolescents with Prader-Willi Syndrome

Barbara J. Goff

Adolescence—ages 13 to 21—can be a trying time for the individuals experiencing it, as well as for their caregivers. This is no less true for the child with a disability. Parents and teachers need to understand typical developmental experiences during this time in order to recognize experiences and behaviors that may be reflective of a child's particular disability.

It's All About Change

The transitions from elementary school to middle school and again to high school pose dramatic changes for any student. The safety and security of being in a single classroom in a familiar building with the same teacher for at least one, possibly several years is suddenly taken away. Classmates who have been together for many years are dispersed. For some students, this provides an exciting opportunity to try out new behaviors and put on a more grown-up persona. For many students with disabilities, it can be the beginning of a special kind of loneliness and rejection. This is especially true for students with significant cognitive and social deficits such as those with Prader-Willi syndrome (PWS).

It is not unusual for a student with PWS to be several academic grade levels behind his/her same-age peers. This may not have been a significant problem in the student's earlier class placement, where there was a single teacher who knew the student's capabilities in all areas and could adapt the environment and curriculum accordingly. The social gap between the student with Prader-Willi syndrome and his/her classmates also may have widened, as peer relationships now require greater sophistication and savvy. For many, an additional significant change involves the loss of the one-to-one paraprofessional assigned to him/her for a good part of the elementary school years.

The student placed in an inclusive program may be involved with several teachers and potentially new groups of students in each class. Each teacher has his/her own style, expectations and, perhaps, beliefs about the student with PWS. Most, if not all, will have had no prior experience working with students with PWS; further they will not have the luxury of getting to know the student every day, all day. Thus, a great deal of collaboration and contact between school and home must occur on a routine basis.

Students placed in self-contained classrooms have a greater possibility of being with at least one or two familiar classmates. Typically in this setting there is one teacher with one or more paraprofessionals. Physical transitions are fewer and a consistent routine is more likely.

Some students may transfer to a special school for students with learning disabilities at this time. Here, too, staff need to be prepared for working with the student with PWS. The student, on the other hand, may need reassurance that the school change is not the result of some kind of failure, but an opportunity to better learn and to make new friends.

In all situations, however, administrative, clinical, and support staff are new, requiring the establishment of new relationships. Further, all who will now interact with the student throughout the day have a great deal to learn. Caregivers must consider and plan for how this learning and relationship-building will occur.

Moving from classroom to classroom throughout the day poses a special challenge for many individuals with PWS. The building is probably much larger than the previous school building, and locations of the various classrooms and offices must be learned quickly. Further, changing classes means moving fairly quickly from room to room so as not to be tardy. This requires prior organization of needed materials for each class.

Larger buildings also mean more food sources, a source of increased stress. It also means bigger and more chaotic cafeterias with food service staff that are unfamiliar with the drives of a student with PWS. It means resisting the many opportunities to obtain food (e.g., open lockers, offices, snack machines). All present significant challenges for the person with Prader-Willi syndrome, challenges that should not be underestimated.

Transportation to and from school will also likely be affected. There may be a new driver and bus monitor, new children on the bus, and a new route requiring a new pick-up and drop-off time. This change alone can be very stressful for the student with Prader-Willi syndrome and should be considered when designing transition plans.

Students with PWS often experience increased stress and anxiety when facing the many changes required, feelings that are sometimes manifested in unacceptable ways. The student may be "in trouble" more often, with the result that the family is getting more and more phone calls from the school. In some instances, the school may decide that it can no longer serve the student within its existing programs and may recommend placement in a special class or school for students with behavior difficulties. Such placements have not proven optimal for the student with PWS. Schools for children and adolescents with "routine behavior disorders" focus on intervention strategies aimed at returning the student to a less restrictive program. These strategies assume that behavior can be substantially altered using traditional behavior modification approaches. Since most such strategies are ineffective for those with Prader-Willi syndrome, placement in traditional "behavior disordered" programs often results in worsened rather than improved behavior for the student with PWS. Programs utilizing environmental modifications and a variety of situation-specific preventative strategies are most effective for managing PWS-associated behaviors. Traditional rewards and consequences strategies—no matter how seemingly powerful—cannot override some of the neurological drives inherent in the syndrome.

With these background caveats in mind, let us consider some transition strategies for assisting the student with PWS in moving to middle and high school.

Transition to Middle or High School: Paving the Way

One of the most successful strategies for an educator is to "be prepared." Training and support in working with the student with Prader-Willi syndrome should be the first priority for middle or high school personnel unfamiliar with the affected student's special needs. Parents may need some additional support as well. When all people who work with the student with PWS are informed and environmental concerns have been addressed, there is a greater chance for a positive learning experience to occur.
Strategies:

- The student's multidisciplinary team should meet as early as possible in the school year prior to the planned transition in order to identify the specific program and services the student may need, and to develop a transition plan.
- If a change to a new program or school is necessitated, parents and school personnel should visit to ascertain the fit between the program's services and the student's needs. One criterion for future success is that the new program or school demonstrates openness to learning about Prader-Willi syndrome from previous teachers, parents, and experts in the field.
- Once a program and setting are determined, visits for the student to tour the building, learn where his/her primary classroom is located, and meet the teacher(s), administrators, and clinical staff should be arranged. If the student will have more than one teacher, a key teacher who will be the decision-maker and primary communicator should be identified. A regular and easily used communication system between other school personnel and the key teacher, and between the key teacher and the family or caretakers, should be established. For caregivers, e-mail may be preferable to sending a notebook back and forth in order to prevent the student from ripping out pages or "losing" the notebook.
- The daily schedule (if known), especially the snack and lunch schedule, as well as where food will be eaten, should be reviewed with the student.

- Similarly, school or classroom policies or procedures impacting the student (e.g., code of conduct, dress codes, behavior management programs) should be reviewed. If the student takes issue with any of these policies or procedures, a meeting with the principal to review the policies often resolves the problem.
- Environmental changes that need to be made prior to the school year should be determined and a plan for their implementation developed. For example, is food kept in classrooms? Where are lunches and snacks kept? Should the student eat in the cafeteria, and if so, what degree of supervision is needed? If there is a food preparation component to any of the student's classes, how will this be managed while considering the needs and abilities of the student?
- Does the student need an adapted physical education program? Will there be opportunities for the student to engage in physical activity beyond the one time per week in a physical education class?
- A meeting should be held with personnel from the transportation staff or company, especially the driver and bus assistant/monitor, to discuss the nature of PWS and the specific needs of the student.
- It is often helpful for the student to view pictures of school personnel prior to starting the school year, particularly if the student is very anxious. Therefore, caregivers may want to photograph the school, the assigned classroom, and the key teacher(s). This way, over the summer months, caregivers can use visuals to review what will be happening and with whom during the next school year.
- Finally, assuming all the basics are in place for the next school year, it is advisable to set up a time in the fall for a training session on PWS. The advantage of waiting until after the student has been in the new program for a few weeks is that school personnel will have a chance to know the student; subsequent training will then have more meaning and student-specific questions can be addressed.

Behavior Challenges

The typical characteristics associated with PWS persist throughout adolescence. Certain behavioral characteristics, such as those associated with obsessive-compulsive disorder (e.g., hoarding and skin picking) often become progressively worse as the child moves into the teen years.[2] However, this is not universal.

Some caretakers report an increased skill, ability, and creativity in obtaining food or other desired items or activities. It is not unusual to hear parents and teachers report that the adolescent is suddenly doing all kinds of food seeking that he/she never did before. Temper tantrums and "shutdowns" that they may have had throughout elementary school now take on a new dimension, with increased intensity and keener awareness of how to obtain desired responses. The adolescent may also experience greater and more exacting demands, which produce greater stress and confusion, resulting in an increased frequency and intensity of difficult behaviors. The majority of individuals with PWS (75% to 90%) do exhibit acting-out behaviors, including temper tantrums, impulsivity, aggression, and stubbornness to a greater degree and with greater intensity than individuals with mild or moderate mental retardation with other etiologies.[1,4] This might cause a new teacher to erroneously conclude that the behaviors of the student with PWS are worsening, when comparing them with those of children with other developmental disabilities.

Research has found that older children with PWS are more likely to experience withdrawal and depression relative to younger children, but these symptoms may abate in adulthood.[2] Weight may also affect behaviors, particularly individual, internal maladaptive behaviors such as mood disorders; those who are thinner demonstrate more maladaptive behaviors than those who are heavier.[3,9] Studies have found more symptoms of psychosis (confused thinking, delusions, and hallucinations), depression (anxiety, fearfulness, and sadness), and characteristics of dependent personality among thinner individuals with PWS.[3,9] While many perceive that individuals with higher IQs have greater and more severe maladaptive behaviors, the evidence suggests no significant differences based on IQ.[3]

Since there is significant individual variation in behavior, behavior management strategies or interventions need to be person-specific. This necessitates involving the interdisciplinary team, including the family, and the individual if appropriate. Behavior management strategies take a different shape in the adolescent and young adult years with an increasing emphasis on self-awareness and self-monitoring. Sticker charts and verbal praise are no longer adequate in providing needed supports. Behavior management strategies have far more success if the student had input in the development of the strategies. Thus, one needed focus is on the individual's recognizing his/her problematic behaviors and agreeing to change them. A common example would be skin picking. Currently there is no known cure, nor is it clear the extent to which individuals can control the urge, but it may be possible to more easily redirect an individual when that person has recognized the problem and agreed to address it. For this reason behavior contracts are generally a more effective strategy than sticker charts for the adolescent and young adult (see Figure 13.1). As with all teenagers, students with PWS react very negatively to any perceived condescension. There must be a *genuine* foundation of respect and appropriate levels of involvement to ensure the maximum effectiveness of any behavioral strategies. While for many teenagers specific behavioral programs can assist with modulating excessive behaviors, others may require ongoing counseling or other forms of support.

Food/Nutrition

Larger schools have a greater number of food sources available. Middle and high school cafeterias often provide numerous food choices in addition to a relatively wholesome school lunch. Many offer soda and bagels, chips and candy, as well as a variety of dessert options. The typically developing adolescent does not choose to spend money on the school lunch (unless it's pizza), opting *instead* for soda and snack

STUDENT CONTRACT

Name_____ Date_____

School_____Class_____

_____agrees to work toward the following personal goals:

GOAL

    1. To let my paraprofessional or teacher know when I am getting upset.

- **WHAT I WILL DO**

When I am feeling upset, I will say to the para or teacher, "I'm getting upset and I need to_____(e.g., go to identified private space, take a 5-minute walk with the para, go outside the classroom and talk to the para…) *For individuals who may not be able to articulate their needs, provide a set of feeling cards (drawings of various facial expression, such as mad, glad, sad, scared. The goal then will be to show the para or teacher the card that represents the student's feelings.*

GOAL

    2. To get to my classes on time.

- **WHAT I WILL DO**

When the para or teacher gives me a five-minute warning that it is time to finish what I am doing and prepare to move on to the next class, I will work for two more minutes, and then put my work into the TO BE DONE folder. I will make sure I have everything I need in my bookbag for the next class. I can complete my work during lunch period, if I choose, or take it home.

I understand that at the end of each day, I will review these goals with the para or teacher.

- If I was able to appropriately express my feelings throughout the day, I will receive_____.

- If I was able to get to my classes on time, I will receive_____.

I AGREE TO DO MY BEST.

Student_____ Family Member_____

Teacher_____ Paraprofessional_____

Principal_____

Figure 13.1. Sample student behavior contract.

foods. Individuals with PWS will choose the school lunch *in addition to* the soda, snacks, and desserts unless closely supervised. Research suggests that individuals with PWS are more likely to choose a larger quantity of a less-preferred food over a lesser quantity of a more-preferred food.[5] This provides an opportunity for the student's caretaker or cafeteria aide to promote healthy choices, such as offering a large bowl of chef salad as an alternative to a lesser quantity of macaroni and cheese. Many parents report that as their children age and their social world expands, so too do their exposure and opportunities to access food. Therefore, it is imperative that educators work with parents in planning for any events that involve food.

Fitting In: The Adolescent Struggle

The social success of a typically developing adolescent depends, in large measure, on how well they can follow fashion, music, and other leisure time trends. Most teenagers are very self-conscious and quite critical of themselves and others. The student with Prader-Willi syndrome may have had some very good relationships with other classmates up until this point; in this period, however, the academic and social gap widens significantly. The typically developing student's tolerance for the immaturity and rigidity of the student with PWS frequently lessens, particularly for episodes of lying, stealing, and occasional aggressiveness, which are not uncommon in those with the syndrome. The student with PWS may not understand why classmates are distancing themselves both physically and psychologically from him, or even teasing or bullying him, and therefore may try even harder to be part of a social group. This lack of a social group often precipitates a transfer of the student with PWS out of inclusive situations and mainstream classes into smaller, less-integrated special education settings. Optimally, other solutions will be employed.

There are many students with social skills deficits who have disabilities other than Prader-Willi syndrome. They may be students with learning or emotional problems, sensory or neurological impairments, or students with medical disorders or physical disabilities. This constitutes a fairly large population of students who would benefit from formal social skills training. While teachers seek to take advantage of teachable moments in developing social skills in their students, this is sometimes inadequate.

It has been shown that the most effective social skills training occurs in small groups with students meeting (at least) weekly throughout the school year, facilitated by an adult proficient in social skills training.[6,7] In this setting, the group can practice many practical skills such as listening, using good manners, asking someone out for a date, expressing anger and frustration, and solving a variety of likely problems, as well as learning relaxation techniques.

Role-play is particularly useful. Students are given situations to act out, first with inappropriate behaviors to be critiqued by the group, and then with appropriate behaviors producing desired outcomes. The supplemental use of videotaping provides a powerful training tool, especially for those students with PWS who are predominantly visual learners. Students, including those who participated in the role-play, are able to see themselves as others see them and critique their own behavior, both positively and negatively. It thus makes the role-play situation more real. It is also important to create situations that mimic real life. For example, a student may want to learn how to join a club, how to ask a friend to go to a dance, or even how to dance!

Additionally, teachers and caretakers need to be up-to-date on current fashions and trends to help the student learn how to dress in style and converse about subjects of general interest. In 1960s jargon, they need to learn how to be "cool"—or, in today's jargon, "wicked," "sweet," or "awesome."

Young people with PWS want what most teens want—friends and a close relationship with another. They can carry this desire to extreme lengths in their search for a boyfriend or girlfriend, including obsessing over a particular individual (sometimes a favorite adult). Dating and relationships are another area where adolescents with PWS need education and guidance. Phone etiquette needs to be taught. Excessive and often inappropriate use of the phone can become an obsessive, problematic situation. Rules and guidelines may be needed that outline when and where public displays of affection can take place. Many need to learn what should be said and done to make sure consent is obtained. It is recommended to use this consent approach for hand-holding, hugging, and other intimate situations. Educators need to make sure they have an understanding of what their students are being told at home. If we rely on students with PWS to learn from their peers, they may see and mimic inappropriate affectionate behavior. Some parents and providers have used written contracts to outline specific dating or friendship behaviors, and when and how they are to be employed.

Sexuality is also part of social development and, while individuals with PWS may not go through a complete puberty, they are still subject to sexual feelings and desires to be romantically involved. Most schools provide some kind of health class that describes the physical changes of adolescence, so the students know what to expect. For students with PWS, these changes may be minimal or nonexistent, unless they are receiving hormone therapy. This is often a very sensitive area for individuals with PWS, who want to know why they are not growing, have an unusually small penis, are not developing facial hair or experiencing a deepening of their voice, or, for girls, are not having a period. Girls frequently want to know about their future as mothers. While reproduction is highly unlikely, it has happened in several documented cases of young women with PWS. Instructors and caregivers must be prepared to deal with these issues in an honest, respectful, and sensitive manner. Collaboration between home and school is essential to convey information and advice in a consistent fashion. See Chapter 21 for further discussion of sexuality issues.

Finally, many individuals with PWS are highly motivated to play the role of nurturer or caregiver. This drive allows for opportunities to have the student assist others who may be younger or less capable (e.g., push a wheelchair, carry someone's knapsack, read to a young child, or be in charge of the classroom pet). There are endless opportunities to use the student's strengths and interests to build social skills. Caring for others can be a big boost to one's self-esteem, a critical component of self-development and socialization.

Looking Ahead

Many parents and caregivers ask, "Is there life after graduation?" The answer is "Yes, but it must be carefully designed." Federal special education laws mandate that transition planning be initiated when the student is 14 years old. This is the time to consider the specific strengths and needs of the student in light of future goals, whether they include continuing in an educational program, securing a job, or participating in a vocational or pre-vocational program. While the student is eligible to remain in school until 21 or 22 years old (varies by school district), not all will; those that do should achieve maximum benefit from available programs. Schools may provide a variety of community-based work experiences whereby the student experiences the world of work, gains a better understanding of work expectations, and determines where his/her interests lie.

Some school districts have a transition specialist who works with the multidisciplinary team in assessing the student's strengths and needs, designs a high school curriculum that addresses these areas, and identifies community resources relevant to the student's goals. Table 13.1 gives an overview of the timetable for various transition activities and events during the adolescent education years.

Making specific plans for entry into the adult world can be particularly stressful for all. The protections of IDEA no longer apply and there are no laws guaranteeing services for the student after 21 years of age. For most caregivers, and even school personnel, this comes as a surprise, often too late to have the optimal transition plan in place.

There are many things to consider in planning for the future:

- If the student is capable of further education in either a 2-year college or specialized training program, how will it be paid for and who will monitor his access to food?
- If the student has proven capable of entering the world of competitive employment, who will assist him/her in locating a job and providing the necessary job coaching? How will access to food be restricted?
- If the student requires a sheltered vocational placement, where does one exist? Is there a waiting list? Does the student meet eligibility requirements? Are the program, environment, and staffing compatible with the needs of the student with PWS? Is the program open to receiving training on the syndrome?
- Do the student and caregiver(s) desire a residential placement after high school? If so, what are the options? Is there a waiting list? Does the student meet eligibility requirements? What is the funding mechanism? If a Prader-Willi designated residence is unavailable, is the available program amenable to making significant environmental changes and to receiving training about PWS?
- Who knows the answers to these questions? Where does one go for help? Does anyone out there understand Prader-Willi syndrome as well as the school personnel?

Table 13.1. Timetable for Transition Services

This is a guide to when to expect certain activities to occur during the transition years in special education. As this is an individualized program, there may be great variation in scheduling from one student to another.

	Ages 14–18	Ages 18–20	Ages 20–22
Academic	• If mainstreamed, continue academic goals; may graduate at 18, which may cease special education services. • If in Special Day Class (SDC), include functional skills in goals.	• If mainstreamed and still in special education, continue academic goals. • If in SDC, continue mix of academic and functional skills goals.	• If mainstreamed, continue academic goals; determine use of post-secondary education and apply for funding and to post-secondary placements (junior college, trade school). • If in SDC, de-emphasize academics, increase focus on domestic and vocational functional skills.
Domestic	• Assess skills in personal hygiene, nutrition, cooking, household maintenance. • Build skills.	• Build skills.	• Continue to build skills using home, classroom, and community environments.
Community	• Assess skills in purchasing transportation, interpersonal communication and relationships, use of community services.	• In community more frequently as learning skills.	• Continue to build skills in all domains in the environment in which those skills will be used.
Vocational	• Rehab counselor assigned, application made, with proof of disability and vocational potential. • School assesses vocational knowledge, interests, aptitudes, skills.	• Begin to meet/-interview vocational agency staff. • Gain a variety of work experiences and skills.	• Continue gaining work experience. • Select vocational agency, if wanted, to serve needs after leaving school.
Site of Educational Activities	• Classroom with some community-based exposure.	• Classroom. • Community. • Vocational sites.	• Classroom may be on junior college campus. • Community. • Vocational sites.
Leaving School	• Some choose to leave (drop out) with incomplete skills.	• Many of non-disabled friends have left school (graduated) and gone on to college, trade training, or work. • May graduate but continue studies.	• Will exit special education sometime during age 21, typically with certificate, not diploma.

Source: Adapted from J.A. Seguin and R.M. Hodapp, *Transition From School to Adult Services in Prader-Willi Syndrome: What Parents Need to Know.*[8] Copyright 1998, The Prader-Willi Syndrome Association (USA).

- Without early planning, the student with PWS may find him- or herself sitting at home with no work or "adult" residence. In the worst case, a parent may have to quit his/her job to provide supervision for the young adult at home.

To facilitate a smooth transition, one of the first services needed is that of case management, typically provided by a county or state agency or contracted out to a community-based agency. The school district should provide this information. Case managers assist the student in linking up with a variety of community resources for which he/she qualifies. Generally, case management services are provided to individuals with mental retardation and/or developmental disabilities. This can be a problem for individuals with PWS, since many exceed the IQ eligibility criterion (usually 70 to 75), and many states do not recognize developmental disabilities unless the person also has mental retardation. When this occurs, help can be obtained from the national Prader-Willi Syndrome Association.

Service providers for adults, once identified, should be brought together no later than when the student is 18 years of age (see Table 13.1, Timetable for Transition Services). Both vocational and residential service providers may be needed. A note of caution is in order: the language of adult service providers is quite different from that of educators, so families and school staff should ask for translations. Also, families seeking residential services for their child should be made aware that residential programs specifically designated for people with PWS are scarce; thus the student may need to be on a waiting list while looking at other residential options.

Vocational services are also difficult to secure for individuals with PWS. Most programs promote competitive employment and cannot provide the degree of supervision required by a person with PWS. Typically, the student is matched with a job and provided with a temporary job coach (someone who assists in training the individual until he/she is able to perform the job independently). In the case of individuals with PWS, the need for a job coach, in the absence of diligent and caring co-workers, is long-term, if not permanent. Parents are encouraged to tap their own resources by calling friends in business or others who can provide some kind of meaningful employment in a safe environment.

Because many individuals with PWS are nurturing and loving, common career choices include working with animals or children. Some individuals are dog walkers or work at shelters or veterinarian offices taking care of the cages. Jobs with children where there is little-to-no food are difficult to find. While many individuals could be very effective in caring for young children in a preschool setting, there would need to be supports and environmental adaptations where food is involved. Some look into volunteer positions at local libraries where they may read to young children.

For a more detailed discussion of the transition process and related issues, see Chapters 14–16.

Conclusion

In short, early planning along with a great deal of networking and creativity is required on behalf of adolescents with PWS. Schools must provide relevant programs to prepare these students for an entirely new set of challenges. Parents must continue to work collaboratively with educators while they advocate for their sons and daughters. Indeed, the adult service system is generally not as "user friendly" as the educational system and requires continued diligence in securing needed services and programs.

Editor's note: Parts of this chapter were adapted from *The Student with Prader-Willi Syndrome: Information for Educators,* by Barbara Dorn, R.N., B.S.N., and Barbara J. Goff, Ed.D., published by PWSA (USA) and PWSA of Wisconsin, Inc., 2003.

References

1. Curfs LMG, Hoondert B, van Lieshout CFM, Fryns JP. Personality profiles of youngsters with Prader-Willi syndrome and youngsters attending regular schools. *Journal of Intellectual Disability Research.* 1995;39(3):241–248.
2. Dykens, EM. Maladaptive and compulsive behavior in Prader-Willi syndrome: new insights from older adults. *American Journal on Mental Retardation.* 2004;109:142–153.
3. Dykens EM, Cassidy SB. Correlates of maladaptive behavior in children and adults with Prader-Willi syndrome. *American Journal of Medical Genetics.* 1995;60:546–549.
4. Dykens EM, Kasari C. Maladaptive behavior in children with Prader-Willi syndrome, Down syndrome, and non-specific mental retardation. *American Journal on Mental Retardation.* 1997;102:228–237.
5. Joseph B, Egli M, Koppekin A, Thompson T. Food choice in people with Prader-Willi syndrome: quantity and relative preference. *American Journal on Mental Retardation.* 2002;107:128–135.
6. Lewis TJ, Sugai G, Colvin G. Reducing problem behaviors through a school-wide system of effective behavior support: investigation of a school-wide social skills training program and contextual interventions. *School Psychology Review.* 1998;27:446–459.
7. Schectman Z. [Group counseling in school in order to improve social skills among students with adaptation problems.] *The Educational Counselor.* 1993;3(1):47–67.
8. Seguin JA, Hodapp RM. *Transition from School to Adult Services in Prader-Willi Syndrome: What Parents Need to Know.* Sarasota, FL: Prader-Willi Syndrome Association (USA); 1998.
9. Whitman B, Accardo P. Emotional symptoms in Prader-Willi syndrome adolescents. *American Journal of Medical Genetics.* 1987;28(12):897–905.

14

Transition from Adolescence to Young Adulthood: The Special Case of Prader-Willi Syndrome

Ellie Kazemi and Robert M. Hodapp

Although all of life involves transitions, within the field of disabilities the term has a specialized meaning. For those with disabilities, "transition" usually refers to two time points: transition into and out of school. Although the transition from early intervention into school is of obvious interest in Prader-Willi syndrome, we focus here on the second transition, the movement of the young adult into the adult world.

In this chapter, we begin by describing transition as it pertains to students with disabilities in general. We discuss the transition from school into work and into community living, examine the role that families play in the transitional process and explore some of the issues families and students face. This first section thus applies to all persons with disabilities and their families.

In the second section, we focus more on the special issues involving Prader-Willi syndrome. In addition to discussing some preliminary findings on transition and adult services for individuals with this syndrome, we also explore Prader-Willi-specific issues and provide suggestions for making transition a smoother and more successful process for adolescents with this syndrome.

From Adolescence to Adulthood: How Does Transition Differ for Persons with Disabilities?

Adolescence is a developmental period that involves biological, physical, and environmental changes. Like most adolescents, adolescents with disabilities struggle with issues of gaining vocational and life skills, building and maintaining friendships, sexuality, obtaining independence from their parents and care providers, and making decisions about their adult lives and living situations.

In contrast to typically developing adolescents, however, most adolescents with developmental disabilities have delayed cognitive and social skills. In order to function as independent or semi-independent adults within their community, these adolescents must therefore continue to learn daily living, adaptive, and social skills. In addition, many adolescents with disabilities show behavior problems that interfere with more independent, adult functioning. Such problems must also be dealt with before adolescents can fully assume their adult roles.

Another difference between typically developing adolescents and adolescents with disabilities involves the long-term preparation for transition. In certain respects, children without disabilities have been preparing for the transition out of adolescence their entire lives. Even preschoolers consider their adult futures when they talk endlessly about being "a firefighter when I grow up," or becoming a doctor, lawyer, teacher, or helicopter pilot. More formally, school-age children begin preparing for their adult futures when they talk with guidance counselors about career planning or to their parents about parental jobs and aspirations or when they take specific classes or perform extracurricular activities needed to get into colleges with good programs in their desired fields.

On the other hand, children with developmental disabilities usually have very different childhood experiences. It is probably rare that a child with a developmental disability is asked, "What do you want to be when you grow up?" Many of the informal and formal preparations also probably differ, even as the child's functioning as an adult gradually becomes *the* important issue for the adolescent with a disability and his or her family.

A further difference even from 20 or 30 years ago involves the lengthening life spans of persons with disabilities, particularly among persons with different genetic syndromes. To take the example of Down syndrome, in 1929 it was estimated that the average life span of persons with Down syndrome was 9 years (see Penrose[10]). Today, individuals with Down syndrome routinely live into their 50s and 60s. Similarly in Prader-Willi syndrome, as recently as a few decades ago most individuals did not live into their middle-age or older years. Nowadays, with increased attention to weight management and the prevention of obesity, more and more persons with Prader-Willi syndrome are living longer lives. As a result, the needs of individuals with disabilities and their families have changed—one therefore must consider the lifelong impacts of changes made as the individual transitions from the adolescent into the early adult years.

Service Delivery: From School to Adult Services

One of the most significant milestones an adolescent with a disability faces is the shift from school to adult-oriented services within the community. In the late 1980s, researchers and advocates began to discuss the dire need for transitional services between school and work environments.[11] Beginning in 1990, the U.S. government recognized the need for transitional services and, by 1997, the main federal special education law, the Individuals with Disabilities Education Act (IDEA), mandated that schools prepare students for the transition to adult life by including transition plans.

Since 1997, transition planning and services have been incorporated in the Individualized Education Program (IEP) of all eligible students as early as age 14 and no later than age 16. Thus, from between age 14 and 16 until leaving school at 21, adolescents, their families, and school and adult-service personnel increasingly focus on successful transition. At least for a 5- to 7-year period, then, everyone involved should be asking the question, "What do you want to be as an adult?"

Based on the individual's current and future needs, such transition services vary widely. In general, schools promote transition by helping students obtain the life skills needed to live in their communities. Some of these basic skills might involve taking the bus or other public transportation, shopping for clothes or food within a certain budget, using the post office, or visiting their doctor or a clinic. Other services might help lessen the adolescent's maladaptive behavior or teach appropriate social skills, thereby making possible more independent movement within the community. Postsecondary education, vocational training, job coaches, or other help in employment might all be included in the mix of support services.

When thinking about transition services more generally, three other issues are also important to consider. The first has been the movement toward so-called "person-centered planning." This term, recently used by policymakers, advocates, and parents of children with disabilities, concerns a shift in who receives—and in who decides upon—appropriate services. In short, for an adult with disabilities, the final say about services is given to the adult him/herself, whereas for a child with disabilities the final say about services is given to the family. Although this issue may sometimes be problematic in Prader-Willi syndrome, the IDEA mandates that transition services and activities should be based on the *young adult's* individual needs, interests, and preferences. The family plays a major role in talking to and counseling their young adult with disabilities, as well as in working with service providers to obtain appropriate services for their offspring. Ultimately, however, unless parents go to court to obtain guardianship or conservatorship rights, the adult with disabilities holds the ultimate say in which services will be provided and in which ways.

A second change from school-related services concerns the responsibility for providing services. During the school years, services come to the child through the school itself as a federally mandated entitlement. But, as young adults with developmental disabilities leave school and enter community life, unlike the school-age years, services are usually not mandatory but are instead provided through state-funded agencies and *at the discretion of the state*. Depending on the state's financial situation, such agencies are often hard-pressed to meet the demands of all of the state's persons with disabilities. Thus, while a particular state may decide that specific services must be provided when individuals have certain needs, the services available may not match the range of services to which the person was previously "entitled" through the school years.

Third, services during adulthood often are not provided in a single place. Whereas schools had earlier operated as one-stop service providers, during adulthood the individual with disabilities must travel to different locations and coordinate different agencies' services. For young adults and parents alike, such changes can sometimes feel overwhelming.

The Transition to Work

Transitioning into the workplace is a process that an adolescent with a disability must prepare for long before the actual placement in a job. As a part of their transitional planning, schools generally teach vocational skills to students with disabilities. While in school, a student may be taught interviewing skills, how to fill out an application, and how to use public transportation. When this training ends, however, most students with disabilities must independently seek out suitable positions. Although schools usually do not directly help with job placements, certain adult-oriented services do aid in this process (e.g., the state's Department of Rehabilitation); the person with the disability, with his or her family, can meet with such agencies and request their participation in the transition meetings.

As shown in Table 14.1, individuals with disabilities and their families may consider various types of paid positions. Such positions range from "least restrictive"—or the most "typical" and within the community—to "most restrictive," or more specialized, segregated from the community, and less often engaged in by nondisabled persons. Since each person differs in severity of disability, abilities, skills, needs, goals, and desires, it is important to know about all of the possible options for employment. Also, no job placements should necessarily be thought of as permanent positions. A person may move across levels of support as needed at different points in time.

The Transition to Community Living

For any person, moving out of their parents' home is a significant step. The decision to move out affects the family as well as the person and

Table 14.1. Paid Work Opportunities from Least Support to Most Support Needed

Term	Description	Examples
Competitive Employment	A job like that of every nondisabled person, with the same pay levels and benefits. This also means that the person is expected to have the same job responsibilities and performance as other employees.	Cashier, stock person, utility clerk, landscaping crew, factory employee
Supported Employment	Same sites as a person with competitive employment but with more support services, offered through rehabilitation service agency. Generally a job coach, or someone specifically trained to offer job support to persons with disabilities, will teach persons the basic skills needed to maintain their jobs.	Same as above but with support
Sheltered Work	Individuals with disabilities are in a more restricted setting, in a segregated environment, and they work in groups with other individuals with disabilities.	Packaging and product assembly of office supplies, jewelry, etc.

this is even truer for persons with disabilities. The appropriate time to move out of the parents' home differs for each individual. The decision is affected by such factors as the young adult's readiness to move out and parental attitudes toward moving out, both of which can be affected by such factors as culture and socioeconomic status. Nevertheless, as in employment opportunities, to make informed decisions one needs to know about the various residential placements available for persons with disabilities.

Just as in school or in employment, residential opportunities vary from least restrictive to most restrictive. In this instance, "most restrictive" generally signifies placements that offer the most support and that are most segregated from the community. Considering residential placements on a continuum from least to most restrictive, a person with a disability may live independently or with some support in a semi-independent living situation (in their own house or apartment), in a large or a small group home (sometimes called Community Care Facilities), in an Intermediate Care Facility, or in a state developmental center or a state hospital (see Table 14.2).

Table 14.2. Living Options from Most to Least Restrictive Environments

Term	Description
State Hospitals or Developmental Centers	Serve individuals who need 24-hour supervision in a structured health facility where they receive programming, training, care, and treatment on site.
Intermediate Care Facilities (ICF)	Offer 24-hour service to 4 to 16 individuals with disabilities. These facilities serve individuals with developmental disabilities who have a primary need for developmental services, as well as some needs for skilled nursing services.
Group Homes or Community Care Facilities (CCF)	Offer 24-hour non-medical residential care to individuals with developmental disabilities who may need personal services, supervision, and or assistance crucial for self-protection or sustaining the activities of daily living. These residential models are popular because of their integration into the community.
Supported Living	The individual with the disability has support systems, typically nondisabled roommates who are trained to help or outside agencies that teach independent living skills. In most situations the person is monitored and supervised because he/she may not yet have mastered basic financial, shopping, and self-care skills. They learn to manage these duties while living either alone or with roommates in apartments or condominiums.
Independent Living	The individual with the disability may choose to live alone in a home that they own or lease in the community. They may have hired staff that aid in some daily activities, but they do not need supervised care and training in basic life skills.

Of special interest here is one type of group home, the so-called "dedicated" or "specialized" Prader-Willi group home (for examples, see Greenswag et al.[8]). These placements are group homes—with six to eight individuals living within a home in a residential neighborhood—in which all residents have Prader-Willi syndrome. In addition, services are tailored to the needs of individuals with this syndrome, with refrigerators and food cabinets oftentimes locked. In addition, most Prader-Willi group homes feature exercise rooms for their residents and daily times during which residents exercise each day. Staff members are specially trained in dealing with individuals with this syndrome. Although we say more about dedicated Prader-Willi group homes below, suffice for now to note that such homes exist and serve as the residential option of choice for many adults with Prader-Willi syndrome.

Transition Issues Specific to Prader-Willi Syndrome

Until now, we have discussed transition with only slight mention of Prader-Willi syndrome. Indeed, many transitional issues are generic: regardless of which type of disability the young adult has, one deals with adult-service personnel around questions of how the young adult will work and where he or she will live. We now address issues more specific to Prader-Willi syndrome.

PWS Behavioral Issues Affecting Transition

Looked at from the perspective of many adult-service professionals, Prader-Willi syndrome is both unknown and irrelevant. Indeed, case managers or vocational rehabilitation counselors are often puzzled by individuals with Prader-Willi syndrome. On one hand, here is a young adult who shows fairly high cognitive skills, while on the other, the individual has maladaptive behaviors that prevent successful independence. Typically, in adults with other disabilities who have such high cognitive abilities, the individual would normally be able to live in a supervised or supported-living apartment and be employed competitively or with some support. For some adults with Prader-Willi syndrome, these options are suitable and accomplishable. In most cases of young adults with Prader-Willi syndrome, however, the individual cannot live independently and often fails in competitive jobs.

Although it is beyond this chapter's scope to discuss behavioral issues in depth, six general issues should be emphasized when considering vocational and residential issues for young adults with Prader-Willi syndrome:

1. Overeating

Overeating and food foraging occur in the large majority of persons with Prader-Willi syndrome. As a result, many individuals are obese or may become obese if allowed open access to food. Overeating is not a trivial issue. In one recent study, Whittington et al.[14] found that most deaths in Prader-Willi syndrome relate to complications of obesity (e.g., Type 2 diabetes, respiratory and circulatory problems). In addition, despite years of research, no drug regimen or behavior modification therapy has successfully curbed overeating for most individuals with this disorder.

2. Obsessions and Compulsions

Many young adults with Prader-Willi syndrome become "stuck" and need particular help during transitions.[6] Such difficulties—which often lead to tantrums and other disruptive behaviors—are rarely tolerated in jobs and group homes. To help counteract such obsessions and compulsions, young adults with Prader-Willi syndrome may need ample warning about upcoming transitions or special auditory or visual cues to prepare them that a change is coming.

3. Temper Tantrums

Many individuals with Prader-Willi have full-blown, disruptive temper tantrums.[7] Such tantrums can arise around food-related or other issues; they also often occur when the individual's routine has been disrupted. Temper tantrums are especially problematic when they occur during work, in a group home, or out in the community.

4. Changes with Age

The age of the adult with Prader-Willi syndrome may be an important predictor of behavior problems. Prader-Willi syndrome has historically been considered a "two-stage" disorder, with problems beginning during the 2-to-5-year-old period.[1,2] More recently, however, it is becoming clear that behavior problems may intensify from childhood into early adulthood before lessening during the later adult years. Cross-sectionally examining a large sample of 3- to 50-year-olds, Dykens[3] found that maladaptive behaviors generally increased throughout childhood up until about age 30, after which problems decreased from age 30 to 50. This "up until 30, then down" pattern held for the overall amount of behavior problems, the amount of problems directed toward others (e.g., such externalizing problems as tantrums or aggression), as well as for skin picking and for the individual's number and severity of compulsive symptoms.

5. Differences in Maladaptive Behaviors Related to Weight

In contrast to what one might expect, thinner—as opposed to heavier—individuals with Prader-Willi syndrome seem to have more psychiatric problems. These problems especially relate to distorted, confused, and delusional thinking, as well as to anxiety, sadness, and distress.[3,4,13] Why such a counterintuitive relationship occurs remains unknown, but simply keeping the weight off will not solve the behavior problems of most young adults with Prader-Willi syndrome.

6. Possible Beneficial Effects of Prader-Willi Group Homes

Many parents and professionals feel that specialized Prader-Willi group homes are particularly beneficial for most adults with this disorder. Several studies find that individuals placed outside the home—especially in Prader-Willi group homes—have lower BMIs (body mass index, a measure of weight and obesity).[3,7] Hanchett and Greenswag[9] also point to the importance of more structured group living for lessening problems with maladaptive behavior for most individuals with Prader-Willi syndrome. Although few studies directly compare Prader-Willi group homes to generic group homes or to home living for adults with the syndrome, the consensus is that Prader-Willi group homes are most beneficial.

In thinking more generally about work and residential issues, then, young adults with Prader-Willi syndrome present special challenges to the service-delivery system. Weight and dietary control of calories are almost always issues in this disorder, much more so than in other adults with disabilities. In addition, individuals with Prader-Willi syndrome show extreme problems with temper tantrums, obsessions and compulsions, skin picking, and other behaviors. At the same time, however, many young adults with the syndrome do well with more structured, and at times more segregated, services.

More segregated, specialized residential settings, however, generally go against the current emphasis on inclusive programming. Clearly, a balance must be reached between the needs of the young adult with Prader-Willi syndrome and the philosophical goal that all persons with disabilities live as typically and as normally as possible (see Dykens et al.[6] for discussions).

Parental Experiences with Transition and Adult Services

Few published studies have yet examined issues of transition and adult services in young adults with Prader-Willi syndrome. Several years ago, however, Seguin and Hodapp[12] performed 30 extended telephone interviews with parents of 18- to 35-year-old adults with Prader-Willi syndrome. We present below some of the major findings of these interviews. As before, we separate our discussions into work/vocational issues and residential issues.

Vocational Issues

The major issue mentioned by many parents in our survey concerned the lack of knowledge of Prader-Willi syndrome. Simply stated, the large majority of vocational staff have little understanding of Prader-Willi syndrome. As one parent noted, "There is a lack of understanding of PWS on the part of the staff, so they don't know how to deal with it." Another parent gave a workshop to her child's staff about Prader-Willi syndrome. But even when such staff learn about the syndrome, their understanding is more intellectual than emotional. Staff members thus continue to be frustrated—and to have a low tolerance for—various Prader-Willi behaviors. Indeed, most staff consider overeating to be simply "willfulness" or a lack of self-discipline on the part of the young adult with Prader-Willi syndrome. As one mother noted, "When he was successful, staff were great," but when her child was having difficulties, "They just didn't want to deal with his problems."

Partly because adult-service personnel know so little about Prader-Willi syndrome, behaviors characteristic of the syndrome frequently lead to job terminations. Thus, young adults who might, cognitively, be appropriate for supported or even competitive employment often have difficulties holding a job. As a result, these individuals often find themselves working in sheltered workshops. In the Seguin and Hodapp[12] sample, over half of all young adults with Prader-Willi syndrome were employed in sheltered workshops; less than one third were in competitive or supported employment. Many had earlier been in these less restrictive employment settings but had since lost their jobs.

But the sheltered workshop experience was also mixed. Looked at positively, many parents considered the sheltered workshop a safe environment in which their child had a few friends who were functioning at the same level. At the same time, however, tasks performed by workers in sheltered workshops were often repetitive and unstimulating. In addition, other (non-PWS) co-workers were often lower functioning.

These are difficult issues, and many parents expressed frustrations and diminished hopes for their young adult's work life. There may, however, be some strategies that help.

A first suggestion relates to *provider education*. In short, parents need to educate adult-service staff about Prader-Willi syndrome and its effects. Staff need to understand the physiological basis of overeating—it is not willfulness or poor self-control. This volume and other publications may help in this process, but parents should be forewarned that few adult-service staff—be they case managers, vocational rehabilitation counselors, job coaches, or (generic) group home staff—will have heard about the syndrome. Fewer still will truly appreciate the difficult management issues presented by young adults with Prader-Willi syndrome. Here we highlight the role of the Prader-Willi Syndrome Association (USA) in increasing knowledge about the syndrome. At each of its yearly national meetings, separate sessions are provided for service providers; PWSA also has available for providers tapes, booklets, and other informational materials.

A second issue concerns *gaining allies from special education* to help work with adult-services staff. Many parents expressed relief when, over time, their child's special education teacher became familiar with their child and their child's disorder. Particularly in small towns where families could interact socially with the teacher, parents felt that they could affect their child's classroom experiences. If parents have one or more special educators who know their child well, that person might then intercede on their behalf with adult-service providers. Many parents felt that, over the years, schools and teachers "were receptive to suggestions and to listening to me"; it may be helpful to use this alliance to get adult services.

A third issue concerns the *pace of change*. In contrast to other young adults with disabilities—and particularly young adults with mild intellectual disabilities—many individuals with Prader-Willi syndrome may not quickly or easily transition to a supported employment setting.

As a result, rehabilitation counselors and job coaches may be dismayed by the increased numbers of tantrums and disruptive behavior shown by young adults with Prader-Willi syndrome during the first few days or weeks on the job. The first reaction of these professionals may be to terminate workers with Prader-Willi syndrome, to automatically assume that the adult with this syndrome cannot possibly handle the demands of a particular work setting. Given time, however, young adults with Prader-Willi syndrome may get used to job demands and routines. In short, parents may need to work hard to help their offspring's case manager, rehabilitation counselor, and job coach expect a slower, more gradual transition into an appropriate work placement.

The fourth and most important issue concerns *parental advocacy.* Unfortunately—and sometimes in contrast to earlier school services— parents may need to be their young adult's best advocate. Parents who were interviewed often noted that they were unaware of what services they could request, and that case managers and rehabilitation counselors were not always forthcoming. As several parents stressed, "You have to know how to work the system, and you have to demand services," and you have to "make it your business to go out there and find out what is available."

A fifth issue relates to offspring becoming their own advocates. In line with the idea that services focus on the adults with disabilities themselves, many individuals with disabilities have been trained to advocate for themselves. Some even attend self-advocacy classes, and parents and professionals can informally teach, probe, and guide their young adults with PWS to speak up for themselves to attain needed services.

To help in thinking about such services, Table 14.3 (from Seguin and Hodapp[12]) lists a variety of services that parents and their offspring can request from case managers. Not every young adult with Prader-Willi syndrome will need every service. Even if needed, state funding and provision practices would probably preclude anyone from getting all these services anyway. But parents and professionals may be able to advocate for those services most necessary to their child's job (and residential) success.

Residential-Community Issues

The other main focus of our interviews with families concerned residential-community issues. Here again, parents often expressed frustration with the adult service system. These frustrations generally fell into the following areas:

Scarcity of Prader-Willi Group Homes

Many parents noted that there simply were not enough Prader-Willi group homes. As a result, waiting lists were common. Once their son or daughter was in a Prader-Willi syndrome group home, however, parents generally agreed that their child's weight was under control and that the staff usually had some experience in dealing with tantrums and other behavioral issues.

Table 14.3. Typical Services Provided for Adults with Disabilities, by Service/Funding Source

	Social Security (SSI/SSDI)	Case Management	Department of Rehabilitation	Personal Expenses
Health Insurance				
"Approved" medical	Monthly living allowance*	Pre-vocational day programs	Center-based sheltered workshop	Food
Dental	Eligibility for government-funded medical insurance	Transportation to day program	Supported employment: —assessments	Community transportation
Optometric		Case management	—placement	Recreation & leisure activities
			—job coaching	
Physical therapy+		Residential fees*		Health club/gym
Speech therapy+				Clothing
Occupational therapy+		Therapies not covered by medical insurance		Therapies not covered by medical insurance

+ Depending on the medical insurance plan coverage, these may not be "covered" services.

* This may be a fee that is shared by Supplemental Security Income (SSI) and by the case management organization.

Adapted from J.A. Seguin and R.M. Hodapp, *Transition From School to Adult Services in Prader-Willi Syndrome: What Parents Need to Know.*[12] Copyright 1998, The Prader-Willi Syndrome Association (USA).

Given that so few Prader-Willi homes exist, many parents were forced to choose among several less-than-perfect options. One parent noted that, while she had a choice as to which program would take her child, "there was only one that had had someone with PWS before." Others worked hard for supported living arrangements, although one added that, "If we had really known we had to demand supported living, we would have demanded it sooner."

Staff Turnover, Inexperience, and Opposition

Many parents complained about the frequent turnover in group home staff. Because of such frequent changes, it is often difficult for parents to develop relationships with staff or for staff to get to know their adult with Prader-Willi syndrome. As one parent put it, "There are quite a few changes in staff. The family is the only stable factor."

Parents have the right to know what is happening with their child at both the group home and the work sites. Several parents recommended that, in order to get a balanced view, parents should speak with both the staff and their child about what is happening. When speaking with staff, parents may want to pick a time when the staff member is free to focus exclusively on the parents' concerns, not trying at the same time to assist several residents. Parents might also want to speak to the counselor for the home so that they get an overview of group home activities and how well their child is managing. One parent noted: "We used to have meetings where the staff told me how she was doing. Now she tells me how she's doing." As another parent concluded, ultimately "Communication with the residential staff is a function of what you make of it."

Difficulties in Their Child's Dealings in the Community

Several parents noted problems that their child had when outside in the community. The most persistent issue concerned manipulation and the ways that peers take advantage of young adults with Prader-Willi syndrome. As the parent of one young man explained, "He's easily manipulated by others. He needs supervision to maintain his diet."

Yet how one stops such manipulation is another matter. In reacting to this issue, several parents found that the solution requires a delicate balance between the parents' or group home staff's desire for protection and the young adult's need for independence. As one family noted, "We've decided now that he can't be too protected; he has to make decisions for himself."

As in discussions of work-related topics, parent interviews again revealed a less-than-flattering picture of adult services and of whether most adults with Prader-Willi syndrome are getting the necessary residential services. For every parent who reported that "Things are much better than we ever hoped for" or that "People by and large are great," equal or greater numbers made comments such as "I wish I had known how difficult it would be to find a group home." Like services related to work, residential services also do not seem adequate in the eyes of most parents of young adults with Prader-Willi syndrome.

Although it is impossible to address certain issues—like state funding or whether one gets a case manager who is helpful or not—we echo here many of the suggestions parents supplied about residential services. Their first and most important suggestion concerned *advocacy*. Virtually every successful parent commented on the importance of advocating for services for their young adult. Second, one needs to have *information*. In contrast to one parent's sense that "I still feel like I don't know anything," others noted the importance of learning all they could. Much of this knowledge came from other parents of young adults with Prader-Willi syndrome. As one parent noted, "All parents need to be aware of services through the Prader-Willi syndrome associations: keep aware, attend conferences, and talk to parents of older kids."

Conclusion

In moving from the childhood into the adult years, young adults with Prader-Willi syndrome and their families are entering a new and foreign world. Like many foreign countries, the language differs—rarely if at all, for example, have school personnel ever talked much about "person-centered planning." Also like in a foreign country, things are simply done differently in the childhood versus the adult years. As opposed to the local school and school district being responsible for services, the state is now responsible. As opposed to the one-stop service provision that schools offer, services are now decentralized and sometimes uncoordinated.

And yet, as frustrating and overwhelming as many of these changes may seem, individuals with Prader-Willi syndrome and their parents can survive—and even thrive—in this new country. As in any new country, "expatriates" from one's old country will be invaluable. Specifically in the Prader-Willi syndrome community, many parents of young adults have been there before and are more than happy to provide their hard-earned advice, guidance, and contacts. Knowledgeable and caring professionals can also help, as can the Prader-Willi Syndrome Association (USA) and the many local, state, and national organizations designed to help parents of individuals with disabilities. Like any change, the transition from childhood into adulthood is never easy. In this case, though, many people, organizations, and services make it possible for the young adult with Prader-Willi syndrome to become an independent or semi-independent individual who is able to fully enjoy their adult years.

References

1. Cassidy SB. Prader-Willi syndrome. *Current Problems in Pediatrics*. 1984; 14:1–55.
2. Dimitropoulos A, Feurer ID, Butler MG, Thompson T. Emergence of compulsive behavior and tantrums in children with Prader-Willi syndrome. *American Journal on Mental Retardation*. 2001;106:39–51.
3. Dykens EM. Compulsive and maladaptive behavior in Prader-Willi syndrome: new insights from older adults. *American Journal on Mental Retardation*. 2004;109:142–153.

4. Dykens EM, Cassidy SB. Correlates of maladaptive behavior in children and adults with Prader-Willi syndrome. *American Journal of Medical Genetics.* 1995;60:546–549.

5. Dykens EM, Goff BJ, Hodapp RM, et al. Eating themselves to death: have "personal rights" gone too far in Prader-Willi syndrome? *Mental Retardation.* 1997;35:312–314.

6. Dykens EM, Leckman JF, Cassidy SB. Obsessions and compulsions in Prader-Willi syndrome. *Journal of Child Psychology and Psychiatry.* 1996; 37:995–1002.

7. Greenswag LR. Adults with Prader-Willi syndrome: a survey of 232 cases. *Developmental Medicine and Child Neurology.* 1987;29:145–152.

8. Greenswag L, Singer SL, Condon N, et al. Residential options for individuals with Prader-Willi syndrome. In: Greenswag LR, Alexander RC, eds. *Management of Prader-Willi Syndrome.* 2nd ed. New York, NY: Springer-Verlag; 1995:214–247.

9. Hanchett J, Greenswag LR. *Health Care Guidelines for Individuals With Prader-Willi Syndrome.* Sarasota, FL: Prader-Willi Syndrome Association (USA); 1998.

10. Penrose LS. *The Biology of Mental Defect.* 2nd ed. London: Sidgewick & Jackson; 1966.

11. Rusch FR, Phelps LA. Secondary special education and transition from school to work: a national priority. *Exceptional Children.* 1987;53(6): 487–492.

12. Seguin JA, Hodapp RM. *Transition from School to Adult Services in Prader-Willi Syndrome: What Parents Need to Know.* Sarasota, FL: Prader-Willi Syndrome Association (USA); 1998.

13. Whitman BY, Accardo P. Emotional symptoms in Prader-Willi syndrome adolescents. *American Journal of Medical Genetics.* 1987;28:897–905.

14. Whittington JE, Holland AJ, Webb T, Butler J, Clarke D, Boer H. Population prevalence and estimated birth incidence and mortality rate for people with Prader-Willi syndrome in one UK health region. *Journal of Medical Genetics.* 2001;38:792–798.

15

Vocational Training for People with Prader-Willi Syndrome

Steve Drago

The Challenge

The standard of care for persons with Prader-Willi syndrome (PWS) has changed dramatically in recent years. People with PWS now are routinely diagnosed at an early age, the majority in the newborn period. Physicians are better educated on the disorder, and effective treatment protocols are rapidly increasing. Hormone replacement therapies are resulting in people who are healthier, more energetic and robust, and who look "normal." Many children with PWS are successfully mainstreamed in school. Effective behavioral management strategies have been developed and are readily available. In contrast to even a decade ago, today residential placements exist that effectively manage the behavioral and weight issues of people with the disorder.

Because of these improvements in health care, education, and behavior management, more and more individuals with PWS of near normal or normal intelligence are living well into adulthood.[2] As a result of the rapid improvements in both the length and quality of life, the bar of parental expectations has been raised. Parents are no longer *merely* concerned that their affected child will live to age 30. Instead, parents want—and expect—to see their son or daughter thrive, live a normal life, and be included in the fabric of their communities. More importantly, the aspirations of people with PWS have changed as well. Many now expect eventually to live independently, in their own home with a spouse and pets (not necessarily in that order), and to be gainfully employed. Expectations for good health and for satisfying adult living situations are no longer unrealistic. In the continuum of effective services for people with PWS, successful job placements have evolved to the status of "the last hurdle" to be cleared. For this population, jobs remain the "final frontier."

Effective vocational placements are still difficult to achieve for people with PWS, despite extensive health and care improvements. There are several reasons why this is so. First, vocational providers have not had to live with the disorder; many are slow to understand the seriousness of the appetite and emotional volatility associated with the disorder.

The truly debilitating nature of PWS is often difficult to grasp until time is spent working or living with an affected individual. One parent uses the analogy of alcoholism or drug addiction to describe her son's disability: "His drive to eat is just as strong and difficult to overcome as any other addiction." Unlike someone addicted to drugs or alcohol, however, her son cannot simply stop imbibing or using. He *must eat*; furthermore he *must eat several times per day*, in small controlled quantities, and *then stop*. Unlike the alcoholic or drug addict, an individual with PWS is unable to simply adopt a new lifestyle by surrounding himself with people who do not use alcohol or drugs, or by avoiding places where these things are used. Everyone eats, and food is everywhere in our society. Thus the stress on an affected individual is ubiquitous and constant. Until an employer achieves this level of understanding and appreciation of the disorder, job placements have a high likelihood of failure.

An employer's effort to achieve this level of understanding is often undermined by first impressions from meeting someone with PWS. Usually prospective employers are initially presented with a mild-mannered, intelligent individual who is motivated to work. Many employers have employed other people with disabilities whose initial presentation did not signify either the capacities or the (apparent) social and intellectual skills initially observed in a person with PWS. Therefore, when the person with PWS disappears from the work site to obtain food, or becomes argumentative with the boss, employers are most likely to see these behaviors as "merely discipline problems" rather than as a natural manifestation of PWS requiring workplace adaptation.

An employer's expectations for individuals in an adult work setting are different from those at home or in school. Schooling is an entitlement; as such, schools must adapt for the individual. By contrast, jobs are earned; the employee serves by the privilege of the employer. Arguing and noncompliance, both frequent typical behaviors for someone with PWS, are not usually tolerated. An initial adjustment period is usually required, during which both employer and employee acquire new learning and skills.

The Continuum of Work Placement Options

Prior to discussing successful work placement strategies for individuals with PWS, a discussion of work placement options and terminology is needed. *Sheltered workshops* are the oldest and most common type of job placement for workers with developmental disabilities. These settings offer job skills training in a *nonintegrated setting*. This means that the entire work or training force is made up of disabled individuals. Individuals in sheltered workshops are paid on a piece-completed rate; that is, they do not receive an hourly wage, but are paid for each piece of completed work at a rate comparable to that paid a nondisabled person. Typical sheltered workshop tasks include packaging and simple product assembly. Workshop placements are generally funded at the state level by programs such as the Medicaid Waiver. Medicaid Waiver programs (available in most states) are state-run programs where the state is able to leverage matching federal dollars specifically dedicated to services for people with a developmental disability. In return, the programs must comply with established work and safety standards and rules for service provision and documentation such as those mandated by the U.S. Department of Labor.

Work enclaves consist of small, nonintegrated groups of disabled individuals who go into the community to perform service type work.[1] Typical work enclave jobs are lawn maintenance, janitorial, and restaurant work. Each enclave has at least one nondisabled supervisor. Individuals in work enclaves are routinely paid an hourly wage. Because individuals in work enclaves are considered trainees, federal wage and hour regulations allow them to be paid at a rate less than minimum wage, as long as they are paid a percentage of the competitive wage equal to the percentage of work they perform (compared with that of a nondisabled individual). Thus, if a typical hourly wage for a nondisabled person is $7.50, and the disabled person averages 60% work productivity compared with the nondisabled person, then the disabled person's salary would be 60% of $7.50, or $4.50 per hour. Funding for work enclaves generally comes from state programs such as the Medicaid Waiver; however, some individuals may also be eligible for funding through the U.S. Department of Vocational Rehabilitation. The Department of Vocational Rehabilitation is a federal government program that provides services to individuals who are trying to work and who have a disability. While those with any type of disability may be eligible for funding through the Department of Vocational Rehabilitation, such funding is limited to only 180 days of employment (referred to as "employment stability"). During the employment stability time frame, the Department will pay for a large variety of supports. If an individual requires support beyond the employment stability time frame, funding must be obtained through other programs.

Supported employment is a community-based job placement in an integrated work force; the other workers performing comparable jobs are nondisabled. Individuals placed in supported employment must receive a wage that is equal to or above minimum wage and comparable to that of nondisabled individuals performing the same job at that location. Supported employment is supervised by a *job coach*, who provides on-the-job training and acts as a liaison with the employer. Ultimately, the goal is for the job coach to become less and less necessary over time until, ideally, the job coach is no longer needed. When this successfully occurs, the individual is said to be "competitively placed." These placements are typically paid for initially by the federal government Department of Vocational Rehabilitation. When successfully trained so that the job coach is no longer necessary, the employee becomes a "regular" employee of the host company. Employers may benefit from tax incentives for hiring disabled workers.

The vocational services just described comprise a natural *continuum of services.* Individuals can progress from a sheltered workshop to an enclave, to supported employment, and finally, to competitive employment. The concept of a continuum of services has been under attack in recent years.[4] Many argue that individuals should be placed initially at the optimal level of employment with needed supports to insure success provided in that setting. One obvious limitation to this approach is cost. Providing sufficient supportive resources to train and maintain a consumer at the optimal level may exceed even the most generous state funding. Further, the assumption that initial assessment strategies are sufficiently reliable and able to predict an individual's optimal work placement is not supported by previous data. Standardized assessment tools frequently fail to assess individual needs, personality strengths and weaknesses, motivations, and individual "quirks" that impact job placement and performance. These individual characteristics frequently determine placement success or failure but often aren't evident until the person is in the work environment. This is particularly true for those with Prader-Willi syndrome.

The Person with Prader-Willi Syndrome in the Workplace: Pitfalls and Successes

Challenges to Successful Vocational Placement

Because those with PWS are far fewer than those affected with other disorders such as Down or fragile X syndromes, most vocational service providers serve no more than one or two adults with PWS. The ARC of Alachua County (Gainesville, Florida) is an exception, currently providing vocational services to more than 50 individuals with PWS. One staff member states: "We pride ourselves in being more knowledgeable than most providers when it comes to working with individuals with PWS. In spite of this experience, we have not been spared the typical pitfalls. In fact, due to the sheer numbers of individuals with PWS we have worked with, we have probably encountered many more." Several typical examples follow.

One of the first issues encountered was getting individuals to work on time. Getting up and out of the house is often difficult for people with PWS. For many affected individuals, the early morning routine consists of one compulsive ritual after another. A parent who was having particular difficulties getting her daughter to work on time attempted to circumvent these rituals by getting her daughter up earlier and earlier. The parent sought help from the vocational staff when she and her daughter were waking up 5 hours before it was time to go work each day, yet the problem continued to worsen! Getting up earlier and earlier was obviously not the solution. Rearranging the contingencies in the morning environment to make it more efficient was. Mom was instructed to place breakfast at the end of the chain of morning requirements rather than in the middle. This hastened her daughter's performance of the morning routine. Some fine-tuning was necessary to streamline the morning ritual, which remained somewhat lengthy but by no means 5 hours.

Food temptations are everywhere for those with PWS, causing some adults to lose their daily specialized transportation due to eating fellow riders' lunches. And, not unexpectedly, behavior issues occur when there is insufficient supervision of the vending machines and at lunch. Further, those in sheltered workshop settings have been known to bring more and more possessions to work. A supervisor observed: "One puzzle book for break time is acceptable; every puzzle book collected *and completed* during the past 5 years is not." Other problems have occurred around returning from bathroom breaks on time and disengaging from one work task and beginning another.

Work enclaves encounter both the previous issues in addition to others that are unique to the nature of the job. The ARC of Alachua County vocational services are located in Florida, which has a warm climate year-round; as a result, lawn care is a major business opportunity. Many individuals with PWS select work on a lawn crew as one of their goals. Lawn work is hard physical labor. Many currently employed adults with PWS reached adulthood before growth hormone replacement therapy was available and thus have not benefited from this treatment. As a result, many lack the physical stamina required for lawn work. In addition, the summer sun may be medically contraindicated for the adult with hypopigmentation. Even those individuals who are normally pigmented may have increased photosensitivity and heat intolerance from psychotropic medications. The issue of job goals obviously has to be negotiated. The time and place for the negotiation are as important as the negotiation itself. It should not be done at a time when emotionally charged behavior is present. It should be done at a calm time in a professional setting and with respect for the choice the individual wants to make. Vocational staff should have a variety of alternatives to present.

Many individuals select and are valued members of janitorial crews. Even so, crew supervisors report being constantly stressed and challenged: "No matter how hard we scour the job sites for food access, individuals with PWS are better at it then we are. People will always have food in their desks, coffee creamer in the break room, and even edible Christmas tree ornaments during the holiday season." Examples include a young man with excellent work skills whose heart was set on working with a janitorial crew. After several failed work trials, a contract to clean a large warehouse belonging to the Department of Transportation seemed the perfect work placement for him. After a 10-pound weight gain in 1 week, it emerged that the young man had located the storage space housing the entire Gatorade drink supply for local road crews. It was not possible to secure this area, nor was it possible to sufficiently supervise the individual; as a result this placement failed.

Another man requested his own removal from an interstate highway rest area crew in spite of excellent work performance and zero weight gain. He reported that watching people throw away bags of partially eaten fast food all day was too much for him; he was afraid that the temptation was becoming too great. He preferred working elsewhere, even if it meant returning to the workshop.

As the previous examples illustrate, the biggest issue with supported employment is the unrestricted nature of the environment along with the lack of supervision. For many in these settings, maintaining dietary restrictions has to rely primarily on self-control. One young woman was given a trial as a filing clerk with the Social Security Administration, despite their initial reluctance to employ her because of her disability. After the first week, her work performance generated rave reviews. Her filing was excellent, she thrived on the repetitive task that quickly bored other workers, and she was making friends and was liked in the office. She was terminated a month later for taking unexpected breaks and for failure to return from lunch on time. An investigation revealed several fast food restaurants in close proximity to the office. Since termination occurred without consulting the job coach, there was no opportunity to provide additional supports that might have made this placement a long-term success.

One applicant for services seemed, at first, inappropriate. At the time of application, his proud parents reported that the young man was currently employed in the mailroom of a large downtown business earning a living wage. Further, he lived alone, rode buses around town visiting friends, and seemed from all outward appearances to have the ideal life. When parents were queried regarding the reason for disrupting what appeared to be excellent supported employment services, they responded, "because he weighs 350 pounds and is near death due to congestive heart failure."

While these examples illustrate that vocational placements for adults with PWS present specific syndrome-related challenges, these challenges by no means preclude successful vocational placements. To assure employment success, however, employers must be aware of and address these issues prior to placement, must provide constant monitoring of these issues after placement, and finally, must be prepared to deal with these challenges if and when they occur.[5] It must be recalled that historically it was thought that the health, weight, and behavioral issues of people with PWS could never be successfully dealt with outside a hospital setting. With a growth in expertise, public awareness, and sensitivity concerning people with disabilities, these challenges have been met. Similarly, successful vocational placements will soon be the norm rather than the exception. The following guidelines may be helpful.

Structuring for Vocational Successes

The average profile of those admitted to ARC of Alachua County's Prader-Willi syndrome program is a person 20 years of age weighing approximately 255 pounds; 50% evidence significant health problems—including Type 2 diabetes, congestive heart failure, and sleep apnea—and all are challengingly intelligent. Characteristically, these individuals cannot be well served in a program that does not successfully occupy their interests. In addition, many first entering the program have physical limitations that present obvious and unique challenges. Unstructured time frequently leads to food stealing, movement violations (individuals leaving the work area), and argumentative, and eventually tantrum, behaviors. Untrained staff may inadvertently increase the possibility that any or all of these behaviors may occur. Nonetheless, many people with PWS possess skills that are uniquely suited to the typical workshop environment. This is one reason that many individuals choose this more restrictive work setting. Among these skills are an ability to perform repetitive, fine-motor behaviors without loss of interest. Experience demonstrates that assembly and packaging tasks frequently serve to direct and focus otherwise variable attention on the task, while precluding engagement in other more problematic behaviors, and that they are comfortably performed by those with limited physical abilities. Prior to undertaking these tasks, however, the person with PWS must be assured of a safe, food-free environment that is sufficiently structured and supervised to prevent or respond to the most dedicated efforts to find the "loopholes."

Structure and expectations must be consistent and clear. At least two basic tracking systems (cash and food) are essential tools in a sheltered setting. In one setting, an individual's cash is tracked at all times with reconciliation prior to and after work. All spending must have a receipt or a staff signature to be valid. Missing cash results in consequences. Lunch box lists are included with every lunch individuals take to work. Residential staff sign these prior to the individual leaving home. Food found in someone's possession at work can be verified against the list, which is turned in to the work supervisor upon arrival.

These tools are only as good as the staff trained to implement them and the consumers' understanding of behavioral expectations under these rules. All direct care staff should be trained and certified in the performance and implementation of these procedures. Competency should be rewarded with salary increases. Prior to initiating work, expectations should be explained and agreed to by the employee and then *consistently* implemented. While this may seem obvious and only common sense, vocational placement success will only be achieved to the degree that consistent implementation is successfully achieved.

Some programs, in cooperation with living-care staff, view food stealing as a choice. Calorie allotments begin on arising and are tracked throughout the day. Individuals who engage in food stealing are viewed as exercising a choice to spend their calories on the items they have procured. Incentives are employed to discourage exercising this type of choice, but it is recognized as a choice when it happens (and it will) and is treated in a dignified manner. Calories are subtracted from the prescribed allotment and the business of work continues. When calories are gone, however, eating is finished for that day. Strong and consistent supervision is necessary to ensure this outcome. Similarly, clear, constant communication and cooperative programming with the living-care staff must be in place for such a system to be successful. When consistently implemented, these procedures minimize workplace problems.

For vocational efforts to succeed with these unique adults, environmental modifications and supports must ensure adequate supervision and consistent use of positive consequences. People with Prader-Willi syndrome are well grounded in the concept of what is "fair." Inconsistencies across settings or even inconsistent application of consequences across settings are common causes of behavior problems. Programs that offer both residential and vocational services are ideal for providing a comprehensive program of continuously integrated supports. Because of the reactive nature of those with PWS, a program with a proven commitment to positive behavioral supports has a greater likelihood of success.[3] In addition, the program must have an administration dedicated to understanding the issues of Prader-Willi syndrome and must provide programmatic support designed by staff versed in behavior analysis.

Table 15.1 suggests questions for families to ask when they are considering a vocational provider for their son or daughter with PWS. If

Table 15.1. Questions for Families To Ask a Potential Provider of Vocational Services

General

1. Is the employment agency known for high-quality services?
2. How long has the agency been in business?
3. What is the depth of the agency's experience supporting people with Prader-Willi syndrome?
4. How many people have been discharged from the program in the past year? Regarding those that have Prader-Willi syndrome, in general terms, what were the reasons for the discharge(s)?
5. With permission, can the prospective provider share names and phone numbers of individuals and their families who currently use the agency's services and would be willing to be a reference for the agency?
6. Does the agency have any written materials about its services that we could take with us?
7. How can I obtain a copy of the most recent licensing survey, if applicable?

Staffing and Consultants

8. Describe the hiring process used when selecting staff.
9. Does the agency have ongoing training and expect that all administrators and staff learn about Prader-Willi syndrome?
10. What initial and ongoing training does the agency require of staff?
11. How are direct support staff supervised?
12. What is the turnover rate for direct support and supervisory staff? What is the average length of employment for a direct support staff at your agency?
13. How does the provider select and access physicians, ancillary medical services, behavioral consultants, etc.?

Policies

14. Does the agency keep records of service and progress individuals with PWS have made? If so, in what form do you display progress?
15. How does the agency handle health limitations of workers?
16. What are the agency's policies regarding medication and its administration?
17. What contingencies are in place for medical and behavioral emergencies?
18. What is your agency's relationship to the individual's family or residential provider? Can or will it change based on the individual's needs?
19. What is the frequency of communications with families? What if a family wants to know more/less than is standard?
20. What issues are you mandated to report to parents/guardians?
21. What is your agency's policy on family observation? Can families visit unannounced? What is the agency's grievance procedure for the individuals and/or their families?

Program Issues

22. How does the agency modify assignments/settings to accommodate the food-related needs of an individual with PWS?
23. Does the agency create individual development plans to manage the behaviors associated with Prader-Willi syndrome?
24. How does the agency handle money issues for an individual with PWS?
25. How does your agency individualize services to meet varying needs?
26. What recreational or social activities are part of the work programming? Are the opportunities for participation individualized, accessible, and consistent? What happens when an individual chooses not to participate with the rest of the group?
27. What kind of transportation is available for the individuals' use?

the agency lacks experience in working with Prader-Willi syndrome, the family should anticipate being very involved in the process to increase the possibility of a successful job placement. Table 15.2 lists questions that agency providers might want to ask families in order to gather important information about the individual with PWS and to establish open communications with the family as a foundation for a positive work placement.

Table 15.2. Questions for Providers To Ask a Family Seeking Vocational Services

General
1. What is your family's expectation of this vocational placement? For example, do you consider this a transitional placement, or do you expect this to be your son's/daughter's permanent job setting?
2. Have you attended preparatory workshops regarding the range of vocational placements that are available for your son or daughter?
3. Where does the family see the individual best suited for success?
4. If you are not your (adult) son's/daughter's guardian presently, do you intend to apply for guardianship?

Background Regarding the Individual with Prader-Willi Syndrome
5. What are the career goals and type of work sought by the individual?
6. How much does the individual want to work?
7. What are the preferred hours of work?
8. Are there benefit limitations to the amount of work an individual may do? That is, will any financial entitlements such as SSI impose limits on the number of hours an individual can work?
9. Are there any medical limitations other than those associated with Prader-Willi syndrome?
10. Has the individual completed any vocational assessments through other agencies such as Vocational Rehabilitation?

Agency Policies and Staffing
11. Do you anticipate that you will have input into the agency's decisions?
12. What are your expectations of how conflicts will be resolved?
13. How much involvement would you like regarding your child's food and finances?
14. For what issues/incidents would you like to be contacted immediately? What is the best manner in which to reach you?
15. Do you prefer written or verbal communications from the agency staff? Barring any major incidents, how often would you like an update?
16. What are your preferred days/times for team meetings?
17. Do you have a relationship with your child's current residential staff (if applicable)?

Program Issues
18. What are your preferred methods of managing the behavioral manifestations of Prader-Willi syndrome exhibited by your son/daughter?
19. What interventions have been consistently successful with your son/daughter? What other interventions have been attempted but proven unsuccessful for your child?
20. Are you using a behavior support plan currently?
21. Will you need assistance from agency staff with transportation?
22. Other than yourself, who else in your extended family can transport your child when needed?
23. Does your son/daughter have a religious affiliation? Which church/synagogue does he/she regularly attend?
24. How will your child's prescription medications be secured, if necessary, and administered? Will the agency need to administer any of the medication?
25. What assistance, if any, would your family like in transitioning your son/daughter to their new work situation?

Toward the Future: A Pilot Community-Based Program Model

Recently, the ARC of Alachua County identified a need for an employment agency that actively serves, trains, and places individuals with severe disabilities in jobs. The data at that point indicated there were at least 200 individuals in the local area whose work applications had been denied by other vocational placement and human service agencies. Most of these individuals required long-term, close monitoring and training to achieve vocational success. Among these were many persons with PWS. As a result, the ARC implemented a specialized employment agency for people with PWS. There are several specialty components in this pilot model, including education and recruitment of employers, a co-coaching training and support model, a temporary labor pool, and specialized skills training. Each component is discussed in more detail in the following paragraphs.

Education and Recruitment of Employers

The business recruitment and education component has been designed to provide potential employers with an understanding of the benefits of hiring people with disabilities. These benefits include, but are not limited to, lower-than-average turnover rates, punctuality and above-average attendance rates,[5] prescreening of employees to help match potential employees with employers, a reduction in the cost to the employer for the employee's salary during training, follow-along support for the employee, and the possible extension of ARC of Alachua County's Sub-Minimum Wage Certificate.

Co-Coaching Training and Support Model

In addition to the traditional coaching model, the ARC overlays a co-coaching strategy, which entails identifying natural supports among current employees. This is based on the assumption that the co-coaching model will provide long-term, continuous support for the employee with significant disabilities. The co-coaches provide support, continuous coaching, and communication assistance to people with disabilities in exchange for a weekly stipend.

As earlier indicated, people with PWS are extremely capable individuals who often fail community-based employment due to excessive overeating and food stealing. The co-teaching model provides an *employment specialist* who identifies and trains co-coaches to provide supervision and support to those with PWS in an effort to alleviate many of these problems. Individuals with PWS are very capable of learning to control these behaviors if they know people are there to help. The co-coaching model has been used very successfully with one participant supportively employed at a day care center for 6 months. During this time, co-coaching provided by several co-workers helped control unauthorized food intake. This successful placement would have continued had not overall staff reductions at the day care center ended the employment.

Traditional placement accompanied by a fading of support services typically fails for individuals with PWS, whose needs are intense and unremitting. The success of this pilot co-coaching model suggests that many people with PWS may have increased vocational success utilizing this continuous support model. This requires that stipend supports be ongoing with no fading. However, this minimal investment may mean the difference between the individual with PWS keeping and losing a job.

Conclusion

Progress in the field of vocational services remains behind progress achieved in medical, behavioral, and residential supports for people with Prader-Willi syndrome. The reason for this lag is obvious: arranging these environments for success is far more difficult, as their very nature lessens the capacity for controlling critical components. In sheltered workshops there is not always consistency in the ownership/management between residential and vocational service providers. Typically, supported employment requires placing adults in someone else's business whose good intentions may be eroded by behavior that adversely affects business. Nonetheless, progress has been achieved, and that progress can be directly attributable to successfully structuring the environment through staff training, physical modifications, and programmatic procedures. The "final frontier" is definitely upon us.

References

1. Gold M. *Did I Say That? Articles and Commentary on the Try Another Way System*. Champaign, IL:Research Press; 1980.
2. James TN, Brown RI. *Prader-Willi Syndrome: Home, School and Community*. San Diego, CA: Singular Publishing Group, Inc.; 1992.
3. Latham G. *The Power of Positive Parenting*. Logan, UT: P&T Inc.; 1998.
4. Lutfiyya ZM, Rogan P, Shoultz B. *Supported Employment: A Conceptual Overview*. Center on Human Policy. Syracuse, NY: Syracuse University; 1988.
5. Wehman P, Kregal J. A supported work approach to competitive employment of individuals with severe handicaps. *Journal of the Association for Persons with Severe Handicaps*. 1985;10(3):132–136.

16

Residential Care for Adults with Prader-Willi Syndrome

Mary K. Ziccardi

Provision of residential services to adolescents and adults with Prader-Willi syndrome was not seriously addressed prior to the late 1970s, primarily because until then most individuals were not expected to live beyond the adolescent years.[3] As a natural outgrowth of earlier diagnosis and appropriate nutritional and weight management, individuals with Prader-Willi syndrome now live well into adulthood. Thus, from the mid-1980s to the mid-1990s, establishment of adult residential services was a major thrust of both local chapters and the national Prader-Willi Syndrome Association (USA). And, while many additional programs are needed to care for this expanding population, nonetheless, this experience has taught us that individuals with Prader-Willi syndrome have unique needs that must be addressed in order to ensure a safe and successful living environment. This chapter addresses the issues involved in both the process of selecting a supported living environment and the necessary elements of that environment. These elements are the minimal requirements needed for the success of any supported living situation, ranging from the single-resident apartment to larger, congregate settings.

For many families, exploring residential placement for their family member is an overwhelming and emotionally charged task.[1] At the most basic level, placements are difficult to locate and secure. Many families face long waiting lists and lack of governmental funding for developing appropriate alternative living opportunities. Families are often forced to wait years for any residential facility that is even remotely willing to assume the challenges of serving someone with Prader-Willi syndrome (PWS).

That said, the family's full participation and disclosure of their family member's strengths and needs is the first step in securing a *successful* residential placement. A residential provider has a significantly improved chance of meeting the individual's needs if those needs are discussed honestly and openly. Many willingly provide behavioral accommodations and environmental modifications to facilitate a lasting placement that includes a clinically sound approach to treatment, an acceptable quality of life, and positive outcomes. The prospective provider and the family will be most successful for the individual with PWS if a spirit of openness and cooperative, respectful communications are established early in the relationship.

Questions Families and Providers Should Ask

For most families, the task of selecting a residential provider is a process and experience unlike any other. While families may not know what areas to address and questions to ask, at the same time providers may be unaware of the potential impact on families of an agency's culture and operational structure. Many families benefit from training workshops that both educate and support them through the process of selecting residential supports and services.[2] In addition to providing factual information regarding funding, eligibility, and application processes, many workshops provide an informational checklist to help families obtain sufficient information for making informed placement decisions.

Some of the questions that a family may find beneficial to ask when interviewing a potential provider are included in Table 16.1. While not all-inclusive, this list can provide a stepping-off point in seeking and sharing information that may result in a positive, long-term relationship between the family and the provider agency. Asking and discussing these questions with a potential provider agency can assist in determining if the agency's program matches the family member's needs. Further, any potentially contentious issues can be openly and thoroughly discussed. Likewise, the set of questions suggested in Table 16.2 might assist the residential service provider in better preparing to serve an individual with Prader-Willi syndrome.

We will discuss a number of environmental and program areas that are critical to the quality and success of a residential placement.

Program Issues

Provider's Manifestation of Philosophy

When a family is considering their residential options, the philosophy of the provider agency should also be considered. Agencies, just as families, may consciously or unconsciously develop a set of beliefs and practice accordingly. Guardianship may be one of the more significant areas that merit early exploration. For example, is the family planning to assume guardianship of their adult son/daughter? Some agencies may demand guardian oversight as well as an active role in the seemingly smallest occurrences. Providers who advocate this philosophy may solicit a total team approach, involving the families in the home purchase, renovations, staff selection and training, behavioral supports and policy setting. Families may need to consider their own relationship dynamics, availability, and degree of interest in participating. Conversely, other providers may prefer that families practice a more "hands off" approach, only notifying a family for the most major life issues.

Table 16.1. Questions Families Should Ask Potential Providers of Residential Services

General

1. How long has the agency been in business?
2. What is the depth of the agency's experience supporting people with Prader-Willi syndrome?
3. How many people have you discharged from your program in the past year? Regarding those that had Prader-Willi syndrome, in general terms, what were the reasons for the discharge(s)?
4. With permission, can the prospective provider share names and phone numbers of individuals and their families who currently use the agency's services and would be willing to be a reference for the agency?
5. Does the agency have any written materials about the services that we could take with us?
6. How can I obtain a copy of the most recent licensing survey, if applicable?

Agency Staffing and Consultants

7. Describe the hiring process used when selecting staff.
8. What initial and ongoing training does the agency require of staff?
9. How are direct support staff supervised?
10. What is the turnover rate for direct support and supervisory staff? What is the average length of employment for a direct support staff at your agency?
11. How does the provider select and access physicians, ancillary medical services, behavioral consultants, etc?

Agency Policies

12. How does the agency handle an individual's finances?
13. Do you provide reports to the individuals and/or their families regarding their finances? If so, how frequently?
14. What are your agency's policies regarding medication and its administration?
15. What contingencies are in place for medical and behavioral emergencies?
16. How can a supervisor be contacted by a family during an emergency situation outside of normal office hours?
17. What is the frequency of your communications with families? What if my family wants to know more/less than the standard?
18. What issues are you mandated to report to parents/guardians?
19. What is your agency's policy on visitation? Can families visit unannounced? Would we have a private place in which to visit?
20. What is the agency's grievance procedure for the individuals and/or their families?

Program Issues

21. How does your agency individualize services to meet varying needs?
22. What is your agency's relationship to the individual's school or vocational programs? Can or will it change based on my family member's individual needs?
23. What recreational, religious, and social activities are available? Are the opportunities for participation individualized, accessible, and consistent? What happens when an individual chooses not to participate and the rest of the group plans to go?
24. What kind of transportation is available for the individuals' use?
25. What is the agency's relationship with neighbors and the community at large? Describe how the individuals belong and have a community presence.

Source: Adapted from C. Norwood, *What Questions Should I Ask?* Center for Mental Retardation, 2002.[2]

An honest self analysis that defines the family's style and comfort level and finding a provider that best matches their beliefs can be critical to a placement's long-term success.

Families and providers also need to openly discuss larger issues regarding their views and practices surrounding the very meaning of normalization and to what depth community integration is expected. Discussing the negotiables of these issues can be far-reaching to many

Table 16.2. Questions Providers Should Ask Families Seeking Residential Services

General
1. What is your family's expectation of this placement? For example, do you consider this a respite or short-term assistance for your family, or do you expect this to be your son's/daughter's permanent home?
2. For what issues/incidents would you like to be contacted immediately? What is the best manner in which to reach you?
3. If you are not your (adult) son's/daughter's guardian currently, do you intend to apply for guardianship?
4. What are your preferred days/times for team meetings?
5. Do you prefer written communications from the agency staff? Barring any major incidents, how often would you like an update?
6. What assistance, if any, would your family like in preparing for the move and transitioning of your son/daughter to the new home?

Agency Policies and Staffing
7. Do you anticipate that you will have input into the agency's hiring decisions?
8. What are your expectations of how conflicts will be resolved?
9. Describe your preferences regarding medical appointments. Will you be scheduling any or all of them? Attending any or all appointments? With or without agency staff?
10. How much involvement would you like regarding your child's finances?

Program Issues
11. What are your preferred methods of managing the behavioral manifestations of Prader-Willi syndrome exhibited by your son/daughter?
12. What interventions have been consistently successful with your son/daughter? What other interventions have been attempted but proven unsuccessful for your child?
13. Are you using a behavior support plan currently?
14. Should extended family visitation be used as a reinforcer for your child?
15. Do you have a relationship with your child's current educational/vocational staff?
16. Will you need assistance from agency staff with transporting your child for routine family visits? Would you like a staff member to assist your son/daughter by accompanying them to other family events, i.e., holidays, birthday celebrations, weddings, etc.?
17. Other than yourself, who else in your extended family can visit your child? Who can take them out of the facility for day/overnight visits?
18. Does your son/daughter have a religious affiliation? Which church/synagogue does he/she regularly attend?
19. How will your child's prescription medications be secured, if necessary, and administered during a visit to your home?
20. During visits to your family home, would you like to have menus and/or food packed to assist with nutritional management?

significant aspects of the individual's daily life. Does the son/daughter ride a sheltered workshop bus or independently travel—that is, become trained to use community transportation? How are food and caloric restrictions handled during family visits and holiday celebrations? Is the agency expected to provide staff to accompany and support the individual at family events? Can the individual carry spending money? A common ground should be realized regarding the extent to which an individual with Prader-Willi syndrome is their "own person," able to make decisions and reap the benefits or suffer the consequences. Or, does the agency advocate that the individual should be supported and afforded freedoms within the confines of the developmental disability, while receiving more intensive direction and supports for issues related to health and safety? In addition to exploring the daily dynamics identified in Table 16.1, all parties would benefit from delving into the reality of the practices manifested from the overall philosophical foundations.

Before starting out, families should establish their "must-have" priorities, whether it is location, number of housemates, behavioral management strategies, parental involvement, or other issues critical to their values and wishes for their family member, and consider conceding some smaller, less important issues. When faced with minor compromises and small disagreements, it's important to be mindful of the greater gains to be achieved for and by the person with Prader-Willi syndrome. Families should ask themselves and the agency how the person's health, their very life, will be improved and enriched by this residential placement.

Once the selection has been made and an individual has been accepted into a program, the family and provider agency must work as a team to provide the best care and support possible. Any obvious disagreements, displayed by either party, can and will most likely be used by the person with Prader-Willi syndrome to create conflict and to manipulate. If an issue that requires discussion becomes evident, the family and provider representatives should make every effort to discuss it out of earshot of the individual receiving services. Remaining committed to respectful and professional interactions will go far in providing a secure environment in which the individual with Prader-Willi syndrome can learn new skills, remain healthy and safe, and achieve a quality of life.

Environmental Requirements

Several environmental factors are key to providing both a safe and successful residential program for people with Prader-Willi syndrome. For most, securing food is primary. The methods used to do so are ultimately of less importance than the simple fact that it is secured. To many with Prader-Willi syndrome, the very knowledge that food is inaccessible is a comfort and reduces a great deal of anxiety. Many residential programs choose to utilize a totally locked kitchen; others elect to lock refrigerator and cupboards. Either method is appropriate. Most importantly, consistency in properly securing these locks must be applied by everyone at all times.

There are a variety of methods to prepare and apportion the meals and snacks. Determining what is best in a particular environment depends on the overall philosophy, the physical structure of the home, and the number of people who live and work there. Constant factors, however, include the need for accurate menus and recipes. In addition, those who cook and prepare food must be adequately trained to measure quantities in accordance with the calories prescribed. The role of the dietitian is later addressed in this chapter under "Medical Care and Ancillary Services."

Individuals with Prader-Willi syndrome enjoy the comfort and perceived sense of control of their own lives associated with the anticipatory knowledge of what foods will be included in meals and snacks. To this end, a menu board posted daily outside the kitchen may help alleviate some of the ever-present questions and concerns regarding food.

To accommodate the short stature of many adults with Prader-Willi syndrome, especially those who have not benefited from growth hormone therapy, providers may want to lower shelves and closet rods when building or modifying a home. This simple adaptation can serve to increase an individual's independence.

Most people with Prader-Willi syndrome "skin-pick"; the frequency and severity of skin-picking episodes can be especially challenging for all involved. Therefore a range of behavioral interventions and treatment approaches may be needed (see Chapter 12). A water-circulating, Jacuzzi-style bathtub can assist with healing open areas and promote skin integrity. Under some funding sources, the Jacuzzi tub may be allowed and reimbursed as medical equipment.

People with Prader-Willi syndrome have unique space needs. A propensity for both collecting and hoarding, accompanied by an extraordinary sense of protectiveness for these collections, particularly from others, dictates that the resident must have ample space. Individual bedrooms, which are locked and accessible only to the occupant and support staff, are ideal to reduce both real and imagined incidents of stealing and property destruction. When single bedrooms are not possible, strict guidelines need to be established and enforced with regard to individual space, property, and privacy. Integrating these expectations into the "house rules" may help to prevent disagreements later.

Furnishings in the home should be durable, sturdy, and easily cleaned and maintained. When making purchases, consider items made of wood that cannot be easily overturned. As an example, purchase a console television rather than an entertainment center, which may be tall, may contain several pieces of audio-visual equipment, and would be easier to topple. Unwieldy pole lamps or ceiling-installed track lighting may be better choices than table lamps that can easily be thrown in the midst of a behavioral episode. For decorative purposes, small wicker baskets, items made of soft material, and hanging quilts may be preferable to metal and glass picture frames, large mirrors, and candlesticks. When hanging large frames, consideration should be given to discreetly bolting the frame to the wall. The individual's preferences, safety, and common sense should be balanced when making these decisions. Further, the intensity of environmental safeguards provided should be commensurate with the behavioral needs of the residents.

Large and open common living spaces, such as a family room and living room may serve multiple purposes such as providing opportunities for family visits, individual and small group activities, and social interactions. However, a common television may require supervision to avoid arguments about "who chooses the next show." Small game tables, for crafts and puzzles, are often used. A computer loaded with games and e-mail access may also be appropriate, when used with some supervision.

The proper utilization of both separate bedrooms and common living areas is essential to containing and managing behavioral tantrums, including possible escalation of physical aggression and property destruction. When the house rule "Leave the room when asked" is enforced, and plenty of space is available, potential chain-reaction behavioral tantrums may be avoided.

Overall, carpeted areas in the bedrooms and general living spaces seem most desirable and comfortable. Wide staircases, with carpet or treads and handrails on both sides, provide additional safety. Due to affected individuals' propensity to hoard items, storage closets can be very beneficial in reducing the general clutter in the bedrooms. These closets can be used for off-season clothing storage, as well as storage of overflow word-search books, jigsaw puzzles, crafts, and videos.

If space is available, it is ideal that key supervisory staff have an office area on-site. This will afford a private space to consult with family members, complete necessary paperwork, and train and counsel staff. Ideally, this office space can be locked and inaccessible to the men and/or women living in the home.

Yet another environmental consideration in program location is access to the community. Neighborhoods with recreation centers (e.g., indoor and outdoor paved walking tracks, swimming pools, basketball courts), bowling alleys, churches, shopping, and movie theaters may help to promote exercise and encourage an overall sense of community belonging. Conversely, building a program next to a restaurant or convenience store might invite food-related incidents and simply provoke already challenging behaviors. While food cannot be completely avoided when in the community, common sense and good judgment should prevail.

No environment is perfect and without pitfalls. Families and providers can work together to prioritize, set guidelines, and define expectations. All involved will need to be prepared to adapt and make changes to meet current and future challenges.

Staff Training and Supervisory Approaches

A well-trained, supported, and empowered staff is the cornerstone of any successful residential program. These attributes are even more critical for supporting people with Prader-Willi syndrome. Creating a successful team of staff begins with the employment application and interview phases. It is helpful to develop some questions specific to the program's structure and needs. Every attempt should be made to assess the applicant's approach to difficult situations, such as being the target of verbal or physical aggression. It is often beneficial for the applicant to be interviewed by more than one supervisor, either in subsequent appointments or by using a team interview approach. Further, requiring the applicant to respond to written questions may help determine literacy and problem-solving abilities, while also providing additional insight as to whether the applicant would be a positive addition to the team of staff.

Once a hiring decision has been made, both initial and ongoing training is crucial. Training sessions on the most basic issues, such as genetics, environmental controls, diet, exercise, and behavior management strategies, is best completed prior to an introduction to the residential setting. Once a new staff member has toured the home and been introduced to the program, an orientation checklist detailing specific requirements and approaches can be an excellent training tool. Further, a new staff member should ideally be trained by an experienced supervisor or lead staff, working side by side for a full 40-hour work week on the shift they will be scheduled to work. The opportunity to witness and participate in a variety of shift responsibilities and events will only serve to strengthen the trainee's overview of the entire program.

Hiring and initial training of new staff is only one aspect of the overall training program. Direct support staff encounter challenges daily. Ongoing support and training are required to ensure that the residents are constantly provided with the best possible responses from staff. When a program is new, or when difficulties arise, weekly staff meetings are beneficial. A group of work partners who are cohesive and empowered can be solidified through an atmosphere of sharing, honesty, and mutual support. Role-playing actual or potential events is a great learning tool for staff.

The program supervisor is key to a successful environment. This leader must earn the respect of staff, individuals served, and parents. The supervisor must establish a strong presence on each shift, demonstrate the ability to mediate the issues of both individuals served and staff, and display expertise in training and coaching others. The direct line supervisor will most likely interact routinely with parents and other professionals and be required to lead an interdisciplinary team to consensus on a variety of difficult issues.

Not only should direct care staff be supported, they must also be empowered to make the multitude of decisions that are needed on any given shift. Therefore, if a staff member needs to seek advice or direction from a supervisor, it is best done out of earshot of the individuals with Prader-Willi syndrome. This simple approach may help to establish the credibility and authority of the direct care staff.

A career supporting people with Prader-Willi syndrome is not the best match for everyone. Successful staff must possess maturity, display consistency, and not need the "last word" in a debate. Finally, staff must demonstrate forgiveness and know that, regardless of what happened today, tomorrow is a new day. Nationwide, the most successful programs have little staff turnover. Staff possess integrity and maturity and are highly trained and competent. Constant staff turnover is a key indicator of programmatic problems.

Programmatic Components: Behavioral Management and Skill Development Issues

The content of the program is as critical as staff training. People with Prader-Willi syndrome require—nay, demand—consistency and structure in all areas including programming.

Generally, a basic set of house rules, applied equally to all, will establish baseline standards for acceptable behavior. Depending on the structure and uniqueness of the residential program, some examples of house rules may include the following:

1. Keep your hands and feet to yourself.
2. Leave the room when asked.
3. No food is to be brought into the house.
4. No trading, loaning, or borrowing items.
5. Keep your bedroom door locked at all times.

Based on overall cognitive abilities, including understanding of cause and effect (i.e., if I do x, then y will occur), a person with Prader-Willi syndrome may function best with a behavior plan that offers rewards and consequences. A young lady with Prader-Willi syndrome was once overheard to say, "We don't do something for nothing!" Capitalizing on that approach may be easier said than done, because, historically, reinforcers used in MRDD (mental retardation/developmental disability) programs have been food-based. While it is indeed a challenge to identify potential nonfood reinforcers, it is not only possible, it is the responsibility of caregivers to do so. Behavior support plans must be individualized and must address the unique needs, strengths, and desires of each person. Many years of successful programming coupled with practical, anecdotal experience of many professionals have established that response cost programs, level systems, and point systems work well. A brief explanation of each follows:

1. Response Cost

The basic premise of this strategy is that, while an individual has the opportunity to earn "tokens" for compliance (e.g., absence of target behaviors or completion of daily expectations), previously earned tokens can also be forfeited for negative behavior. The reinforcement schedule of the tokens can be for task-based compliance or at predetermined intervals. This may work best if the token is a tangible item so that it can be visibly presented and retained for desirable behavior, as well as functionally returned for negative, less desirable activity.

2. Level Systems

This program strategy requires that privileges be earned and available relative to the type and amount of compliance and positive, desirable behaviors. More coveted and difficult-to-attain privileges are rewarded at the highest level; that is, when the greatest amount of compliance is shown the most desired reinforcers are available. This approach may work best with an individual who can cognitively grasp future goals, plan accordingly, and make informed choices about willingness to comply in order to earn desired rewards.

3. Point System

This approach can be designed much like the strategy used with the level system. Various expectations are assigned a point value, and points translate into privileges and reinforcers. This plan may work best with someone who doesn't require a concrete "token" as a symbol of their compliance but for whom the level system is too abstract.

Behavior support plans for individuals with Prader-Willi syndrome have the best chances for success when they are soundly designed, supported by parents and significant others, and *consistently* implemented. Because individual needs can be constantly changing, behavior support plans should be regarded as perpetual works in progress. Accurate and complete data recording of behavior plan compliance is an essential responsibility of each and every shift. Many of the most common behavioral strategies used with people with Prader-Willi syndrome build on each other. For example, if a staff member who works on a Tuesday afternoon shift leaves without documenting program compliance, the staff working on Wednesday may not have all of the information needed to determine the individual's privilege status. While this certainly does not qualify as an acceptable clinical practice, lack of documentation is also problematic because it leads to inconsistent program implementation, staff not working as a team, and possible manipulation of the behavior plan by the person with Prader-Willi syndrome. Conversely, if this lack of documentation has resulted in the loss of an earned reward, which subsequently precipitates a behavioral incident, it may appear that the behavior plan is unsuccessful, when in fact it was working. Staff must be held accountable for consistent and thorough completion of basic program documentation. The uncanny ability of some individuals to determine the weakest point of any behavior support plan further compels the interdisciplinary team to continuously assess the plan's strengths and weaknesses and modify accordingly.

Many people with Prader-Willi syndrome possess and can display adequate skills in completing personal care and household tasks. While compliance may be an issue, skills are often at acceptable levels. Therefore, creating meaningful skill development program ideas can be challenging. A comprehensive functional assessment could assist in determining need areas. The interdisciplinary team can use this information to determine priority areas and may want to consider integrating compliance into the reward system of a behavior support plan. While most people with Prader-Willi syndrome require high levels of supervision throughout their lives, greater independence and autonomy can be achieved through increasing skills and abilities in many areas.

Medical Care and Ancillary Services

Locating, informing, and working cooperatively with physicians is a major responsibility of a residential provider. In some cities where residential services for people with Prader-Willi syndrome exist, physicians are knowledgeable, experienced, and well versed in treating the multifaceted needs arising from the syndrome. Unfortunately, this is often the exception rather than the rule. More often, the provider and parents must work to develop relationships with physicians willing to provide primary care as well as specialists for psychiatric oversight, podiatric care, dental treatment, and a host of other specialty areas.

A nurse employed by the residential provider can have a positive impact on the relationship between physicians and the individuals requiring medical care. In addition to ensuring that the regulations of each state are met, the nurse can be the "eyes and ears" for the physicians. The nurse's knowledge of Prader-Willi syndrome, familiarity with each individual, and excellent assessment skills may save unnecessary trips to the physician's office. Although faced with many significant medical conditions and risks, people with Prader-Willi syndrome also regularly create or embellish medical complaints. A skilled nurse can assist the physician by decreasing time spent on hypochondriacal complaints, thereby assuring that the person with PWS will receive proper attention and care when brought to the physician for a true area of need.

The nurse also plays a critical role in educating others. As earlier indicated, a well-trained staff is imperative for a successful program. The agency's nurse should be involved in both the initial and ongoing staff training. This training can include descriptions of the physical manifestations of the syndrome, as well as standards and expectations for dietary needs, weight control, and exercise. The nurse's role in educating others may extend to neighbors, teachers, and vocational support staff. A nurse who enjoys an overall case management approach of coordinating, assessing, and educating would likely succeed in this role.

In addition to nursing, other ancillary services are important to a residential program supporting people with Prader-Willi syndrome. An approachable, informed dietitian is essential to create menus, discuss food preferences, and determine what food items are permissible for special occasions. Oftentimes, several people living together may have differing caloric requirements. A creative dietitian can assemble menus that are tasty, healthful, and follow the physician's prescriptions. The successful dietitian will be well versed in all aspects of Prader-Willi syndrome, in part to avoid potential manipulations by the individuals regarding their diet plans. Further, it is beneficial that staff receive training related to the food aspects of the syndrome directly from the dietitian. This training can include basic dietary guidelines and requirements, discussion of specific diets, practice with weighing and measuring portions, and role-playing that addresses the inevitable "what ifs." Staff armed with knowledge and confidence are more likely to make competent and reasonable decisions when unusual situations present themselves. While a dietitian may not always be immediately accessible, the training he/she has provided may enable the staff to alleviate the immediate concern until official clarification is obtained.

Lastly, a dietitian may make regular house visits to monitor weights, make necessary dietary modifications, and adapt menus. Because food is of utmost importance to the person with Prader-Willi syndrome, a strong relationship with the dietitian may help to provide assurance that the prescribed diets are indeed being followed.

An exercise physiologist also has a major role in a residential program. At the onset of a new program, the exercise physiologist's role is to assess each individual's overall status. Once the assessment is completed, an appropriate exercise schedule can be created and individualized. Periodic review of progress and necessary modifications in type and amount of exercise can assist individuals to safely achieve weight loss goals. Similar to the nurse and dietitian, the exercise physiologist can play a supportive and teaching role to the direct support staff. This can be achieved by answering staff's questions about therapeutic exercise and supporting staff in their quest to encourage the individuals with their required exercise regimen.

A residential program's nursing, dietary, and exercise physiology staff play key roles in the success of the individuals receiving services. The education of community members and support to individuals and staff can be invaluable tools in the multidisciplinary approach toward achieving safe, healthy, and enriched lives for people with Prader-Willi syndrome.

Current Residential Options Across the United States

The implementation and operational practices of residential programs throughout the country are diverse. This is due, in part, to the requirements of the funding sources and philosophies of the provider agencies. While environmental controls remain paramount, the physical structure of the program, whether an apartment, single family home, or dormlike setting, may be less important than the care, consistency, and treatment that takes place inside those walls.

Segregated programs, which provide services only to people with Prader-Willi syndrome, can be managed in several ways. One example is a small (i.e., three-to-five-person) single-family home that blends into a residential neighborhood. Additionally, residential homes exist in which larger numbers of individuals (i.e., up to 16) are supported. In this type of program, general house rules and overall philosophies are the same for all, while medical care and behavior support plans can be individualized. This model promotes and capitalizes on community presence and belonging, yet provides the environmental structure and treatment that the syndrome demands.

Similarly, some individuals with Prader-Willi syndrome are best served in a segregated program on a large, campuslike setting. Individual or shared sleeping rooms are contained in a cottage or dorm-style building on the grounds. Many children and young adults who live in this setting attend school or a vocational training program that is on the campus; others are enrolled in school or work settings in the community. In this type of program, services are also provided that address exercise, nutrition, activities of daily living, and behavioral supports.

Some persons with Prader-Willi syndrome live in homes with others who do not share the syndrome. This may occur as a result of the family's or individual's choice. Unfortunately, however, this residential option may often occur out of necessity or an emergency placement need, and when no segregated programs exist or are readily available. This arrangement presents particular challenges for those with Prader-Willi syndrome as well as for those with whom they live who have other developmental disabilities. While this situation may indeed be successful for some, significant attention must be paid to food access as well as to the unique privacy and property issues and behavioral manifestations of the syndrome.

Due to tight state and federal funding and budget constraints, it is rare to see a residential program that requires 24-hour staffing in which three or fewer individuals live. While this may be philosophically preferred, it is simply not deemed cost-effective by many funding sources. As a result, many states now encourage families to keep their adults with developmental disabilities in the family home with supportive services provided as needed. This option, even when desired by families, is rarely successful since most families find that, by the time the individual with Prader-Willi syndrome reaches adulthood, the family is no longer able to adequately meet the demands of 24-hour vigilance and care.

A variety of approaches occur within the walls of a residential program supporting people with Prader-Willi syndrome. All existing programs that were researched for this chapter provide dietary management, food restrictions, and education, as well as exercise, access to ancillary services, and behavior support. One unique approach to dietary management involves a base calorie amount to be consumed each day, with the opportunity for earning additional caloric incentives for compliance in areas like exercise, attending work, and exhibiting appropriate behavior. Ultimately, an argument can be made that there is no absolutely "right" or "wrong" approach, but the key to a successful residential placement may lie in the matching of the individual's needs to the program strengths.

Conclusion

Many people with Prader-Willi syndrome will face significant challenges and difficult decisions throughout their entire lives. Deciding on the appropriateness and type of residential supports is one of the most important considerations facing any person with the syndrome and their family. Guidance from family and significant others, national research of the options available, and seeking answers to probing questions will greatly assist with making the best possible decision. Successful residential programs, equipped to manage and support the unique needs of individuals with Prader-Willi syndrome, have continued to be developed over the past two decades. It will be through the relentless efforts of both parents and professionals that all people who desire a residential program will have a safe home in which to learn and achieve a quality of life.

References

1. Greenswag L, Singer S, Condon N, et al. Residential options for individuals with Prader-Willi syndrome. In: Greenswag L, Alexander R, eds. *Management of Prader-Willi Syndrome*. 2nd ed. New York, NY: Springer-Verlag; 1995: 214–247.
2. Norwood, C. *What Questions Should I Ask?* Cleveland, OH: Center for Mental Retardation; 2002.
3. Thompson D, Greenswag L, Eleazer R. Residential programs for individuals with Prader-Willi syndrome. In: Greenswag L, Alexander R, eds. *Management of Prader-Willi Syndrome*. New York, NY: Springer-Verlag; 1988: 205–222.

17

Inpatient Crisis Intervention for Persons with Prader-Willi Syndrome

Linda M. Gourash, James E. Hanchett,[†] and Janice L. Forster

Editor's comment: The authors have summarized their clinical experience in caring for several hundred persons with Prader-Willi syndrome (PWS) who were hospitalized for crisis intervention. It is emphasized that this experience was with a referral population in crisis and does not reflect the population of persons with PWS in general. Crises in persons with PWS appear to be most often associated with extrinsic factors.

Inpatient care of persons with PWS has afforded a unique opportunity to obtain detailed historical information, to review extensive medical records including prior hospitalizations, and to observe firsthand and in depth the complex phenomenology of PWS. While clear patterns of PWS crisis have emerged and are described here, the interventions and recommendations are far more variable and must be individualized based on each patient's unique personality and circumstances. Inpatient hospitalization is at times the only way to evaluate in adequate detail the circumstances leading to crisis in PWS and to identify or create the resources necessary for crisis resolution.

At the time that this work was done at The Children's Institute of Pittsburgh, Pa., the clinical leadership consisted of a developmental pediatrician who functioned as attending physician and team leader, a developmental neuropsychiatrist, an internist (nephrologist), a head nurse, a psychologist, and case manager(s). There was an integrated collaboration of these disciplines meeting as often as three times per week in addition to a more typical patient-centered staffing model, which included other rehabilitation disciplines such as physical therapy, speech and language therapy, and occupational therapy.

Extended hospitalization (usually 1 to 6 months in duration) provided a unique venue to evaluate treatment interventions in situ. Outpatient follow-up through office visits, telephone, and e-mail communications enabled the clinical evaluation of treatment results across time for a large number of patients manifesting the more severe problems associated with PWS.

Crisis intervention requires involvement in complex scenarios and problems that do not lend themselves easily to systematic investigation. The relative contribution of patient factors and environmental factors leading to crisis is

[†] Deceased.

an area worthy of study. The authors offer their experience as an aid for clinicians addressing these complex problems perhaps for the first time and for researchers who may choose to investigate the sources and resolution of crises in individuals with PWS.

The Nature of Crisis in Prader-Willi Syndrome

Crises in persons with Prader-Willi syndrome (PWS) involve deterioration in level of functioning across medical, behavioral, and/or psychiatric domains, or in the person's support system. Crises most often appear to be the result of extrinsic factors interacting with the typical features of PWS, while a smaller number of cases appear to involve the more severe behavioral spectrum of the disorder or frank psychiatric illness. Usually several areas of deterioration must be addressed simultaneously in order to stabilize the patient. Frequently outpatient interventions may not be comprehensive or sufficiently intensive to effect a sustained improvement.

A crisis for an individual with Prader-Willi syndrome may be abrupt or, as often occurs, the culmination of multiple contributory events. Because of the unique and complex problems associated with PWS, these situations may benefit from a specialized approach implemented by a multidisciplinary team experienced with the disorder.

The goals of crisis intervention include the following:

- Reversal of medical, psychiatric, or behavioral deterioration
- Restoration of an existing support system or prevention of breakdown in a potentially overwhelmed system
- Development of a realistic and comprehensive post-hospitalization plan that supports the above goals for the foreseeable future

Reasons for Referral

Medical Crises

Medical crises in persons with PWS stem primarily from the consequences of a fundamental disorder of satiety.[21] Rarely, stomach rupture from overeating has occurred (see Chapter 6). More often there is an unrelenting weight gain leading to morbid obesity. Persons with Prader-Willi syndrome usually have *exceptionally low caloric needs*, resulting from decreased muscle mass and physical activity, as well as varying degrees of excessive food-seeking behavior. The latter is sometimes extraordinary. Overestimates of actual caloric needs by caretakers and professionals can add to the problem. Persons with Prader-Willi syndrome are at risk for sleep-disordered breathing (SDB)[14,15,19,20,30,33] (see discussion in Chapter 5). Clinical experience also indicates that obese persons with PWS and a body mass index (BMI) greater than 35 appear to be at increased risk for complications resulting from SDB (BMI = weight in kg/height in meters^2). These breathing abnormalities, if untreated, can eventually lead to obesity hypoventilation and right heart failure *(cor pulmonale)*, which may first present as sudden critical illness, prolonged hospitalizations, chronic disability, and sometimes death. Children appear to tolerate obesity less well than adolescents and adults; persons of all ages, including young children with Prader-Willi syndrome, can die of obesity-related complications.

Food-Related Behavioral Crises

Food-seeking behavior may include foraging for food in and out of the home and consuming spoiled, raw, frozen, or otherwise inedible foods. Older children and adults may steal food or money or use their own money to buy additional food. They may "phone out" for food, elope in search of a restaurant, or enter a stranger's home to seek food. Persons with PWS may pick and break locks or steal keys to enter a locked kitchen. They may display violent outbursts or aggression related to food acquisition or to attempts to set appropriate limits. The more success/notoriety they experience, the more persistent they are in attempting to obtain additional food. In some cases, once this process has begun, the more families attempt to intervene and the more persons with PWS may exhibit behaviors that become intolerable. Families often find themselves in a situation that is spiraling out of control. Law enforcement may become involved, further complicating management.

Behavioral and Psychiatric Crises

Individuals with PWS appear to be more vulnerable to stress. Cognitive deficits (specific learning disabilities) often diminish their ability to adapt to change. Persons with PWS are exceptionally dependent on others to provide structure and control of their environment. Transitional periods such as family moves, changing schools, or moving from school to work environments can all result in a loss of structure and consistency. Limited coping mechanisms may result in extreme maladaptive behaviors that are uncommon in persons without PWS. Dangerous, aggressive, destructive, disruptive, and otherwise intolerable behaviors are sometimes symptoms of psychiatric illness. Self-endangerment, self-mutilation, and rectal-picking can all result in additional medical complications. Stress appears to be a contributing factor in the onset of major psychiatric disorders such as adjustment disorders, anxiety disorders, mood disorders, psychoses, and impulse-control disorders such as intermittent explosive disorder. Accurate psychiatric assessment and diagnosis is greatly aided by a familiarity with the common personality features of the syndrome, including the propensity of persons with PWS to engage in manipulation and falsification of facts.

Medical Problems Requiring Intervention

Editor's note: Selected medical terms are defined in the Glossary at the end of this chapter.

Morbid Obesity

Morbid obesity is defined as a degree of excess body fat that is associated with a high risk for obesity-related complications. In the general adult population, a BMI of 40, or 200% of ideal body weight is often used as a boundary for defining morbid obesity, although obesity-associated morbidities certainly occur at lower levels of BMI. In PWS, the additional factors of substantially reduced lean body mass at any given level of BMI, inactivity, and behavioral characteristics may further exacerbate or interfere with the management of some types of obesity-related morbidities. Such morbidities include the following:

- Sleep-disordered breathing, hypoventilation, pulmonary hypertension and cor pulmonale
- Leg edema, skin breakdown, cellulitis, and venous stasis disease with risk for thrombotic events
- Type 2 diabetes and associated complications
- Hypertension
- Intertrigo (skin breakdown, yeast and bacterial infections in deep fat folds)

Obesity Hypoventilation and Cor Pulmonale

Disordered breathing and respiratory compromise are the most commonly encountered PWS-associated medical crisis requiring inpatient management. Virtually every kind of sleep-disordered breathing has been described in persons with PWS,[13,19,20,32,33,36,37] including individuals who do not meet BMI criteria for obesity. In our experience, morbidly obese persons with PWS at any age may be particularly susceptible to nocturnal hypoxia and edema. The onset may be rapid or slow. In some patients, especially the young, respiratory failure can develop quickly in a situation of steadily worsening obesity. More typically, gradually increasing weight is accompanied by decreasing stamina, reduced activity, and the insidious onset of cardiopulmonary abnormalities. In older persons, long-standing mild obesity can result in chronic dependent edema and venous and lymphatic damage. Recognition is aided by a high index of suspicion for sleep-disordered breathing in obese patients.

Koenig[25] describes the clinical picture of hypoventilation in obese persons as oxyhemoglobin desaturation in the absence of abnormalities in the pattern of breathing. Hypoventilation is characterized by "constant or slowly diminishing oxyhemoglobin desaturation without the cyclic, episodic or repetitive changes in oxygen saturation associated with apneas and hypopneas or the arousal that terminates these abnormal breathing events." This sustained hypoxia can be seen on sleep pulse oximetry and is the typical pattern that we have observed in obese patients with PWS, but is relatively uncommon in non-PWS obese persons.[25,38] Hypoxia has long been known to cause increased pulmonary vascular resistance which, over time, leads to right heart overload.[10]

In PWS, cardiomegaly on chest radiograph (Figure 17.1) is a late finding in the course of obesity hypoventilation and may indicate right heart failure. In these patients, the right ventricular failure (cor pulmonale) of obesity hypoventilation usually occurs in the presence of healthy, *asymptomatic* lungs. Even when daytime and nighttime hypoxemia is profound, there may be wheezing in some, but frank pulmonary edema is usually absent and the left ventricle is generally healthy. Diagnostic modalities that are clinically useful in identifying left heart failure are less helpful in this condition. Right ventriculomegaly or evidence of increased pulmonary artery pressures on echocardiography depend on a good view of the right ventricle, which is often difficult to obtain in a very obese person.

The clinical picture of cor pulmonale is of shortness of breath, worsening daytime sleepiness, leg swelling and cardiomegaly. Unrecognized right heart failure undoubtedly contributes to some cases of sudden death, pneumonia, and reactive airway disease in obese patients with PWS. Healthy persons with PWS sometimes have a low normal hematocrit of 33% and a hemoglobin of 11g per dl. For those with obesity hypoventilation and whose hematocrit and hemoglobin usually fall in the low normal range, levels of 36% and 12g per dl or greater, respectively, may represent an elevation in response to hypoxia. The usual carbon dioxide (CO_2) combining power is 25mmol/L or less, but in those with early CO_2 retention it is 29mmol/L or higher. These subtle changes may help identify individuals at significant risk for CO_2 narcosis if given amounts of supplemental oxygen (O_2) sufficient to normalize oxygen saturations. Overuse of oxygen causing iatrogenic CO_2 narcosis has led to intubation and admission to critical care with consequent deconditioning, markedly worsening and prolonging this crisis. Recovery and reconditioning become especially difficult if a tracheotomy is performed in a PWS-affected person.

The physiology of obesity hypoventilation syndrome (OHS) has not been fully illuminated.[4,25] Many variables have been studied in compar-

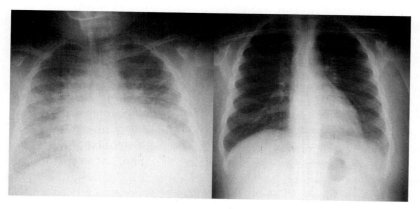

Figure 17.1. A young patient (shown in Figure 17.4) had cardiomegaly on radiograph (left), which resolved to normal heart size (right) after 5 months of rehabilitation. Normalization of pulmonary hypertension was demonstrated by echocardiogram shortly afterwards.

ing persons with OHS with obese persons who do not develop OHS. One common finding appears to be that persons with OHS ultimately develop reduced lung volume.[3,22] It is not clear why some obese persons (non-PWS) develop the disorder and others do not,[31,34] implying multiple mechanisms.[4] It is likely due to a combination of factors causing hypoventilation including congenitally decreased musculature,[24,35] increased work of breathing due to decreased chest wall compliance and, in some cases, partially obstructed airway anatomy.[32] There is some evidence for decreased central or peripheral ventilatory response to hypoxia and hypercapnia,[2,28] an abnormality also reported in non-PWS persons with obesity-hypoventilation.[27,31,39] Other factors may exist. Obstructive hypopneas or apnea during sleep may be present but are clearly not necessary.[4,24]

The late stages of obesity-hypoventilation have been termed "Pickwickian syndrome"[8] after the boy "Joe" who appears in Chapter 4 of Charles Dickens' *The Pickwick Papers*. Numerous authors have suggested that Joe was modeled on a child with Prader-Willi syndrome. Despite its historical and literary interest, the term is best avoided as it has confused parents and other caretakers by suggesting that the patient has acquired yet another "PW" syndrome.

Clinical Presentation

The clinical presentation of OHS in PWS has been delineated from a large number of patients with PWS cared for by the authors with various degrees of hypoventilation and right heart failure. The sequence of events leading to morbidity, disability, and critical illness from obesity hypoventilation is fairly typical and can be observed in reverse during rehabilitation. The sequence can develop over a period of months in the face of rapid weight gain and severe SDB or slowly in patients whose weight has been more stable but in the obese range for many years:

Stage 1—asymptomatic nocturnal hypoxia (detectable only by pulse oximetry study during sleep)

Stage 2—fluid retention (clinical edema, nonpitting increase in tissue turgor); decreased endurance

Stage 3—daytime hypoventilation and hypoxia; edema may be massive

Stage 4—respiratory failure, which may be subtle or brought on suddenly by illness or overuse of oxygen therapy

Stage 1. Nocturnal Hypoventilation and Hypoxia: A maximum "safe" weight for obese persons with PWS has not been defined but, based on our experience, appears to be something less than 200% of IBW (ideal body weight) based on the 50th percentile weight for height. Hypoxia first appears during REM phases of sleep.[19] This finding is subclinical and not evident unless detected by specific testing with pulse oximetry. Full sleep studies are needed to recognize obstructive and apneic events, however. Not every sleep lab is aware that significant hypoxia may occur in the absence of apneas and arousals.

Stage 2. Fluid Retention: In the PWS population, edema often is the earliest clinical sign of obesity-hypoventilation in this population; *it is frequently missed.* The reason for this appears to be the visual subtlety of edema in the obese child or adult. One useful way to describe this type of edema is that "the fat gets hard" as the turgor (firmness) of dependent tissues increases. In our experience, *pitting is usually absent.* Manual comparison (not compression) of tissue in the lower part of the body to the upper extremities will demonstrate an increased density of the tissue in the lower part of the body to the level of the knees, thighs, hips, waist, or higher. This finding is not always appreciable in children. In the absence of diuretic use, the level of edema correlates fairly well with the severity of hypoxia. Therefore detection of a lesser degree of edema to the knees or thighs is especially valuable as an early sign of increased pulmonary artery pressures. These patients typically have normal resting oxygen saturations during the day but pulse oximetry testing during exercise will sometimes demonstrate desaturation. In the presence of *any* recognizable edema, nocturnal oxygen desaturations are usually quite extensive, especially in children and adolescents, and may be present throughout the night without arousals (Figure 17.2).

Decreased exercise tolerance can also be a sign of obesity hypoventilation. However, decreased tolerance is difficult to differentiate from the noncompliance with exercise often displayed by persons with the syndrome. Families do not always perceive the symptom because young children are adept at appearing to carry out their usual activities while conserving their energy. Similarly, orthopnea (sleeping with extra pillows or sitting up) and symptoms of OSA (obstructive sleep apnea) are only sometimes present. Rapid weight gain in an individual with PWS that is not explained by increased access to food may also be a sign of fluid retention.

Stage 3. Daytime Hypoxemia, Clinical Cardiopulmonary Compromise: Even in this late stage, obese patients with PWS sometimes come to medical attention with only complaints of reduced exercise tolerance. Ambulatory patients with daytime oxygen desaturations often have edema (nonpitting) to or above the level of the thighs and hips. Extensive nonpitting edema to the level of the chest can still be subtle enough to be missed (Figure 17.3), but other patients visibly display massive edema, especially in the lower extremities, causing secondary morbidity: weeping sores, cellulitis, and most ominously, impaired ambulation. Some patients are quite sedentary and increased daytime sleeping may be a prominent symptom. Inactivity further impairs the quality of ventilation both during the day and at night. Oxygen saturations when the patient is awake and sitting quietly may be well below 85%, dropping still lower with activity. Cardiomegaly on chest X-ray sometimes still appears "mild." As persons reach the stage of daytime hypoxemia, they will typically increase their resting respiratory rate, but this tachypnea >25/min is not readily appreciated since it is not accompanied by a visible increase in respiratory effort. At rest the tidal volume is small.[31] Resting breath sounds are often barely audible with the stethoscope.

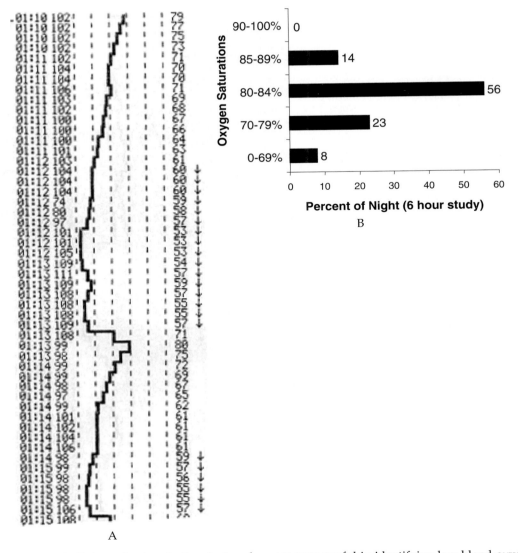

A

Figure 17.2. A. Pulse oximetry studies during sleep are very useful in identifying low blood oxygen in obese patients with PWS. These studies are relatively inexpensive and may be arranged at home while waiting for a formal sleep study. This 5-minute sample from a 6-hour study shows the severe hypoxia with oxygen saturation dipping below 60% (normal 93%–96%), which took place in a severely obese teenager who was only slightly symptomatic. She had mild shortness of breath with exertion and a subtle increase in tissue turgor of her lower body (fluid retention from early right heart failure). **B.** The bar graph depicts the data from a 6-hour night time study of the same teenager. The graph shows that once asleep she spent the entire night with oxygen saturations below 90%, most of the night (56% of the 6-hour study) with oxygen saturations in the 80%–84% range, and 8% of the time under 70% (0%–69% range). These abnormalities in oxygenation may take place without obstructive or non-obstructive sleep apnea and are primarily due to poor ventilation. Hypoxia is sometimes overlooked on sleep studies because there are no "events" in the form of apnea or arousals.

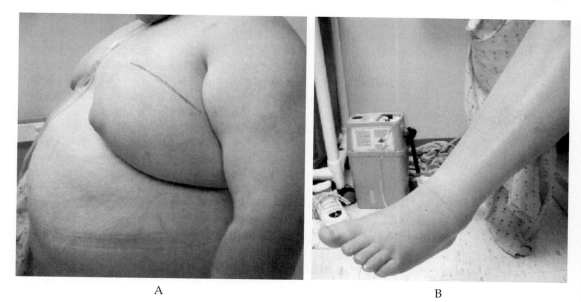

A   B

Figure 17.3. A. This young man with PWS had severe nocturnal and daytime hypoxia and palpable edema (increased tissue density) to the upper chest (marker line). **B.** Lower extremity edema may not be visually impressive (same patient) and is easily overlooked.

With activity, however, increased respiratory effort is more evident and often reported by family members as shortness of breath.

Stage 4. Respiratory failure: Respiratory failure with CO_2 retention is a life-threatening condition that may be acute or chronic. Obese persons with PWS may continue to survive in a compensated state in Stage 3 for years without evidence of respiratory failure *if* their obesity is stable and *if* they remain active. However, they will deteriorate eventually or they may suddenly become critically ill when decompensation is precipitated by an intercurrent respiratory illness or an injury resulting in decreased ambulation. Our clinical experience indicates that both inactivity and overuse of oxygen therapy may cause worsening daytime and nighttime hypoventilation with worsening CO_2 retention. (See Management section, below.)

We have rehabilitated a number of patients from very late and chronic obesity hypoventilation and cor pulmonale and believe that as long as a patient can be made to be ambulatory and be calorically restricted so that weight loss occurs, the condition is usually reversible. Clearly the younger the patient and the earlier the intervention, the better the prognosis for full recovery from critical illness.

Management of Obesity Hypoventilation and Cor Pulmonale

Activity and Diet

Two primary modalities are effective in reversing the cardiopulmonary deterioration of obesity hypoventilation. These are calorie restriction and ambulation. Rehabilitation to a higher level of physical activity is essential for recovery. Even the most seriously ill patients, if conscious, will benefit from this process immediately upon hospitalization. Patients who are critically ill, edematous, and short of breath are understandably reluctant to move. The typical Prader-Willi behavioral traits of stubbornness and manipulation may become immediately life-threatening when patients refuse to cooperate. Therefore skilled therapists working in teams of two or three may be needed to initiate activity in a nonambulatory patient (Figure 17.4).

Every effort should be made to encourage ambulation and physical activity, and patients with PWS should not be fed in bed unless absolutely required by their medical condition. Nurses and therapists will need to work together; consistent behavioral rewards and consequences should be used by the nursing and other hospital staff. Tactics may include delaying meals until modest therapy goals are achieved. Communication with the patient should take into account the personality traits typical of the syndrome to avoid nonproductive efforts. Rehabilitation consists of gradually increasing demands for physical activity beginning, if necessary, with walking a few steps to a chair for meals. Physical and occupational therapists should be given adequate support from nursing staff and adequate time to wait out the inevitable PWS

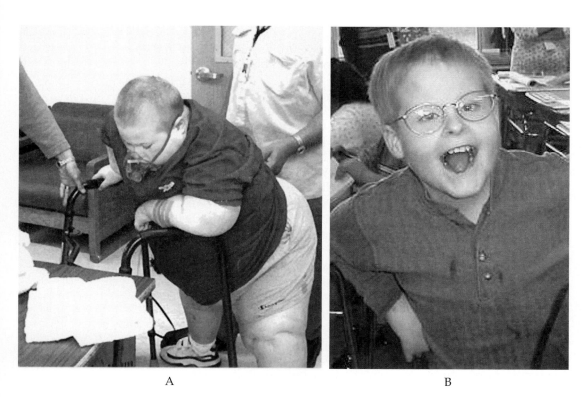

A   B

Figure 17.4. A. Nine-year-old boy with massive edema, profound hypoxia, and dangerous levels of CO_2 retention from obesity hypoventilation and cor pulmonale. Physical activity is an essential component of management and recovery, even in the critically ill patient. **B.** Six months after presenting in critical condition this same 9-year-old boy was active and healthy in school.

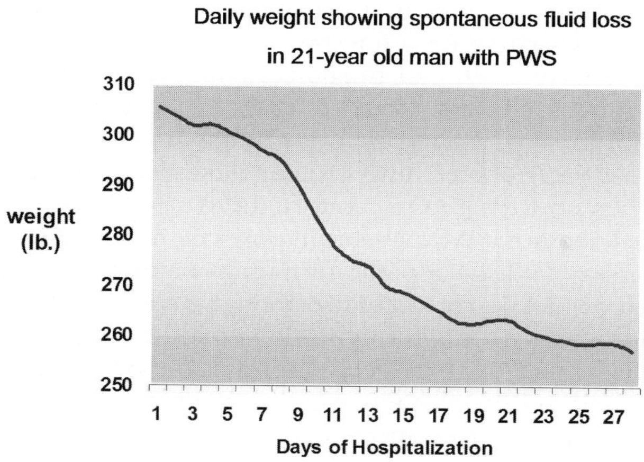

Daily weight showing spontaneous fluid loss

in 21-year old man with PWS

Figure 17.5. This graph shows the daily weight measurements of a patient (pictured in Figure 17.3), beginning at the time of his admission to a rehabilitation unit. He lost 48 pounds in his first 28 days of rehabilitation, with rapid improvement in his endurance and hypoxia. This massive fluid loss is achieved without the use of diuretic medications and appears to be a response to calorie restriction and increased physical activity, which improves ventilation.

behaviors such as whining, crying, delaying, manipulation, and frank refusals. Additional behavioral incentives also may be required. A crisis intervention team approach can help to facilitate this process.

Our experience is that a daily intake of 600–800 kcal provides adequate early nutritional maintenance in the obese, ill patient with PWS. Higher caloric intake is not needed, even for healing of self-injury wounds or decubiti (bedsores). The reader is referred to Chapter 6 for additional discussion of diet and nutrition in PWS.

Spontaneous Diuresis

Spontaneous diuresis of edema fluid and improving oxygen saturation are hallmarks of recovery. The known natriuretic effects of a low-calorie diet[23,26] combined with increased activity[29] result in a diuresis that is at times dramatic (Figure 17.5). In our experience, patients have eliminated as much as 2 liters of fluid per day without diuretics during the initial stages of rehabilitation. Half a kilogram weight loss per day is typical. In the absence of drugs that alter renal function, such as diuretics or topirimate, this rapid diuresis has not been associated with electrolyte abnormalities. The time frame for this diuresis is variable. While it is usually seen within days, in several severely affected individuals, diuresis was delayed many weeks despite early ambulation and strict diet. Further, we have observed that in this patient population, use of diuretics appears to delay this diuresis rather than assist it, even if azotemia is avoided.

Use and Misuse of Oxygen Therapy

The fundamental pathophysiology of OHS should be kept in mind in order to avoid mismanagement. Patients are hypoventilating, day and night. Underlying this condition is congenitally weak respiratory musculature. Adiposity thickens the chest wall thereby further restricting excursion (Figure 17.6), and in some cases there is additional tissue stiffness with loss of compliance brought on by edema fluid, as seen in Figure 17.3. In sum, the respiratory drive is inadequate to overcome the increased work of breathing. Oxygen therapy and inactivity worsen the condition of decreased respiratory drive.

In the edematous PWS patient, hypoxemia may be assumed to be chronic (present at least at night for months or years) and need not be corrected too quickly. It is sometimes thought that oxygen therapy used to treat a young person with hypoxemia and no lung disease can do no harm. However, our experience indicates that hypoventilation may gradually and subtly worsen over hours or days in patients with hypoxemia who are given more than 1 liter/minute of oxygen (24%). Hypoventilating patients who are deteriorating do not appear distressed; the only indications of excessive use of oxygen are dropping O_2 saturations and lethargy with seemingly increased O_2 "requirements" to maintain "normal" saturation. Although it may appear clinically reasonable to increase the rate of oxygen flow such action can lead to worsening CO_2 retention, with worsening lethargy and potential respiratory arrest. An alternate response to this situation is to decrease the oxygen flow and get the patient up and moving. If needed, assisted external positive airway ventilation (BiPAP and CPAP) is preferable to intubation because it is more compatible with keeping the patient awake and mobile.

In our experience, patients in Stages 3 and 4 appear to benefit from 1 liter/minute of oxygen without worsening their hypoventilation, provided increased activity is also demanded of them. One liter of O_2 per minute by nasal cannula may raise oxygen saturations to the high

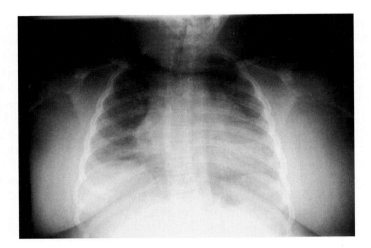

Figure 17.6. This chest radiograph illustrates the thickening of the chest wall in a severely obese patient with PWS. Excessive subcutaneous fat changes the energy requirements for breathing, causing hypoventilation in obese persons with the disorder.

70s or low 80s and this appears to be adequate while rehabilitation takes place. Oxygen saturations will be lower during activity, and by our observation, this is neither harmful nor does it hamper recovery. Patients should have their electrolytes checked especially if they are on oxygen therapy. After 2 to 3 days on oxygen therapy, bicarbonate levels may rise as the kidneys produce a compensatory metabolic alkalosis. This finding confirms a respiratory acidosis from CO_2 retention; furosemide and other diuretics may mask this effect. Unfortunately, prior to referral, some patients who refuse the BiPAP are then treated with nocturnal oxygen alone without being checked to see if their bicarbonate is rising. Prior to hospitalization, external positive airway ventilation definitely improves nocturnal ventilation in some patients, but this benefit should not be allowed to delay definitive therapy of rehabilitation and weight loss.

Many patients who initially refuse these modalities (CPAP and BiPAP) have successfully accepted these treatments with behavioral training. Sleep studies show improved ventilation and some patients report more comfortable sleep. Other patients who have completely refused therapy have also been fully rehabilitated despite ongoing profound nocturnal hypoxia. Nocturnal ventilation improves (sometimes rapidly) once rehabilitation with increased activity is underway. We do not recommend tracheostomy; the majority of persons (6 of 7 in our experience) with PWS given tracheostomies have pulled out their own tubes or otherwise endangered themselves by injuring the stoma and airway.

Use and Misuse of Diuretics

If used at all, diuretics must be administered with a clear understanding of their benefits and risks. Excessive use of diuretics decreases intravascular volume with little impact on the interstitial edema of the lower body. Further, diuretic use risks the development of renal and hepatic hypoperfusion. Pulmonary edema is not characteristic of right heart failure and should not be cited as a reason to give diuretics unless left heart dysfunction has been established.

The doses of diuretics that produce azotemia also produce hepatic ischemia with rising levels of transaminase. Diuretics should be tapered and ACE inhibitors and non-steroidal anti-inflammatory drugs (NSAIDs) should be discontinued immediately. In our experience, the typical teenage and older person with PWS who has not been on growth hormone (GH) or testosterone has a serum creatinine of 0.5–0.7 mg/dl; if the creatinine is greater than 0.9 mg/dl then renal hypoperfusion/injury should be considered.

Positioning

Ventilation is well known to deteriorate in the reclining position.[25] Very obese, edematous individuals may further compromise their own ventilation if the abdominal mass is resting on their thighs in a hospital bed. Patients benefit from sleeping in a recliner rather than a hospital bed so that their legs can be supported from below on a footstool and the abdomen can remain pendulant so as not to impinge on lung volume. These measures generally apply only to persons in Stages 3 or

4 of obesity hypoventilation. Obese persons with PWS who prefer sleeping at night in an upright position (orthopnea) can be assumed to be in danger from their obesity.

Obesity-Associated Conditions

Although the following medical conditions are not inpatient crises per se, they may be commonly observed in patients with PWS who are hospitalized for medical or psychiatric problems.

Leg Edema and Cellulitis

Leg edema in a person with PWS may accompany sleep-disordered breathing. Longstanding edema results in chronic tissue changes of the lower body including legs and lower abdomen (Figure 17.7). The result-

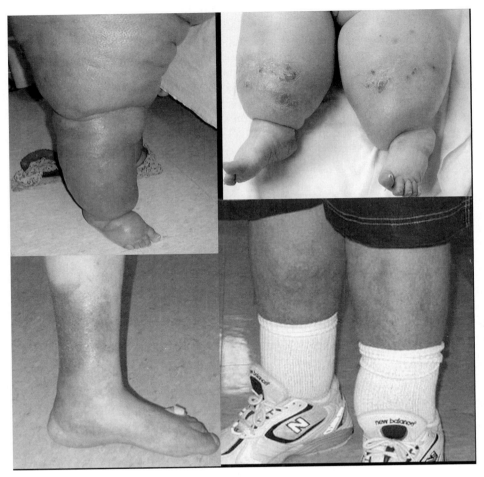

Figure 17.7. While some edematous patients retain fluid throughout their subcutaneous tissue as in Figure 17.3, others demonstrate severe leg swelling. In either case, longstanding (probably 10 or more years) and often unrecognized obesity hypoventilation results in changes in the lower extremities that are irreversible. These include dilated veins, damaged lymph vessels, and chronic stasis changes of the skin.

ing venous stasis and lymphatic damage predispose tissue to ulcers, thrombosis, and cellulitis. Intervention to prevent the prolonged condition of obesity hypoventilation is essential to avoid irreversible damage to the lymphatic and venous systems of the lower legs. There is no question that skin-picking behavior in those with PWS, while not usually resulting in infection in other parts of the body, is a major contributor to some episodes of leg cellulitis.

Direct pressure techniques to the legs or use of support hose appear to be of limited use unless they are part of a medical protocol for treatment of lymphedema. In some cases, support stockings may be counterproductive, causing tissue breakdown from pressure or constriction of fluid outflow.

Signs of cellulitis can be difficult to ascertain in the very obese individual since the legs are often already chronically swollen, indurated, and discolored. A high index of suspicion and close daily examination of the legs by caretakers seeking changes in feel or appearance is essential. Patients do not always exhibit fever or pain. Limited cases of cellulitis, diagnosed early, can be managed with oral antibiotics sometimes in combination with an antifungal agent (such as fluconazole). Preventative use of antibiotics is discouraged to prevent development of resistant strains of bacteria. However, intravenous antibiotic may be necessary in severe cases, especially if there is evidence of systemic infection. In all cases, an attempt should be made to identify the causative organism.

Maintaining and increasing physical activity and leg elevation when the patient is sitting have proven useful adjuncts in the management of these difficult conditions. Cellulitis and superficial venous thrombosis are not reasons to limit activity; rather, the reverse is true. Patients who have ceased to walk for any reason are at high risk for thromboembolic events and prophylactic anticoagulation should be considered. Rehabilitation to some level of ambulation is the highest priority.

Diabetes

Elevated blood glucose or Type 2 diabetes, seen in 25% of adults with PWS,[9] is frequently indicative of excessive calorie intake. In our experience, the vast majority of patients achieve normal glycemic control without medication when exercise and diet are implemented. In patients requiring diabetes medications, increasing dose requirements may be a sign of a crisis with the patient's intake and weight given the rapid rate of weight gain possible in persons with this syndrome. Patients who are on insulin are at risk for hypoglycemia if their access to excess calories is suddenly interrupted by hospitalization or other intervention.

Hypertension

Most hypertension in PWS is directly related to obesity and requires definitive intervention with diet and exercise. As with diabetes, the need for medication usually indicates inadequate intervention for the patient's deteriorating clinical condition. In our experience, the hypertension is usually labile and resolves with adequate weight loss. The vasodilating effects of most antihypertensives only add to the amount of accumulating edema, and in some individuals their use delays the onset of the diuresis that occurs with increasing activity and caloric restriction.

Intertrigo

In the very obese individual, the deep fat folds are prone to monilial and bacterial infections with occasional severe ulceration. The physical effort required to cleanse and dry deep fat folds may be beyond the capability of a single caretaker; sometimes two persons are needed to support the adipose tissue while another person performs hygiene. Obviously this is beyond the capabilities of most families, especially if the patient is resisting care.

Management of severe intertrigo includes daily or twice-daily cleansing of nonulcerated skin using a dilute (1:3 ratio) vinegar and water spray and air-drying with a heat lamp or hair dryer. Clotrimazole or powdered nystatin applied two or more times per day is also useful, especially if there is evidence of fungal infection. On rare occasions, oral antifungal or antibiotic agents may be needed if the area of involved skin surface area is great or there is ulceration. Severe ulceration may require aggressive inpatient nursing care.

PWS Medical Issues Not Related to Obesity

Renal Dysfunction

There is, at present, little evidence of any kidney dysfunction directly related to PWS. Renal problems seen in this population have included obstructive uropathy of congenital origin, glomerulonephritis post-infection, "diabetic nephropathy" (i.e., persons with longstanding hyperglycemia and elevated glycohemoglobin levels but without the retinopathy), renal tubular acidosis from psychotropic medication (topiramate) and Syndrome of Inappropriate Antidiuretic Hormone (SIADH) with hyponatremia from psychotropic medications, especially oxcarbazepine and carbamazepine (Trileptal and Tegretol), but also from selective serotonin reuptake inhibitors (SSRIs). Patients have developed azotemia from the volume depletion of aggressive diuretic use that at times may be greater than 70mg/dl without postural symptoms. Renal biopsy has been rarely attempted, possibly due to concerns about both patient cooperation and massive overlying fat. Persons with PWS who have not had GH treatment have a low serum creatinine due to low muscle mass. For these patients creatinine of 0.9 to 1.0 should be considered abnormal in this syndrome where typical values are 0.5–0.7.

Skin Picking

The "skin picking" behavior of PWS has a wide range of severity from patient to patient and may vary in the same patient over time.[17] Some patients have occasional minor skin picking while others maintain large open wounds.

Skin picking is an activity that goes on continuously, intermittently, or clandestinely when the patient is calm; typical skin picking in an individual with PWS does not appear to be an expression of emotional distress. It has been related to boredom and anxiety but objective evidence for this is difficult to establish. No specific intervention has been uniformly effective. In some cases extremely frequent picking behavior itself interferes with the patient's other activities. Severe disfigurement, recurrent infection, and anemia are reasons to consider more intense efforts to modify the behavior with hospitalization or a medication trial.

The behavior often is extinguished, at least temporarily, if healing of the wound(s) is achieved. There has been limited success using protective dressings and an intense program of alternative activity until wound healing occurs. Some wounds have healed with frequent application of an antibacterial ointment, which functions as a lubricant and interferes with picking.

Behavioral interventions targeted at the activity itself are difficult to implement. Some parents have reported success by attaching a major reward to the healing of a lesion. This approach is compatible with the basic principle that no attention, positive or negative, should be paid to the behavior itself other than to require the patient to observe social conventions and good hygiene. Spontaneously or in search of a reward, patients may cease the behavior or substitute another area of skin to pick. In rare cases, primary reinforcers have been used effectively.

Self-Mutilation

Sudden self-injury or self-mutilation, with or without an emotional outburst, is a different phenomenon representing more serious pathology. Self-mutilation does not differ from skin picking in severity of the wound but in the circumstances under which it is inflicted. The actual injury may be mild (hitting self in the head) or severe (gouging a deep wound with a pen, knife, or fingers). The patient may verbalize distress, anger, self-hatred ("I'm so stupid!") or appear to be in a dissociative state. Psychiatric assessment and treatment of underlying conditions is indicated. Some patients have reported auditory hallucinations as triggers for these behaviors.

Rectal Self-Injury

Rectal self-injury is a problematic behavior occurring in some persons with Prader-Willi syndrome. The behavior is not well understood. Fortunately it becomes serious in only a small number of persons with PWS. This includes patients with frequent rectal digging/injury that results in medical problems such as bleeding, infection, or fecal incontinence.[5]

We have made the following observations with respect to this behavior in individuals with PWS:

1. Indirect evidence of the behavior can aid in the diagnosis of unexplained medical symptoms, even if the behavior has never been observed.
   - Excessive time spent in the bathroom (a PWS trait without rectal picking)

- Feces or bloody smears on hands, toilet, fixtures, shower, bath tub, or bed linens
- Bowel incontinence, urgency, or diarrhea (due to reflex bowel emptying following rectal stimulation)
- Rectal bleeding

2. Anemia from chronic blood loss, while rare, can be severe. Acute hemorrhage or perforation has not been reported. Chronic ulceration is well documented by endoscopy and has been misdiagnosed and treated as inflammatory bowel disease (IBD). Clinical experience in the referral population has been that all cases of rectal bleeding or ulceration have proved to be from self-trauma and not from IBD.
3. The behavior appears to be obsessional and compulsive; however, medications targeting OCD have not been helpful.
4. Rectal picking/digging does not appear to be a sexual behavior, nor is it a behavioral sequela of sexual abuse.
5. Rectal picking/digging does appear to be a nonspecific stress symptom. Marked increases in the behavior have been observed in association with ongoing interpersonal conflict and punitive disciplinary approaches, as well as during episodes of psychosis and mood instability. Diminution and/or elimination of the behavior has occurred when the underlying illness or stressor is addressed effectively.

Special Considerations

There are a number of unusual characteristics in patients with PWS. Familiarity with these issues is necessary for the clinician to respond appropriately.

Unreliable Self-Report

Abnormal pain awareness and *unpredictable fever response* can lead to underreporting of pain and missed diagnoses of serious conditions that would normally be expected to produce severe pain or fever. Diagnoses of acute surgical abdominal conditions, fractures, and serious infections have been delayed due to the failure of the patient to report pain or to show a fever. Therefore, any fever or a refusal to walk (after an injury or fall) should be taken seriously, and a good clinical examination and close follow-up are indicated until serious illness or injury has been ruled out. Patients who show decreased interest in food should be considered potentially seriously ill, medically or psychiatrically. Manipulative refusal to eat has also been observed.

Somatic complaints are frequent among individuals with PWS who often report numerous symptoms for which objective evidence is lacking. When there is a marked discrepancy between objective findings and subjective stress, discomfort, or functional impairment, malingering should be considered. Malingering is the intentional production of false or grossly exaggerated physical or psychological symptoms motivated by external incentives.[1] For the individual with PWS, external incentives may include obtaining food or medication, escaping structure in order to procure food, or avoiding demands for exercise or other tasks.

Factitious disorder is differentiated from malingering by the absence of external incentives. Individuals with factitious disorder need to maintain the sick role because of intrapsychic needs.[1] They may engage in pathological lying about their symptoms or their personal history. Sometimes for no apparent reason an individual with PWS will provide an inaccurate medical history, reporting events or procedures that never happened. Therefore clinicians should try to verify presenting symptoms and history independently.

Suggestibility is common among patients with PWS, and they may falsely endorse complaints when asked during a routine interview about symptoms of disease. The symptoms may be variable in their occurrence (present only when asked or examined). This suggestibility in persons with PWS complicates both their medical and psychiatric evaluation and management.

Medication seeking is prominent with some patients (usually higher-functioning individuals) and can result in massive polypharmacy for miscellaneous complaints. These are usually conditions for which objective evidence is difficult to obtain: pain, allergies, reflux disease, constipation, and urinary symptoms. Complaints of pain and requests for analgesia are common. Consequently adult patients in the referral population were frequently on an excessive number of prescription drugs and over-the-counter and topical preparations. Persons with PWS are often very attached to their medication regimens and will sometimes vigorously resist change. They sometimes request liquid medications that are sweet to the taste. Prescriptions for symptomatic relief should therefore always be time limited and dose limited. Patients with a clear pattern of medication seeking should not have unsupervised visits to a physician. Persons with the disorder cannot be allowed to dose themselves with "as needed" medications. Nondrug therapies are preferred: heat, cold, massage, sympathy, or reassurance.

On the other hand, individuals with PWS can be truthful and reliable historians. A careful interviewer takes seriously reports of abuse, serious physical symptoms, delusional thinking, or the experience of hallucinations.

Abnormal Temperature Regulation

As discussed in Chapter 5, temperature dysregulation may occur more frequently in individuals with PWS, perhaps related to a defect in thermogenesis. In two well-described cases,[16] hypothermia began with a change in behavior followed by decreasing activity proceeding to near coma. The patients did not want to eat, did not complain of being cold, felt cool to the touch, and became ashen. Hypothermia (81°F–94°F), decreased blood pressure, bradycardia, and slow respirations were observed. Laboratory studies revealed decreased hemoglobin, low white blood cell count, decreased platelets, hyponatremia without acidosis or hyperkalemia, and elevated renal and liver function tests. All of these changes returned to normal levels over several days as the patients were rewarmed. Sepsis was suspected but blood cultures were negative. Neuroleptics may have played a role.

Hypothermia has usually occurred when the outside temperature was cool. In our experience, some patients have repeated episodes during the winter and experience relief during the summer months, but have recurrence during the following late fall. Bray et al.[7] have made similar observations.

Hypersomnia, Daytime Sleepiness

Abnormalities in the sleep of persons with PWS have been documented at all ages with and without obesity.[11] Excessive daytime sleepiness appears to be a characteristic of PWS, but excessive medication and obesity hypoventilation should be ruled out as contributing to the symptom. Some patients meet criteria for narcolepsy and benefit from appropriate medications.[18] In our experience, stimulant medications and modafinil have been useful in selected cases, keeping in mind the risk of mood activation in susceptible patients. It is certain that some patients use their capacity for short sleep latency as an escape-avoidance mechanism at will and can respond to behavioral incentives to stay awake in the classroom or workshop.

Behavioral Problems and Psychiatric Crises in PWS Requiring Intervention

Stress Sensitivity

Individuals with PWS are *stress sensitive;* they rely on the predictability, consistency, and stability of their environment. A stress response will occur if the integrity of the environment deteriorates for any reason. The stress response can be nonspecific (increase in typical PWS behaviors), specific (emergence of atypical behaviors), or outrageous (individuals with PWS who are stressed have the potential for outrageous behavior, which should not be attributed automatically to severe psychopathology). Because the individual with PWS is dependent on environmental structure, and because their problem-solving ability is limited by impaired judgment that exceeds the deficit predicted by their intellectual deficiency and/or learning disability, individuals with PWS are usually not candidates for complete independence. Their life goal is *maximal function with support.* Among the referral population, the most common cause of functional deterioration in adults with PWS has been a misunderstanding of this principle.

Caring for an individual with Prader-Willi syndrome exhausts the resources of many families. Controlling food access and dealing with other difficult behaviors is an ongoing challenge. The price tag of "keeping the peace" or "giving in" is the inadvertent reinforcement of disruptive behaviors and the spiraling pattern of obesity and its morbid complications. The mainstay of behavior management for this syndrome is an environmental buffer in the home, school, workshop, and community that prevents rather than reacts to behavioral problems. On an inpatient unit, this environmental buffer is called the *therapeutic milieu*. Crisis intervention provides an opportunity to have an impact on both the individual and the home environment by reestablishing the structure and consistency of rules and mutual expectations needed to effectively manage the most common behavioral issues.

The basic premise of intervention acknowledges that *at no time can a person with Prader-Willi syndrome be expected to voluntarily or independently control his/her own food consumption.* Extensive experience with a referral population has demonstrated that once food access is controlled and the principles of food security are implemented, acceptance of restrictive programming is usually excellent. Appropriate social skills and compliance with an exercise program, activities of daily living, chores, and personal hygiene are all supported by the therapeutic milieu. The experience of success is built into the program; patients receive rewards for their progress, and families are rewarded by a decrease in their level of stress.

Psychiatric Crises

Among persons with Prader-Willi syndrome the proportion who develop serious psychiatric illness requiring hospitalization is unknown. The literature is replete with case reports of serious psychiatric illness concurrent with PWS. Among the referral population, psychiatric crises are not limited to those who have severe psychiatric illness. Persons with PWS have limited coping mechanisms due to cognitive rigidity, variability in adaptive functioning and sensitivity to stress that can result in a precipitous deterioration in mental status. Sometimes precipitating events can be identified, such as life stressors associated with leaving school, failure in a workshop, change in educational or caretaking staff, or death or illness in the family. Some presenting symptoms leading to hospitalization include depression, suicidality, aggression, property destruction, elopement, increased irritability and explosiveness, hypomania, delusional thinking and hallucinations, and self-mutilation or self-endangerment. Persons with PWS are prone to a variety of mood disorders often presenting with psychotic features.[6,12] In addition, psychiatric illness in this population is often accompanied by a range of behaviors usually restricted to severe and chronically mentally ill persons: attempts to swallow inedible objects, severe self-mutilation, running into traffic, fecal smearing and throwing, coprophagia, rectal self-injury, psychogenic water intoxication, deliberate property destruction, hunger strikes, refusal to move, etc. Despite the seriousness of the presenting symptoms, the prognosis is quite variable: many patients stabilize in a structured, low-stress environment either with or without psychotropic medication. At the other end of the spectrum, a small number of patients have refractory symptoms despite appropriate environmental management and multiple trials of medications.

Individuals with PWS who develop serious psychiatric crises may display behaviors that are difficult to manage even in psychiatric facilities. The therapeutic milieu of most general psychiatric units is not designed for and often does not meet the needs of persons with PWS. Subtle but significant cognitive deficits may not be recognized, and individuals with PWS may not benefit from traditional group and individual psychotherapeutic interventions. Some patients undergo weight gain because they have access to too much food. They may forage freely from other patients. Access to recreational and exercise facilities is often limited. Among the referral population, reports of weight gains of as much as 20 pounds have occurred in a matter of days during the course of a psychiatric hospitalization. Syndromal behaviors are often misunderstood as manifestations of more severe psychopathology, or they are not adequately addressed with potentially beneficial behavioral techniques.

Even in the absence of a specialized program, most patients with PWS show an immediate improvement in their behavior upon admission to a hospital. They respond to the predictable regimen. They often have no demands made on them, and they receive too much food. This response should not be interpreted as an end to the behavioral crisis.

A Specialized Therapeutic Milieu

A specialized therapeutic milieu, an essential element in inpatient crisis management, is dependent upon a trained, experienced staff. *Consistency* is the gold standard; every health care worker, nurse, physician, and therapist should understand the behaviors typical of PWS and respond to their occurrence in a low-keyed manner. For example, behaviors such as screaming and verbal abuse are best ignored. All statements made to patients should be framed in a positive rather than punitive context. It is important to assure that the patient's attempts to manipulate or triangulate personnel are unsuccessful. Even housekeeping and dietary staff members can be trained so that they will not be tricked or manipulated. Important questions about length of stay, family visits, dietary orders, and disposition plans should be answered only by designated staff. Clear lines of communication with the patient, the family, and the staff are critical. Families and other caretakers should be incorporated into the milieu through training that may involve "homework." These assignments may involve observation of direct care in the hospital or the design of a daily schedule, meal plan, or behavioral contract for use in the home. Telephone contact after discharge can assist in the maintenance of structure and the ongoing monitoring of treatment goals and interventions.

The ideal inpatient treatment plan for a patient with PWS has six major components: (1) milieu management, (2) behavioral interventions, (3) psychological therapies, (4) psychotropic medication, (5) family/staff intervention, and (6) disposition/systems intervention.

1. Milieu management

Milieu management is an essential tool for the inpatient treatment of individuals with PWS. The milieu contains several components:

- Rules of conduct
- Daily schedule of activities

- Psychological food security
- Mandatory supervised exercise
- "The Day Stops Here"

Rules of Conduct: The rules of conduct for patients on the unit are explained and posted next to the patient's bill of rights in the individual's room.

Daily Schedule of Activities: This modality consists of a predetermined time line of activities including wake up, grooming, therapies, meal times, exercise, leisure, rest, and bedtime. It defines the *flow* of the day and establishes structure, consistency, and predictability. This timeline is best presented to the patient in a concrete form; a wall chart may be supplemented by a written schedule which the patient carries with him through the day. The planned flow of activities is continuously reinforced by the verbal prompts of staff and therapists to "check the schedule."

It should be assumed that the patient with PWS could have difficulty with the timely completion of ADLs (activities of daily living). Some will need assistance with grooming activities depending upon the degree of obesity as well as the level of dyspraxia.

Because individuals with PWS have hypothalamic hypogonadism, their physical and emotional sexual maturation may not be age appropriate. Many of them appear to be younger than their stated age, and their gender may appear to be ambiguous because of obesity and poorly developed secondary sexual characteristics. Among our referral population, both males and females tend to be immodest, and they may require verbal reminders to maintain privacy for grooming and dressing. Also, this privacy is essential to prevent other patients who do not have PWS from making destructive comments leading to discomfort.

Psychological Food Security: Psychological food security is an essential component of the management of individuals with PWS; the goal is to maintain food security across all settings both in and out of the hospital. Food security is one of the most basic skills taught to the patient's family or other caretakers.

Food security is achieved when food access is controlled to the extent that three criteria are established:

1. There is *no doubt* when, what, and how much the person with PWS will eat;
2. There is *no hope* of receiving any more; and
3. There is *no disappointment* due to false expectations.

   Food security = No doubt + No hope + No disappointment

When food access is restricted, individuals with PWS require *no doubt* about their meals and snacks. Menus are planned ahead and posted; calories are controlled, but the amount of food presented can still be generous. Although the timing of the meals and snacks remains fixed, it is not focused on the clock; it is set by the sequence of activities across the day. This concept is critical to the achievement of *flow* through the day.

The quantity of food presented aids *psychological satisfaction*. The Red, Yellow, Green Diet (see www.amazingkids.org/) is an excellent meal plan that provides large quantities of "green foods" (lettuce, green beans, tomatoes, broccoli, cauliflower, peppers), moderate amounts of "yellow foods" (lean meats and complex carbohydrates such as fruits, grains, fat free dairy, mushrooms), and extremely infrequent or minimal amounts of "red foods" (high-fat, high-calorie foods such as fats, oils, most deli meats, cheeses, desserts). The food is presented with controlled access to low-calorie, fat-free seasonings and dressings as well as salt, pepper, and hot sauce. (Individual packets eliminate discussions about the quantity of these condiments.) Sugarless gum and diet beverages including Crystal Light®, coffee and tea, and diet soda are limited, and measured quantities are used as reinforcement for exercise compliance and effort.

Mandatory Supervised Exercise: Mandatory supervised exercise must be scheduled into the daily plan. Walking is by far the best aerobic exercise for individuals with PWS. Two different periods for exercise are scheduled through the day.

The Day Stops Here! This concept is *the* most effective milieu management program for hospitalized individuals with PWS. It requires 1:1 staff for implementation. If the flow of the daily schedule is interrupted by a behavioral situation (refusal, "shutdown," or outburst), the daily schedule stops at that point until the individual has regained motivation and behavioral control to return to programming. Individuals are encouraged to return to scheduled therapies; if the therapies have ended, prescribed "make-up work" must be accomplished before moving to the next activity. *It is essential to delay meals until make-up work has been completed.* Sometimes individuals with PWS stage behavioral shutdowns that can last for as long as a full day. However, nourishment is never withheld and an alternative calorie source is provided even if the privilege to attend mealtime is lost.

2. Behavioral interventions

Behavioral interventions are an essential tool for a successful treatment plan. Noncontingent reinforcement (NCR) is the delivery of reinforcers independent of an individual's response. It is a powerful tool for establishing rapport. Typical noncontingent reinforcers include talking to the individual (nonspecific topics such as orientation, daily schedule, current events, weather; specific topics such as clothing, grooming, leisure interests) and providing leisure activities (puzzles, magazines, paper, markers).

Contingent reinforcement is the delivery of reinforcers dependent upon the individual's response. It is a powerful tool for shaping appropriate behaviors. Extinction, selective attention, praise, and differential reinforcement of other behaviors (DRO) are examples of contingent reinforcement. For example, the patient is praised for all appropriate behaviors, especially those facilitating daily transitions such as the timely completion of ADLs, grooming, exercise effort, social skills, following unit rules and directives, and working toward psychological treatment goals. A token economy works well as a structured system for delivering contingent response to behavior. For oppositional behavior among individuals with PWS, a response cost intervention is recommended. It will be necessary to post the rules and expectations; delineate rewards (tokens) for achieving expectations, and define the cost (loss of rewards) for not achieving the desired results.

3. Psychological therapies

Throughout inpatient hospitalization, individual and group therapy sessions focus on acceptance of and building expectations for a continuation of the therapeutic milieu following discharge. An appreciation for intellectual level and learning style is essential; adapting the daily plan to meet the individual's unique pattern of strengths and weaknesses can lead to better compliance during the hospital stay. For example, an individual with receptive and expressive language disability will not perform well in a group context. Psychological testing should be available by history to ascertain overall intellectual ability. Neuropsychological testing can be requested to elucidate learning disorders, and strategies for adaptation can be tested during the hospital stay and taught to the family or other caretakers before discharge.

The utility and effectiveness of psychological interventions is based entirely upon an individual's verbal and intellectual abilities. High-functioning individuals with PWS may benefit from all psychotherapeutic and behavioral modalities. Clinical experience suggests that young adults with PWS have the capacity to realize that their syndrome may limit their potential for independence and that many of the life goals that they share with typical peers and siblings may never be actualized. They need support as they grieve the loss of a "normal" life. Social and family situations that are usually joyful, such as college graduations, marriages, and births, may be reminders of unachievable milestones and precipitate dysphoric responses. These situational crises can be addressed through traditional psychotherapy using interpersonal and cognitive strategies with the individual and the family.

Lower-functioning individuals with limited insight may require individual therapy with supportive and psychoeducational goals such as minimizing stress, enhancing coping abilities, and improving participation and compliance with the inpatient program. Strategies and modalities for relaxation (progressive muscle relaxation, deep breathing, and occasionally visual imaging), anger management, social problem solving, and social skills training may be prescribed, taught to the individual with repetition and drill, woven into the fabric of the daily plan, and implemented with prompts, cues, and supervision.

4. Psychotropic medication

The use of psychotropic medication is determined by (1) the patient's response to behavioral and eco-environmental interventions, (2) psychiatric diagnosis, (3) the nature of targeted symptoms, and (4) severity of impairment. There is no syndrome-specific medication. Effort is directed toward eliminating unnecessary medications that may be producing unwanted side effects as well as selecting the best medication for improving adaptive function.

Several observations about the use of psychotropic medication in the referral population of individuals with PWS are offered. First, dose titrations should be made in small increments because individuals with PWS may respond to lower doses of psychotropic medication. Second, clinical experience with the referral population suggests that activation as well as discontinuation syndromes occur with selective serotonin reuptake inhibitors (SSRIs), nonselective serotonin reuptake inhibitors (NSRIs), and even some psychostimulants at a rate that exceeds the neurotypical population. Third, it is assumed that the sedating side effects of some agents could increase respiratory compromise due to hypoventilation in obese individuals. Finally, the risk of hyponatremia associated with the use of certain classes of medication (anticonvulsants, SSRIs, and atypical neuroleptics) should be kept in mind, especially during upward titration of the dose.

Pharmacokinetic variations that should be considered among individuals with PWS are summarized below:

- High fat/lean ratio in body composition (even in normal weight individuals with PWS) may affect metabolism and bioavailability of fat-soluble medications.
- Low muscle mass may alter the presentation of extrapyramidal side effects of typical and atypical neuroleptics.
- Doses of medication and schedule of administration should take into consideration persistent prepubertal status due to hypothalamic hypogonadism.
- Concurrent administration of calcium for osteoporosis may affect absorption and action of the psychotropic medication.

5. Family/group home staff involvement

Preparation for discharge begins before admission. Family education is an intense process that continues throughout hospitalization. Training sessions are attended by parents, step-parents, and siblings, as well as involved extended family members (aunts, uncles, and grandparents). School personnel, in-home support personnel, local mental retardation and developmental disabilities (MRDD) staff, as well as supervisors, school nurses, nutritionists, group home staff, and mental retardation program administrators also are invited.

The goal of this training is to teach the caretakers how to devise, within the limits of their own situation, a therapeutic milieu similar to the one shown to be effective during hospitalization. The plan mimics the hospital program by providing the same special structure, expectations, consequences, and schedule. The hospital exercise program to which the patient has become accustomed is duplicated or modified as needed for the home environment.

Psychoeducational intervention is the keystone of relapse prevention. Eco-environmental interventions (i.e., the treatment plan) can be developed only after a thorough review of predisposing, precipitating, and perpetuating factors. This information provides a basis for understanding the strengths and weaknesses in the system of care. The results of the patient evaluation and the comprehensive formulation are provided to the family/staff to assist their understanding of the acute and chronic aspects of the crisis that precipitated hospitalization. In this way they are prepared to make the changes that will be required to meet the individual's needs. Finally, the family/staff work with the team to develop an action plan for how to change the environment, the daily structure, and the existing behavioral patterns in order to meet the individual's needs. The degree of additional support necessary to implement the treatment plan is assessed and a therapeutic network of professional and extended family supports may be identified for recruitment as needed.

6. Disposition planning

Disposition planning includes an examination of all available short-term and long-term resources and living arrangements. Outpatient treatment, school- and/or workshop-based interventions, and wrap-around or habilitative services help families to provide for the needs of their child in the home. Residential treatment or group home placement is explored when appropriate, especially in the case of patients with an exceptionally high need for structure and supervision. Persons with PWS are rarely well managed in facilities that do not have prior experience with the syndrome or a commitment to develop a program specifically suited for individuals with the syndrome. The Prader-Willi Syndrome Association (USA) is a valuable resource for families, and membership is encouraged.

Psychiatric follow-up for individuals with PWS can be challenging. "Finding Psychiatric Help for Your Child" is a monograph pertaining to this topic available at the PWSA (USA) Web site (www.pwsausa.org).

Conclusion

Prader-Willi syndrome is a genetic disorder with neurologic abnormalities affecting cognition, behavior, and energy balance. The complications of the disorder, when environmental controls are not in place, are chronic disability and cardiopulmonary deterioration from morbid obesity in addition to the complications of uncontrolled diabetes. Inpatient crisis intervention is briefest when done early on during the process of deteriorating food control and weight gain and prevents a pattern of multiple acute-care hospitalizations.

Behavior problems complicate the management of persons with PWS and sometimes necessitate inpatient hospitalization. Psychiatric vulnerability appears to be a direct result of the stress sensitivity associated with this condition. A specialized team with experience working with Prader-Willi syndrome can provide successful intervention for patients in medical or behavioral crises who are not responding to outpatient interventions. The therapeutic milieu is the definitive intervention for both food and behavioral management. Although it is typically implemented during an inpatient hospitalization, it can be translated through training to other settings including the home, school, and community living situations.

The use of psychotropic medication is based on psychiatric diagnosis and the nature and severity of psychiatric symptoms. The diagnosis of Prader-Willi syndrome does not determine the selection of psychotropic medication but rather informs the interpretation of symptoms and the resulting diagnostic formulation, which in turn guide the psychiatrist in the choice of medication.

Glossary

ACE inhibitors—(angiotensin converting enzyme inhibitors) medication used in treating left heart failure.

adipose—fat tissue.

asymptomatic—without symptoms.

azotemia—a build-up of nitrogen waste products in the blood when kidney function is deficient.

BiPAP and *CPAP*—Bi-level Positive Airway Pressure and Continuous Positive Airway Pressure.

cardiomegaly—enlarged heart.

cellulitis—infection of tissue, often a complication of venous stasis and lymphedema (serum).

creatinine—a measure of kidney function.

CO_2 narcosis—toxicity of the brain from excessive amounts of retained CO_2, resulting in decreased breathing.

dissociative state—a mental condition causing an incomplete awareness of pain or other stimuli.

echocardiography—sonar exam of the heart; gives dynamic information about heart function.

edema—excess fluid in tissue.

deconditioning—weakness and loss of flexibility from inactivity.

glycemic—pertaining to blood sugar.

heart failure—the pump function of the heart is not adequate to meet the body's needs.

hematocrit—measures the proportion of red blood cells to serum in the blood.

hemoglobin—measures the oxygen carrying capacity of the blood.

hepatic—having to do with the liver.

hypercapnia—high CO_2; very high levels of CO_2 cause coma and cessation of breathing.

hyperkalemia—high potassium in the blood.

hyperglycemia—high blood sugar.

hypoglycemia—low blood sugar.

hyponatremia—low sodium in the blood.

hypoperfusion—undersupply of blood to tissue due to low pressure.

hypothermia—low body temperature.

hypoxemia, hypoxia—low oxygen in the blood.

iatrogenic—caused by medical intervention.

interstitial—the space in tissue between the cells.

intubation—breathing tube placed in windpipe, enables the use of a ventilator.

ischemia—inadequate oxygen to tissue, usually due to inadequate blood flow.

lymphatic—the vessels in the body that carry tissue fluid back to the heart.

lymphedema—fluid build-up in one part of the body as a result of damage to the lymphatic vessels.

monilial—yeast, produces red rashes and skin breakdown in areas of skin that are allowed to remain moist.

pitting—finger pressure on edema will often produce a depression that remains visible for several seconds or longer; physicians often use this sign to help detect edema.

pulmonary edema—excess fluid in the lungs.

pulmonary hypertension—increased blood pressure in the lungs.

pulmonary vascular resistance—roughly positively related to the blood pressure in the lungs.

radiograph—X-ray picture.

REM—rapid eye movement, one of the deeper stages of sleep associated with dreaming.

renal—having to do with the kidneys.

respiratory failure—breathing is inadequate to rid the body of CO_2, which builds up in the blood.

superficial venous—pertaining to the veins near the surface.

thrombosis—blood clot formation.

thromboembolic—pertaining to blood clots that travel to another part of the body.

tracheotomy—surgical opening in neck to bypass the upper airway.

transaminase—a measure of liver function.

venous stasis—poor return of blood through the veins in the legs.

ventilation—breathing.

ventricular failure—the right or left ventricle of the heart is unable to meet the body's demands.

ventriculomegaly—enlarged ventricle.

References

1. American Psychiatric Association. *Diagnostic and Statistical Manual of Mental Disorders*. 4th ed. Washington, DC: American Psychiatric Association; 1994.

2. Arens R, Gozal D, Aomlin K, et al. Hypoxic and hypercapnic ventilatory responses in Prader-Willi syndrome. *Journal of Applied Physiology*. 1994; 77:2224–2230.

3. Bedell GN, Wilson WR, Seebohm PM. Pulmonary function in obese persons. *Journal of Clinical Investigation.* 1958;37(7):1049–1060.

4. Berger KI, Ayappa I, Chatr-Amontri B, et al. Obesity hypoventilation as a spectrum of respiratory disturbances during sleep. *Chest.* 2001;120: 1231–1238.

5. Bhargava SA, Putnam PE, Kocoshis SA, Rowe M, Hanchett JM. Rectal bleeding in Prader-Willi syndrome. *Pediatrics.* 1996;97(2):265–267.

6. Boer H, Holland A, Whittington J, Butler J, Webb T, Clarke D. Psychotic illness in people with Prader-Willi syndrome due to chromosome 15 maternal uniparental disomy. *The Lancet.* 2002;359:135–136.

7. Bray GA, Dahms WT, Swerdloff RS, Fiser RH, Atkinson RL, Carrel RE. The Prader-Willi syndrome: a study of 40 patients and a review of the literature. *Medicine.* 1983;62(2):59–80.

8. Burwell CS, Robin ED, Whaley RF, et al. Extreme obesity associated with alveolar hypoventilation: a pickwickian syndrome. *American Journal of Medicine.* 1956;21:811–818.

9. Butler JV, Whittington JE, Holland AJ, Boer H, Clarke D, Webb T. Prevalence of, and risk factors for, physical ill-health in people with Prader-Willi syndrome: a population-based study. *Developmental Medicine & Child Neurology.* 2002;44(4):248–255.

10. Carroll D. A peculiar type of cardiopulmonary failure associated with obesity. *American Journal of Medicine.* 1956;21:819–824.

11. Cassidy SB, McKilliop JA, Morgan WJ. Sleep disorders in the Prader-Willi syndrome. David W. Smith Workshop on Malformations and Morphogenesis, Madrid, Spain; 1989:23–29.

12. Clarke D, Boer H, Webb T, et al. Prader-Willi syndrome and psychotic symptoms: I. Case descriptions and genetic studies. *Journal of Intellectual Disability Research.* 1998;42:440–450.

13. Clift S, Dahlitz M, Parkes JD. Sleep apnoea in the Prader-Willi syndrome. *Journal of Sleep Research.* 1994;3(2):121–126.

14. Eiholzer U, Nordmann Y, l'Allemand D. Fatal outcome of sleep apnoea in PWS during the initial phase of growth hormone treatment. *Hormone Research.* 2002;58(Suppl 3):24–26.

15. Hakonarson H, Moskovitz J, Daigle KI, Cassidy SB, Cloutier MM. Pulmonary function abnormalities in Prader-Willi syndrome. *Journal of Pediatrics.* 1995;126:565–570.

16. Hanchett JE, Hanchett JM. Hypothermia in Prader-Willi syndrome. Presentation at 14th Annual Scientific Conference, Prader-Willi Syndrome Association (USA), San Diego, CA; July 1999.

17. Hanchett JM. Treatment of self-abusive behavior in Prader-Willi syndrome persons. 3rd Prader-Willi Syndrome International Scientific Workshop and Conference, Lido di Jesolo, Italy; May 1998.

18. Helbing-Zwanenburg B, Kamphuisen HA, Mourtazaev M. The origin of excessive daytime sleepiness in the Prader-Willi syndrome. *Journal of Intellectual Disability Research.* 1993;37:533–541.

19. Hertz G, Cataletto M, Feinsilver SH, Angulo M. Sleep and breathing patterns in patients with Prader-Willi syndrome (PWS): effects of age and gender. *Sleep.* 1993;16(4):366–371.

20. Hertz G, Cataletto M, Feinsilver SH, Angulo M. Developmental trends of sleep-disordered breathing in Prader-Willi syndrome: the role of obesity. *American Journal of Medical Genetics.* 1995;56(2):188–190.

21. Holland AJ, Treasure J, Coskeran P, Dallow J. Characteristics of eating disorder in Prader-Willi syndrome: implications for treatment. *Journal of Intellectual Disability Research.* 1995;39(Pt 5):373–381.

22. Holley HS, Milic-Emili J, Becklake MR, Bates DV. Regional distribution of pulmonary ventilation and perfusion in obesity. *Journal of Clinical Investigation.* 1967;46(4):475–481.

23. Katz A, Hollingsworth D, Epstein F. Influence of carbohydrate and protein on socium excertion during fasting and refeeding. *Journal of Laboratory and Clinical Medicine.* 1968;72:93–104.

24. Kessler R, Chaouat A, Schinkewitch P, et al. The obesity-hypoventilation syndrome revisited. *Chest.* 2001;120(2):369–376.

25. Koenig SM. Pulmonary complications of obesity. *American Journal of Medical Science.* 2001;321(4):249–279.

26. Kreitzman SN. Total body water and very-low-calorie diets. *The Lancet.* 1988;1(8585):581.

27. Leech JA, Onal E, Aronson R, et al. Voluntary hyperventilation in obesity hyperventilation. *Chest.* 1991;100:1334–1338.

28. Livingston FR, Arens R, Bailey SL, Keens TG, Ward SL. Hypercapnic arousal responses in Prader-Willi syndrome. *Chest.* 1995;108(6):1627–1631.

29. Nicholls DP, Onuoha GN, McDowell G, et al. Neuroendocrine changes in chronic cardiac failure. *Basic Research in Cardiology.* 1996;91 (Suppl. 1):13–20.

30. Nixon GM, Brouillette RT. Sleep and breathing in Prader-Willi syndrome. *Pediatric Pulmonology.* 2002;34:209–217.

31. Pack A, Kubin L, Davies R. Changes in the cardiorespiratory system during sleep. In: Fishman AP, ed. *Fishman's Pulmonary Diseases and Disorders.* 3rd ed. New York, NY: McGraw-Hill; 1998.

32. Richards A, Quaghebeur G, Clift S, Holland A, Dahlitz M, Parkes D. The upper airway and sleep apnoea in Prader-Willi syndrome. *Clinical Otolaryngology.* 1994;19:193–197.

33. Schluter B, Buschatz D, Trowitzsth E, Aksu F, Andler W. Respiratory control in children with Prader-Willi syndrome. *European Journal of Pediatrics.* 1997;156:65–68.

34. Teichtahl H. The obesity-hypoventilation syndrome revisited. *Chest.* 2001;120(2):336–339.

35. Van Mil E, Westerterp K, Gerver W, et al. Energy expenditure at rest and during sleep in children with Prader-Willi syndrome is explained by body composition. *American Journal of Clinical Nutrition.* 2000;71:752–756.

36. Vela-Bueno A, Kales A, Soldatos C, et al. Sleep in the Prader-Willi syndrome: clinical and polygraphic findings. *Archives of Neurology.* 1984;4:294–296.

37. Vgontzas A, Bixler E, Kales A, Vela-Bueno A. Prader-Willi syndrome: effects of weight loss on sleep disordered breathing, daytime sleepiness and REM sleep disturbance. *Acta Paediatrica.* 1995;84:813–814.

38. Von Boxem T, de Groot GH. Prevalence and severity of sleep disordered breathing in a group of morbidly obese patients. *The Netherlands Journal of Medicine.* 1999;54:202–206.

39. Zwillich CW, Sutton FD, Pierson DJ. Decreased hypoxic ventilatory drive in the obesity-hypoventilation syndrome. *American Journal of Medicine.* 1975;59(3):343–348.

18

Social Work Interventions: Advocacy and Support for Families

Barbara Y. Whitman

It has been well established that the psychosocial impact of Prader-Willi syndrome (PWS) is ubiquitous and involves the entire family, as well as schools, community, and society as a whole. For the family, from the moment of the birth and diagnosis of the child with PWS, the added stress usually disrupts the dynamic balance (at least temporarily) at both the family and personal levels. In addition, any personal, family or environmental concerns present *before* the birth of the affected child may make it harder to maintain adequate family functioning and meet the infant's special needs. Targeted social service intervention by the social worker can (1) help the family reestablish a working balance, while (2) providing education and guidance, and (3) providing support in obtaining and coordinating multiple medical and early intervention services for the affected child.

In Time of Need—Resource and Advocate

For most families, this initial stressful period—full of grief, confusion, and often overwhelming challenges—is just the beginning of a journey through a lifelong maze of unique experiences, educational challenges and needs, and behavioral differences and uncertainties. Moreover, they face a complex, multilayered service system that often may seem designed more to deny than to deliver services.

As noted, individuals with Prader-Willi syndrome present unique needs to their families, schools, and communities. These special needs are not always fully appreciated outside the family, and sometimes not even by the extended family. Too many families continue to report that professionals and systems fail to recognize the physiologic etiology and unyielding nature of the food-related constellation of behaviors, often implicitly if not explicitly citing poor parenting or purposeful bad behavior as the source of difficulty. In the face of this interpretation, the process of obtaining appropriate services can become adversarial, with the parents often feeling alone, isolated and exhausted. Indeed, in a study of stress and social support in families of children with PWS,

Hodapp et al.[3] found that parents of children with Prader-Willi syndrome reported higher levels of parent and family problems, suffered greater pessimism, yet got less support from professionals when compared with parents of similarly aged children with cognitive deficits from other causes. This was despite a major thrust in the last decade for "family-focused" and "family-driven" intervention strategies.

As a member of an interdisciplinary team, the social worker identifies the frustrations, demands, and conflicts encountered by families and functions as an advocate for both the family and the team. This chapter will address many family issues and needs in which the social worker plays a supportive role, beginning with the confirmation of diagnosis and continuing throughout the child's life.

What Is a Family?

While many currently frame this question as a political, moral, and ultimately a legal issue, the question is raised here in order to understand how to define a "family unit" for purposes of support and intervention. Conceptually, systems theory[8] provides a framework for identifying and understanding family strengths and limitations and can provide guidance in identifying the need for additional support.

Briefly, a systems theory perspective defines the family as a small social unit of interconnected persons that reciprocally influence one another, over time. Two distinct properties distinguish families from other social groups. First, unlike work, task, or interest-focused groups, *families are formed on the basis of reciprocal emotional ties.* While legal contracts—such as marriage—may be struck as a result of this emotional connectedness, the primary bindings in a family are affectional; the emotional bonds of attachment, loyalty, and positive regard are paramount. Secondly, *family membership is virtually permanent.* Other social groups have recognized means for joining and leaving; resignation or being fired from a family is virtually impossible. For many, that permanency is a source of support, a foundation for growth, and a recognition that, no matter what, the family will always be there. For others, it is an imprisoning, stultifying entrapment that suppresses growth and from which there is no escape. Whatever the feeling, the ties are permanent. Parents may divorce, but the emotional ties, whether positive or negative, inextricably bind that child to both parents in perpetuity.

Many studies document that family "processes"—or how, and with what emotional tone, a family interacts—have greater impact on developmental outcomes than does family composition. From the viewpoint of family processes as primarily determinative, systems framework is neutral on the political/moral issues of family composition, recognizing viable family structures beyond the traditional two-parent family including single parent families, extended family caretakers, and homosexual couples raising children. When a family member has PWS, understanding family processes, and the impact of PWS on these processes, is key to management. Yet, despite its importance, research in this area is virtually nonexistent. Thus, much of this chapter is based on collective clinical wisdom from families served and data and experience from the national Prader-Willi Syndrome Association of the United States, combined with general research findings regarding families and developmental disabilities. The purpose of this chapter is to provide parents, grandparents, and those multidisciplinary teams who would work with these families a basis for understanding the many complex, and often contradictory, feelings reported by families and to provide some examples of the multitude of creative coping strategies developed by these families.

Initial Sources of Stress for Parents: Birth

With rare exception, families begin interacting with multiple service-providing systems at the moment of birth, when the characteristic hypotonia and accompanying feeding difficulties signaling the presence of PWS usually demand extended hospital stays in special care nurseries accompanied by multiple, often painful diagnostic testing. Few require "hard-core scientific data" to acknowledge both the extraordinary levels of parental stress specifically associated with the situation and the extraordinary levels of parental anxiety specifically for their baby and for the possible outcome. In addition, frequently the baby is transported to a specialty pediatric facility that may be quite distant both from the delivery hospital, where the mother may remain for a while, and from the family home. Distance, parental health, and even travel expenses or transportation difficulties may prevent parents from being with their infant at this time, further increasing stress and anxiety. Moreover, parents may encounter, and be overwhelmed by, multiple medical specialists, noisy and unfamiliar biomedical equipment, constantly changing staff, and confusing communication processes. The stress of radically altered, unfamiliar, and unpredictable circumstances, combined with extraordinary levels of anxiety, change the dynamics and processes within the family unit, which, without support and intervention, may foreshadow a shift into a less functional system. Proper health care for the infant dictates provision of support and care for the parents during this time, along with a life-care plan based on an accurate assessment of family stresses, family processes and dynamics, finances, coping mechanisms, extended family and community resources, and the ability to make future plans. The social worker is the logical choice as a communication liaison for the care team, an emotional support for the family, and an investigator of the needed facts for developing a long-term care plan for both the child and family.

Reactions to Diagnosis

Full genetic confirmation of Prader-Willi syndrome has only been available for the past decade. Prior to that time, diagnosis was suspected but often delayed until the emergence of hyperphagia and the accompanying obesity. Prior to confirmatory genetics, the average age at diagnosis for those with a deletion was 6 years and for those with uniparental disomy, 9 years.[1] Many families who experienced such a diagnostic process reported that "having a name for it" was an enormous relief. In addition to having some framework for knowing what to expect, many parents said that finally having a diagnosis relieved an often unspoken fear and guilt that they had somehow caused or were causing the problem. At the same time, many reported experiencing an active grief process as they realized the unreality of another, often unspoken hope that eventually the problem would be named *and fixed.*

By contrast, today when the profound hypotonia of affected newborns signals a genetic work-up, it is often followed by a confirmed diagnosis while the infant is still in the special care nursery. Thus many families now must concurrently absorb and comprehend that their newborn "isn't quite normal" *and* the implications of PWS. With rare exception, parents experience an acute stress associated with learning of the genetic abnormality. Each parent may silently wonder if they were somehow to blame; conversely, they may each (silently, it is hoped) blame the other. Whether the etiology is a deletion or uniparental disomy, parents need to be regularly assured that there is nothing they did to cause the disorder, nor anything they could have done to prevent the disorder. Mother needs to hear that *nothing* she did (or didn't do), drank (or didn't drink), medications she took (or didn't), or activities and exercises she engaged in (or didn't) was causative.

Research indicates that, faced with extraordinary and overwhelming stress (earthquake, fire, accident, birth of a child with disability), people respond in one of three predictable ways. Fifteen percent of the people can and do take positive action, 15% are emotionally paralyzed and unable to act, and 70% exhibit odd behavior. Even beginning the adjustment to a diagnosis can take 9 to 12 months. Family members often experience the classic stages of grief, progressing from (1) shock and numbness to (2) yearning and searching, (3) disorientation and disorganization, and (4) resolution, reorganization, and integration. But these stages are not accomplished in a vacuum; nor do parents have the luxury of coming to terms with their feelings before they must provide day-to-day care for their child, make decisions, and seek services. Parents may take their baby home feeling anxious and unprepared to handle the special feeding techniques. Infants with PWS have a weak cry and little energy for crying, so a parent may be trying to provide for the infant's needs in the absence of normal infant cues signaling hunger, distress, and discomfort. They may immediately face the need for understanding what early intervention services are required, how they are obtained, and finding the financial resources for the services. Further, participating in multiple services can be time consuming. If a family was planning on, or needs, two incomes, lifestyle adjustments may be needed. If the family already has other children, their needs cannot be put on hold. Previously unresolved conflicts regarding child rearing and discipline will be exacerbated and magnified under the stress, to the detriment of the family and marriage. Restrictions imposed by the syndrome often cause other family members to feel trapped and helpless, and emotional ties become distorted. Research documents that, without adequate support through this period, the level of reorganization may be fragile and insufficient to adequately meet the challenges facing the child and the family.

It is important at this point for the social worker to fully assess, monitor, and when necessary, intercede in family interaction patterns so that, over time, the resources and capacity of family members are not overtaxed to the point that parents doubt their capacity to fulfill their prescribed roles. If the previously developed habilitation and management plan provides inadequate support, the social worker must act as advocate for the family with the team, reconvening the necessary professionals to design a more comprehensive habilitation plan. At the same time, the social worker must establish and maintain an ongoing supportive relationship with the family so that he/she can act as a link to the team and community resources. In this role, he/she may review, reinforce, and where necessary, clarify team recommendations to the family and work to establish the necessary linkages in the community for obtaining needed services. Although most families reach grief resolution, several authors describe the chronic sorrow that can be found when families address this lifelong disorder and its consequences.[4,5]

In addition, many parents find it helpful to connect with other parents. This can be done formally through the parent mentoring program of the PWSA (USA) or informally through state or local support groups or one of the Internet-based discussion groups. It is often recommended that parents avoid Internet groups until they are through some of the more vulnerable periods of grief and are more able to cognitively evaluate the validity of the information.

Moving Forward

As the intensity of the initial reactions lessens, parents can more purposefully focus on integrating this child's special needs into the daily fabric of family life. Early in their educational process, many parents will request as much information about the syndrome as possible, attempting to reach and absorb some understanding about PWS. Others simply want to know "what to do next," perhaps acknowledging at some pre-conscious level that true understanding is illusive. There are inherent emotional traps in each request to which the social worker must remain alert. For the family requesting as much information as possible, the social worker needs to monitor the emotional impact of the information. The information presented must be factually and sensitively presented. However, despite the request for information, for some families the trauma of learning about PWS may distort their ability to listen and absorb fact, creating a backlash of anger directed at what they perceive as insensitivity in the presentation of the facts. Questions such as, "Are you telling me that my child won't be able to go to the same parish school that my other children have gone to? It's the same one I went to," indicate that the impact of this disorder in this family may reach into many sacred, well-loved areas whose constancy over time has always offered the family a sense of safety and stability.

Thus a tempered, yet realistic answer should be provided, indicating that we cannot yet predict the severity of this child's learning deficits, nor do we know how easily and extensively necessary modifications will be provided by the school. Further, it can be indicated that research is ongoing regarding the management of some of the more difficult areas of the syndrome, so that by the time this child reaches school age, school placement issues may be quite different from those currently encountered, and may or may not preclude attendance at any particular school.

By the same token, those who request, "Just tell me what to do next," may be realistically relating that they wish to act appropriately based on what is now needed and will seek further guidance as the situation changes—a perfectly valid coping strategy. The team, however, must be alert for the family that is so overwhelmed that they need further supports and for the family that picks and chooses what is comfortable to hear, acting on what is convenient. In all instances, it is recommended that written information be available to be taken home and reviewed.

Long-Term Family Concerns

Marital Relationship

The stressors of everyday life can greatly affect both general family functioning and specifically the ability of the family to adequately provide for their child with PWS. In addition, having a child with PWS places further stress on marriages. Harried schedules, work overload, financial difficulties, and other "external" precipitants of stress may be so overwhelming that they increase the need for social work intervention. Within the family, the marital relationship is particularly vulnerable to stress, even those relationships of longer duration. Frequently, marriages are in their early stages and spouses are learning how to rely on each other when their children are born. Areas of difficulty and unresolved conflicts will be exacerbated at this time. Further, it is not unusual for each spouse to be at a different stage of absorbing and adjusting to the presence and implications of PWS. As a result, each spouse may be unable to respond to the emotional needs of the other. Early on, special feeding demands may limit access to qualified babysitters, so there is little or no time for private couple interactions. As the child gets older, potential behavior problems may continue limiting childcare options. The needs of other children add to these concerns. Both spouses may feel confused, angry, and trapped. Sometimes, the depth of frustration and unhappiness does not emerge for some time. As a result, expression of feelings toward their child and each other may be blunted or erratic, unpredictable, and even episodically volatile. Van Lieshout et al.[7] compared the impact of family stress on the marriage and on behavior of spouses toward the affected child in parents of children in three syndrome groups: Prader-Willi syndrome, fragile X syndrome, and Williams syndrome. In all groups, higher perceived family stress was related to more marital conflict and less parental consistency. However, parents of children with Prader-Willi syndrome more often expressed this stress as anger directed toward the child. This child-directed anger was not without consequence; in those families evidencing the highest parent anger toward the child, the child was noted to have a more negative personality and poor behavior when compared with those from families in which parental anger was less. These findings extend the earlier findings of Holmes and associates[4] and Whitman,[9] who demonstrated that if intervention is not carried out immediately after diagnosis and whenever problems arise, behavior problems are more likely to develop or compound. A high level of family conflict acts as a predictor of behavior problems, reinforcing the need for social work intervention.

In addition, for most families, financial concerns are ongoing until the child reaches age 18. Worries about payment of medical bills and the cost of other services (e.g., respite) can become overwhelming. To the extent that the marital relationship is strained, the energy each spouse expends on coping with those strains drains that available for dealing with their child(ren), and in those families coping with Prader-Willi syndrome may be manifested by greater anger directed toward the affected child. Thus, the social worker should continually assess the status of the marital relationship, obtaining the appropriate interventions as needed. This may be as simple as finding funding and trained respite providers to allow the couple some time alone, or as intense as providing couples therapy.

Parenting Issues

Many families are able to maintain a stable marital relationship, but fundamental differences in parenting styles serve as a constant source of conflict. Clearly there are many "right ways" to parent. However, since children and adults with PWS have an extraordinary need for sameness and consistency, parenting style differences must be identified as early as possible and sensitively addressed to prevent later intractable behavior problems. Nutritional management remains a primary issue for affected individuals. How well food is managed has critical implications for all other areas of behavior management, yet it remains an area that parents often find difficult to discuss. For example, even when parents are in perfect accord over the need for dietary vigilance, particularly at home, they may be at odds over how to handle food-related behavioral incidents in public. Thus, while one parent may respond to an impending incident over a denied treat by preparing to leave a restaurant, the other may advocate giving in "just this once" in order "to avoid an incident here." Such differences are quickly apparent to the affected child, who soon learns to manage his/her environment through bad behavior. Mutually acceptable methods for handling these differences need to be developed as soon as the differences arise. If the couple is unable to come to some decision rules on their own, counseling help should be sought.

Parents may also be in conflict about the division of labor regarding nutritional and behavioral management, with one parent feeling burdened with too much responsibility while perceiving that the other is insufficiently so. Thus, if a child gains excess weight, one spouse often attacks the parenting style of the other, diverting attention from solving the problem of how to prevent further weight gain. Leaving such issues unresolved leads to families becoming dysfunctional, if they are not already so.

Even when there are no differences between parents regarding nutritional management, defensiveness is a natural first response when a child gains excess weight. Many families worry that the professional staff with whom they work will perceive them as "bad parents." Many express guilt that in "stealing a few moments for themselves," they failed to lock a food storage area and found their child "indulging" vigorously. Over time, many families have developed a number of ways to cope with weight and behavior control issues. A survey of 293 parents/ guardians of children with PWS aged 25 years and younger identified a number of strategies for coping with food and eating behaviors, and a second set of strategies for controlling access to food, as well as the strengths and weaknesses of each.[2] The most consistently effective are listed in Table 18.1, while those that are often effective (given the right conditions) are listed in Table 18.2. The effectiveness of many strategies depended on family characteristics. For example, many respondents

Table 18.1. Effective Strategies for Nutritional Management

Strategy	Respondents* (N)	Effectiveness Score** (Mean ± SD)
Lock away food	116	4.5 ± 0.9
Always supervise the child, particularly around food	143	4.2 ± 0.9
Keep lots of low-energy foods on hand for snacks	125	4.2 ± 1.0
Keep lots of low-fat foods on hand for snacks	120	4.2 ± 1.0
Count kilocalories and have less energy at a meal to allow a snack later	104	4.1 ± 1.1
Try to provide more nonfood-related rewards and treats	124	4.1 ± 1.1
Give the child small portions	141	4.0 ± 1.0
Give other children treats when the child with Prader-Willi syndrome is not around	101	4.0 ± 1.1
Only have snacks when the child with Prader-Willi syndrome is not around	105	4.0 ± 1.1
A special diet	85	4.0 ± 1.1

* Not all respondents rated every strategy; a total of 154 families provided ratings.
** Strategies were rated on a scale from 1 to 5, with 1 being least effective and 5 being most effective.
SD = standard deviation.
Source: Goldberg et al., "Coping with Prader-Willi syndrome," *Journal of the American Dietetic Association*, 2002;102:537–542.[2]

Table 18.2. Moderately Effective Strategies for Nutritional Management

Strategy	Respondents* (N)	Effectiveness Score** (Mean + SD)
Keep snack food in your bedroom or other special locked area	94	3.9 ± 1.2
Only serve food from the kitchen, not from the table	86	3.9 ± 1.2
Send special snacks to school	114	3.8 ± 1.2
Give the child half a portion, and let the child ask for seconds	109	3.8 ± 1.2
Keep strict meal times	91	3.8 ± 1.3
Everyone eats only low-fat and low-energy foods	76	3.6 ± 1.2
Put food on smaller plates so that it looks like more	97	3.6 ± 1.2
Discuss the menu before going to a restaurant	106	3.6 ± 1.3
Keep limited amount of food in the house	69	3.6 ± 1.4
Meet with caregivers and teachers to explain the syndrome	141	3.5 ± 1.2
Allow the child to be part of menu planning and preparation, making him or her aware of energy and fat content	103	3.5 ± 1.2

* Not all respondents rated every strategy; a total of 154 families provided ratings.
** Strategies were rated on a scale from 1 to 5, with 1 being least effective and 5 being most effective.
SD = standard deviation.
Data source: Goldberg et al., "Coping with Prader-Willi syndrome," *Journal of the American Dietetic Association*, 2002;102:537–542.[2]

indicated that lack of effectiveness was related to (1) the time, energy, and stress involved in maintaining a strictly regimented diet; (2) other caregivers (such as school personnel) not understanding the importance of dietary restrictions and how to implement them; (3) difficulty making appropriate food choices at restaurants and parties; (4) difficulty limiting access to food (a difficulty that increased as the child got older); (5) the impact of limiting access on the other children and general family life; (6) a busy lifestyle; and (7) general difficulties in managing behavior problems. A number of respondents highlighted the need for unwavering consistency of approach both between parents and across time. Thus, while the particular strategy that is effective may differ between families, the need for consistency is vital for all families.

Siblings

Perhaps no area of research is more neglected than the impact on, and adjustment of, siblings living in a family coping with PWS. When parents are obliged to constantly focus on the needs of the affected child, the risk for siblings having adjustment difficulties, although not inevitable, is clearly increased. Anecdotal evidence suggests that siblings of a child with Prader-Willi syndrome often have conflicting and confusing feelings about their brother or sister. They may be embarrassed by their abnormal eating behaviors and resent the need for controlling access to food. The constant risk of behavior problems in public, particularly around these issues, are a chronic source of stress and anxiety, frequently leading to a "Does he/she have to go?" query. Siblings often express reluctance to invite friends to their home, while at the same time indicating a lot of guilt associated with this reluctance. Jealousy at the attention given the affected child is often noted, along with a conflicting feeling of being glad they can "hide" and get away with doing what they want while the family is preoccupied with the affected sibling. Often siblings are pressured into parenting roles (e.g., "I have to run to the store; make sure he/she doesn't get into the pantry while I'm gone"), given additional responsibilities, or forced to "grow up" too quickly. Even feelings of guilt that the sibling does not have food restrictions often emerge.

It is important to give siblings permission to express all feelings regarding their brother or sister and the syndrome. Parents may need to be taught how to encourage this expression, and how to hear them in a nonjudgmental way. Parents can model for their children by inviting them to medical or counseling appointments where the parents identify and talk about their own feelings. Attending siblings should be encouraged to ask questions; the tone of their questions can serve as a conduit for asking, "That makes me wonder if you are feeling . . . ," and pave the way for the siblings to air concerns and feelings.

Younger siblings may wonder if they can "catch" Prader-Willi syndrome, while older siblings may worry about having a child of their own with PWS. Most geneticists encourage genetic counseling for the maturing adolescent so that all questions can be honestly and sensitively answered.

Many adult siblings retrospectively report that they knew their family was different and had to be more careful about food, but they didn't see it as "abnormal." Research documents that many adult siblings feel they have developed a greater understanding and acceptance of differences and a sense of empathy for those who are disenfranchised.[6]

The Extended Family

Many families have extended family members from whom they usually get support. Grandparents, parents' adult siblings, aunts, uncles, and even cousins often can be called on for child care, emotional support, and occasionally financial relief. The extended family can also, unintentionally, become an additional source of stress. Many grandparents have stated, "You can keep him on a diet at home, but at my house it's my grandparent's right to give him what he wants!" In family systems that regularly or primarily socialize with and draw support from each other, a family meeting of all concerned should be convened as soon as possible after the diagnosis has been made. The purpose of this meeting is to insure that all members have the same information and are provided the opportunity to ask questions and to seek understanding. Concerns about a recurrence risk in their own families can be addressed. The importance of controlling access to food, limiting intake, and establishing consistent behavioral approaches across environments can be underscored. The special pain of grandparents, both for their adult child who is the parent and their grandchild, can be aired and supported. Periodically, the social worker will need to inquire how these relationships are functioning and whether additional meetings are necessary.

Additional Major Concerns

Over time, the relationship of the social worker with the family is such that multiple additional concerns may be raised and addressed. Among these are behavior management, community resources, education, guardianship, and adult living and working arrangements. Since these areas are addressed at length elsewhere in this volume, they will be dealt with only briefly here.

Behavior

In addition to nutritional management, behavior is the other major challenge for families, and one that increases as the affected individual gets older. It is important to identify and intervene in areas of concern as soon as they arise and before a self-reinforcing pattern is established. While consistency is the hallmark of behavior management, it must be remembered that the constant food vigilance and behavior management needs of the child with PWS can drain family emotional and financial resources. Families are often unable or reluctant to admit their exhaustion; it may, however, become evident in clinic visits when a family appears to ignore inappropriate behavior. Clinic staff may misinterpret this lack of response as chronic inadequate parenting and may respond to the family in ways that prevent a full understanding of the level of exhaustion. In addition, acting out and tantrums make families reluctant to go out in public or invite others to their home, encouraging isolation. The social worker needs to be alert to and acknowledge this exhaustion. Realistic recommendations for behavior management and additional family support may temporarily be needed to allow the family to regain sufficient stability to deal adequately with the behavior difficulties.

Community Resources

The family's need for community services will vary according to the age of the child and the services provided through federally mandated intervention programs. Across all ages, it is important for the child to have access to special recreation programs and social activities. Many younger children may want to access therapeutic horseback riding, dance, and gymnastics programs as a part of their physical therapy program. Older youngsters and adults may want to participate in sports activities through such programs as Special Olympics or attend specially designed summer camps. Social activities such as coed dances and movie outings are favorites of older adolescents and adults.

As previously indicated, all parents need an occasional break; therefore, local respite services are critical to how well families function. Parents of a child with PWS are chronically fatigued and stressed by the constant vigilance required and the demands of behavioral challenges. Unfortunately, respite is increasingly limited for parents of children with special needs and even more so for those with PWS. Respite can take the form of a few hours, overnight, or even a period of days. When funded respite is unavailable, the social worker may want to help parents organize a parent-based respite cooperative. This serves not only the respite needs of parents, but also the need for several children with PWS to socialize in each others' homes.

Education

All U.S. children with Prader-Willi syndrome are eligible for federally mandated early intervention programs, usually delivered through the local school district, although this varies by state and school district. Parents should be encouraged to enroll their children in such a program as soon as they receive a diagnosis. Physical and occupational therapy, speech and language services, and behavior interventions are available through these programs and should be instituted as soon as possible. In addition to specific therapies and activities, these programs are a source of support for parents. School personnel not familiar with PWS may call on the social worker to arrange an in-service training to familiarize school staff with the syndrome and to answer questions regarding management. A day care for children with special needs may also be needed at this age. Special education services through the Individuals with Disabilities Education Improvement Act (IDEA) serve the child from ages 3 through 21; Individualized Educational Plan (IEP) goals should anticipate and program for special services (see Chapter 11).

For adolescents, transitional planning should begin on entering high school. Issues center around job training and the transition from high school to a workshop or supported employment position. Transition planning should begin with prevocational and vocational planning during the first years in a high school setting. Vocational service staff will need considerable education about the special aspects of PWS (see Chapters 13–15).

Discussion regarding the ultimate possibility of alternative living arrangements should be raised at this time as well. Making the decision to allow the young adult with PWS to move out of the family home to another living arrangement is often a very difficult decision for the family, and one that may take considerable time to make. The social worker should gently bring up the subject regarding future living plans early with the family. The family should be encouraged to discuss their feelings and fears with regard to this. Many parents need time to absorb the fact that it is "normal" for young adults to move out of the family home, even a young adult with PWS. Feelings of relief and feelings of guilt at that relief are not uncommon, along with fear that the staff of the home won't care for and about their child as well as the family does. Families will need assurance that the placement will provide complete control of food, constant supervision, and structured opportunities for social activities. The social worker will need to assist the family in contacting the local support person for alternative living arrangements. See Chapter 16 for more information on residential services.

Guardianship and Wills

While all parents with children under age 18 should have wills, it becomes even more critical when there is a child with a disability. At this point the issue is not dispersal of financial resources, if any; rather it is declaring who should assume guardianship of the child in the event that the parents become deceased. This requires a very open and intimate conversation with those being asked to assume that responsibility. Often the social worker can help the family rehearse ways to approach this issue or can join the family during the conversation, supporting them through some of the more emotionally difficult aspects such a discussion inevitably generates.

Guardianship and long-term financial planning become issues as the young adult approaches age 18. The laws of each state vary but are usually readily accessible on the Internet. For both wills and the guardianship issues, the social worker can provide parents with basic information and direct them toward the appropriate legal resource.

Conclusion

When a child has PWS, in addition to the usual parenting concerns, the family faces a number of additional and often emotionally difficult issues requiring special information and services. The social worker is in a unique position to aid and support the family as they face the challenges and joys of parenting this child. Targeted social service intervention by the social worker can (1) help the family reestablish a working balance, while (2) providing education and guidance, and (3) providing support in obtaining and coordinating multiple medical and early intervention services for the affected child.

References

1. Butler MG, Meaney FJ, Palmer CG. Clinical and cytogenetic survey of 39 individuals with Prader-Labhart-Willi syndrome. *American Journal of Medical Genetics*. 1986;23:793–809.
2. Goldberg D, Garrett C, Van Riper C, Warzak W. Coping with Prader-Willi syndrome. *Journal of the American Dietetic Association*. 2002;102:537–542.
3. Hodapp R, Dykens E, Masino L. Families of children with Prader-Willi syndrome: stress-support and relations to child characteristics. *Journal of Autism and Developmental Disorders*. 1997;27(1):11–24.
4. Holmes CS, Yu A, Frentz J. Chronic and discrete stress as predictors of children's adjustment. *Journal of Consulting and Clinical Psychology*. 1999;67: 411–419.
5. Olshansky S. Chronic sorrow: a response to having a mentally defective child. *Social Casework*. 1962;43:190–193.
6. Powell TH, Ogle PA. *Brothers and Sisters: A Special Part of Exceptional Families*. Baltimore, MD: Paul H. Brookes; 1985.
7. Van Lieshout C, De Meyer R, Curfs L, Fryns JP. Family contexts, parental behavior, and personality profiles of children and adolescents with Prader-Willi, fragile-X, or Williams syndrome. *Journal of Child Psychology and Psychiatry*. 1998;39(5):699–710.
8. Von Bertalanffy L. *General Systems Theory*. New York, NY: George Braziller; 1968.
9. Whitman BY. The prevalence and type of behavior problems and the contribution of extended family health histories and current family stress to such problems in persons with Prader-Willi syndrome: a cross cultural study. Abstract presented at 2nd Prader-Willi Syndrome International Scientific Workshop and Conference, Oslo, Norway; June 1995.

19

A National Approach to Crisis Intervention and Advocacy

David A. Wyatt

Prader-Willi syndrome (PWS) often places great stress on the family and all persons who come into contact with the person with PWS. The Prader-Willi Syndrome Association (USA) national office provides crisis intervention in the areas in which our member families often need advocacy and/or educational information to obtain the local support and assistance they need. Requests for this service come through e-mails and telephone calls to a toll-free number. The Crisis Intervention Counselor and/or the Executive Director respond personally to these requests and with appropriate educational materials about the syndrome. A "crisis letter" that addresses the issue(s) at hand often accompanies a "crisis packet" containing the appropriate information. The Association maintains an account of the types of crisis calls received and reported the following summary of calls for one recent year:

36%—Legal and advocacy: court system, schools, insurers, Supplemental Security Income, etc.
17%—Residential placement cases
16%—Medical cases (other than morbid obesity)
11%—Sexuality issues in residential placement or school
10%—Behavior management in the home, school, or residential placement
10%—Morbid obesity

Following is a description of the various crisis issues that are presented by families with a child, adolescent, or adult with the syndrome.

Medical Crisis Issues

While parents initiate a majority of our medical crisis calls, the national PWSA office also receives calls from physicians, nursing personnel, emergency room personnel, and intensive care units when they are in the process of treating a person with PWS because of medical issues related to the syndrome. These are usually critical emergencies. The situation is exacerbated when the medical personnel are not familiar with PWS.

PWSA immediately sends acute medical information written for these situations, such as a general medical alert document or articles on respiratory concerns and anesthesia considerations. The information is faxed or e-mailed directly to the nursing station and/or the attending physician. The national office may notify one or more physicians on the Scientific and Clinical Advisory Boards about the situation, then inform the medical staff on the scene that they can obtain a medical consultation from the physician who was contacted.

PWSA can often give input and get feedback from several expert sources, thanks to e-mail. The following is an e-mail sent to the medical boards. This case demonstrates how an individual with PWS is often affected in quite different ways, and with more complexity, than other patients because of the syndrome.

11-year-old girl, 4 foot 5 inches and 109 pounds. She had surgery for scoliosis January 29th. The surgery was uneventful, but she has had dramatic cognitive changes since then. She repeats things 60–100 times as if she does not remember just saying it, can barely hold a fork, drops things, thinks she sees people in the room that are not there, and is wetting herself. She is also sleeping a dramatic amount of time. She was on antibiotics, then off and is on again because they discovered Sunday she has pneumonia. No change while off and the only pain med she is taking is Tylenol. Her long-term memory appears to be still OK. She had a CT of the head, and an EEG. Nothing showed up. She cannot get an MRI right now due to all of the new hardware in her back. The only thing (besides the pneumonia) that shows up is a slightly elevated ammonia level. She went home from the hospital 5–6 days post surgery (they were hoping that the cognitive issues were the result of the anesthesia and would be temporary) and ending up back on Sunday due to the pneumonia, which seems to be responding to the antibiotics, but her cognitive functioning is no better.

Several members of the PWSA medical boards responded to this e-mail with advice on further testing, and two agreed to consult with this girl's physician. The consultations had two results: First, the medical treatment was modified in addressing the pneumonia according to the consulting physician's clinical experience with antibiotic medication and oxygen impairment in patients with PWS. Secondly, specific psychotropic medications and dosages were given, again related to the clinical experience with individuals with PWS, and medication changes began to clear up her cognitive difficulties.

Morbid obesity is always a threat for the person with PWS. When the obesity is leading to life-threatening medical problems, crisis intervention is needed to reverse the gaining of weight. The initial response to a call from the parents, who are aware and frightened that the eating is out of control, is to support them in instituting means of controlling food access. If this is already being done, the Crisis Counselor can help the parents solve the mystery of where the individual with PWS is obtaining food and support them in getting this source shut down. In some cases, it is more appropriate to assist the parents in getting the person with the syndrome into an inpatient rehabilitation program, which is often successful in reversing the most life-threatening weight gain and/or managing the most difficult behaviors (see Chapter 17). Often, the weight and behavior problems occur concurrently. The following case is a typical example:

Call from mother: Son, 16yrs., 320lbs. Just officially diagnosed with PWS. Hides food everywhere, e.g., under his pillow, bed, etc. Becoming aggressive towards her. Took out a knife and tried to hurt himself. Embarrassed about his small penis and his obesity. Others stare and point at him. He has no social life. He cried and said, "Momma, why am I like this? I don't want to be this way." Has diabetes and sleep apnea and is beginning to demonstrate right-sided heart failure.

Education Issues

Often, major problems arise when the child with PWS begins his/her educational program, even as early as preschool. One issue is obtaining an Individualized Education Plan (IEP) that is appropriate for the person with PWS. An IEP meeting can seem overwhelming and intimidating to the parents without experience or thorough preparation. Parents are encouraged to go to the Internet and locate NICHCY (www.nichcy.org), a federally-funded information clearinghouse on disabilities. NICHCY has abundant information on the Individuals with Disabilities Education Act (IDEA), the U.S. federal law governing public schools in the area of programming for persons with disabilities. They also have information about the IEP process. The national PWSA office can provide parents with educational materials to help prepare them for the IEP meeting and, perhaps, put them in touch with other parents of children with PWS who have had successful IEP results.

One area of the IEP that can cause parents consternation is that of requesting a one-on-one aide to assist the child with PWS. When the child's eruptive behavior is interrupting the educational process and the school's response is suspension or expulsion, an experienced aide can often defuse the environment. An aide is also needed when the child with PWS is a serious food forager. This request is often met with resistance because of the added expense for the school and school district. The national PWSA office has been successful in assisting parents to obtain this aide through the crisis packet and a letter from the Executive Director that points out that the expense of not providing an aide can far exceed the expenditure of having one.

Another problem area of the IEP is in the development of a behavioral plan that is appropriate for a person with PWS. Children with PWS usually have difficulty in meeting the requirements of a classroom because of the effects of their syndrome. They may disrupt a class by having a "meltdown" because of their lack of emotional control. Often, the response to this behavior is based on a false assumption that the child with PWS has a control mechanism or can learn to gain control by receiving a consequence such as a threatened suspension. Another typical behavior exhibited may be called "stubbornness." This may play out as being oppositional, sitting down, refusing to move, or throwing something. This often occurs because people with PWS have difficulty identifying priorities. When they are confronted with stimuli from several different sources, it is difficult for them to prioritize or sort out the major from minor issues. This produces anxiety and confusion that leads to the behavior.

To assist the administrative staff and the teacher to understand the syndrome and offer suggestions for addressing the various behaviors that may be occurring, the national office provides written material, audiotapes, and videos that give insight and facts about PWS. When this approach does not help, the Crisis Counselor can contact an experienced advocate in the area to go with the parents to the next IEP meeting. If this is not possible, or does not succeed in establishing an appropriate IEP for behavior, PWSA (USA) can provide, through a grant, an educational consultant who is experienced with persons who have PWS. This person usually talks with the school representatives and the parents by telephone to obtain a clear picture of the major issues. This is typically followed by a "speaker phone" conference in which the parents, school personnel, and the educational consultant work toward a suitable behavioral and food control plan.

Legal Issues at School

In the U.S., most children with disabilities are now integrated into regular schools. At the same time, due to the rash of school shootings by students, there is now a "zero tolerance" discipline policy in many schools, which also includes disruptive behavior in general. This is creating many problems for our children with PWS. Placement in a mainstream class brings a faster pace of learning and higher expectations, which heighten the pressure and performance anxiety. It is also much more difficult to write a behavior management program into the IEP that is practical for school staff to implement. The response on the part of the school is now weighted toward longer suspensions and, if the behavior continues, may lead to expulsion. Reports of arrest for stealing food, hitting teachers, or destroying property are not uncommon. In one case, for example, a teacher filed sexual battery charges against a 9-year-old child with PWS because he grabbed her breast. In another case, a 10-year-old with PWS was taken to jail in handcuffs for kicking a teacher during a tantrum. And, of course, there are the many, many cases where students are suspended from school for taking food from the lunchroom, from other students' lunch boxes, or from the teacher's desk.

If an IEP does not have an appropriate behavioral plan for the child with PWS, it increases the likelihood of legal problems. The scenario may be that when the student with PWS has a "meltdown" the school calls the local police department. Then the child kicks the police officer! The person with PWS might be placed in handcuffs and taken to jail. We have known children as young as 7 years old being subjected to this procedure. Criminal charges may be consequently filed because of the destruction of property or physical assault. The PWSA Crisis Counselor will respond to this incident, when contacted by the parents or the school, with a telephone call to the law enforcement representative or a crisis letter addressed to the defense attorney, or to the court if there is a scheduled hearing of the case, and a crisis information packet. If the incident does not go to court, the option to return to writing an appropriate IEP is usually productive.

There have been a few occasions when the school staff person who has been struck by the person with PWS files an assault and battery charge. There have also been instances in which a sexual assault charge is filed because the person with PWS has touched an adult or another student in an inappropriate place. None of these cases was pursued after PWSA (USA) responded with calls, educational materials, and a crisis letter.

In PWSA's experience, these cases can be resolved out of court, and both the parents and the school can work for a more suitable response. Everyone must be made aware that the individual with PWS may be driven by the insatiable appetite and may not have the usual mechanisms to exercise control over their eating and emotions. It is always emphasized that individuals with PWS usually do not have the ability to stop anti-social behavior by themselves, but with assistance, they can be helped to learn socially acceptable means of dealing with the classroom stress. For further discussion of school discipline issues, from the perspective of an educational consultant, see Chapter 20.

Legal Issues in the Public Arena

On rare occasions, crisis support may be needed because the young adult or adult with PWS may get into a major legal situation related to sexual behavior. Court cases are complicated because the adult with PWS may appear to know the difference between right and wrong sexual behavior. The prosecution usually believes that the individual is competent to stand trial and is often not willing to consider leniency due to mental and social capacity. This results in a long, drawn-out legal process and months of agonizing worry for the adult with PWS and his/her family. To date, all of these cases have eventually been dismissed and the records expunged. See Chapter 21 for further discussion of sexuality issues.

Young adults with PWS can also be charged with theft. This occurs when they have tried to walk out of a store with merchandise that has not been paid for or have eaten food merchandise in the store. In these instances, the national office has been able to intercede with crisis intervention, resulting in the charges being dropped and an equitable resolution obtained. A recent example was a 21-year-old male with PWS who entered a clothing store, picked up two shirts at the back of the store, and brought them to the cashier. He explained they were gifts and he wanted to return them and get a refund. His plan was that he could then use the money for food. The clerk refused since there was no receipt and called the police. A charge of shoplifting was filed. The case was dropped after the prosecuting attorney received a crisis letter and information packet from PWSA (USA).

Supported Living Issues

The national crisis intervention program can also be helpful at a time when the family of an individual with PWS is desperate to find either residential services or a rehabilitation facility. The national office keeps contact with group homes and other facilities that are designed to provide a safe environment for persons with PWS. National office staff have written letters about the individual and the syndrome that have assisted families to obtain funding necessary for an appropriate placement, where the individual with PWS can live a happy and successful life within the parameters of the syndrome.

What has become a major dilemma in residential placement is: What is the least restrictive environment for an individual with PWS? Consider the following:

- *Under age 18*—parents can be charged with neglect for allowing a child with PWS to become so obese that it is a life-threatening situation.
- *Over age 18*—a judge can rule that a person with PWS can have the right to unrestricted access to food.
- *Mixed supportive living homes* in the United States often cannot lock up food due to the rights of other clients in the home who do not have PWS.

PWSA's position is that the least restrictive environment for a person with Prader-Willi syndrome is a home with restricted food access. The rationale is clear:

- It frees their mind from the constant striving to secure food.
- If everyone in the home has the same restrictions, they now feel more "normal" and less singled out, which reduces frustration. Less frustration equals fewer behavior problems.
- An extremely obese body with related health problems is very restrictive!
- Would you hand a person who is suicidal a gun? Food will kill our children and adults with PWS just as assuredly as a gun.
- The children's right to life is more essential than their right to free access to food.

The following case is an example of how complex some of the problems can be:

A young woman called in urgent crisis. Her sister with PWS weighs 346 lbs., is 37 years old and 5'2" tall. She was in a respite program but now is in the hospital. The case manager is skeptical of the PWS diagnosis, although she tested positive for PWS. The agency stated, "She has the right of unrestricted access to food." Her sister stated, "Her 'rights' are killing her." In a series of calls over the next 6 months, the sister reported that the case manager manipulates her sister and is making up charges of abuse against their mother. The sister with PWS has been in and out of the hospital. In May, she weighed 440 lbs., having gained 140 lbs. in 2 months in a placement. She now has a tracheotomy. Sister stated, "It's out of my hands and I have to watch it happen. It is really horrible." In July, her sister was in the hospital on a ventilator. Crisis packets, referrals, and counseling were provided to the family, hospital personnel, and physicians.

Advocacy for Services

Children with the syndrome are eligible for various services offered at the local level by national and state governments or their designees. When the child is diagnosed with PWS, it is important to obtain the assistance of a speech therapist, occupational therapist, and physical therapist to intervene in the area of low muscle tone, or hypotonia. A majority of the states offer these services through a local ARC organization or through the state Division of Disabilities/Mental Retardation. Usually it is not a problem to establish eligibility for these services, but in some cases services are denied. However, if the service is denied, it is always possible to appeal this decision. When making an appeal, the resources of the national PWSA office can often be very helpful in resolving the issue and gaining the services. A call from the Crisis Counselor to the agency social worker or administrator can provide that person with the information they need to qualify the applicant. In most instances, a letter and a crisis packet are also sent to the agency, often through the family making the application, and this provides specific details about PWS and the reasons the child is eligible for the services requested.

There have been instances in which the services have been provided in a timely manner, yet when an evaluation of the results occurs, usually at the age of 3 or 4 years, the agency proposes to stop the services because the child is doing so well. Families have been successful in achieving the continuance of these services after the agency receives appropriate educational materials and their knowledge of the syndrome is broadened.

Supplemental Security Income (SSI) is available to persons with qualifying disabilities who have income and resources that do not exceed certain limits set by the U.S. Social Security Administration. This usually means that the person with PWS who qualifies for SSI payments is also eligible to receive Medicaid or a similar state medical insurance coverage. Sometimes the national PWSA office assists in an SSI appeal because this is another instance where benefits have been denied because the person with PWS is not officially "mentally retarded"; that is, their IQ is 70 or above. To win an appeal, one must demonstrate that PWS is lifelong and very debilitating. A family with a young person with PWS should obtain information about these services even if they feel that they are not financially eligible. When the child reaches the age of 18, the income used for evaluation of eligibility is based on the individual's income only and not the family's income. See Appendix D for further guidance on SSI application and appeal.

Other organizations can sometimes provide assistance to families of children with PWS, such as special equipment that may be needed. The Shriners' Hospitals have been helpful in providing medical care for orthopedic issues, and the March of Dimes, Elks, Kiwanis, and other service organizations have assisted with equipment and other specific needs.

Conclusion

The PWSA (USA) Crisis Intervention Program began in January 2001, thanks to a 5-year grant from the Alterman Family Foundation. It has become an invaluable service to hundreds of PWS families and professionals, as reflected in the following quotes:

- "I called the PWSA office today. They are sending me information and a video. The counselor was very helpful with issues about guardianship and educating the school staff. I feel a little bit better; actually it's unbelievable, and I am overwhelmed. Looks like there may be a light at the end of the tunnel after all."
- "First of all, I can't thank you enough. Your work in helping me deal with this matter—or shall I say, nightmare—has been so supportive and informative, even at times when I just wanted to give up."
- "Thank you so very much for your recent letters and information you sent on behalf of our daughter and our SSI 'battle.' They were very nice and hopefully will make our problems with the SS Administration go away."
- "We would like to thank all of you for the time and assistance you offered us to achieve our goal of getting our son appropriately placed. We will always remember you in our thoughts and prayers."
- "Thank you so much for your letter. It helped more than you could know."
- "Thank you so much for the excellent material you sent me recently. It was a great help to us and to the judge who was hearing our case."

20

Advocacy Issues: School Discipline and Expulsion

Barbara J. Goff

Editor's note: Barbara J. Goff is a disabilities consultant specializing in Prader-Willi syndrome. The following information is based on Dr. Goff's personal experience in advocating for appropriate educational services and placement for students with PWS.

Sara was 7 years old when I received a call from her mom. It seems Sara had been receiving home-bound instruction for the previous 6 months as a result of disciplinary action taken in response to her classroom behaviors. These behaviors included: wetting herself; throwing objects; property destruction; removing socks, shoes, and other clothing; engaging in self-injurious behavior (SIB); and demonstrating aggressive behavior toward her classmates. In the meantime, the school was seeking an appropriate alternate placement. Sara's parents felt that the placement that was identified and recommended was not at all suitable for their daughter. Mom had appealed the decision and called to ask if I could assist her, as an expert in PWS, in formally assessing the proposed alternative and make recommendations to the district as to the most appropriate placement. At this point, the family had engaged a lawyer with whom I began to work.

Sam was 14 years old when I received a call from his school. He had been receiving home-bound instruction for the previous 6 months following several acts of aggression toward school personnel and attempting to leave school grounds. The school had made a few recommendations which were unacceptable to Sam's parents. The next step was mediation or a due process hearing. In an effort to avoid that and to maintain a good working relationship, Sam's parents agreed to have the school engage me as an expert in PWS to review the options presented and make recommendations for an appropriate educational program.

Introduction

The Individuals with Disabilities Education Act (IDEA, Public Law 101–476, 1990; amended 1997, 2004) appears to be straightforward regarding the rights of students with disabilities to a free and appropriate public education and the provisions of multiple necessary supports to enable that education. For many students with disabilities, the process works smoothly and allows the student to optimally achieve his/her educational goals. Not infrequently, however, families of students with Prader-Willi syndrome (PWS) report far more difficulty obtaining necessary supports and services for their child than a reading of the law would suggest. This is particularly true when the student is demonstrating challenging behaviors in the classroom. Many families indicate an apparent unwillingness on the part of school personnel to understand the behavioral triggers and to modify the program in order to minimize the occurrence of those triggers. They often feel that, instead, the school employs ineffective and academically counterproductive disciplinary actions. Some parents go so far as to suggest a purposeful inducement of these behaviors in order to effect a change in placement.

On the other hand, school personnel frequently report that, in spite of their best efforts, the student with PWS has become far too disruptive and/or aggressive to be managed in the current placement. In some cases, the school feels that the student is putting him/herself and others in a potentially harmful situation. For example, the student may be throwing objects and/or hitting others or running out of the school building into busy roads. All too often, the school's management of these behaviors is not rooted in a thorough understanding of the syndrome, and comprehensive training and additional supports are needed. However, some schools simply do not have the resources to appropriately and safely serve such a student in the present setting and seek alternative, more protective settings.

This chapter reviews a number of typical scenarios reported by parents and school personnel along with options under IDEA for improving the educational environment for students with Prader-Willi syndrome.

Causes and Process for Disciplinary Actions and Alternate Placement

The national office of PWSA (USA) received 35 calls during a recent year from families regarding disciplinary actions (up to and including alternate placements) being taken against their children. The reasons for such actions included property destruction, food stealing, disruption of the class, hitting other students, and physically abusing personnel.

In most cases, the events that elicited the aggressive acts resulting in expulsion from the placement related directly to characteristics of PWS and, as such, were largely preventable. Following are some examples:

- A male teen was told that he couldn't go on a particular field trip as a consequence of recent unacceptable behavior. The field trip involved a picnic lunch. Upon receiving this news the young man had an outburst culminating in grabbing the phone out of the teacher's hand and attempting, with some success, to hit her with it as she was calling for assistance. While no serious physical harm resulted, the teacher was quite upset with the severity of the student's reaction and pressed charges. Those familiar with people with PWS would have predicted such an extreme response to that which amounts to the taking away of food and would have handled the situation differently.
- A paraprofessional kept insisting that a student with PWS "hurry up" and put an (unfinished) worksheet away and move on to the next class. The child with PWS ripped up his paper, screamed at the paraprofessional, and stormed out of the room.
- A teacher failed to reassure a child who was obsessing about an upcoming event. The child continuously interrupted the class to repeat her concern and, given an inadequate response, she left the classroom and marched directly to the counselor's office where she insisted on using the phone to call her mother to get the reassurance she needed.

In analyzing the specifics of the reports on disciplinary action made to the PWSA (USA) office, some interesting patterns emerge. First, there appears to be no relationship between gender and disciplinary action in general in terms of students with PWS; however, males were more likely to be expelled from their current school-based program and be recommended for alternate placements. In most cases the precipitating events involved acts of aggression. In 7 cases of alternate placements reported to PWSA (USA) in one year, 5 involved physical aggression toward a teacher, paraprofessional, or administrator; 1 involved an act of inappropriate touching of a teacher; and 1 had to do with possession of drugs on school grounds. All of the perpetrators were male, ranging from 12 to 18 years of age.

Being in a separate special education class does not appear to have preventive or protective value with regard to recommendations for alternate placements. Six of the 7 students expelled were full-time participants in special education classes, with 1 participating in both special education and general education classes.

There was no discernable pattern with regard to cognitive ability, which isn't surprising since the majority of students with PWS fall into the mild mental retardation range. While the grade level at which such serious action is taken is typically in middle or high school, even very young children have been recommended for alternate placements, as Sara's story indicates.

There does appear to be a pattern in the target of the students' transgressions. In all cases of alternative placements, school personnel—not other students—were the targets of aggression. These data suggest, and experts in the field have observed, that the teacher or other adult personnel are most often the source of frustration. Hence, the notion that the student with PWS will hurt other children may be unfounded in many instances. This can be seen as a positive. Most parents can easily understand another parent's concern that their child is being assaulted or bullied in school and would be hard put to defend such behavior. However, where the target of aggression is a professional adult, there is hope and opportunity for prevention.

The typical story goes something like this: The child has engaged in a series of disruptive incidents over time and/or has engaged in a life-threatening behavior towards self or others. Some examples of disruptive behavior include consistent refusal to complete assignments; sleeping through classes; refusal to transition; routinely interrupting the teacher or other students; severe skin picking requiring intervention; periods of screaming, yelling, or crying; and frequently leaving the classroom to address some concern or to escape the situation. Examples of life-threatening behavior include physical aggression directed toward teachers or students, repeated attempts to leave the school building and grounds, and placing oneself in dangerous situations such as heavy traffic or unstable neighborhoods.

What is important for parents to understand is that schools may discipline but not totally expel (cease providing educational services in *any* setting) a student if the student's unacceptable behavior was a "manifestation" of the student's disability. This is often the case for children with PWS who have disciplinary problems at school. However, students with disabilities can be disciplined in the same manner as nondisabled students if their behaviors are not disability-related.

Whether a child with PWS is suspended or expelled from the school site, he/she is still entitled to a free and appropriate education under IDEA.

Suspension

Suspension from school can be used as a disciplinary measure for up to 10 days per school year before the school is required to provide an alternative academic program to a student with disabilities. This is justified as a "cooling off" period. Longer-term suspension or removal to an interim alternative education setting for more than 10 days (but not more than 45 school days) can occur when the child has carried a weapon to school, possesses or uses illegal drugs, has "inflicted serious bodily injury" on another person, or if remaining in the current placement presents a substantial likelihood that injury will result to the child or others (the latter requires approval by a hearing officer).

Should a long-term suspension or change in placement be recommended by the school, there must be a meeting to determine whether the behavior that led to the suspension is a *manifestation* of the student's disability or the "direct result" of the school's failure to implement the IEP. IDEA requires that a "manifestation determination review" hearing be held within 10 school days of any decision to change the placement of a child with a disability because of conduct violations. The Manifestation Review Team typically consists of the members of the Individualized Education Program (IEP) team and "other qualified personnel."

In the case of Prader-Willi syndrome, "other qualified personnel" may include an expert in Prader-Willi syndrome—someone who can review the records, observe the child, conduct interviews, and make a determination as to whether or not the cause for suspension was a result of the child's having PWS. Parents may want to consider contacting an educational advocate as well to assist them in understanding the laws and process related to a manifestation hearing. Parents do have the right to appeal the decision of the Manifestation Review Team, during which time the child remains in his present placement unless other options are agreed upon by the school and family. (Exceptions to this include removal for behaviors involving weapons or illegal drugs on school grounds, and those resulting in "serious bodily injury" to another person or substantially likely to cause injury to self or others.)

Assessment and Behavior Plans

When a student is subject to any kind of discipline beyond the short-term (up to 10 days) suspension, the school must conduct an assessment and develop a plan to address the student's behavior. Such an assessment is referred to as a functional behavioral assessment (FBA), with the plan referred to as a behavioral intervention plan. If none existed prior to the disciplinary action, the IEP team must act immediately (within 10 school days). If an FBA was done previously and a behavioral plan exists, the IEP team must review and modify it as appropriate to address the behavior.

In either case, an expert in Prader-Willi syndrome may be useful in assisting the school personnel in understanding the functions of the child's behaviors, which may not be easily identifiable but are intrinsic to the nature of the disability.

It is highly unlikely that a school would attempt to expel a student with PWS from receiving educational services altogether, but it is quite common for a school to recommend alternate placements.

Parents' Right to Due Process

Following any decisions by the school regarding a child's identification, evaluation, placement, or other provisions of a free appropriate public education for children with disabilities, the parent has the right to appeal. The parents or their attorney may file a complaint against their state education agency (SEA) or their local education agency (LEA), specifically stating the reason for such action. Prior to a due process hearing regarding the complaint, the SEA must allow the parents and the SEA or LEA to resolve their disputes through mediation. This is a desirable option as it avoids the cost and time of a court trial and may serve to improve communication between parties. Generally, when cases make their way to the courtroom, the relationship between the school and the family has suffered a great deal of damage, with the ultimate victim being the student.

If mediation is unsuccessful, the parents can pursue their complaint through an "impartial due process hearing" conducted by the SEA or

LEA. The participants in the hearing would include an impartial hearing officer, representatives from the LEA, the parents or guardians, attorneys for either party, if desired, and any individuals with special knowledge regarding children with disabilities who can present evidence or opinion pertinent to the case (e.g., an expert in Prader-Willi syndrome). Parents can continue the appeal process up to the federal court level, if necessary.

Alternate Placements

Generally, the process of alternate placement begins with the student persisting in low levels of unacceptable behavior, such as work refusal, speaking out of turn, perseverating, skin picking, etc. Meetings are held, recommendations made, consultants called in, and behavior plans developed; yet the behaviors persist, perhaps even worsen. Finally the child engages in some kind of aggressive or otherwise dangerous behavior that becomes the "last straw." The child is suspended indefinitely, provided with approximately 2 hours per day of home-bound instruction and any special services identified in his/her IEP. During this period, the school investigates more appropriate placement options and parents become increasingly frantic.

When PWS specialists become involved in such situations, it is usually because the school recommended a program that is unacceptable to the family. This is what occurred with Sara and Sam (see beginning of this chapter). In both cases, as the Prader-Willi syndrome expert involved, I agreed with the parents (although this may not always be the case). Typically, the alternative placements recommended are special schools for children with severe behavior problems. The intent of these schools is to teach the child how to control his/her behaviors so that he/she may return to a public school placement. These programs may have moderate success for children diagnosed with emotional and behavioral disorders. However, it is certain that children with Prader-Willi syndrome do not respond to the behavior modification programs these schools utilize. While the programs are in place, there may be behavioral improvement, but these changes are not internalized as part of a permanent behavioral repertoire. When the program is withdrawn, the characteristic behaviors of PWS return.

One young man with PWS was in a residential school program where aversive therapy was used. The therapy included electric shocks, noxious odors, irritating sounds, water sprayed in the face, and other aversive consequences to reduce or eliminate undesirable behaviors. The students at this school typically exhibited severe aggression and/or severe self-injurious behavior (SIB) over time. The school accepted only those students for whom every other effort at behavior change had failed. The targeted behaviors for the young man with PWS included lying, food seeking, and skin picking. The records indicate that he did demonstrate improvement during his stay at the school; however when he was transferred to a group home for individuals with PWS, all of the targeted behaviors returned with just as much frequency and intensity (if not more) as existed prior to the use of aversive therapy.

Are the school districts being negligent or punitive in making these recommendations? I don't believe so. What I find to be the case is that school personnel simply do not understand the syndrome and view the student as a child with an eating disorder and behavior problems. Subsequently, they seek out the best programs serving children with similar characteristics. The professional staff at these special schools may be excellent and well-meaning in their willingness to work with children with PWS, but they lack an accurate understanding of the syndrome.

Indeed, every effort had been made to maintain Sara and Sam in their current public school placements, including employing consulting psychologists to develop a functional behavioral assessment and behavior plan, increasing staff support, decreasing the length of the school day, and a myriad of other unsuccessful efforts. The missing ingredient was the failure to engage the services of someone with an expertise in PWS, both to train personnel and to consult. Most parents do a fabulous job at working with their school staff and providing any information available, but they may be less able to articulate broader concerns that affect the educational experience of a child with PWS.

Making the Best Decision for the Child

So, what now? What happens when the parents and the PWS specialist agree that the options presented are inappropriate? Where will the child be educated? What can the school and family agree upon?

Some schools are eager to provide the most appropriate education for the child with PWS, no matter what it takes. That was the case with Sam. With school personnel, family, and the PWS specialist working closely together, a carefully crafted program housed within the school itself was created. They have made a commitment to pursue specialized training in Prader-Willi syndrome in order to ensure the success of this young man.

For Sara, and many other children, an alternative setting was and is a reasonable option and should be pursued. It may be that the child's current placement team has lost enthusiasm and commitment for working with the child and it would be better to let go, as was true for Sara. However, schools designed to serve children with severe behavior problems with the goal of returning the child to his/her previous setting usually are not the answer. Both the school and family need to look at programs within public schools that can provide the academic and behavioral supports needed and have staff who are willing to be trained in Prader-Willi syndrome. Where such programs do not exist, alternative schools designed for children with severe learning disabilities may be the right choice. These children tend to have concomitant behavior problems which the school personnel are trained to understand and address. And since they are not schools specifically designed for children with behavior disorders, their approaches to behavior management are often more individualized and flexible.

Eventually, Sara was placed in a school for children with special needs and flourished. She still exhibits many of the behaviors that caused the transfer in the first place, but they are managed quite differently and are much less frequent and intense as a result. For the first time, Sara is experiencing success, a sense of community, and a feeling of safety. She looks forward to school and works diligently at being a well-rounded student. This same experience holds true for many other students with PWS who exit inclusive, or even self-contained, programs in public schools and find themselves in schools for children with a variety of learning problems.

While inclusion is the educational philosophy of today, and rightfully so, some children find it extremely difficult to negotiate all the demands of a public school setting, even with special supports and services. Not all children with Prader-Willi syndrome are alike and require identical programs. Not all children can handle the emotional experience of being so different from their classmates. On the other hand, some are successful in inclusion classes throughout their school career and wouldn't have had it any other way. Others do fine for the first several years of schooling and then signs of unhappiness, perhaps regression, on the part of the child (and the teachers) begin to appear. This typically starts in third or fourth grade, when the academic and social gaps between the child with Prader-Willi syndrome and those without disabilities widens dramatically.

If a school cannot create a positive and effective learning environment for a child, for whatever reasons, it may be time to look elsewhere and/or create something that does not yet exist.

Recommendations and Conclusion

Preventing disciplinary action for a student with PWS necessitates the good faith of all concerned: parents, teachers, administrators, clinicians and anyone else involved in the student's education. Many of these youngsters go through school with only minor bumps along the way, while others seem to encounter major obstacles at every turn. For these students, specific attention must be paid to the unique characteristics of PWS and how they interact with the learning environment. Where parents and professionals choose to collaborate in good faith, there are amazing outcomes.

For those who are parents, here are some tips for developing effective partnerships with school personnel so that disciplinary action may be prevented:

- Genuinely try to connect with the educators to help them in getting to know your child.
- Be a proactive participant on your child's team.
- Be an advocate for your child *and* the school.
- Validate the teacher's efforts; send written compliments with copies to their administrators.

- Ask the teacher when (daily, weekly, afternoons, evenings) and how (notes, phone calls, e-mail, personal visits) he/she would like to communicate.
- Volunteer to provide supervision on field trips to ensure that your child can successfully participate.
- Communicate, collaborate, and coach—even compromise—instead of criticizing and complaining.
- Pay attention to the early warning signs of discontent, either from the teachers, school, or your child.
- Assist in developing behavior management strategies that work for your child; clearly identify those that do not.
- Recommend specialized training (in addition to the training and materials you routinely provide) for personnel involved with your child before negative feelings and attitudes develop. (The national office of PWSA-USA can assist you in locating qualified consultant/trainers.)
- Search your school district for alternative programs and visit them. Should the school decide that they can no longer serve your child in the current setting, they will tell you what they believe to be an appropriate alternative—not necessarily apprise you of all the programs that are available. Consider out-of-district placements if the distance is not prohibitive.

Finally, you may need the help of an educational advocate, attorney, and/or education specialist in PWS if you decide to appeal any decisions the school has made. Educational advocates are very helpful in interpreting laws and ensuring that you are aware of your rights as a parent. Attorneys with a specialty in education and disability law can direct the appeal process in the most beneficial manner. Education specialists with an expertise in PWS can provide testimony and recommendations regarding the syndrome in general and your child in particular with regard to the most appropriate programs and services.

As the parent, you are vulnerable and should not expect yourself to carry the entire burden of contentious placement decisions. Seek support in a timely fashion. The decisions made about your child's program will have an impact on his entire school career.

For those who are teachers and professionals, you are entitled to adequate training, supports, and services to appropriately educate children with special needs.

As a cohesive team, we are most likely to achieve the mandate of a free and appropriate public education for children with disabilities in the least restrictive setting.

21

Advocacy Issues: Sexuality

Janalee Heinemann, David A. Wyatt, and Barbara J. Goff

In the past, the primary caregiving concern for those with Prader-Willi syndrome was preventing life-threatening obesity. Although, of necessity, this remains the first priority, better understanding and management techniques leading to significantly lengthened life spans now allows for an additional focus on the quality of their lives. The joint focus on preventing life-threatening obesity through restrictions and limitations, while at the same time attempting to mainstream affected individuals into society, forces many to deal with previously latent ethical and legal issues. The primary issue of sexuality has placed many parents and professionals at odds and piqued new levels of discomfort.

Although sexual maturation in those with PWS is rarely complete, it does not mean that a person with Prader-Willi syndrome is asexual. Indeed, sexuality is a "state of being" that incorporates gender identity, relationships, intimacy, and self-esteem as well as physical development and hormones.[1] We are all sexual beings, including persons with PWS—who sometimes need intimate contact, sometimes dream about a romantic relationship, and always are influenced by what is on TV, read, or observed.

Until recently, it was believed that infertility in this population was universal. Parents, particularly mothers, of individuals with PWS have quipped, "We are the only parents who worry more about our teen and adult children going into their dates' kitchen than into their bedroom!" A new reality has emerged. Published reports now document three pregnancies: two young women with the chromosome 15 deletion form of PWS have given birth to children with Angelman syndrome (see Figure 21.1); the third young woman, whose PWS originated from maternal disomy, had a child who appears to be typically developing.[2,8,9] Since the ability to adequately parent is unlikely in even the most capable of persons with PWS, most parents and providers do not wish to deal with either of these situations.

This then raises a number of questions: What are the parameters of appropriate and allowable expressions of sexuality? Who decides—parents or professional caregivers? If sexual activity is allowed, what

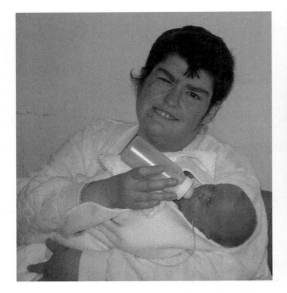

Figure 21.1. A young woman with Prader-Willi syndrome and her daughter with Angelman syndrome (shown in infancy at left and at 11 months of age at right). Both mother and daughter were diagnosed by the FISH test (fluorescence *in situ* hybridization), which identifies chromosome deletions. A mother with the classic PWS deletion on her paternal chromosome 15 has a 50% chance of contributing a defective chromosome 15 to her child; since the child receives it as a *maternal* chromosome deletion, the child develops Angelman syndrome instead of PWS (see Part I of this volume for details on the genetics of PWS and AS).

kind of protections should be taken? Who does the teaching? Who does the monitoring? When an adult child with PWS lives away from home in supported care, how are parents assured that protection is really being utilized since many young women with PWS express a longing to get pregnant?

A logical corollary issue is that of marriage or cohabitation. When couples with PWS want to get married or live together, what is their right to have this type of relationship? Is it so wrong for young adults with PWS to want a romantic/sexual life? Some parents ask, "Isn't it bad enough that they have to spend every day of their lives watching everyone around them enjoying the pleasure of eating? Do they also have to be deprived of this other basic pleasure and need in life?" On the other hand, others ask, "What is their potential for maintaining a long-term relationship?" These are a few of the many questions facing parents and caregivers for which there are no easy answers.

An even more troublesome situation might be when an individual with PWS is romantically involved with someone who does not have PWS. This is not an unusual situation for two reasons. Many individuals with PWS do not want to be identified by their disability and establish relationships with people who are not as readily identifiable. The other reason is that the person with PWS knows that their non-PWS partner will have greater access to food and money. One parent reported that her daughter was dating a young man she knew through her workshop. He did not have PWS and routinely provided her with snacks, either brought from home or purchased with his money. The mother liked the young man and spoke directly to him about the harm he could be causing by supplying her already obese daughter with food. He replied that he understood and he cared very much about the well-being of his girlfriend, but if he didn't bring her what she wanted, she wouldn't be his girlfriend anymore. He was quite correct.

While cognitive or emotional capacity is never the defining criterion for marriage or parenthood, cognitive capacity does impact the ability to understand right and wrong and allowable and disallowable moral and sexual behavior. A lack of understanding has led to problems of sexual charges against some young people with PWS. One instance involved a mid-30s couple, both with PWS, who had been dating two years. Although they were mutually consenting adults, when the young man's fondling of his girlfriend caused slight bleeding in her vagina, she was taken to the hospital emergency room, and he was charged with "felony, first-degree sexual assault and battery." A much better response would have been to provide counseling to the couple with regard to their sexual expression.

Like most issues, the answer to what is acceptable sexual behavior is not a simple one. Some of the intertwining issues are listed below:

- *Family's religious and moral beliefs*—The family may be strongly opposed to sex outside of marriage, yet marriage may be out of the question due to lack of maturity and the questionable ability to maintain a long-term relationship.
- *Sexually transmitted diseases*—How do you protect a vulnerable adult from picking the wrong partner? How can you be assured that they are adequately protected from sexual diseases?
- *What is appropriate sexual behavior?* Help from the Crisis Intervention Counselor at PWSA (USA) has been sought for situations in which poor behavioral choices have led to legal charges. Who is at fault when a person with PWS observes through peers, movies, and TV that others are getting affection and pleasure in more sexual ways? Then, wishing to act normally, they subsequently make similar attempts that are unsophisticated—and perhaps socially unacceptable—and that unexpectedly result in legal trouble.

The Role of the Service Provider

Because residential service providers have a mandated responsibility to address issues of sexuality for adults with PWS, the following are responses, from a provider's perspective, to the questions posed at the start of this chapter.

What are the parameters of appropriate and allowable expressions of sexuality? This is dependent on the individual and his/her desires and capabilities. Where there is a question regarding competency to knowingly engage in sexual activity, a provider is responsible to conduct an assessment to determine competency and provide supports and supervision accordingly.

Who decides—parents or professional caregivers? Parents, quite naturally, wish to guide the lives of their children according to specific religious or moral beliefs; providers, while respecting those beliefs, must also adhere to established agency and state standards of care, which include matters of sexuality and sexual expression. While there may never be complete agreement as to what is best for the individual with PWS, there must always be a process and a dialogue that keeps in focus the health, happiness, and fulfillment of the individual with Prader-Willi syndrome.

Again, if the individual is an adult living in a residential program, competency must be established. If the person is deemed competent, then the policies of the agency and/or state must be adhered to. In some cases, sexual activity is prohibited within the residence itself, but other options may be explored. In other instances, individuals who are deemed incompetent to consent to sexual activity may be required to receive a certain level of supervision if sexual activity is suspected. In most all cases, the agency is obligated to provide sex education commensurate with the capacity of the individual(s) to understand and benefit. This includes the use of protective devices.

What if the parent has guardianship? Having a guardian does not mean that an individual has to abide by their demands regarding sexual expression or lack thereof. The specific allowable interventions from a guardian vary from state to state and may, or may not, include influences into the areas of sexuality and sexual choices.[10]

If sexual activity is allowed, what kind of protection should be used? Any type of protection to be utilized during sexual activity would need to be medically approved and recommended and be of the type that the individual is most likely to use consistently and effectively.

Who does the teaching? The teaching of sexual information is delivered by the person(s) deemed to be most appropriate. For example, if a person has severe cognitive deficits, a direct-service staff person with whom the individual has a close relationship and who possesses some level of expertise might be most appropriate. It might be prudent to have an established protocol wherein the staff person would function primarily as a conduit to introduce the individual to a sex educator/counselor who specializes in these issues. In this way, staff are protected so that accusations of inappropriate behavior may be avoided. The staff member would reassure the individual that this counselor is safe to talk with in private. A counselor is needed to work on building relationships, not just to discuss sexual activity.

Who does the monitoring? Generally, it is the residential staff that monitors any developing romantic or sexual relationships. They do so in order to ensure the safety and well-being of the individuals involved. Typically, there are agency guidelines to follow and supervisory staff to provide oversight with regard to sexual activity. If a sexual relationship is perceived as harmful, it most certainly will be stopped and investigated immediately. If it is assessed to be consensual and not harmful physically or emotionally, then most agencies would seek to provide education and counseling as appropriate.

When an adult child with PWS lives away from home in supported care, how are parents assured that protection is really being utilized, since many young women with PWS express a longing to get pregnant? If a sexual relationship is being permitted after recommended assessments are completed and appropriate education provided, then the provider agency can only depend on the willingness of the individual to be honest and forthcoming in their discussions of safe sex. If agency staff suspects failure to follow recommended and agreed-upon protective measures, they may prohibit sexual activity until or unless the situation is rectified.

These responses are typical but not reflective of the practices of every provider agency. It is suggested that parents inquire as to any policies, procedures, or prohibitions a particular agency may have with regard to sexual activity prior to their child's placement.

The issue of cohabitation or marriage is, again, one to be made on a case-by-case basis. Many agencies will design a residential program that would support a couple choosing to marry. However, such a serious decision would not be supported without a great deal of discussion, counseling, and education. Parent support for such relationships is a very powerful indicator of whether or not it will be successful.

Sexual Abuse

Even when parents and caregivers agree on allowable sexual expression, the issue of sexual abuse remains a critical concern that is uncomfortable to discuss but a reality with which we all must deal. Current statistics indicate that

- In the general population, at least 1 in 4 girls and 1 in 10 boys are molested before age 18.[12] The risk is significantly increased for those with disabilities.
- The average age of all molestation is around 8 years. Because males (and less obviously females) with PWS have more childlike genitalia, even adults with the syndrome remain at higher risk for abuse by pedophiles.
- The offender is often the person one would least suspect. National data indicate that the majority of offenders are not psychopathic and many are considered "pillars of the community."[3,6]

Adult Sexual Abuse

In a study conducted in Connecticut over a 5-year period and published in 1994,[5] researchers reviewed 461cases involving alleged sexual abuse of adults with intellectual disabilities. Of the 171 (37%) confirmed cases, 48% of those abused had mild mental retardation, 28% had moderate mental retardation, 15% had severe mental retardation, and 9% had profound mental retardation. Most were able to communicate verbally and had few, if any, secondary disabilities (e.g., blindness, physical disabilities). Based on this description, we can see that a higher IQ is not protective.

Other findings of the study included the following:

- 72% of cases involved female victims, and 28% involved males.
- Average age of the victim was 30.

- 88% of perpetrators were men.
- Most abuse occurred in the victim's residence: institution 25.2%, group home 26.9%, family home 14.6%, community 7%, supervised apartment 6.4%, sheltered workshop 5.9%, other settings 14%.
- 92% of the victims knew their abuser.

Childhood Sexual Abuse

In one study investigating sexual abuse of females with mild mental retardation, 80% reported having been sexually abused at least one time.[11] Estimates of sexual abuse against males with developmental disabilities have been as high as 31%.[4]

In another study of 43 children with global developmental disabilities who had been sexually abused, the following statistics were reported[7]:

- 46% were abused by a family member (compared with 38% for non-disabled children).
- 36% were abused by a nonfamily member (compared with 52% for nondisabled children).
- 18% were abused by both a family member and a nonfamily member (compared with 10% for nondisabled children).

The perpetuation of child sexual abuse is aided silently by two factors: (1) Most children don't tell, due to fear, guilt, shame, and concern for others; and (2) most parents and professionals don't suspect—or don't believe the child when they do tell.

The failure to report occurs with increased frequency in those with cognitive disabilities. Many have insufficient language or understanding to adequately report. They may not even perceive or experience that what occurred was wrong. Just as often, the perpetrator threatens harm if the abuse is reported. As a result, many demonstrate an acute behavioral deterioration. Possible sexual abuse as the etiology of this deterioration is frequently overlooked resulting in misdirected treatment. Often the facts surrounding the sexual abuse emerge much later and frequently after others have also been victimized.

Why Children and Adults with PWS Are More at Risk

Individuals with PWS are especially vulnerable to being sexually exploited and abused for several reasons:

1. An uncritical trust of everyone—They generally have an emotionally and socially innocent nature.

2. An increased dependency on others—They lack control over their environment and thus are forced, due to their situation, to place an inordinate amount of trust in those who care for them.

3. An increased vulnerability to anyone who befriends them—Often it is difficult for them to socialize and communicate, thus they are more vulnerable to anyone expressing kindness. Many molesters use the "special friend" and "shared secret" approach.

4. An increased vulnerability to bribes—Anyone who would give food secretly to a child or adult who has Prader-Willi syndrome would

be considered a friend for life, even if this friend was sexually abusive.

5. A normal need for physical attention, often coupled with a lack of adequate sexual education or appropriate outlets—Children with PWS typically receive and give many hugs and kisses. However, many students with PWS are not included in school-based sex education programs, as it is assumed the information is not relevant or will not be understood. As they grow into adolescence, the typical socializing and dating experiences are not available to them, and there may be problems during this period and adulthood with inappropriate sexual behavior.

6. An increased fantasy life—Being more socially isolated than the average child, people with PWS often build up great romantic fantasy lives of which a molester can take advantage. Also, their inability to think abstractly in many cases means they will be less able to make appropriate moral and social decisions.

7. An increased likelihood to be in settings where other adults have control over them—Foster homes, group homes, institutions, camping programs, and other supervised settings are all potential settings for molestation.

Why Parents and Providers Are Also at Risk

The previous section addressed the issue of why children and adults with PWS are at increased risk for sexual abuse. These same characteristics frequently cloud the issue of believability when a person with PWS reports such abuse, since a well-documented characteristic of children and adults with PWS is purposeful lying when angry, in an effort to hurt or get revenge against the perceived offender. And, as mentioned, many have active fantasy lives to compensate for the lack of social and emotional stimulation in their real lives, which can lead to made-up stories about adults that may seem real in their minds. As a result, some parents and providers have been unfairly charged with having committed physical or sexual abuse. It is this combination of increased vulnerability and increased propensity to assert an imagined, self-serving "truth" that makes it difficult to sort out the facts when allegations are made.

Conclusion

Providing a quality life for a person with PWS has layers of complexity and ethical issues that are not easily resolved. The rights and responsibilities regarding sexuality will continue to require a combination of an open mind, common sense, and ongoing dialogue among parents, caregivers, and most importantly, the person with Prader-Willi syndrome.

Perhaps the primary question remains: Who has the right to make these decisions? As with the general population, when considering the issue of sexuality for those with PWS, every parent, every caregiving agency, every ruling judge, and indeed, every culture and every involved person has a different opinion regarding what is fair and appropriate and correct. Ultimately the decision must hinge on "what is best" for the person with PWS. And, just as often, what is best may not be what is "wanted" by that person or easiest for those providing care.

This chapter has dealt briefly with several dimensions of the sexuality issue: the need to recognize the facts and fictions surrounding our beliefs about sexuality and people with PWS; the increased risk for abuse associated with those with disabilities; and the complexity and challenges of supporting individuals with PWS in fulfillment of their hopes, dreams, and wishes. There are many ethical and legal questions surrounding the issue of healthy sexuality and sexual expression for those with PWS. Where care is shared between families and providers, both parties must come together with the individual to determine what makes sense for that person at a particular point in time. There are no easy or right or wrong answers—no simplifying blacks or whites—just shifting shades of gray.

References

1. Ailey SH, Marks BA, Crisp C, Hahn JE. Promoting sexuality across the life span for individuals with intellectual and developmental disabilities. *The Nursing Clinics of North America*. 2003;38(2):229–252.
2. Akefeldt A, Tornhage CJ, Gillberg C. A woman with Prader-Willi syndrome gives birth to a healthy baby girl. *Developmental Medicine and Child Neurology*. 1999;41(11):789–790.
3. Armstrong L. *Kiss Daddy Goodnight*. New York, NY: First Pocket Books; 1979:179.
4. Finkelhor D. *A Sourcebook on Child Sexual Abuse*. Newbury Park, CA: Sage Publications, Inc.; 1986.
5. Furey E. Sexual abuse of adults with mental retardation: who and where. *Mental Retardation*. 1994;32:173–180.
6. Justice B. *The Broken Taboo: Sex in the Family*. New York, NY: Human Sciences Press; 1979:16.
7. Mansell S, Sobsey D, Moskal R. Clinical findings among sexually abused children with and without developmental disabilities. *Mental Retardation*. 1998;36:12–22.
8. Prader-Willi Syndrome Association (New Zealand), Inc. *Pickwick Papers*. 2004;15(3):3.
9. Schulze A, Mogensin H, Hamborg-Petersen B, et al. Fertility in Prader-Willi syndrome: a case report with Angelman syndrome in the offspring. *Acta Paediatrica*. 2001;90(4): 455–459.
10. Schwier KM, Hingsburger D. *Sexuality: Your Sons and Daughters with Intellectual Disabilities*. Baltimore, MD: Paul H. Brookes Publishing Co.; 2000.
11. Stromsness MM. Sexually abused women with mental retardation: hidden victims, absent resources. *Women and Therapy*. 1993;14:139–152.
12. U.S. Department of Health and Human Services. *Child Sexual Abuse: Intervention and Treatment Issues* [user manual online]. 1993. Available from: National Clearinghouse on Child Abuse and Neglect Information (NCCAN), http://nccanch.acf.hhs.gov. Accessed January 18, 2005.

Appendices

Appendices

Appendix A

First Published Report of Prader-Willi Syndrome

Note: The first published description of the condition that we now know as Prader-Willi syndrome was authored by Swiss doctors Andrea Prader (Figure A.1), Alexis Labhart (Figure A.2), and Heinrich Willi (Figure A.3) and appeared in the journal *Schweizerische medizinische Wochenschrift* (*Swiss Medical Weekly* [*SMW*]) in 1956. Thanks to Dr. Urs Eiholzer, a protégé of Dr. Prader, and with copyright permission from the publisher of *SMW*, we are able to reproduce that first report in the original German as well as in an English translation with footnotes, commentary, and historical photographs. Dr. Phillip D.K. Lee assisted with editing the translation and footnotes.

Figure A.1. Andrea Prader, 1919–2001.

Figure A.2. Alexis Labhart, 1916–1994.

Figure A.3. Heinrich Willi, 1900–1971.

Ein Syndrom von Adipositas, Kleinwuchs, Kryptorchismus und Oligophrenie nach myatonieartigem Zustand im Neugeborenenalter*

A. Prader, A. Labhart, und H. Willi
(Universitätskinderklinik, Zürich)

Es handelt sich um ein Syndrom von Kleinwuchs, Akromikrie, Adipositas und Imbezibillität, dem im Säuglingsalter regelmässig eine extrem schwere Muskelhypotonie vorausgegangen ist. Neben variablen kleineren degenerativen Merkmalen findet man beim Knaben regelmässig ein hypoplastisches, flach verstrichenes Skrotum mit inguinaler oder abdominaler Hodenretention.

Bisher haben wir dieses Syndrom bei 5 männlichen und 4 weiblichen Patienten beobachtet. Der älteste Patient ist 23jährig und die älteste Patientin 15jährig. Die übrigen sind 5–10 Jahre alt. Jüngere Patienten haben wir vorläufig nicht miteinbezogen, da sie noch nicht das volle klinische Bild erkennen lassen.

Alle diese Patienten hatten als Neugeborene eine extreme Muskelhypotonie, die sich darin äussert, dass die Kinder fast ganz bewegungslos und schlaff daliegen und weder schreien noch saugen können, so dass einen längere Hospitalisierung notwendig ist. Die Sehnenreflexe sind in diesem Zeitpunkt nicht oder nur schwach auslösbar. Die Diagnose lautet regelmässig "Lebensschwäche" oder "Myatonia congenita". Nach einigen Wochen macht sich wider Erwarten eine leichte Besserung bemerkbar, doch dauert es Monate, bis die Säuglinge schreien und sich kräftig bewegen können.

Wohl als Folge dieser sich nur ganz allmählich bessernden Muskelhypotonie lernen die Kinder erst mit 1 Jahr sitzen und erst mit 2 Jahren gehen. Während die Hypotonie und Adynamie zusehends bessern, tritt ungefähr um das 2. Jahr die Adipositas auf, und gleichzeitig werden der Wachstumsrückstand und die Oligophrenie deutlich bemerkbar.

Neurologisch findet man nach dem 5. Jahr noch eine geringfügige Muskelhypotonie und eine gewisse motorische Unbeholfenheit, jedoch ein normales Reflexbild. Der Kopf ist im Verhältnis zur Körpergrösse eher klein. Im Röntgenbild fehlen signifikante Sellaveränderungen. Die dreimal durchgeführte Luft-und Elektroencephalographie ergab unauffällige Befunde.

Stoffwechseluntersuchungen konnten leider nur bei der Hälfte der Patienten durchgeführt werden. Der Grundumsatz ist normal. Mit Ausnahme des ältesten Patienten, bei dem mit 17 Jahren ein Diabetes mellitus aufgetreten ist, ergibt die Prüfung des KH-, Elektrolyt- und Wasserstoffwechsels mit den üblichen Untersuchungen normale Befunde. Zeichen einer Hypothyreose fehlen. Die Pubertätsentwicklung scheint verzögert und unvollständig zu sein. Die 17-Ketosteroide der älteren Patienten sind auffallend tief. Die Gonadotropinausssscheidung des 23-jährigen Patienten ist erhöht, d.h. es besteht wohl als Folge des Kryptorchismus ein hypergonadotroper Hypogonadismus. Der Vaginalabstrich des 15jährigen Mädchens zeigt eine deutliche Östrogenwirkung. Es scheint also keine Hypophyseninsuffizienz, sondern eher noch eine Hypothalamusstörung vorzuliegen. Bezüglich Ätiologie konnten wir bis jetzt weder für die Heredität noch für eine Embryopathie genügend Anhaltspunkte finden.

Zusammenfassend glauben wir, dass es sich um ein nicht so seltenes, gut abgegrenztes, einheitliches klinisches Syndrom handelt. Beim Säugling und Kleinkind erinnert es an die Myatonia congenita Oppenheim. Im Schulkindalter und später an die Dystrophia adiposogenitalis Fröhlich, an das Laurence-Moon-Biedl-Syndrom und an den hypophysären Zwergwuchs. Trotz mancher Ähnlichkeit lässt es sich aber von allen diesen Syndromen deutlich unterscheiden.

English translation:

A Syndrome Characterized by Obesity, Small Stature, Cryptorchidism and Oligophrenia Following a Myotonia-like Status in Infancy

A. Prader, A. Labhart and H. Willi
(Zürich Children's Hospital)

The syndrome to be described is characterized by small stature, acromicria,[1] obesity and imbecility, regularly preceded by extreme muscle hypotonia in infancy. Apart from variable minor degenerative characteristics, one generally finds in boys a hypoplastic, flat scrotum with inguinal or abdominal retention of testicles.

So far, we have found this syndrome in 5 male and 4 female patients. The oldest patient is 23 years old, the oldest female patient 15 years. The others are between 5 and 10 years old. For the time being, we have not included younger patients, since they do not present the entire clinical picture.

As neonates, all these patients had suffered from extreme muscle hypotonia, leading to the children lying almost entirely motionless and floppy, not being able to either cry or suck, resulting in prolonged stays in hospitals. Hardly any tendon reflexes can be found at that stage. Typically, "Congenital Myotonia" or "Lebensschwäche"[2] were diagnosed. Unexpectedly, some improvement was generally seen after several weeks, but it takes months before the infants are able to cry and move with ease.

Probably as a consequence of the very slowly improving muscle hypotonia, the children are only able to sit at 1 year of age and to walk at the age of 2 years. While the hypotonia and adynamia[3] gradually improve, obesity sets in around the second year of life. At the same time, growth retardation and oligophrenia[4] become distinct.

The neurologic findings persist after age 5 years. Despite some motor clumsiness, reflexes are normal. The size of the head is rather small in relation to body height. X-rays do not reveal any disturbances in the sella[5] area. The pneumo- and electro-encephalograms, performed three times, yielded normal results.

Metabolic tests could be conducted in only half of the patients but resulted in normal basal metabolic rates. Apart from the oldest patient, who had developed diabetes mellitus at the age of 17 years, tests of the carbohydrate, electrolyte and water metabolism yielded normal results when measured with conventional methods. No signs of hypothyroidism were found. Puberty seems to be delayed and incomplete. In the older patients, urinary 17-ketosteroid[6] excretion measurements were very low. The gonadotropin secretion of the 23-year-old patient was increased, the cryptorchidism probably led to a hypergonadotropic hypogonadism. The vaginal smear of the 15-year-old girl revealed a distinct effect of estrogens, which makes a hypothalamic disorder more likely than pituitary insufficiency. Regarding aetiology, we were not able to find sufficient evidence for heredity or for embryopathy.

In summary, we believe that this syndrome is not all that rare, clearly distinguishable, and well defined. Whereas in infants, it shows some similarity to amyotonia congenita of Oppenheim,[7] from school age on and later, it resembles Fröhlich's syndrome (adiposogenital dystrophy),[8] the Laurence-Moon-Biedl-Bardet syndrome,[9] and later, pituitary small stature.[10] Despite all the similarities, it can be clearly distinguished from the syndromes mentioned.

Footnotes to translation:

1. Acromicria: small hands and feet.
2. Lebensschwache: literally, life-weak (e.g., listless, moribund).
3. Adynamia: lack of physical movement.
4. Oligophrenia: a type of mental retardation leading to social incompetence; "feeble-mindedness."
5. Sella, *or* sella turcica: the area of the skull that contains the pituitary gland.
6. 17-ketosteroids: a urinary test, commonly used in the past as a marker for androgen production.
7. Amyotonia congenita of Oppenheim: a condition of severe, usually nonprogressive neonatal hypotonia described by Hermann Oppenheim in 1900. It appears that this is not an actual condition, but a description of signs and symptoms that are seen in a number of neonatal neuromuscular conditions, most notably spinal muscular atrophy.
8. Fröhlich's (adiposogenital dystrophy) syndrome is usually used to describe a condition in which adolescent boys are noted to have obesity and hypogonadotropic hypogonadism. The original case was due to a pituitary tumor and subsequent cases have involved a similar etiology, whereas other cases may have had a variety of conditions. This term is not commonly used in current medical practice.
9. Laurence-Moon-Biedl-Bardet syndrome was actually described by Bardet and Biedl in the 1920s and is currently known as Bardet-Biedl syndrome. It is characterized by obesity, short stature,

moderate mental retardation, retinal dystrophy, polydactyly, hypo-gonadism in males, and a variety of other abnormalities.

10. Pituitary short stature: e.g., growth hormone deficiency.

Commentary

The first description of the Prader-Willi syndrome—as it is now called—consisted of only 21 lines. The paragraphs above constitute the entire article—not just the abstract. The completeness and accuracy of this description and its pathophysiological implications meet with much admiration. Considering the limited methodological techniques of the time, this achievement becomes even more impressive. The description was so comprehensive that up until the 1980s, no substantial new knowledge was added.

Urs Eiholzer, MD
Head, Foundation Growth Puberty Adolescence
Zurich, Switzerland

Appendix B

A Comprehensive Team Approach to the Management of Prader-Willi Syndrome

Note: This document represents the consensus of an international meeting of PWS specialists on April 24, 2001, sponsored by the International Prader-Willi Syndrome Organization (IPSWO) and funded through a grant from Pharmacia Corporation. The consensus statement was originally edited by Dr. Urs Eiholzer in 2001 and subsequently revised with the assistance of Dr. Phillip D.K. Lee in 2004.*

Meeting Participants

Moderator and Presenter: Urs Eiholzer, MD, Head, Foundation Growth Puberty Adolescence Zurich, Switzerland

Other Presenters:

Margaret Gellatly, BSc (Hons), SRD, Hon Dietary Adviser, Prader-Willi Syndrome Association (UK), Chelmsford, UK

Phillip D.K. Lee, MD, Director, Division of Endocrinology and Metabolism, Children's Hospital of Orange County, Orange, California, USA [*Note:* Dr. Lee is currently Chief Scientific Officer, Immunodiagnostic Systems Ltd., Bolden, Tyne & Wear, UK]

Martin Ritzén, MD, Professor, Department of Pediatric Endocrinology, Karolinska Hospital, Stockholm, Sweden

Barbara Y. Whitman, PhD, Professor, Department of Developmental Pediatrics, St. Louis University, Cardinal Glennon Children's Hospital, St. Louis, Missouri, USA

---

*In 2003, a warning label was added to Genotropin/Genotonorm (Pfizer), the growth hormone preparation labeled for treatment of PWS. The warning label includes recommendations for evaluation of sleep and breathing disorders and screening for morbid obesity prior to initiation of therapy. The Clinical Advisory Board of the Prader-Willi Syndrome Association (USA) subsequently issued its own guidelines regarding sleep, breathing, and respiratory evaluation. These subjects are discussed extensively in Chapters 5 and 7 of this volume and are not specifically addressed in the Comprehensive Care guidelines.

Panelists:

Giuseppe Chiumello, MD, Professor, Università degli Studi di Milano, Clinica Pediatrica III Ospedale San Raffaele, Milan, Italy

Yukihiro Hasegawa, MD, Chief, Endocrinology, Metabolism and Genetics Unit, Tokyo Metropolitan Kiyose Children's Hospital, Tokyo, Japan

Priv-Doz Dr med Berthold P. Hauffa, Abt. F. pädiatrische Hämatologie/Onkologie und Endocrinologie, Universität GHS Essen, Essen, Germany

Maïthé Tauber, MD, Professor, Children's Hospital Toulouse, Toulouse, France

Introduction

The treatment of children with Prader-Willi syndrome (PWS) represents a challenge, particularly in the field of pediatric endocrinology. The handicaps and problems of affected children are manifold, more so than in any other typical disease of pediatric endocrinology, perhaps with the exception of craniopharyngioma. Therefore, management of children with PWS may be most successful with a team approach to comprehensive care. We thank Pharmacia Corporation for organizing a workshop on such an approach in St. Julians, Malta, on April 24, 2001.

The reader will notice that the development of a comprehensive professional team approach to PWS has only just begun. Much work remains to be done, primarily to define what, exactly, a "comprehensive team approach" to PWS means. For example, it appears necessary for one highly experienced specialist team member to assume leadership, to allow patients and their families to interact with one single professional. Furthermore, growth hormone (GH) treatment has become a very important tool in the management of PWS. Nevertheless, it must be emphasized that without a comprehensive team approach, especially to control weight gain, optimize dietary intake, and provide family psychosocial support, children with PWS may continue to suffer from excessive weight gain and major behavioral problems despite the beneficial effects of GH.

Some centers have a great deal of experience and know-how in managing PWS. This know-how, however, is most often attributable to the experience of a single person. Through intensive study of the experience and strategies of such centers and individuals, a professional comprehensive team approach can be developed that will allow centers all over the world to offer optimum care to their patients with PWS.

Prader-Willi syndrome (PWS) is characterized by infantile hypotonia; short stature; small hands and feet; increased body fat; decreased muscle mass; scoliosis; reduced resting energy expenditure (REE); reduced bone mineral density (BMD), which may lead to osteopenia and osteoporosis; hypogonadism; hypothalamic dysfunction; and a particular facial appearance. These clinical features are accompanied by hyperphagia, cognitive disabilities, and behavioral problems, including skin picking. In approximately 70% of affected individuals, the syndrome is the phenotypic expression of a complex genetic disorder resulting from a de novo deletion of the "PWS region" located on the proximal long arm of the paternal copy of chromosome 15 (at bands 15q11.2–15q13). Maternal disomy 15 (inheritance of both chromosome 15 copies from the mother, with no paternal copy) is seen in about 25% of individuals with PWS. Chromosomal translocations involving the paternal "PWS region" of 15q and imprinting center defects account for a small percentage of cases.[1] Prader-Willi syndrome and Angelman syndrome (an entirely different clinical syndrome involving defects in the maternal copy of 15q) were among the first examples in humans of genomic imprinting, or the differential expression of genetic information depending on the parent of origin. Prader-Willi syndrome is one of the more common conditions seen in genetics clinics worldwide, occurring in an estimated 1:10,000–25,000 individuals, and is the most common syndrome associated with morbid or life-threatening obesity. For affected individuals, the various clinical manifestations of PWS are major causes of morbidity and social limitation. Learning ability, speech and language, self-esteem, emotional stability, social perception, interpersonal functioning, and family dynamics, in addition to cognition and behavior, are all adversely affected in PWS.

A panel of international experts on PWS was convened to share their clinical experience and to identify strategies for managing PWS. The panel agreed that, because PWS produces various adverse functional as well as metabolic effects, individuals with PWS require a variety of interventions to optimize their growth and development. These include growth hormone (GH) replacement; dietary management; physical and occupational therapy; speech, language, and learning disability services; behavior management; and family interaction, support, and care. Children with PWS should be evaluated and treated in a multidisciplinary clinic that is managed by a nurse coordinator and staffed by a physician specialist in PWS, geneticist, psychologist, and dietitian. Ancillary resources should include support by neurology, physical therapy, social services, and educational services, as well as readily available facilities for measuring body composition (including whole-body DEXA) and exercise physiology. The Table (B.1) lists the components of the initial evaluation and testing. Follow-up visits are recommended at 6-month intervals for patients receiving GH therapy. In the majority of patients puberty will not occur, and gonadal steroid replacement therapy should be considered for them on the basis of clinical and DEXA findings.

GH Effects on Physical Parameters in PWS

Dysregulated GH secretion associated with deficient GH responses is a principal cause of short stature in the majority of individuals with PWS. It is probably also an important contributor to the decreased muscle mass and osteopenia in patients with PWS, whereas hypogonadotropic hypogonadism is the probable primary cause of osteopenia and osteoporosis in these patients.[2] Evidence is mounting that GH

Table B.1. Recommended Components of the Initial Visit to a Multidisciplinary Prader-Willi Syndrome Clinic

Evaluation
- **Confirmation of diagnosis, genetic counseling**
- **Complete examination**
- **Dietary evaluation and counseling**
- **Physical therapy evaluation (developmental, neuromuscular)**
- **Psychological evaluation and recommendations**
- **Educational evaluation and recommendations**
- **Initial discussion of growth hormone therapy and approval process**

Testing
- **DNA studies**
- **IGFBP-3, IGF-I, thyroid panel, lipid panel (other lab tests as clinically indicated)**
- **Screening for glucose intolerance if patient is obese (fasting glucose, glycated hemoglobin, oral glucose tolerance test, if indicated)**
- **Body composition analysis (DEXA, anthropometry, or other method)**
- **Psychological and/or educational testing**
- **Strength and endurance testing**

DEXA = dual-energy X-ray absorptiometry; IGF-I = insulin-like growth factor-I; IGFBP-3 = IGF binding protein-3.

deficiency may contribute not only to the abnormal growth pattern but also to the excess of body fat and the lean body mass deficit.[3,4] Growth hormone treatment of children with PWS normalizes linear growth,[5–10] promotes growth of lean body mass,[6–8,11,12] and decreases fat mass.[6–8,11,12] However, the benefits associated with GH therapy can be optimized and maintained only in conjunction with a multidisciplinary approach that emphasizes comprehensive care for the complex neurobehavioral and endocrine needs appropriate for the patient's age.

The role of GH as a component of the overall management of PWS has been studied extensively in the United States, Switzerland, and Sweden.

American Experience

Parra and co-workers observed in 1973 that a deficient GH response to pharmacologic stimuli appeared to be related to the abnormal growth pattern in patients with PWS.[13] In 1987, Lee and colleagues reported for the first time that GH therapy led to significant increases in the linear growth rate of patients with PWS.[6] Patients in their study initially had low serum levels of GH and insulin-like growth factor-I (IGF-I); during GH therapy, levels of IGF-I normalized. These results indicated that the low GH levels observed in these cases were not an artifact of obesity and supported the premise that the poor linear growth in patients with PWS might be caused by a true deficiency of GH. In 1993, Lee and collaborators reported the results of an uncontrolled trial of GH therapy in 12 obese children with PWS and associated chromosome 15 abnormalities.[14] All 12 children initially had low serum levels of GH, IGF-I, IGF-2, IGF binding protein-3 (IGFBP-3), and osteocalcin. These levels normalized and height velocity increased during GH therapy. Dual-energy X-ray absorptiometry (DEXA) at baseline revealed increased fat mass, normal (not weight-corrected) BMD, and very low lean body mass. Within 3 months of the patients' beginning GH therapy, DEXA revealed variable changes in fat mass and increased BMD and lean body mass, with redistribution of fat mass from the trunk to the thighs. The majority of parents reported improved behavior and appetite control. The decreased GH secretion commonly seen in children with PWS had been considered by some to be an effect of obesity, but reduced GH secretion had also been found in non-obese children with PWS.

Angulo and colleagues studied 33 obese and 11 non-obese children with PWS to determine whether the suboptimal GH secretion was an artifact of obesity.[15] Spontaneous GH secretion was measured over 24 hours, and GH secretion was provoked by insulin, clonidine, and levodopa. Of the 44 subjects, 40—including 10 non-obese children—failed to respond to at least two of the stimuli, and 43 had reduced spontaneous 24-hour GH secretion. The investigators concluded that the GH deficiency seen in PWS is not a consequence of obesity but rather a significant contributor to the decreased growth velocity and increased adiposity typical of the syndrome.

In a controlled trial reported in 1999, Carrel and associates assessed the effects of GH therapy on growth, body composition, strength and agility, respiratory muscle function, REE, and fat utilization in 54 children with PWS, all of whom had low peak stimulated GH levels at baseline.[6] Thirty-five children received GH at a dose of $1 \, mg/m^2/day$ and 19 were untreated. After 12 months, the GH-treated children showed significantly increased height velocity, decreased percentage of body fat, and improved physical strength, agility, and respiratory muscle function, although there was no significant increase in REE. The investigators concluded that GH therapy, in addition to its effect on growth and body composition, may have value in improving some physical disabilities experienced by children with PWS. After 24 months of GH therapy, patients had experienced sustained decreases in fat mass, increases in lean body mass,[7] and improvements in physical strength and agility.[8] Height velocity remained significantly higher than at baseline ($P < 0.01$), although the growth rate slowed between 12 and 24 months. To achieve these encouraging results, the investigators suggested, GH therapy should be started early; GH therapy started in middle to late childhood may not be capable of normalizing the percentage of body fat in patients with PWS.[7]

At baseline, 70% of subjects had mild to moderate scoliosis on spine films.[8] During the first year of the study, no significant difference in scoliosis progression was seen between the GH-treated group (from a mean of 9.2° at baseline to 12.1°) and the control group (from 14.7° to 16.6°). During the second year, the mean change in curve measurement in the GH-treated group also was not significant.

Swiss Experience

Disturbed satiation and energy expenditure are basic defects in PWS. Reduced muscle mass appears to be the consequence of decreased physical activity, which is probably caused by the central nervous system defects. Reduced muscle mass, in turn, is the cause of the decreased energy requirement. The benefit of GH therapy for children with PWS, according to Eiholzer's group, is an increase in lean body mass and a subsequent increase in REE. If energy intake is not increased, these alterations lead to a reduction of energy stores, mainly of body fat, and a dramatic change in phenotype. However, even though height and weight are normalized during GH treatment, children with PWS must maintain their energy intake at about 75% of the intake of healthy children to stabilize their weight for height. Such a reduction of food intake is possible only through close, strict parental supervision, and this is a major reason why families caring for a child with PWS need psychosocial support. Following is a short summary of the Swiss experience with GH therapy.

Eiholzer and l'Allemand described 23 children with genetically confirmed PWS and divided them into three groups: group 1 comprised young children who were not yet obese; group 2, prepubertal overweight children; and group 3, pubertal overweight children. All were treated with GH 24 U/m²/week (~0.037 mg/kg/day) for a median of 4 years (range, 1.5 to 5.5 years).[9] In group 1, weight and weight for height were lower than normal before treatment and continuously increased up to the normal range during treatment. In group 2, a dramatic height increase and drop in weight for height showed clearly that these obese children had become not only taller but also slimmer with treatment. In group 3, however, the effect of GH on growth and weight was rather limited. The investigators concluded that if treatment is instituted early enough, growth becomes normal and height predictions reach the parental target height. This effect of exogenous GH on growth has so far been described only in children with GH deficiency.

Most importantly, although loss of fat mass, as determined by DEXA,[12] in the older children (group 3) was considerable with exogenous GH administration, fat mass was still in the upper-normal range. The influence of exogenous GH on muscle mass in PWS was found to be limited. Catch-up growth in muscle mass, as estimated by lean mass, was observed only during the first 6 months of therapy; thereafter, muscle mass increased in parallel with height. Therefore, it was deduced that muscle mass remained relatively decreased.

Improvement in body composition is the main goal in the treatment of children with PWS. According to the Swiss experience, the changes in body composition during GH therapy result from several therapeutic interventions. It is critical to maintain control of nutrient intake during GH treatment, in accordance with the reduced energy requirements in PWS. In children with PWS, energy requirements are about 50% below those of healthy children.[17] Growth hormone treatment does not change the feeling of satiety but increases the energy expenditure resulting from the increase of lean mass by an estimated 25%, as shown by another Swiss study.[18] Weight for height and BMI decrease during GH treatment only if energy intake is not increased at the same time. It is therefore imperative that parents continue to keep patients' food consumption under control with the same rigidity as before the start of GH treatment.

Hypothesizing that increased muscle mass in infants may positively influence motor development, Eiholzer and colleagues used the Griffith test[19] to study psychomotor development in 10 young, underweight children with PWS during the first year of GH treatment.[20] At baseline, the children were significantly more retarded on the "locomotor" and "hearing and speech" scales than on the other scales. During GH therapy, locomotor capabilities increased significantly, whereas hearing and speech remained unchanged. The treated children started walking unassisted at an average age of 24.1 months, about 4 to 6 months earlier than untreated children with PWS. Motor development thus seems to be improved by GH therapy.

In older children, improvement in physical performance is—in the opinion of the parents—the most important therapeutic effect of GH.[21] After 1 year of GH therapy, physical performance, as assessed by ergometry, significantly increased in peak and mean power in four prepubertal 7-year-old obese children. Such improvement in physical performance leads to an increase in activity, which, together with the disappearance of the obese phenotype, may relieve patients and their families of a major stigma that accompanies PWS, improving their quality of life.

The Swiss group was also able to show for the first time that insulin secretion in children with PWS is delayed and lower than that shown in otherwise normal, nonsyndromal obese children and in children without PWS on GH therapy.[22] In addition, the increase in fasting insulin and insulin resistance seen in children with PWS during GH therapy is transient.[22] Three years of GH therapy did not impair carbo-hydrate metabolism, but rather counteracted the potential GH-induced insulin resistance by decreasing fat mass and increasing lean mass. Since normal insulin sensitivity remains preserved, the investigators speculated that the primary mechanism for the development of dia-betes in PWS is a reduced secretory capacity of pancreatic beta cells that persists despite GH administration.

According to the Swiss researchers, certain aspects of lipid metabo-lism differ in PWS and non-PWS obesity. In PWS, triglyceride levels are normal (although still correlated with abdominal obesity), but LDL cholesterol levels are elevated and HDL cholesterol levels are decreased.[23] These lipid levels normalize during GH therapy, but the changes are not associated with changes in body fat and probably are caused by the direct effects of GH deficiency and exogenous GH administration on cholesterol metabolism, as described in adult patients with GH deficiency.[24]

Swedish Experience

Despite the evidence from uncontrolled trials that GH therapy is ben-eficial in PWS, a number of pediatric endocrinologists continued to believe that the GH deficiency seen in the syndrome was a result of the characteristic obesity, and they were concerned that treatment with exogenous GH would negatively affect endogenous GH secretion. For this reason, a controlled study was conducted to assess the effects of GH therapy on growth, body composition, and behavior in prepubertal children with PWS.

Lindgren and co-workers reported preliminary results[25] of this study in 1997 and 5-year results[10] in 1999. After a 6-month evaluation period, patients with PWS between the ages of 3 and 7 years were randomized into group A (n = 15), which received GH 0.1 IU/kg/day (0.033 mg/kg/day) for 2 years, or group B (n = 12), which received no treatment for the first year and GH 0.2 IU/kg/day (0.066 mg/kg/day) during the second year. After 2 years, all children stopped GH therapy for 6 months and then restarted GH therapy at a dose of 0.1 IU/kg/day (0.033 mg/kg/day). The 6-month GH-free interval was included to prove that the effects of GH therapy were reversible and to compare the effects of the low and high doses.

Before GH therapy, all patients had low 24-hour levels of GH and IGF-I and low levels of insulin. During the first year of the study, IGF-I levels increased rapidly to supranormal values in group A (GH therapy) but remained essentially unchanged in group B (no treatment). With respect to growth, height velocity standard deviation scores (SDS) increased from −1.9 to 6.0 during the first year of GH therapy in group A, followed by a lower rate of increase during the second year. In group B, height velocity SDS decreased slightly during the first year of the study (no treatment) but increased rapidly from −1.4 to 10.1 in the second year of the study (GH therapy). When GH therapy was stopped for 6 months, height velocity declined dramatically in both groups; height SDS followed a similar pattern. Growth hormone therapy reduced the percentage of body fat and increased the muscle area of the thigh; isometric muscle strength also increased. In addition, parents reported that GH therapy seemed to have psychological and behavioral benefits, which were reversed after treatment was stopped.

Five-year follow-up data on 18 of the children were published in 1999.[10] Following resumption of GH therapy after the 6-month discontinuation, height SDS again increased. Body mass index SDS stabilized at 1.7 for group A (n = 9) and 2.5 for group B (n = 9). In 16 children, levels of fasting insulin, glucose, and the A_{1c} fraction of glycated hemoglobin remained within normal ranges. The remaining two children developed non-insulin-dependent diabetes mellitus following a rapid weight gain, but glucose homeostasis returned to normal when GH was discontinued. Unpublished 7-year follow-up data show that height has been normalizing with prolonged treatment.

Clinical Management of PWS-Associated Behaviors

By adolescence, behavioral problems characteristically have evolved as a major issue for patients with PWS and their families. Adolescents with PWS have been described as stubborn, impulsive, manipulative, irritable, mood-labile, angry, perseverative, egocentric, demanding, and prone to rage episodes when frustrated. Transitioning from one activity to another becomes increasingly difficult, and there is a tendency to confuse day with night. Thus, the food-related behavior constellation, although dramatic, is just one of many neurobehavioral abnormalities characterizing this disorder, and the food behavior often is the easiest to manage.

These behavioral traits are frequently accompanied by depression, obsessions, or even frank psychoses, and they ultimately are responsible for the inability of adults with PWS to succeed in alternative living and work placements. Interestingly, many of the characteristic behaviors of patients with PWS, including cognitive rigidity, hoarding behavior, impaired judgment, denial of deficits, inability to self-monitor behavior, and interpersonal conflicts, are also seen in patients with traumatic brain damage. In patients with PWS, however, the brain damage is genetic and, unlike traumatic brain damage, appears to affect the entire brain. Prader-Willi syndrome may thus be characterized as a pervasive developmental neurobehavioral syndrome whose behavioral manifestations reflect a distributed central nervous system dysfunction that has yet to be fully described either anatomically or biochemically.

In addition to behavioral problems, four cognitive difficulties have been identified in patients with PWS: global mental retardation, language processing problems, learning disability associated with short-term memory and sequencing deficits[27] and failure to develop the ability to apply knowledge in new situations (metacognitive ability). Most patients with PWS score between 60 and 80 on IQ tests, and at least some have IQ scores in the 90s or somewhat higher. Functional aptitude, however, is entirely independent of test scores and appears to be related more to the degree of cognitive rigidity. Impaired metacognitive ability prevents patients with PWS from utilizing their typically extensive compendium of facts in a practical or productive manner. Difficulty with sequencing and language deficits underlies most of the behavioral problems and the inability to change some behaviors. Sequencing difficulty extends beyond simple numerical applications and includes an inability to recognize cause-and-effect sequences. This particular problem necessitates an entirely different approach to traditional behavior management, since patients with PWS fail to link punishment or reward with an antecedent behavior.

Many patients with PWS who frequently exhibit problem behaviors are able to alter these behaviors when environmental changes are instituted. These changes require creativity, hard work, and, often, many months before a behavior is altered, and some environmental and family situations are unalterable. It is particularly difficult when parents disagree about the management approach. Children with PWS who have the worst behavior in terms of depression and anxiety come from families in which parents report the highest level of conflict over child rearing. Although this is also true for normal children, children with PWS do not have the flexibility seen in normal children. Therefore, family therapy is recommended as soon as the diagnosis of PWS is made in an infant or young child.

For many patients with PWS, problem behaviors are resistant to most attempts at behavioral management, and pharmacologic interventions are often considered when this becomes clear. Unfortunately, psychopharmacologic agents frequently worsen problem behaviors in these individuals. A survey of parents of children with PWS conducted between 1989 and 1993 revealed that almost every available psychotropic agent had been prescribed to manage behavioral problems.[28,29] Most agents either were ineffective or increased the occurrence of targeted symptoms; only three—haloperidol, thioridazine, and fluoxetine—were effective.[30] More recently, it has been found that all serotonin specific reuptake inhibitors seem to have a nonspecific behavior-stabilizing effect, characterized by fewer outbursts, a marked reduction in irritability, and less perseveration, but with no specific antidepressant effect.[31] Other psychotropic drugs, such as the antipsychotic agent olanzapine and the anticonvulsant agent divalproex sodium, may have an effect.[31] It must be emphasized, however, that any single agent may produce a dramatically beneficial response in some patients with PWS and a dramatically adverse response in others, and many patients with PWS have idiosyncratic reactions to psychotropic drugs. Those with PWS require only one fourth to one half the standard dose of a psychotropic drug to achieve a benefit; increasing the dose to "normal" often results in toxicity and a return of the problem behavior.[31] In general, psychotropic medication should be used only when all other interventions, including behavioral modification and environmental changes, have failed.

It should also be noted that appetite-suppressing medications have been ineffective in controlling food-seeking behavior and overeating.[1] Pharmacologic agents, including the amphetamines and agents that block nutrient absorption, which are often effective for weight control in non-PWS obese population, do not appear to alter the brain signals, or perhaps peripheral signals, that drive patients with PWS to seek food and overeat. Until a medication is discovered that can accomplish this goal, good management depends entirely on environmental control, protection against overeating, and an understanding caregiver who recognizes that the constant feeling of hunger experienced by these patients underlies some of their irritability and other behavior characteristics.

With regard to the effect of GH therapy on PWS behavior in the setting of behavioral difficulties and refractoriness to psychopharmacologic agents, surveys of parents indicate that some behaviors improve and none deteriorate.[26,32] Since the behavior of children with PWS tends to deteriorate over time, the absence of deterioration is, in fact, a positive outcome. Specific behavioral benefits of GH therapy, as reported anecdotally by parents, included increased energy, increased activity without the need for encouragement, improved personal hygiene, less "annoying" behavior, increased assumption of responsibility, and less perseveration.[26,32] In addition, attention span and compliance seemed to improve and anxiety, depression, and obsessive thoughts decreased, although there was no impact on obsessive-compulsive behavior or improvement in school performance. Growth hormone therapy also produced positive effects on physical appearance, usually within 3 to 6 months of patients' starting treatment. Appearance of the hands, feet, and trunk normalized in all GH recipients, and appearance of the head normalized in 81%. Such changes may positively affect patients' social interaction. Furthermore, 97% of patients had more energy and 83% spontaneously increased their level of physical activity without parental prodding.

Improving Quality of Life in Patients with PWS: Diet, Exercise, and Lifestyle Changes

Surveys performed in the United Kingdom in 1989 and 1999 have provided useful information about the impact of lifestyle changes on PWS. From the standpoint of diet, two distinct phases of PWS are apparent: initial failure to thrive and subsequent obesity.

Failure to thrive results primarily from hypotonia, which makes sucking difficult during infancy. Nasogastric tube feeding may be necessary for as long as 2 months to meet energy requirements. Signs of poor feeding in infants with PWS include changes in the voice or cry, coughing while swallowing, excessive drooling, frequent vomiting, constipation, respiratory infections, irritability during feeding, slow intake, and poor weight gain. For infants who are able to suck, specially designed nipples can reduce the energy expenditure. Early weaning to soft food will reduce energy requirements; introduction of solids is accompanied by a lessening of appetite for milk.[33] However, some 33% of older infants with PWS are unable to eat soft food normally acceptable at 1 year, and children with PWS typically lag far behind children without PWS in their transition to solid food, with 42% of children with PWS unable to chew some solid foods at the age of 5 years.[34]

The change from failure to thrive to excessive weight gain generally occurs between 2 and 4 years; there seems to be a recent shift toward the younger age. Despite their reduced energy requirement, these children are obsessed with food and engage in food seeking and food stealing. Overeating may be due to the prolonged eating drive that results from their disturbed feelings of satiety.[35] The vast majority of parents of children with PWS have attempted to control their children's weight, but dietary compliance is poor. Severe caloric restriction for short periods at home or for longer periods in the hospital setting may be helpful, but most families feel that no intervention will help.

Increased physical activity can increase energy expenditure, promote negative energy balance, raise the post-exercise metabolic rate, build muscle mass, prevent osteoporosis, improve scoliosis, and enhance the overall sense of well-being. However, very few patients with PWS seem to participate in a structured exercise program. Aerobic exercise, toning and strengthening, flexing and stretching, and formal physiotherapy are all useful for patients with PWS. Activities they may find acceptable include bicycling, skating, jumping on a trampoline, dancing, and ball playing.

Lifestyle changes that can be implemented certainly include control of food-seeking and food-stealing behaviors but also must encompass social integration and independence. Specific environmental controls designed to limit hyperphagia include locking places where food is stored, restricting access to money or credit cards, and prohibiting participation in food preparation. Unfortunately, many of these impositions and limitations may actually discourage social integration and independence.

Summary and Conclusion

Because of its many physical and behavioral manifestations, PWS should be managed in a multidisciplinary setting that emphasizes comprehensive care. Clinical trials confirm that GH treatment of children with PWS normalizes linear growth, promotes an increase in lean body mass, and decreases fat mass. However, due to the complex nature of the syndrome, the long-term benefits of GH can be optimized and maintained only in conjunction with dietary control and counseling, physical therapy, and psychological and educational evaluation and support.

References

1. Butler MG, Thompson T. Prader-Willi syndrome: clinical and genetic findings. *The Endocrinologist.* 2000;10(suppl 1):3S-16S.
2. Lee PDK. Effects of growth hormone treatment in children with Prader-Willi syndrome. *Growth Hormone and IGF Research.* 2000;10(suppl B): S75-S79.
3. Eiholzer U, Bachmann S, l'Allemand D. Growth hormone deficiency in Prader-Willi syndrome. *The Endocrinologist.* 2000;10(suppl 1):50S-56S.
4. Eiholzer U, Bachmann S, l'Allemand D. Is there growth hormone deficiency in Prader-Willi syndrome? Six arguments to support the presence of hypothalamic growth hormone deficiency in Prader-Willi syndrome. *Hormone Research.* 2000;53(suppl 3):44–52.
5. Lee PDK, Wilson DM, Rountree L, Hintz RL, Rosenfeld RG. Linear growth response to exogenous growth hormone in Prader-Willi syndrome. *American Journal of Medical Genetics.* 1987;28:865–871.
6. Carrel AL, Myers SE, Whitman BY, Allen DB. Growth hormone improves body composition, fat utilization, physical strength and agility, and growth in Prader-Willi syndrome: a controlled study. *Journal of Pediatrics.* 1999;134:215–221.
7. Carrel AL, Myers SE, Whitman BY, Allen DB. Prader-Willi syndrome: the effect of growth hormone on childhood body composition. *The Endocrinologist.* 2000;10(suppl 1):43S-49S.
8. Myers SE, Carrel AL, Whitman BY, Allen DB. Sustained benefit after 2 years of growth hormone on body composition, fat utilization, physical strength and agility, and growth in Prader-Willi syndrome. *Journal of Pediatrics.* 2000;137:42–49.
9. Eiholzer U, l'Allemand D. Growth hormone normalizes height, prediction of final height and hand length in children with Prader-Willi syndrome after 4 years of therapy. *Hormone Research.* 2000;53:185–192.

10. Lindgren AC, Ritzén EM. Five years of growth hormone treatment in children with Prader-Willi syndrome: Swedish National Growth Hormone Advisory Group. *Acta Paediatrica Supplement.* 1999;433:109–111.

11. Lindgren AC, Hagenäs L, Müller J, et al. Growth hormone treatment of children with Prader-Willi syndrome affects linear growth and body composition favourably. *Acta Paediatrica.* 1998;87:28–31.

12. Eiholzer U, l'Allemand D, van der Sluis I, Steinert H, Gasser T, Ellis K. Body composition abnormalities in children with Prader-Willi syndrome and long-term effects of growth hormone therapy. *Hormone Research.* 2000;53:200–206.

13. Parra A, Cervantes C, Schultz RB. Immunoreactive insulin and growth hormone responses in patients with Prader-Willi syndrome. *Journal of Pediatrics.* 1973;83:587–593.

14. Lee PD, Hwu K, Henson H, et al. Body composition studies in Prader-Willi syndrome: effects of growth hormone therapy. In: Ellis KJ, Eastman JD, eds. *Human Body Composition. In Vivo Methods, Models, and Assessment.* New York, NY: Plenum Press; 1993:201–205.

15. Angulo M, Castro-Magana M, Mazur B, Canas JA, Vitollo PM, Sarrantonio M. Growth hormone secretion and effects of growth hormone therapy on growth velocity and weight gain in children with Prader-Willi syndrome. *Journal of Pediatric Endocrinology and Metabolism.* 1996;9:393–400.

16. Boot AM, Bouquet J, de Ridder MA, Krenning EP, deMuinck Kaizer-Schrama SM. Determinants of body composition measured by dual-energy X-ray absorptiometry in Dutch children and adolescents. *American Journal of Clinical Nutrition.* 1997;66:232–238.

17. Stadler DD. Nutritional management. In: Greenswag LR, Alexander RC, eds. *Management of Prader-Willi Syndrome.* New York, NY: Springer-Verlag; 1995:88–114.

18. Eiholzer U. *Prader-Willi Syndrome. Effects of Human Growth Hormone Treatment.* In: Savage MO, ed. Endocrine Development Series, Vol. 3. Basel: Karger; 2001:51–54.

19. Brandt I. *Griffiths Entwicklungsskalen zur Beurteilung der Entwicklung in den ersten beidenLebensjahren.* Weinheim and Basel: Beltz Verlag; 1983.

20. Eiholzer U, Malich S, l'Allemand D. Does growth hormone therapy improve motor development in infants with Prader-Willi syndrome? [letter]. *European Journal of Pediatrics.* 2000;159:299.

21. Eiholzer U, Gisin R, Weinmann C, et al. Treatment with human growth hormone in patients with Prader-Labhart-Willi syndrome reduces body fat and increases muscle mass and physical performance. *European Journal of Pediatrics.* 1998;157:368–377.

22. L'Allemand D, Schlumpf M, Torresani T, Girard J, Eiholzer U. Insulin secretion before and under 3 years of growth hormone (GH) therapy in Prader-Willi syndrome (PWS) [abstract]. *Experimental and Clinical Endocrinology and Diabetes.* 2000(suppl 1);108:127.

23. L'Allemand D, Eiholzer U, Schlumpf M, Steinert H, Riesen W. Cardiovascular risk factors improve during 3 years of growth hormone therapy in Prader-Willi syndrome. *European Journal of Pediatrics.* 2000;159:835–842.

24. Vahl N, Jorgensen JO, Hansen TP, et al. The favourable effects of growth hormone (GH) substitution on hypercholesterolaemia in GH-deficient adults are not associated with concomitant reductions in adiposity. A 12 month placebo-controlled study. *International Journal of Obesity and Related Metabolic Disorders.* 1998;22:529–536.

25. Lindgren AC, Hagenäs L, Müller J, et al. Effects of growth hormone treatment on growth and body composition in Prader-Willi syndrome: a pre-

liminary report. The Swedish National Growth Hormone Advisory Group. *Acta Paediatrica Supplement.* 1997;423:60–62.

26. Whitman BY, Myers S, Carrel A, Allen D. A treatment/control group study of growth hormone treatment: impact on behavior—a preliminary look. *The Endocrinologist.* 2000;10(suppl 1):31S-37S.

27. Dykens EM, Hodapp RM, Walsh K, Nash LJ. Profiles, correlates, and trajectories of intelligence in Prader-Willi syndrome. *Journal of the American Academy of Child and Adolescent Psychiatry.* 1992;31:1125–1130.

28. Whitman BY, Greenswag L. The use of psychotropic medications in persons with Prader-Willi syndrome. In: Cassidy S, ed. *Prader-Willi Syndrome and Other Chromosome 15q Deletion Disorders.* Berlin: Springer Verlag in cooperation with NATO Scientific Affairs Division; 1992:223–231.

29. Greenswag LR, Whitman BY. Long term follow-up of use of Prozac as a behavioral intervention in 57 persons with Prader-Willi syndrome. Proceedings: 2nd Prader-Willi Syndrome International Scientific Workshop and Conference. 1995 [abstract 20].

30. Whitman B, Greenswag L. Psychological issues in Prader-Willi syndrome. In: Greenswag L, Alexander R, eds. *Management of Prader-Willi Syndrome.* 2nd ed. New York, NY: Springer Verlag; 1995:125–141.

31. Whitman B, Greenswag L, Boyt M. The use and impact of psychotropic medications for managing behavior in persons with Prader-Willi syndrome. Proceedings: 13th Annual Prader-Willi Syndrome Association (USA) Scientific Conference. July 22, 1998; Columbus, OH.

32. Whitman B, Myers S, Carrel A, Allen DB. The behavioral impact of growth hormone treatment for children and adolescents with Prader-Willi syndrome: a two-year controlled study. *Pediatrics.* 2002;109(2):E35.

33. Shaw V, Lawson M. Principles of paediatric dietetics. In: Shaw V, Lawson M, eds. *Clinical Paediatric Dietetics.* 2nd ed. Oxford, UK: Blackwell Science; 2001:Chap 1.

34. Morris M. Feeding the young child with PWS. *The Gathered View.* 1993; 18(1):6–7.

35. Lindgren AC, Barkeling B, Hagg A, et al. Eating behavior in Prader-Willi syndrome, normal weight, and obese control groups. *Journal of Pediatrics.* 2000;137:50–55.

Appendix C

Growth Charts of Individuals with Prader-Willi Syndrome

Data from the United States

Data in Figures C.1 through C.5 are based on measurements of 71 Caucasian U.S. subjects with PWS between the ages of 0 and 24 years, including 42 males and 29 females, reported by Butler and Meaney. Under high-resolution chromosome analysis, 37 subjects had an apparent chromosome 15 deletion, 26 had normal-appearing chromosomes, and 8 had an unknown chromosome status. Approximately half of the subjects were on a calorie-restricted diet, and none were treated with growth hormone. No significant differences were found between those with a chromosome deletion and those with normal-appearing chromosomes, but there were significant variations by gender.

Data source: Butler MG, Meaney FJ. Standards for selected anthropometric measurements in Prader-Willi syndrome. *Pediatrics,* 1991:88(4);853–860. Reproduced by permission of *Pediatrics,* 1991:88;853–858. (Charts were modified by Dr. Merlin Butler to add standard measure equivalents to the original metric units.)

Data from Germany

Data in Figures C.6 and C.7 are based on measurements of 100 subjects of German descent between the ages of 0 and 20 years, including 51 males and 49 females, reported by Hauffa et al. All subjects had genetically confirmed PWS by molecular genetics testing; 76 had deletions, 14 had maternal uniparental disomy, 3 had imprinting mutations, and 7 were of undetermined molecular class. None of the subjects had received a growth-promoting therapy. In comparison with the U.S. data described above, the researchers found that "Height centile curves of the German patients fall in the tall range of standards derived from American patients . . . mainly due to an elevation of the lower centile ranges in both sexes." They also found that after age 14 "German girls with PWS are heavier than their American counterparts."

Data source: Dr. Berthold P. Hauffa provided combination height and weight charts based on PWS data reported in Hauffa BP, Schlippe G, Roos M, Gillessen-Kaesbach G, Gasser T. "Spontaneous growth in German children and adolescents with genetically confirmed Prader-Willi syndrome." *Acta Paediatrica,* 2000:89:1302–1311. These modified clinical charts were prepared by Pharmacia Corporation, substituting German reference data for the Dutch reference data in the original article. Reprinted with English labels by permission of Pharmacia Corporation.

Data from Japan

Data in Figures C.8 through C.15 are based on measurements of 252 Japanese individuals with PWS between the ages of 0 and 24 years, including 153 males and 99 females, reported by Nagai et al. The subjects were diagnosed with PWS by clinical, cytogenetic, and/or molecular genetic methods; 198 were found to have a chromosome 15q abnormality (deletion), 26 had maternal uniparental disomy, and in 28 no chromosome analysis was available. Approximately one third of the subjects were on a calorie-restricted diet. The researchers found that "Growth patterns are not different between Japanese and Caucasian children with the syndrome" but that "the degree of overweight appears much more severe in Caucasians."

Data source: Nagai T, Matsuo N, Kayanuma Y, et al., Standard growth curves for Japanese patients with Prader-Willi syndrome," *American Journal of Medical Genetics,* 2000:95;130–134. Original growth charts from this report, courtesy of Dr. Toshiro Nagai, are reprinted with permission of Wiley-Liss, Inc., a subsidiary of John Wiley & Sons, Inc.

Height

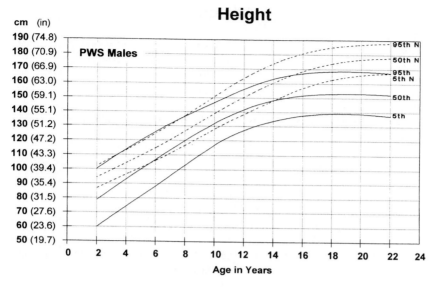

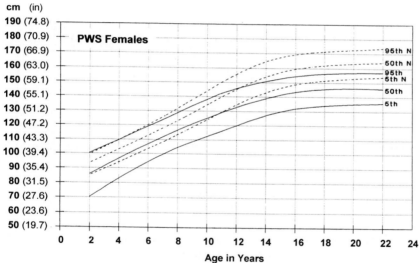

Figure C.1. Data from USA. Standardized curves for height of Prader-Willi syndrome (PWS) male and female patients (solid line) and healthy individuals (broken line). Modified from Butler and Meaney, 1991. Reproduced by permission of *Pediatrics*, Vol. 88, p. 854, Copyright © 1991.

Weight

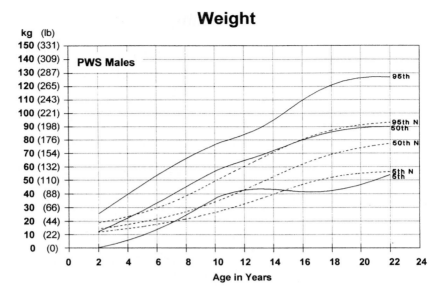

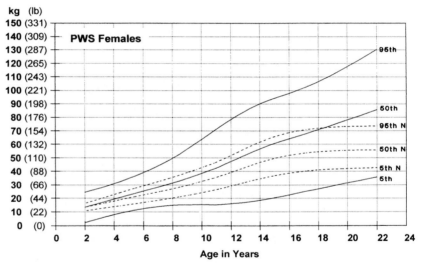

Figure C.2. Data from USA. Standardized curves for weight of Prader-Willi syndrome (PWS) male and female patients (solid line) and healthy individuals (broken line). Modified from Butler and Meaney, 1991. Reproduced by permission of *Pediatrics*, Vol. 88, p. 853, Copyright © 1991.

Head Circumference

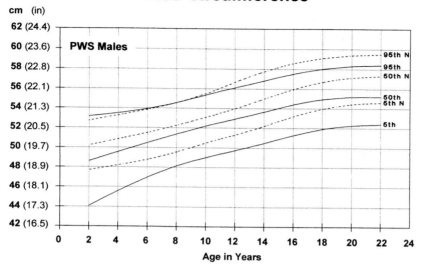

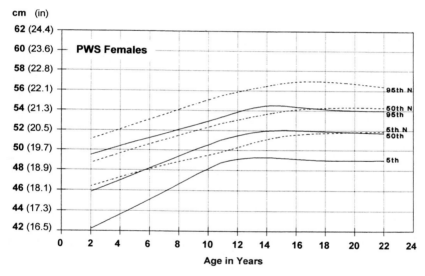

Figure C.3. Data from USA. Standardized curves for head circumference of Prader-Willi syndrome (PWS) male and female patients (solid line) and healthy individuals (broken line). Modified from Butler and Meaney, 1991. Reproduced by permission of *Pediatrics*, Vol. 88, p. 855, Copyright © 1991.

Hand Length

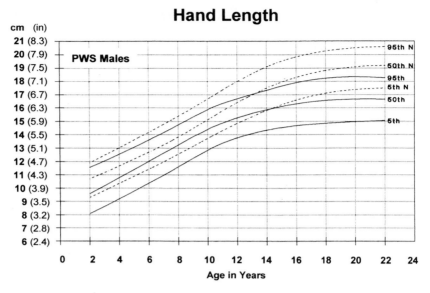

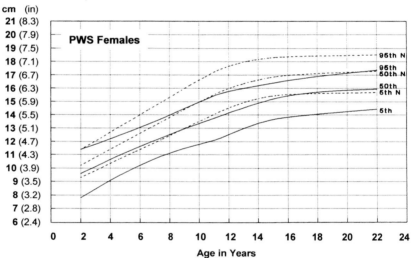

Figure C.4. Data from USA. Standardized curves for hand length of Prader-Willi syndrome (PWS) male and female patients (solid line) and healthy individuals (broken line). Modified from Butler and Meaney, 1991. Reproduced by permission of *Pediatrics*, Vol. 88, p. 856, Copyright © 1991.

Foot Length

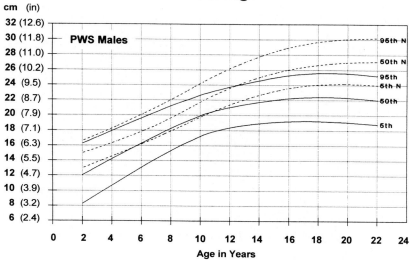

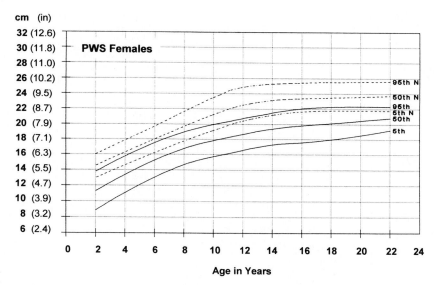

Figure C.5. Data from USA. Standardized curves for foot length of Prader-Willi syndrome (PWS) male and female patients (solid line) and healthy individuals (broken line). Modified from Butler and Meaney, 1991. Reproduced by permission of *Pediatrics*, Vol. 88, p. 858, Copyright © 1991.

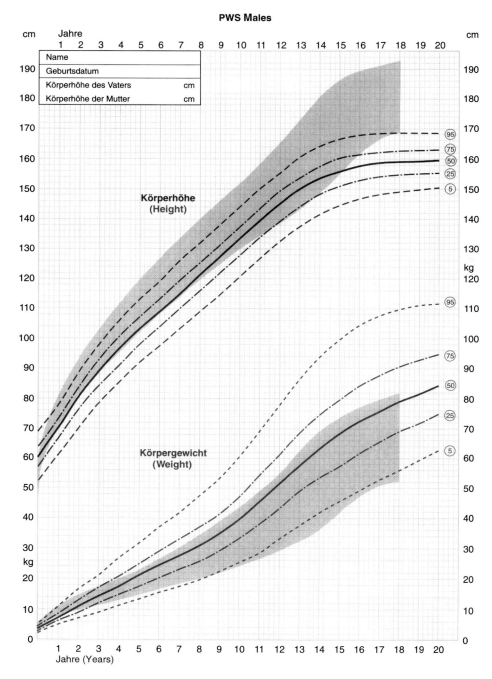

PWS Males

Figure C.6. Data from Germany. Centile curves (5th, 25th, 50th, 75th, 95th centile) for length/height (top) and for weight (bottom) of male German PWS patients, compared with reference growth standards of normal children (shaded area representing the 3rd to 97th centile range). Modified clinical chart based on Hauffa et al., *Acta Paediatrica*, 2000, Vol. 89, pp. 1302–1311. Reprinted with permission from Pharmacia Corp. Chart courtesy of Dr. Berthold P. Hauffa.

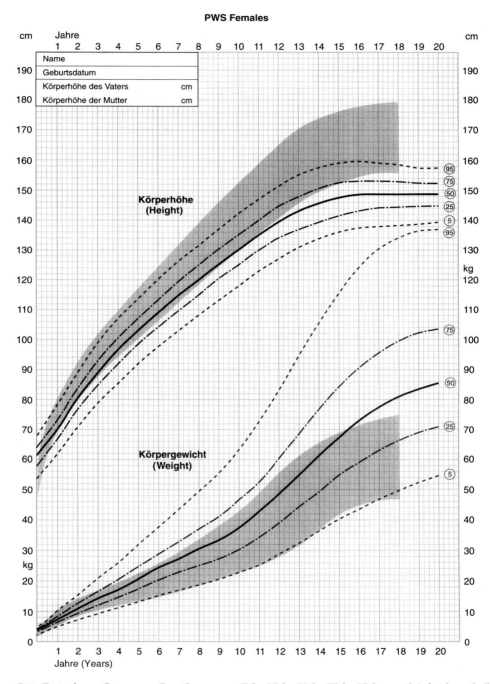

Figure C.7. Data from Germany. Centile curves (5th, 25th, 50th, 75th, 95th centile) for length/height (top) and for weight (bottom) of female German PWS patients, compared with the reference growth standards of normal children (shaded area representing the 3rd to 97th centile range). Modified clinical chart based on Hauffa et al., *Acta Paediatrica*, 2000, Vol. 89, pp. 1302–1311. Reprinted with permission from Pharmacia Corp. Chart courtesy of Dr. Berthold P. Hauffa.

Length in PWS Males

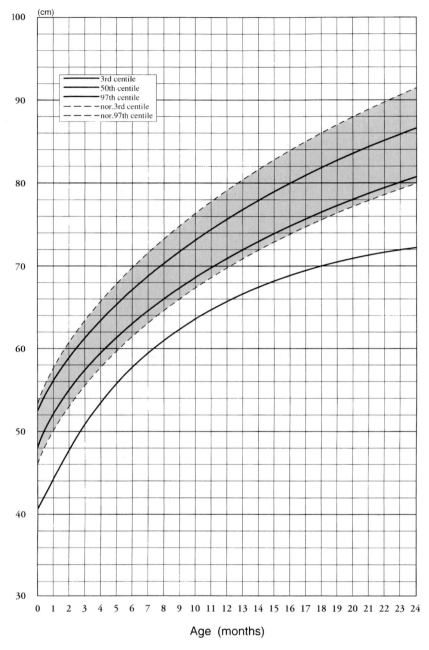

Figure C.8. Data from Japan. Body length of male Japanese PWS patients from birth to age 24 months. Solid lines show 3rd, 50th, and 97th centile values for PWS patients, and dotted lines 3rd and 97th centile values for normal children. From Nagai et al., *American Journal of Medical Genetics*, Vol. 95, p. 131, Copyright © 2000. Reprinted with permission of Wiley-Liss, Inc., a subsidiary of John Wiley & Sons, Inc. Chart courtesy of Dr. Toshiro Nagai.

Length in PWS Females

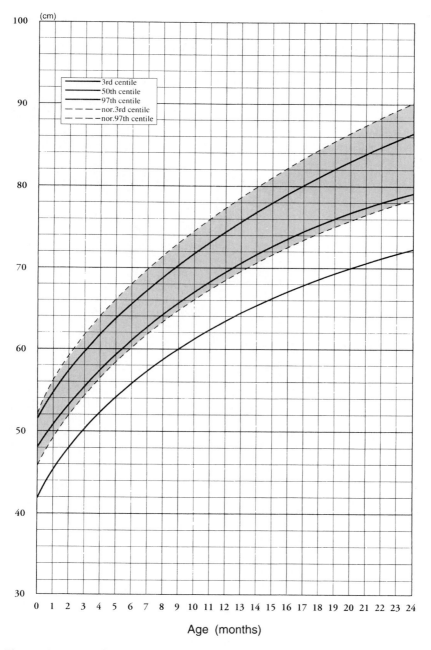

Figure C.9. Data from Japan. Body length of female Japanese PWS patients from birth to age 24 months. Solid lines show 3rd, 50th, and 97th centile values for PWS patients, and dotted lines 3rd and 97th centile values for normal children. From Nagai et al., *American Journal of Medical Genetics*, Vol. 95, p. 131, Copyright © 2000. Reprinted with permission of Wiley-Liss, Inc., a subsidiary of John Wiley & Sons, Inc. Chart courtesy of Dr. Toshiro Nagai.

Height in PWS Males

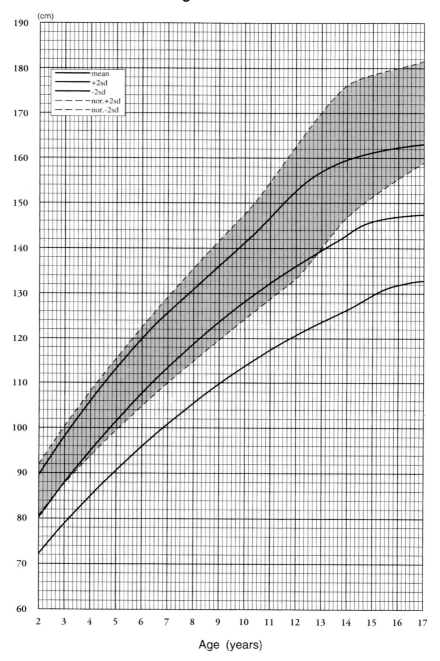

Figure C.10. Data from Japan. Height of male Japanese PWS patients from ages 2 to 17 years. Solid lines show 3rd, 50th, and 97th centile values for PWS patients, and dotted lines 3rd and 97th centile values for normal children. From Nagai et al., *American Journal of Medical Genetics*, Vol. 95, p. 132, Copyright © 2000. Reprinted with permission of Wiley-Liss, Inc., a subsidiary of John Wiley & Sons, Inc. Chart courtesy of Dr. Toshiro Nagai.

Height in PWS Females

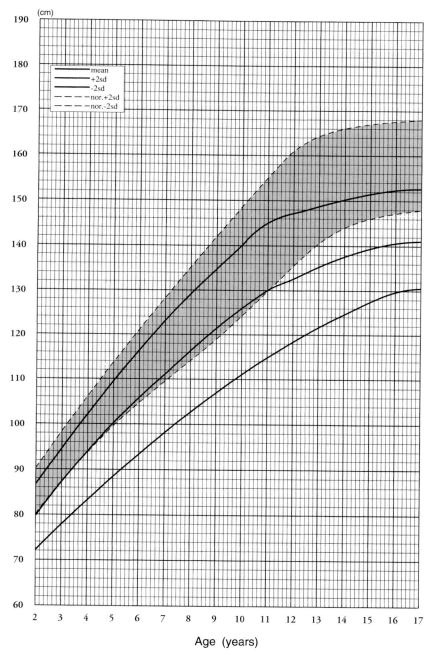

Age (years)

Figure C.11. Data from Japan. Height of female Japanese PWS patients from ages 2 to 17 years. Solid lines show 3rd, 50th, and 97th centile values for PWS patients, and dotted lines 3rd and 97th centile values for normal children. From Nagai et al., *American Journal of Medical Genetics*, Vol. 95, p. 132, Copyright © 2000. Reprinted with permission of Wiley-Liss, Inc., a subsidiary of John Wiley & Sons, Inc. Chart courtesy of Dr. Toshiro Nagai.

Weight in PWS Males

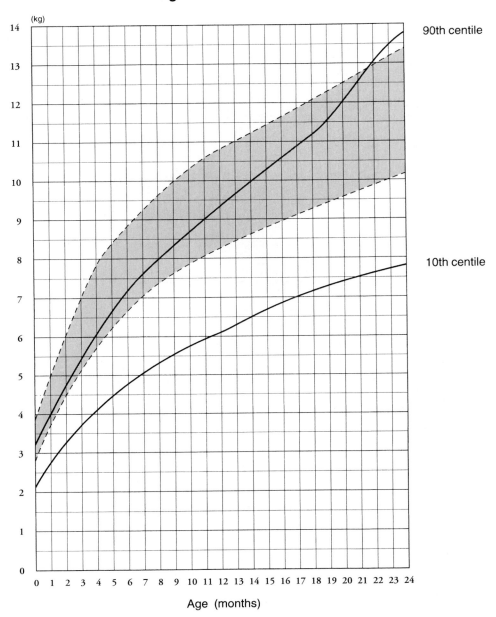

Age (months)

Figure C.12. Data from Japan. Body weight of male Japanese PWS patients from birth to age 24 months. Solid lines show 3rd, 50th, and 97th centile values for PWS patients, and dotted lines 3rd and 97th centile values for normal children. From Nagai et al., *American Journal of Medical Genetics*, Vol. 95, p. 133, Copyright © 2000. Reprinted with permission of Wiley-Liss, Inc., a subsidiary of John Wiley & Sons, Inc. Chart courtesy of Dr. Toshiro Nagai.

Weight in PWS Females

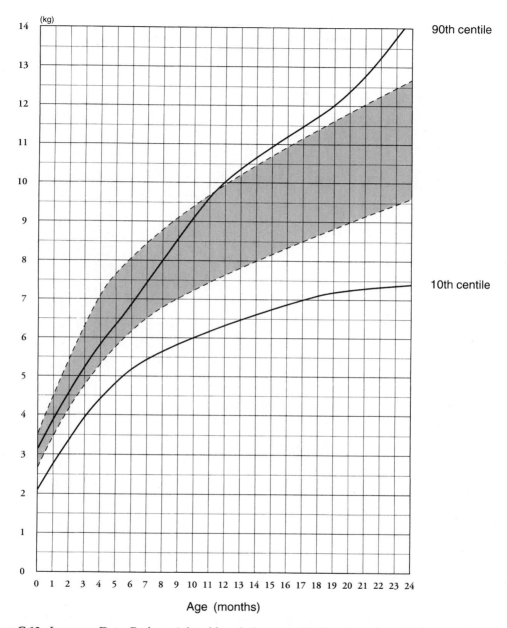

Figure C.13. Japanese Data. Body weight of female Japanese PWS patients from birth to age 24 months. Solid lines show 3rd, 50th, and 97th centile values for PWS patients, and dotted lines 3rd and 97th centile values for normal children. From Nagai et al., *American Journal of Medical Genetics*, Vol. 95, p. 133, Copyright © 2000. Reprinted with permission of Wiley-Liss, Inc., a subsidiary of John Wiley & Sons, Inc. Chart courtesy of Dr. Toshiro Nagai.

Weight in PWS Males

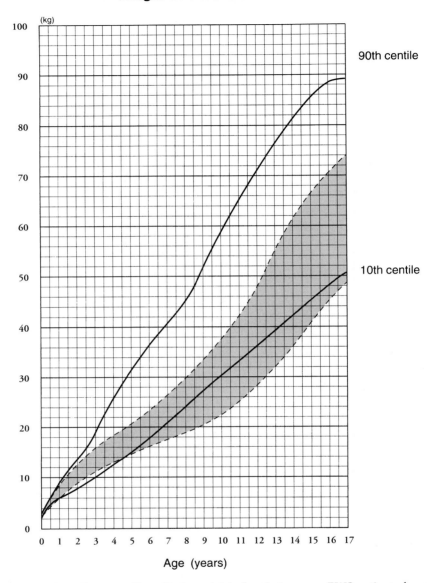

Age (years)

Figure C.14. Japanese Data. Body weight of male Japanese PWS patients from ages 2 to 17 years. Solid lines show 3rd, 50th, and 97th centile values for PWS patients, and dotted lines 3rd and 97th centile values for normal children. From Nagai et al., *American Journal of Medical Genetics*, Vol. 95, p. 133, Copyright © 2000. Reprinted with permission of Wiley-Liss, Inc., a subsidiary of John Wiley & Sons, Inc. Chart courtesy of Dr. Toshiro Nagai.

Weight in PWS Females

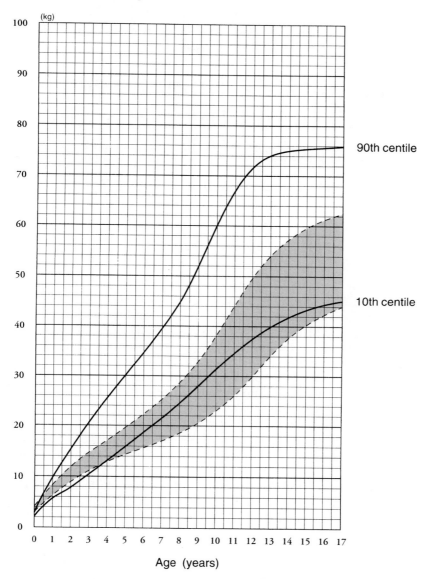

Age (years)

Figure C.15. Japanese Data. Body weight of female Japanese PWS patients from ages 2 to 17 years. Solid lines show 3rd, 50th, and 97th centile values for PWS patients, and dotted lines 3rd and 97th centile values for normal children. From Nagai et al., *American Journal of Medical Genetics*, Vol. 95, p. 133, Copyright © 2000. Reprinted with permission of Wiley-Liss, Inc., a subsidiary of John Wiley & Sons, Inc. Chart courtesy of Dr. Toshiro Nagai.

Appendix D

Eligibility for U.S. Social Security and SSI Benefits for Individuals with Prader-Willi Syndrome

Barbara R. Silverstone, Staff Attorney, National Organization of Social Security Claimants' Representatives

If your child suffers from a chronic condition such as Prader-Willi syndrome, you may have concerns about affording the care he currently needs and his well-being when you are no longer able to support him. If your child is sufficiently disabled and will be unable to work, you need more information about the disability benefits programs administered by the U.S. Social Security Administration (SSA).

The Social Security Administration administers two types of programs for disabled individuals: Social Security Disability Insurance (SSDI), also known as Title II and DIB, and Supplemental Security Income, also known as SSI and Title XVI. Although they have the same definition of disability, there are different requirements for each benefit program.

Social Security Disability Insurance is paid to adults who become disabled and who have a sufficient work history. SSDI also provides income for their dependents. Dependents include their minor children, disabled adult children, and sometimes their spouse. Medicare coverage is available after the disabled individual has received benefits for 2 years. The other benefit, SSI, is means-tested. These benefits are paid to disabled adults who have limited income and resources (most adults with Prader-Willi syndrome receive SSI) and to disabled children whose parents have limited income and resources. Due to the "deeming" of parents' income and resources to minor children, children can receive SSI benefits only if their parents have limited income and resources. Medicaid coverage is available upon receipt of SSI benefits. Many people receive benefits under both the SSI and Medicaid programs. There are special rules for SSI eligibility for noncitizens. If you are not a U.S. citizen, contact your attorney or SSA for more information.

An important benefit to be aware of is called Disabled Adult Children's (DAC) benefits. If a child is unmarried and his disability has continued uninterrupted since before his 22nd birthday, and one of his parents is either retired, disabled, or has died after working enough quarters to qualify as insured, he may be eligible for Social Security disability benefits. Disabled Adult Children's benefits are available to your child even if he or she has never worked and can be used to support a disabled individual whose parents are no longer able to do so. Although these benefits are called "children's" benefits, they are paid to the adult child of a former wage earner. When a recipient of DAC benefits marries, these benefits will end. If you think your child may be eligible for this benefit, you should contact the Social Security Administration for an application. Visit a local office or call 1-800-772-1213. For a referral to a private attorney who is familiar with these benefits, you can call the National Organization of Social Security Claimants' Representatives at 1-800-431-2804.

Determining Eligibility

Social Security Disability Insurance benefits are available to disabled workers who meet two conditions: (1) they are too disabled to work at *any* job, not just the jobs which they held in the past; and (2) through their employment, they have contributed enough FICA tax over the years to be covered. In general, workers who have worked at least 5 out of the 10 years just before the disability began are covered; the rules are different for workers under age 30. An individual's wage history determines the monthly benefit amount.

Remember, even an individual who has not worked, but whose disability began before age 22, may be eligible for Social Security disability benefits, as a Disabled Adult Child.

Supplemental Security Income (SSI) benefits are available to disabled individuals whose income and resources are very limited. Generally, to be eligible for SSI benefits, an individual may have no more than $2,000 in resources ($3,000 for a couple) and income that is less than the SSI benefit amount ($579 per month for an individual and $869 for a couple in January 2005). The income levels change slightly each year. There are several items, such as a primary residence, car, and certain income, that SSA will not count. Income and resources from a spouse or the parents of a minor child are deemed available to the claimant. Parents' income will be deemed to a minor child even if he or she resides at a school, if the parent has parental control (guardianship) over the child.

Claimants who are eligible for Social Security disability benefits but whose payment amount is very low may also be eligible for SSI benefits.

Who Is "Disabled"?

Eligibility for benefits depends on a child's limitations resulting from physical, mental, and behavioral impairments. SSA decides whether a child has been, or is expected to be, disabled for at least 12 months. The

SSA definition of disability for a child is: "An individual under the age of 18 shall be considered disabled for purposes of this title if that individual had a medically determinable physical or mental impairment which results in marked and severe functional limitations, and which can be expected to result in death or which has lasted or can be expected to last for a continuous period of not less than 12 months."

Eligibility for disability benefits depends on the limitations an individual has as a result of *both* physical and mental impairments. For example, SSA will consider the effects of obesity on the individual's heart and ability to walk, as well as other conditions it may cause, such as diabetes or sleep apnea. SSA will also consider mental limitations, including IQ scores and limitations in social functioning and activities of daily living and maladaptive behavior. For example, temper tantrums, obsessive-compulsive behavior, frustration, and need for routine can limit available jobs and can be a "marked and severe functional limitation" in a child's case.

SSA follows a Sequential Evaluation Process to determine whether a claimant meets the disability criteria. For children, the sequential evaluation is a three-step process:

1. Is the child working?
2. Does the child have a medically determinable impairment or combination of impairments that is severe?
3. Does the child's impairment meet or medically equal the requirements of a listed impairment; or are the functional limitations caused by the impairment(s) the same as the disabling functional limitations of any listing, and therefore, functionally equivalent to that listing?

If the child is not working, SSA will compare the child's condition to its criteria in the "Listings of Impairments." There are several different listings under which an individual with Prader-Willi syndrome may be evaluated.

One listing that SSA may refer to is section 110.00, Multiple Body Systems. Those impairments that SSA classifies as "multiple body systems" are "life-threatening catastrophic congenital abnormalities and other serious hereditary, congenital, or acquired disorders that usually affect two or more body systems" and are expected to either produce long-term significant interference with age-appropriate major daily activities or result in early death. SSA will find a child disabled who suffers from multiple body dysfunction due to any confirmed hereditary, congenital, or acquired condition with either persistent motor dysfunction; significant interference with communication due to speech, hearing, or visual impairments; or mental, growth, or cardiac impairments.

SSA may also evaluate a child with Prader-Willi syndrome under the listings for mental disorders, section 112.00. These listings provide specific requirements for each age group. The conditions in this section that are most likely to be present in a child with Prader-Willi syndrome include: organic mental disorders (112.02); mood disorders (112.04); mental retardation (112.05); anxiety disorders (112.06); somatoform, eating, and tic disorders (112.07); pervasive developmental disorders

(112.10); or developmental and emotional disorders of newborn and younger infants (112.12).

If your child's condition precisely meets any of these criteria, SSA will find that he is disabled. But meeting these criteria is not the only way to qualify for benefits. Even if your minor child does not have the exact test results required, SSA will continue to evaluate the claim by determining whether his or her impairments cause the same type of limitations as any of the listings (medically or functionally equal the listing). Keep in mind that your child's disability can be based on a combination of several impairments that may not be disabling when considered separately, but when evaluated together show that he cannot work.

Functional equivalence is shown when the impairment or a combination of impairments causes the same disabling functional limitations as those of a listed impairment. SSA recognizes that some impairments or combination of impairments can be just as disabling even if the impairment itself is not the same as a listed impairment. SSA will also consider other factors such as the effect of medications, or the effects of structured settings, and school attendance. SSA may recognize that your child functions well in his supportive, special class in school or with the family and friends he knows, but may not be able to function well in an unknown setting or without constant support and supervision.

Although there is no longer a listing for obesity, SSA will also rely on the guidance found in Social Security Ruling 02-1p, which explains how obesity can affect an individual's musculoskeletal, cardiovascular, and respiratory systems.

Remember that SSA reviews and sometimes changes the specific requirements of any listing. Be sure you are relying on the current listing when gathering the necessary medical evidence.

SSA will consider the combined effects of all impairments to determine if the child is disabled under this category. Although an individual's physical or mental impairments considered independently may not be found to be disabling, when SSA considers the combined effect of all impairments, the condition can be functionally equal to a listed impairment.

Claimants over 18 years old are evaluated under adult standards. The adult listing for Multiple Body Systems (Section 10.00) is currently limited to Down syndrome, but this listing explains that SSA will evaluate "other chromosomal abnormalities [that] produce a pattern of multiple impairments but manifest in a wide range of impairment severity . . . under the affected body system." An adult can also be evaluated under the adult listings for Mental Disorders (Section 12.00) and Social Security Ruling 02-1p. As in a child's claim, SSA will consider the combined effect of all impairments on the individual's ability to work.

An adult whose condition does not meet the exact criteria, or who does not have the exact test results required may still be found disabled. SSA will continue to evaluate the claim by considering vocational factors (age, educational background, and work history) along with physical and mental residual functional capacities to decide whether a claimant is disabled or whether there are jobs that he can do.

What Can You Do to Help Show SSA That You Are Disabled?

SSA will rely on the results of medical tests to determine whether your child can be found disabled, so it is important that he or she has been properly examined by a doctor. SSA wants to see that a doctor has laboratory tests, including chromosomal analysis, where appropriate, and has diagnosed your child with Prader-Willi syndrome. In addition to medical tests, SSA will consider the opinions of the treating physicians, the child's parents, and teacher's notations in school records. Both adults and children can assist their claim by keeping a diary documenting symptoms and how these symptoms affect the ability to function during a typical day.

You can also provide the Administrative Law Judge (ALJ) with medical information about Prader-Willi syndrome *before your hearing*, so the ALJ can be familiar with the nature of the condition. The Prader-Willi Syndrome Association has valuable information that can be provided to the ALJ. (See Appendix F.)

Applying for Benefits

Application forms for disability benefits from the Social Security Administration are obtainable by calling 1-800-772-1213. When the forms are complete, application for both Social Security Disability Insurance and SSI benefits can be filed at any Social Security office. Some people can also apply online at www.socialsecurity.gov/apply-forbenefits. It is important to complete the form with as much information as possible. Give the full names and addresses of all doctors, and the dates of any hospitalizations. Make a list of the medications or other treatments used, their side effects, and any medications and treatments which have been tried but which no longer work. Describe the child's daily activities, and mention whether his/her behavior or weight prevent him from performing certain activities. Tell the child's doctors that he/she is applying for disability benefits, and that the doctor should expect to receive a request for more information from SSA. Many claimants wait until a hearing is scheduled before hiring an attorney to represent them, but you can choose to be represented at any stage of the application process.

Application and Appeals Process

Only about 30% of disability applications will be approved at the first step of the process. If your application is initially denied, there are several steps in the appeal process.

When your claim is denied initially, you should appeal by completing a reconsideration form. You have 60 days to request reconsideration. A different person from SSA will evaluate your claim. You may have to wait a year before you receive a decision at this stage.

You should be aware that SSA is experimenting with eliminating the reconsideration step. Therefore, in some parts of the country, instead of requiring reconsideration of a denial, you would request a hearing before an ALJ. In most areas, you do not request a hearing unless your claim has been denied at the reconsideration level. Your local SSA office or your representative will know which type of appeal you should file.

If you are denied at the reconsideration level, you will have 60 days to appeal to an Administrative Law Judge for a hearing. Over half of the claimants who request a hearing before an ALJ will receive favorable decisions awarding benefits. At the hearing the ALJ will ask you about your child's condition and how it affects your child's ability to work and perform activities of daily living. The ALJ will ask what your child can do during the day and how the child feels after doing certain activities. Family members may also tell the ALJ about your child's condition. The ALJ may also ask a medical expert to explain Prader-Willi syndrome. You can help your claim by providing information to the ALJ before the hearing. The ALJ may also call a vocational expert who will talk about what jobs your child could perform and whether a significant number of these jobs exist. If your child has worked in the past, the ALJ will also ask the vocational expert whether your child can perform the work that he/she did before.

If the ALJ denies your claim, you have 60 days to appeal to the Appeals Council. In some parts of the country, SSA has eliminated the Appeals Council, so you would appeal directly to federal court. An Appeals Council appeal is a written form. If you disagree with the ALJ's decision, you or your representative must explain exactly what parts of the decision you think are wrong. Occasionally, the Appeals Council will ask your representative for oral argument. But this is very rare. Unfortunately, it is not uncommon to wait 1-1/2 years before the Appeals Council makes a decision. The Appeals Council will usually either send the claim back to the ALJ for another hearing or deny your claim altogether. Eighteen percent of the claims are sent back for another hearing. The Appeals Council finds that claimants are disabled without another hearing in only 2% of the appeals.

If the Appeals Council denies your claim, you can appeal to federal court. You have 60 days to file a claim in federal court. You will want representation at this level, as only an attorney can file an appeal in federal court; a non-attorney representative cannot appeal to federal court.

How Long Will the Application Process Take?

It is not uncommon for a claimant to wait 6 to 12 months for a decision on an application for disability benefits. Claims that must be appealed administratively (to an Administrative Law Judge and the Appeals Council) or to federal court will take much longer. (To provide some perspective on the program, consider that almost 3 million applications for disability benefits were filed in a recent year.) When a case is finally approved, benefits will be paid to cover the months during which the claimant was waiting for a decision. The amount of time and effort it takes to pursue an appeal is definitely daunting. Perseverance and persistence are crucially important.

What Determines the Benefit Amount?

The amount of Social Security Disability Insurance, or Title II, benefits paid depends on the former worker's earnings throughout his/her work history, the number of years worked, his/her age, as well as the number of people in the family and its composition (including divorced spouses). For a person whose earnings, averaged over his/her working life, were $20,000, Title II disability benefits in 2005 would be approximately $835 monthly. A spouse and child would receive an additional $415. For a person whose earnings averaged $50,000, the monthly benefit amount could be approximately $1,570. The family maximum, a cap on the monthly benefits payable on the account of a particular worker, limits the amount of Social Security benefits payable to an entire family when all benefits are based on the account of one wage earner. The effect of the family maximum, however, is that families with more children may not necessarily receive more benefits than families with fewer children, even if the Personal Income Account figures are the same.

Note that there is no family maximum for SSI benefits because each person who receives benefits does so based on his/her own impairment. It is possible, then, that a very poor family with several disabled children will receive more SSI benefits than a wage earner and dependents receiving SSDI (Title II) benefits.

The SSI benefit amount is based on the income, resources, and living arrangement of the disabled individual, and sometimes, his family. There is a federal SSI monthly benefit rate: in 2005, $579 for an individual and $869 for an eligible couple. Many states add a state supplement to this amount. An individual's monthly SSI benefit will be reduced by other resources and income he receives, based on SSA's formula for counting "earned" and "unearned" income. The amount of the SSI check can differ each month if the amount of income changes each month. The formula can be quite complicated. Although SSA is required to provide an explanation of how the benefits amount is calculated, it is still recommended that you ask your representative to explain the decisions in your specific case.

SSA uses a "Retrospective Monthly Accounting" method. This means that the amount of SSI benefits in any given month is determined by the income received in the second previous month. For example, the November SSI amount is based on income received in September.

In certain situations, SSA will count a portion of *another person's income* as the unearned income of an SSI recipient. This is called "deeming." It does not matter whether the other person's income is actually available to the SSI recipient. For example, if one spouse is disabled and not working, the income of the other spouse will be deemed to the disabled spouse and will affect his/her eligibility for SSI benefits. A parent's income is deemed to a minor child and will affect the child's eligibility for SSI and benefit amount, if found eligible. This is why the disabled child in a middle class working family will not be eligible for SSI benefits. The income of a parent's spouse will also be deemed to the child, even if the spouse has not adopted the child. Deeming does not always apply if the child does not live with the parent. Special deeming rules exist if the child lives in a Medicaid-funded institution, or lives with another relative. Parent-to-child deeming ends when the child turns 18 years old. Even without deeming, this does not mean a child will be eligible for SSI. SSA looks at living arrangements also. In some states, a child who is not financially eligible for SSI may still be eligible for Medicaid. Certain types of income, such as other types of welfare benefits, are not deemed.

When determining the monthly SSI benefit amount, SSA will also consider the individual's living arrangements. If an SSI recipient lives with another person and receives food, clothing, or shelter from that person, SSA calls this "in-kind support and maintenance" and includes the value as income. If the SSI recipient can show that he has been *loaned* the support and maintenance and has an obligation to repay it, it will not affect the amount of assistance.

If an SSI recipient is living in another person's house and receiving both food and shelter from that person, SSA will apply what is called a "one-third reduction rule." However, if the recipient is paying a pro rata share of household expenses, or buying his own food, then he is not considered to be living in the household of another and is not subject to the one-third reduction rule. Instead, he is subject to the "presumed value rule."

The main difference between the one-third reduction rule and the presumed value rule is that, under the one-third reduction rule, SSA will reduce the federal monthly SSI benefit by one third ($193, based on the 2005 monthly benefit of $579). The actual value of the support does not matter. Under the presumed value rule, SSA starts with a presumption that the value of the in-kind support is worth one third of the federal benefit rate. However, if you can show that the actual value is less than one third of the federal benefit rate, SSA will use the lower number and will reduce the SSI check by the lower amount.

In addition, SSA will deduct other income, as well as disregard certain amounts of earnings before determining the monthly benefit amount.

Once Approved, Can I Work and Continue to Receive Social Security or SSI Benefits?

SSA has many work incentive programs, which allow recipients to work for a limited amount of time, or under special circumstances, without losing their benefits. Most people who receive Social Security Disability benefits can earn up to $590 per month (in 2005) for 9 months while receiving their benefits. This is called a Trial Work Period. After the 9 months are completed, a beneficiary can work during the following 3 years. Benefits will not be paid for any month in which you earn over $590, but you will receive benefits for any month you do not work. If you are still working at the end of the 3-year period, benefits may be terminated.

Another work incentive available is called Impairment Related Work Expenses (IRWE). If you have certain expenses because of your disability that permit you to work, you can deduct the cost of these expenses from your income. For example, if your seizures prevent you from driving, and there is no public transportation available, or if you are unable to take public transportation, the cost of a taxi to and from work can be deducted from your earnings. SSA will deduct the difference between the cost of the taxi and the cost of the bus that you cannot use. Deducted expenses will not be counted as earnings. This can be used to bring your earnings below the monthly Substantial Gainful Activity level ($830 for 2005). The amount of money considered an IRWE will not be counted as income and will not cause a reduction in your SSI check.

You can also create a PASS Plan. This is a Plan for Achieving Self Support. A PASS plan is a written plan that must be approved by SSA in advance. It permits you to set aside some money earned towards an educational or occupational goal. The money set aside will not be counted as income for Title II or SSI purposes. This can be used to bring your earnings below the monthly substantial gainful activity level, and SSA will then find that you are not "working." The amount of money set aside under a PASS plan will not be counted as income and will not cause a reduction in your SSI check.

SSA has recently started a "Ticket to Work Program," which also permits recipients of disability benefits do to some work while receiving benefits and continued Medicare coverage. People can use their Ticket to get free vocational rehabilitation, job training, and other employment support services. SSA's Web page has information on the Ticket to Work Program and other work incentives that are available. Look at www.socialsecurity.gov or call SSA at 1-800-772-1213.

It is not advisable to return to work before you have received a favorable decision on disability applications. Recipients who are considering trying to work should look at SSA's Web page, and contact SSA at 1-800-772-1213 or an attorney who is familiar with Social Security programs for specific guidance.

How Can I Get Help or Additional Information?

Additional information can be obtained from SSA by calling 1-800-772-1213 or looking at their Web page at www.socialsecurity.gov. Most people apply for benefits on their own but often want assistance in pursuing an appeal. If you need legal representation to assist you in obtaining Social Security disability or SSI benefits, contact your local legal services program or your local bar association referral office. Or you can get a referral to a private attorney in your area from the National Organization of Social Security Claimants' Representatives by calling 1-800-431-2804.

Appendix E

The International Prader-Willi Syndrome Organization

The International Prader-Willi Syndrome Organization (IPWSO) is a global organization of 60 member and associate countries committed to enhancing the quality of life for people with Prader-Willi syndrome and their families, giving our children the best possible opportunities for living their lives to the fullest. This international community of parents, friends, and professionals forms a dedicated network. It connects families and professionals, provides emotional and educational support, spreads general awareness, educates, and encourages scientific research. Regional and international conferences are especially helpful in giving families the occasion to come together for educational workshops and lectures, while also providing a forum for scientists.

Most significantly, IPWSO helps families—even in the most remote corners of the world—understand that they are not alone in dealing with the challenges of this complex syndrome. We provide a window to support and services that already exist within a member country. Where no support exists, we help fledgling associations with their development.

Since education promotes the possibility for early diagnosis and early interventions in medical and behavioral management, spreading awareness is a major goal of IPWSO. Our educational packets ("General Awareness," "Crisis," and "Medical Awareness") cover a wide range of essential topics and are distributed throughout the world in many languages.

IPWSO is an organization without borders—open to people of all origins and cultures. Families, researchers, clinical physicians, and other professionals from all over the world are a part of our network and our family. Please check our Web site (www.ipwso.org) to see if your country has a national association. If it doesn't, contact us and we will assist you in forming an association, and we will connect you to other resources in your region, as well as throughout the world. We invite you to join our global family. With nations working together and sharing our resources and goals, IPWSO provides a beacon of hope for a better life for children with PWS and their families. With love and determination, all people with PWS can have a brighter future!

Please feel free to contact us for information or assistance:

Pam Eisen
IPWSO President
E-mail: pam@ipwso.org

Giorgio Fornasier
IPWSO Director of Program Development
E-mail: g.fornas@libero.it

Main office:
IPWSO
c/o Baschirotto Institute of Rare Diseases (B.I.R.D.)
Via Bartolomeo Bizio, 1—36023 Costozza (VI)—Italy
Tel/Fax: +39 0444 555557
E-mail: ipwso@birdfoundation.org
Web site: www.ipwso.org

IPWSO Members
The following are member countries and contacts as of March 2005:

ARGENTINA
Web site: www.praderwilliARG.com.ar
Professional delegate: Dr. Hector Waisburg, Larrea 1474, (1117) Buenos Aires, Argentina
Tel: +54 14806 4187
Fax: +54 14807 4773
E-mail: waisburg@fibertel.com.ar
Parent delegate: Elli Korth, Forly 680 Loma Verde 1625 Escobar, Pcia. De Bs. As. Argentina
Tel/Fax: +54 3488 493499, +54 155 376 9541
E-mail: ellik@datamarkets.com.ar

AUSTRALIA
Web site: www.pws.org.au
Professional delegate: Dr. Ellie Smith, 23-25 Neil St., Bundeena NSW 2230 Australia
Tel: +61 2 9527 9795 (home), + 61 2 9845 3237 (work)
E-mail: ellies@nch.edu.au
Parent delegate: Ms. Vanessa Crowe, 1 Barron Street, DEAKIN ACT 2600 Australia
Tel: +61 2 6282 1167
E-mail: one.barron@bigpond.com

AUSTRIA
Web site: www.prader-willi-syndrom.at
Professional delegate: Dr. Barbara Utermann, Humangenetische Beratungsstelle, Schöpfstraße 41, A-6020 Innsbruck, Austria

Tel: +43/512/507-3451
Fax: +43/512/507-2861
E-mail: Barbara.Utermann@uibk.ac.at
Parent delegate: Dr. Verena Wanker-Gutmann, Schloß Frohnburg, Hellbrunner Allee 53, A-5020 Salzburg, Austria
Tel: +43 662 624 142
Fax: +43 662 620 826-75
E-mail: frohnburg@salzburg.co.at
E-mail: frohnburg@moz.ac.at

BELGIUM
Professional delegate: Dr. Annick Vogels, C.M.E., KU Leuven, Herestraat 49 B-3000, Belgium
Tel: +32 1634 5903
E-mail: Annick.Vogels@uz.kuleuven.ac.be
Parent delegate: Wilfried De Ley, Prader-Willi Vereniging vzw Belgium, Boechoutsesteenweg 54, B-2540 Hove, Belgium
Tel: +32 3455 6691
Fax: +32 3221 0723
E-mail: wilfried.deley@pandora.be

BOLIVIA
Professional delegate: Carlos A. De Villegas Córdova, M.D., Pediatra Intensivista Av. Pablo Sanchez N° 6763 (Irpavi) Bolivia
Tel: +591-2-2430123, +591-2-2433933, +591-2-71521099 (cell)
E-mail: cadevico@hotmail.com
Parent delegate: Luz Elizabeth Tejada Velez, Casilla N.6861, La Paz, Bolivia
E-mail: eliteve19972@hotmail.com

BRAZIL
Web site: http://geocities.yahoo.com.br/prader_willi_br/
Parent delegate: Gelci Galera, Rua Dom Wilson Laus Schmidt, 303 Córrego Grande, Florianópolis SC CEP 88037-440 Brasil
Tel: +55 48 2330753
E-mail: gelci@estadao.com.br
E-mail: assocnacspw@yahoo.com.br

CANADA
Web site: www.pwsacanada.com
Professional delegate: Dr. Glenn Berall, BSc, M.D., FRCP(C), 4001 Leslie Street, Toronto, ON M2K 1E1 Canada
Tel: +1-416-756-6222
Fax: +1-416-756-6853
E-mail: gberall@nygh.on.ca
Parent delegate: Diane Rogers, P.O. Box 786 Kensington, PEI COB IM0 Canada
Tel: +1-902-836-4452
E-mail: gdrogers@pei.sympatico.ca

CHILE
Web site: www.prader-willi.cl
Professional delegate: Dra. Fanny Cortes, Tabancura 1515 Of 210, Vitacura, Santiago, Chile
Tel: +56 2 678 1478
Fax: +56 2 215 2582
E-mail: fcortes@prader-willi.cl
Parent delegate: Carlos Molinet, 1a Transversal 10252, Dpto 41, El Bosque, Santiago, Chile
Tel: +56 2 559 2343, +56 9 276 9068
E-mail: cmolinet@prader-willi.cl
E-mail: info@prader-willi.cl

CHINA
Professional delegate: Jinghua Chai, M.D., M.Sc., Research Associate, Department of Psychiatry, University of Pennsylvania, Clinical Research Building, Room 530, 415 Curie Blvd. Philadelphia, Pennsylvania 19104-6140, U.S.A.
Tel: +1-215-898-0265
Fax: +1-215-898-0273
E-mail: jinghuachai@yahoo.com

COLOMBIA
Professional delegate: Dr. Alejandro Velàsquez, endocrinologo pediatra, Calle 2 sur # 46-55 Fase 1. Clinica Las Vegas, Consultorio 223, Medellin, Colombia
Tel: +2683763, +3108398091 (cell)
E-mail: alejandrov@doctor.com
Parent delegate: Jorge Restrepo, Apartado Aèreo 11667, Medellin, Colombia
E-mail: jerestrepo@epm.net.co

CROATIA
Professional delegate: Prim. Dr. Jasenka Ille, KBC Zagreb, Rebro, Kispaticeva 12, 10000 Zagreb, Croatia
Tel: +385 1 2388 331
Parent delegate: Ivka Cop and Davor Matic, Bukovacka 158, 10000 Zagreb, Croatia
Tel: +385 9 1 5190 661
E-mail: ivkacop@net.hr

CUBA
Parent delegate: Ilieva Vazquez Bello, 9525 SW 24 St. Apt. D-102, Miami, FL 33165, U.S.A.
Tel: +1-305-207-5739
E-mail: ilimigue@hotmail.com
Address in Cuba: F.lia Vazquez, Calle 4ta #10810, E/7MA Y 9NA Casino Deportivo Ciudad Habana, Cuba
Tel: +41-89-06, +880-27-20 (cell)

DENMARK
Web site: www.prader-willi.dk
Professional delegate: Susanne Blichfeldt, M.D., The Danish PWS Association, Kildehusvej 12, DK-4000 Roskilde, Denmark
Tel: +45 4637 3204, +45 4637 3203
E-mail: s.blichfeldt@dadlnet.dk
Parent delegate: Børge Troelsen, Agervej 23, DK-8320 Mårslet, Denmark
Tel: +45 8629 2141
Fax: +45 8629 2191
E-mail: b.troelsen@email.dk

DOMINICAN REPUBLIC
Professional delegate: Dra. Cristian López Diaz, Neurólogo-Neuropediatra Grupo Médico Naco, C/ Fantino Falco No. 12, Suite No. 5 2do piso, Esanche Naco, Santo Domingo Rep. Dom.
Tel: +809-685-5544, +809-759-8800 (cell)
E-mail: crislopezdiaz@hotmail.com
Parent delegate: Julia Bonelly, Continental Express, SA 5171, P.O. Box 25296, Miami, Florida 33102, U.S.A.
Tel: +1-809-582-3417
E-mail: jbonnelly@navierasbr.com

ECUADOR
Parent delegate: Jorge Oswaldo Zúñiga Gallegos, Urbanización Carlos Montufar, Calle Isla Puná No. 8, San Rafael, Casilla Postal No. 17-23-159, Quito Ecuador
Tel: +593 2 2860-985, +593 2 2865-047
E-mail: jzuniga@espe.edu.ec
Other contact: Armando Castellanos, Reina Victoria 17-37 y La Niña, Quito-Ecuador
Tel: +593 2 2683-647
E mail: iznachi@yahoo.com.mx

EGYPT
Professional delegate: Osama K. Zaki, M.D., Cytogenetics Unit. Dept Of Pediatrics, Ain Shams University, 8, Kamal Raslan St. Heliopolis, Cairo 11771 Egypt
Tel: +20 2 2717445
Fax: +20 2 2731933
E-mail: ozaki@medical-genetics.net
Parent delegate: Omnia Mourad, One Mohamed El Nadi Street, Makram Ebeid, 6th Zone, Nasr City, Egypt
E-mail: omneya@medical-genetics.com

EL SALVADOR
Professional delegate: Dr. Billy Fuentes, 3ª Calle Poniente y 79 Av. Norte, Local N° 110 Condiminio Las Alquerias, Colonia Escalon, San Salvador, El Salvador, Centro America
Tel: +503-264-6990 (work), +503-274-4936 (home)

E-mail: bimifu@hotmail.com
Parent delegate: Mario A. Méndez, Colonia Pórtico San Antonio #13—A Calle San Antonio Abad., San Salvador, El Salvador, Centro America
Tel: +503-284-4032
E-mail: mariomenra@hotmail.com

FINLAND
Web site: www.prader-willi.dk
Professional delegate: Dr. Ilkka Sipilä, M.D., HUCH Hospital for Children and Adolescents, (Stenbäckinkatu 11, 00290 Helsinki), P.O. Box 281, FIN-00029 HYKS, Finland
Tel: +358 50 427 2898
Fax: +358 94 717 5888
E-mail: ilkka.sipila@hus.fi
Parent delegate: Tiina Silvast, Teekkarinkatu 17 B 15, 33720 Tampere, Finland
Tel: +358 40 084 7576
Fax: +358 3 213 3301
E-mail: tiina.s@pp1.inet.fi

FRANCE
Web site: http://perso.wanadoo.fr/pwillifr/
Professional delegate: Prof. Raphael Rappaport, Hôpital Necker, Département de Pédiatrié, Unitéd'endocrinologie et croissance, 149 Rue de Sèvres, F-75743 Paris Cedex 15, France
Tel: +33 144 494 801
Fax: +33 144 494 800
E-mail: raphael.rappaport@nck.ap-hop-paris.fr
Parent delegate: Jean-Yves Belliard, 10 Rue Charles Clément, F-02500 Mondrepuis, France
Tel/Fax: +33 323 987 904
E-mail: jean-yves.BELLIARD@wanadoo.fr

GERMANY
Web site: www.prader-willi.de/
Professional delegate: Dr. Gillessen-Kaesbach, Humangenetisches Institut, Hufelandstr. 55D-45112 Essen, Germany
Tel: +49 201 723 4563
Fax: +49 201 723 5900
E-mail: g.gillessen@uni-essen.de
Parent delegate: Monika Fuhrmann, Weiherstr. 23, D-68259 Mannheim, Germany
Tel: +49 621 799 2193
E-mail: monikafuhrmann@web.de

GREECE
Professional delegate: As. Professor Dr. Christina Kanaka-Gantenbein, MD, Pediatriciac Endocrinology and Diabetes, 1st Department of Paediatrics, University of Athens, 52 Kifissias Ave., 115 26 Athens, Greece

Tel: +30 1 777 9909
E-mail: ganten@hol.gr
Parent delegate: Maria Papaiordanou, Ploutonos 1 & Themistokleous
St., 17455 Kalamaki—Alimos, Athens, Greece
Tel: +30 210 981 8179
Fax: +30 210 413 4648
E-mail: harmar@hol.gr

GUATEMALA

Professional delegate: Licda. Q.B. Mayra Urízar, Laboratorio Bioana-
lisis 18 Av. 4-50 zona 3, Quetzaltenanago, Guatemala, C.A.
Tel: +502 77674597, +502 54070806 (cell)
E-mail: mayraurizar@intelnet.net.gt
E-mail: mayraurizar@cabledx.tv
Parent delegate: Luis Barrios Izaguirre, 6a Calle "C" 4a-25 zona 9, Los
Cerezos I Quetzaltenango, Guatemala, Centro America
Tel: +502 7677202, +502 54033260 (cell)
E-mail: luisbarrios@cabledx.tv

HONDURAS

Professional delegate: Dr. Jose Armando Berlioz Pastor, Colonia
Altos de Miramontes, Diagonal Aguan # 2751 Tegucigalpa, M.D.C.
Honduras, C.A.
Tel: +504-232-4429 (home), +504-221-1939 (office), +504-992-4569 (cell)
E-mail: aberlioz@quik.com

ICELAND

Professional delegate: Stefan Hreidarsson, M.D., Medical Director,
State Diagnostic and Counseling Centre, Digranesuegur 5, 200
Kopavogur, Iceland
Tel: +354 510 8400
E-mail: STEFAN@Greining.is

INDIA

Professional delegate: Dr. Arun Kumar, Asst. Professor of Human
Genetics, Indian Institute of Science, Bangalore 560 012, India
Tel: +91 80 2293 2998 (office), +91 80 2346 5523 (home)
Fax: +91 80 2360 0999
E-mail: Kumarkarun@mrdg.iisc.ernet.in
Parent delegate: Shikha Metharamani, 16\1,Loudon Street, 3D Loudon
Park, Kolkata-700 017 W. Bengal, India
Tel: +91 33 2247 2765, +91 98 3100 8191 (cell)
E-mail: shikha_harlalka@hotmail.com

ISRAEL

Web site: www.pwsil.org.il
Professional delegate: Varda Gross-Tsur, M.D., P.O. Box 2210, Mevase-
ret Zion 90805, Israel
Tel: +972 2 5341193
Fax: +972 2 6422481

E-mail: gros_fam@netvision.net.il
Parent delegate: Urith Boger, P.O. Box 1332, 31 Aya St., Ramat Hasha-
ron 47226, Israel
Tel: +972 3 5409882
Fax: +972 3540 5271
E-mail: koshi1@netvision.net.il

ITALY
Web site: http://digilander.libero.it/praderwilli/main.htm
Professional delegate: Dr. Laura Bosio, c/o Pediatric Dept, Endocrine
Unit, St.Raphael Hospital, Via Olgettina 60, 20132 Milano, Italy
Tel: +39 0226 432 625
Fax: +39 0226 432 626
E-mail: bosio.laura@hsr.it
Parent delegate: Giuseppe Quaglia, Via Cascina Maffeis, 72 24052
Azzano S. Paolo (BG), Italy
Tel: +39 035 530646
Fax: +39 035 531040
E-mail: beppe@quaglia.it

JAPAN
Professional delegate: Dr. Tomoko Hasegawa, M.D., Genetic Support
& Consultation Office (GeSCO), 1-3-5-102 Kami-ashi-arai, 420-0841
Shizuoka, Japan
Tel: +81 5 4248 0457
Fax: +81 3 3944 6460
E-mail: hasemoko@aol.com
Parent delegate: Kazue Matsumoto, Sanjo-cho 33-11, Asiya-city, Hyogo-
prefecture, Japan
Tel: + 81 7 9738 3087
E-mail: takumama@kcc.zaq.ne.jp

KOREA
Parent delegate: Jang Eunju, R# 303, 2FL, 165-13, Seokchon-dong,
Songpa-gu, Seoul, Korea, 138-844
Tel : +016 733 4497 (cell)
E-mail: jenovia2000@yahoo.co.kr

MALAYSIA
Professional delegate: Dr. Yew Sing Choy, Pediatric Institute Kuala
Lumpur Hospital, Jalan Pahang, 50586 Kuala Lumpur, Malaysia
Tel: +60 3 2615 5555, ext. 6889
Fax: +60 3 2694 8187
E-mail: choyo@tm.net.my
Parent delegate: Azmi Baba, Royal Malaysia Police Air Unit, Old
Airport Road, 50460 Kuala Lumpur, Malaysia
Tel: +60 3 2282 5868
Fax: +60 3 2282 4535
E-mail: abhb57@hotmail.com

MEXICO

Web site: http://praderwilli.es.mn
Professional delegate: Carlos Alberto Meza Miranda, Av de los Balsones No. 68 Fracc. Villas del Tey, Mexicali Baja California, Mexico c.p. 21380
Tel: +52 686 5 58 37 01
E-mail: cmezamiranda@yahoo.com.mx
Parent delegate: Diana Cota (same address as Carlos)
E-mail: dianaoliviacota@hotmail.com

MOLDAVIA

Professional delegate: Dr. Victoria Sacara, Apt. 144, 21 Dacha str., Kishinev, Moldova MD 2038
Tel: +373 719670
E-mail: vsacara@mednet.md
Parent delegate: Baranova Tatiana, Apt.12, 23 Renashterey str., Kishinev, Moldova 2005
Tel. +226386
E-mail: tatiaserg@araxinfo.com

MOROCCO

Parent delegate: Mme Fatima Mandili, 6 bis Rd doukkala rue Soulaimane Azzmi Appt. 4 Qu Hopitaux 20000 Casa, Maroc
E-mail: fate\lkorno@hotmail.com

THE NETHERLANDS

Web site: www.praderwillisyndroom.nl
Professional delegate: Prof. Dr. Leopold M.G. Curfs, University Maastricht/Academic Hospital Maastricht, Department Clinical Genetics, P.O. Box 1475, 6201 BL Maastricht, The Netherlands
Tel: +31 43 3877850, +31 43 38775899
E-mail: curfs@msm.nl
Parent delegates: Gerard Meijwaard, Louis Pasteurpad 28, 6216 EV Maastricht, The Netherlands
Tel: +043 3432371
E-mail: Gerard@praderwillisyndroom.nl
Tamara Stranders, Prader-Willi/Angelman-Vereniging, Postbus 85276, 3508 AG Utrecht
Tel: +31 30 2363763
E-mail: t.stranders@fvo.nl

NETHERLANDS ANTILLES

Parent delegate: Marisol Punín, Wolfstraat #4, Noord Aruba, Antillas Holandezas
Tel: +297 9 931869, +297 5 837175 (work)
E-mail: solibeca@hotmail.com

NEW ZEALAND

Web site: www.pwsa-nz.co.nz
Professional delegate: Dr. Esko Wiltshire, Paediatric Endocrinologist, Wellington Hospital, Wellington, New Zealand

E-mail: esko@wcmeds.ac.nz
Parent delegate: Linda Thornton, P.O. Box 143, Masterton, New Zealand
Tel: +64 6306 8424
Fax: +64 6306 8425
E-mail: pwsanz@wise.net.nz

NIGERIA
Parent delegate: Charles Ch. Mayberry, Erdkampsweg 6, 22335 Hamburg, Germany
Tel: +49 40 5204217
Fax: +49 40 5203317
E-mail: Maybec3@aol.com

NORWAY
Web site: www.praderwilli.no
Professional delegate: Christian Aashamar M.Ed., Frambu, Sand-bakkveien 18, N-1404 Siggerud, Norway
Tel: +47 64 85 60 80
Fax: +47 64 85 60 99
E-mail: caa@frambu.no
Parent delegate: Ragnhild Øverland Arnesen, Kollbulia 22, N-5124 Morvik, Norway
Tel: +47 55 18 48 42, +47 48 09 67 09 (cell)
E-mail: ragnhoa@online.no

PAKISTAN
Professional delegate: Dr. Jamal Raza, Associate Professor, National Institute of Child Health, Rafiquee Shaheed Road, Karachi 75510, Karachi, Pakistan
Tel: +9201261 4, ext. 223, +0333 2184376 (cell)
E-mail: jamalraza@yahoo.com
Parent delegate: Ghazala Nomani, 26 Cedar St., Bergenfield, New Jersey 07621, U.S.A.
Tel: +1-201-244-9026
E-mail: gnpwspak@aol.com

PANAMA
Parent delegate: Kathia Díaz Arias, San Miguelito Samaria, Sector 4b, casa 161, Panama
Tel: +507-273-9591
E-mail: angeldavid134@LatinMail.com

PARAGUAY
Professional delegate: Dra. Maria Beatriz N.P. de Herreros, Domingo Portillo 1508 C/ Prof. Fernandez Asuncion, Paraguay
Tel: +595 21 298564
Fax: +595 21 223738
E-mail: mara@cmm.com.py

Parent delegate: Ing. Ubaldo Gonzalez Franco, Villa del Agronomo, Lote Guazù, San Lorenzo, Paraguay
Tel: +595 981 932064
Fax: +595 981 584546
E-mail: danny_winner5@hotmail.com

PERU
Professional delegate: Dr. Alberto Teruya Gibu, Av. Iquitos 293 La Victoria, Lima, Peru
Tel: +51 1 4248007
E-mail: aatg2002@yahoo.com
Parent delegate: Rosalva Espino Moscoso, José Pardo 159, Urb. Astete, San Miguel, Lima, Peru
Tel: +51 1 4206367
E-mail: rosalves@yahoo.com

PHILIPPINES
Parent delegate: Regina Infante, Blk.3 Lot 10 Phase 5 Adalia Street, Elvinda Village, San Pedro, Laguna, Philippines 4023
Tel: +63 917 5011915
E-mail: regina_infante@hotmail.com
Other parent contact: Vienne Go Ang, #45 11th Street, New Manila, Quezon City, Philippines
E-mail: vienne@i-manila.com.ph

POLAND
Professional delegate: Ewa Obersztyn, M.D., Ph.D., Department of Medical Genetics, National Institute of Mother and Child, 01-211 Warsaw, ul. Kasprzaka 17a, Poland
Tel: +48 22 32 77 490
Fax: +48 22 32 77 152
E-mail: eobersztyn@imid.med.pl
Parent delegate: Maria Libura, 02-132 Warsaw, ul. Baleya 4/11 Poland
Tel: +48 22 65 98 778
E-mail: marialibura@op.pl

PORTUGAL
Professional delegate: Prof. Luis Nunes, Conselho de Administração, Hospital D. Estefania, Rua Jacinta Marto, 1169-045 Lisboa, Portugal
Tel: +351 914 907 550
E-mail: luis.nunes@sapo.pt
Parent delegate: Paula Costa, c/o "Rarissimas," Associação Nacional de Deficiencias Mentais e Raras, Rua dos Bons Amigos, Lote 348 Casal do Bispo, 1685-843 Famoes, Portugal
Tel: +351 217 956 205
Fax: +351 217 969 777
Web site: www.rarissimas.org
E-mail: cdlspaulacosta@yahoo.com

PUERTO RICO

Professional delegate: Alberto Santiago Cornier, M.D., Ph.D., Ponce School of Medicine, Genetic Division, P.O. Box 7004, Ponce PR 00732, Puerto Rico
Tel: +1-787-840-2575, ext. 2218/2213/2156
Fax: +1-787-840-5698
Parent delegate: Elsa Alago, Calle El Cerezal F-5 , Jardines de Miramar, Isabela PR 00662 Puerto Rico
Tel: +1-787-830-0439 (home), +1-787-834-8000, ext. 2214 (work), +1-787-645-5483 (cell)
E-mail: ealago@tld.net

ROMANIA

Professional delegate: Szekely Aurelia, Endocrinologist—doctor in medicine, Zalau, str. Oborului, nr.10, Romania
Tel: +40 260 662672, +40 074 2070123 (cell)
E-mail: aurelia_szekely@k.ro.
Parent delegate: Dorica Dan, Str. Simion Barnutiu, nr.97, bl. SB 88, apt.14, Loc. Zalau, Judet Salaj, Cod 4700 Romania
Tel: +40 60 616585, +40 726 248707
E-mail: doricad@yahoo.com

SAUDI ARABIA

Professional delegate: Dr. Mazin S. Fakeeh, M.D., FRCP, Dr. Soliman Fakeeh Hospital, Palastine St. Jeddah, 21461. P.O. Box 2537, Saudi Arabia
Tel: +9662 6655000
E-mail: mazin@drfakeehhospital.com

SINGAPORE

Professional delegate: Denise Li-Meng Goh, The Children's Medical Institute, National University Hospital, Assistant Professor, Dept. of Pediatrics, National University of Singapore, 5 Lower Kent Ridge Road, S 119074, Singapore
Tel: +65 6772 4420
Fax: +65 6779 7486
E-mail: paegohlm@nus.edu.sg
Parent delegate: Eric and Lina Khoo, Blk 721 Pasir Ris St 72 #07-115 S510721 Singapore
Tel: +65 9818 0582
E-mail: linakhoo@gmail.com

SLOVENIA

Professional delegate: Prof. Ciril Kržišnik, University Medical Center Ljubljana University Children's Hospital, Vrazov trg 1, 1000 Ljubljana, Slovenia
Tel: +386 61 320 887
Fax: +386 61 310 246
E-mail: ciril.krzisnik@mf.uni-lj.si

Parent delegate: Mirjana Zokalj, Mirje 25, 1000 Ljubljana, Slovenia
Tel: +386 61 126 23 10
Fax: +386 61 125 24 86
E-mail: vojteh.zokalj@siol.net

SOUTH AFRICA
Web site: www.praderwilli.org.za
Professional delegate: Dr. Engela M. Honey, Department of Human Genetics, University of Pretoria, P.O. Box 2034, 0001 Pretoria, South Africa
Tel: +27 82 5795315
E-mail: ehoney@medic.up.ac.za
Parent delegate: Rika du Plooy, 267 Middelberg Street, Muckleneuk, 0002 Pretoria, South Africa
Tel: +27 12 3440241
E-mail: rikadup@mweb.co.za

SPAIN
Web site: www.prader-willi-esp.com/
Professional delegate: Dr. Fernando Mulas, Jefe de Neuropediatria, Hospital Universitario "LA FE", Avda. Campanar, 21, 46020 Valencia, Spain
Tel: +34 96 386 2700, ext. 50481
Fax: +34 96 362 3194
E-mail: fmulasd@meditex.es
Parent delegate: Maria Helena Escalante, Ramón Lujan, 41-3A, 28026 Madrid, Spain
Tel: +34 91 500 3761, +34 61 925 5813 (cell)
E-mail: osandre@inicia.es

SWEDEN
Professional delegate: Professor Martin Ritzén, Pediatric Endocrinology, Karolinska Hospital, S-171 76 Stockholm, Sweden
Tel: +46 8 5177 2465
Fax: +46 8 5177 5128
E-mail: Martin.Ritzen@kbh.ki.se
Parent delegate: Jean Phillips-Martinsson, Farthings 44 Warwick Park, Tunbridge Wells, Kent TN2 5EF, United Kingdom
Tel/Fax: +44 1892 549492
E-mail: jeanpws@compuserve.com

SWITZERLAND
Web site: www.Prader-Willi.ch
Professional delegate: PD Dr. med. Urs Eiholzer, Foundation Growth Puberty Adolescence, Möhrlistr. 69, CH-8006 Zürich, Switzerland
Tel: +41 1364 3700
Fax: +41 1364 3701
E-mail: mail@childgrowth.org

Parent delegate: Doris Bächli, Im Vogelsang, CH 9477 Trübbach, Switzerland
Tel: +41 81 783 2601
E-mail: Mail@prader-willi.ch

TAIWAN
Web site: http://home.pchome.com.tw/health/twpws
Professional delegate: Dr. Shuan-Pei Lin, Division of Genetics, Dept. of Pediatrics, McKay Memorial Hospital, 92 Chung-San N. Road, Sec. 2, Taipei, Taiwan
Tel: +886 2 2543 3585/3089
Fax: +886 2 2543 3642
E-mail: zsplin@ms2.mmh.org.tw
Parent delegate: Welly Chan, 7th Fl. We Sheng Building, No. 125 Nan-King East Road Sec. 2, Taipei, Taiwan
Tel: +886 2 2508 6626
E-mail: wellyvivian@yahoo.com

THAILAND
Professional delegate: Duangrurdee Wattanasirichaigoon, M.D., 270 Rama VI Rd., Faculty of Medicine, Ramathibodi Hospital, Mahidol University, Department of Pediatrics, Division of Medical Genetics, Bangkok 10400 Thailand
Tel: +66 2201 1488
Fax: +66 2201 1850
E-mail: radwc@mahidol.ac.th

UNITED KINGDOM
(England, Scotland, Wales, Ireland)
Web site: www.pwsa.co.uk
Professional delegate: Prof. A.J. Holland, 2nd Floor, Douglas House, 18b Trumpington Road, Cambridge CB2 2AH, U.K.
Tel: +44 1223 746112
Fax: +44 1223 746122
E-mail: ajh1008@cam.ac.uk
Parent delegate: Jackie Waters, 125a London Road, Derby DE1 2QQ England
Tel: +44 1332 365676
Fax: +44 1332 365401
E-mail: JWaters@pwsa-uk.demon.co.uk

UNITED STATES OF AMERICA
Web site: www.pwsausa.org
Professional delegate: Suzanne B. Cassidy, M.D., Clinical Professor, Department of Pediatrics, Division of Medical Genetics, UCSF, 533 Parnassus Avenue, Rm U100A, San Francisco, CA 94143-0706, U.S.A.
Tel: +1-415 476-2757
Fax: +1-415 476-9976
E-mail: scassidy@uci.edu

Parent delegate: Janalee Heinemann, Executive Director of PWSA (USA), 5700 Midnight Pass Road, Suite 6, Sarasota, Florida 34242, U.S.A.
Tel: +1-800-926-4797, +1-941-312-0400
E-mail: execdir@pwsausa.org

URUGUAY
Professional delegate: Dra. Maria Cristina Suarez, Torre Artigas Ap 507, Maldonado (20000), Uruguay
Tel: +598 42 232946
E-mail: aldebarr@adinet.com.uy
Parent delegate: Fanny Acosta, Calle Ventura Alegre 818, 20000 Maldonado, Uruguay
Tel: +598 42 225531
E-mail: paulag@adinet.com.uy

VENEZUELA
Professional delegate: Imperia Brajkovich, Hospital de Clinicas Caracas, Avenida Los Proceres-San Bernardino, Caracas, Venezuela
Tel: +58 212 978 0462
E-mail: vamilo@cantv.net
Parent delegate: Alicia M. Turio de Borga, Calle Icabaru con Calle Chulavista, Residencias Sierra Nevada, Apto 13-B, Colinas de Bello Monte, 1060 Caracas, Venezuela
Tel: +58 212 754 0353 (home), +58 212 951 0542/0538 (office)
E-mail: familiaborga@cantv.net

Appendix F

The Prader-Willi Syndrome Association (USA)

PWSA (USA)
5700 Midnight Pass Road, Suite 6
Sarasota, Florida 34242 USA
Telephone: 941-312-0400
Toll-free telephone (U.S. and Canada): 800-926-4797
Fax: 941-312-0142
E-mail: pwsausa@pwsausa.org
Web site: www.pwsausa.org

The Prader-Willi Syndrome Association (USA)—PWSA (USA)—is the only national membership organization for children and adults with Prader-Willi syndrome and their families in the United States. PWSA (USA) has been serving children and adults with the syndrome for nearly 30 years. At the time of this writing, the Association also has 33 state and regional chapters, which carry out a range of activities to serve local families and support the mission of the national organization. PWSA (USA) became incorporated in 1977 and was approved for tax-exempt 501(c)3 status as a charitable organization by the U.S. Internal Revenue Service. By 1988, the Association had become a multifaceted international organization, and it currently serves members from 32 other nations in addition to its U.S. members.

Educational Materials—No other resource in the world provides as extensive a range of educational and syndrome management publications as those provided by PWSA (USA). Thanks to thousands of hours of donated time and skills by professionals and parents (often parents who are professionals in a related field), PWSA (USA) provides 35 educational books and/or booklets, 14 brochures, and several videos that cover various topics on medical, emotional, and behavioral management. PWSA (USA) mails, at no cost to the recipients, thousands of awareness and educational packets yearly. Association members are regularly kept up to date through PWSA's bimonthly newsletter, *The Gathered View*, and through articles posted on the PWSA (USA) Web site.

National Conference—The annual national conferences sponsored by PWSA and hosted by the state chapters are the largest in the world (averaging over 1,000 attendees) specifically for education and support for those dealing with Prader-Willi syndrome. This conference is actually several conferences occurring simultaneously—for scientists and other medical professionals, for adult service providers and teachers, for parents and relatives, for those with PWS, and for siblings.

Research—PWSA (USA) has two active medical advisory boards. Its Scientific and Clinical Advisory Boards are comprised of volunteer professionals from around the nation who are researchers and/or medical specialists who treat individuals with the syndrome. PWSA (USA) grants small start-up funding for research projects. The Association also impacts research in many other ways such as advocating for government and private funding of research projects, networking researchers, and through its Clearinghouse Project for research data on PWS.

Medical Intervention Support—Hospitals, physicians, and parents from all over the world consult with PWSA (USA) for medical emergencies and questions daily. Through phone, fax, and e-mail, PWSA (USA) consults with its medical boards and responds to all through a Triage Support System.

Crisis Intervention and Prevention Program—A significant role of the national office is to assist with crisis situations. This is done through the support of a qualified crisis counselor via phone and e-mail consults and networking with specialists on the syndrome from around the nation, PWSA publications, and individualized crisis packets, each containing very specialized letters to address the crisis at hand (see Chapter 19). Our executive director, the crisis counselor, and several of the medical board members consult with attorneys and teleconference with schools and courts.

New Parent Mentoring Program—Coordinated and supported by parent volunteers, this program has the greatest impact on early intervention and prevention, as experienced parents work one-on-one with newly diagnosed families. Early diagnosis, education, and awareness are the keys to prevention of life-threatening obesity and years of isolation and emotional trauma to the family of the child with Prader-Willi syndrome. Early intervention can also save thousands of dollars in medical expense and greatly reduce emotional stress on families.

Bereavement Follow-up Program—This program consists of four separate mailings to bereaved families within the first year after the death of their child. This support program also includes a one-time packet to PWS parents who lose a spouse or another child. Bereavement phone support is also offered.

Technology and International Support—Traffic to PWSA's Web site averages over 38,300 visits a month. Thanks to e-mail, PWSA (USA) has also been able to more effectively support families in the United States and provide support for those in many other nations who have no support system. As stated in a recent e-mail from Giorgio Fornasier, a parent from Italy who is the immediate past president of the International Prader-Willi Syndrome Organization (IPWSO), with which

PWSA (USA) is affiliated: "Technology is also a 'means,' not something to show we're different, inferior, or superior. E-mail is a fantastic way to communicate and assist people, and the Internet a window in the world which any desperate family can open and realize the sun is shining."

The Executive Director of PWSA (USA), Janalee Heinemann, states: "Our short-term goals are to support and educate parents and professionals and to save the lives of our children in crisis. Our long-term goal is to continue to enact our mission statement with a particular emphasis on *preventing* our children from getting into a crisis state. Our mission is not only to educate the families and professionals working with the syndrome, and to save the lives of our children, but also to foster the emotional well-being of our children and young adults with PWS. To give them a sense of worth in a society that shuns anyone different and views obesity as a psychological weakness is not an easy task. Today, understanding, acceptance, and wholeness are only in the dreams and wishes of our children and their families—but our ultimate goal."

PWSA (USA) Educational Materials

Following is a selected list of educational products available through PWSA (USA) as of April 2005:

Publications in English

The Child With Prader-Willi Syndrome: Birth to Three, by Robert H. Wharton, M.D., Karen Levine, Ph.D., Maria Fragala, P.T., Deirdre C. Mulcahy, M.S., CCC–SLP. This booklet discusses the common concerns of the first 3 years and offers specific recommendations for early intervention strategies. A helpful and positive resource for families, physicians, early intervention worker, and other care providers. 34 pages (revised 2004).

Prader-Willi Syndrome: Handbook for Parents, by Shirley Neason, with subsequent revisions by members of the PWSA publications committee. A comprehensive booklet with pictures that covers birth to adulthood. Parent-to-parent handbook for understanding and managing issues related to PWS. 75 pages (revised 1999).

Nutrition Care for Children with PWS, Infants and Toddlers, by Janice Hovasi Cox, M.S., R.D., and Denise Doorlag, OTR. Provides answers to frequently asked questions about nutrition and feeding of infants and toddlers with Prader-Willi syndrome. 62 pages (revised 2004).

Nutrition Care for Children with PWS, Ages 3-9, by Karen H. Borgie, M.A., R.D. Covers calorie needs, supplements, diet planning and food management, and explains food exchange lists. 12 pages (2003).

Nutrition Care for Adolescents and Adults with PWS, by Karen H. Borgie, M.A., R.D. Covers essential diet information for families, caregivers, and residential service providers. 24 pages (2003).

Low-fat, Low-sugar Recipes for the Prader-Willi Syndrome Diet, by Donna Unterberger. Cookbook for the PWS diet filled with recipes designed for use by the whole family. Great substitution list, fun snack recipes, mealtime tips, and full nutritional values calculated for each recipe. 156 pages (2003).

Physical Therapy Intervention for Individuals with PWS, by Maria Fragala, P.T. This booklet provides general information about physical therapy intervention. Includes copies of articles by Janice Agarwal, a physical therapist and mother of a young child with PWS. 11 pages.

Exercise and Crafts & Activities—A Collection of Articles. Contributions by Jennifer C. Deau, M.S., exercise physiologist, and other articles on muscle tone, upper body strength, exercise, and crafts and activities for the individual with Prader-Willi syndrome from infancy to adulthood. 44 pages (1998).

Behavior Management—A Collection of Articles. This booklet includes articles on behavior management and specific concerns, such as use of psychotropic medications, management of skin picking, toilet training, social skills teaching, and more from PWSA's newsletter, *The Gathered View*, and other sources. 79 pages (revised 2003).

Educator's Resource Packet, including the booklet *Information for School Staff: Children with Prader-Willi Syndrome*, by Barbara Dorn, R.N., and Barbara J. Goff, Ph.D. This packet is a resource for educators that includes a teacher's handbook for the student with PWS, an accompanying worksheet about PWS-related issues and interventions for school staff, as well as related articles from PWSA's *The Gathered View*. (2003).

Health and Medical Issues for the Individual with Prader-Willi Syndrome—A Collection of Articles. From the pages of PWSA's newsletter, *The Gathered View*, and other sources, this booklet brings together articles on many aspects of PWS written primarily for the layperson. Covers management of obesity, various medical conditions associated with the syndrome, vision and dental issues, sexual development and sexuality, genetics of PWS, and more. 121 pages (revised 2004).

Prader-Willi Syndrome Medical Alerts. Important resource for parents to give to their child's doctor, emergency room staff, caregiver, etc. Briefly presents cautions regarding aspects of PWS that could lead or contribute to life-threatening situations. A useful pocket-sized handbook written by PWSA's medical professionals. 20 pages (2005).

*Growth Hormone and Prader-Willi Syndrome—A reference for families and care providers, by Linda Keder in consultation with both medical and parent advisors. Covers growth patterns in PWS, research on the effects of growth hormone treatment, and details on using GH therapy in children with PWS. 52 pages (2001).

Prader Willi Syndrome Is What I Have, Not Who I Am! A book of "feelings" written by children and young adults with PWS, collected by Janalee Heinemann, Executive Director of PWSA (USA). This book gives insights into the lives and thoughts of people dealing with PWS on a daily basis. A portion of the book opens the door to journal writing and an opportunity for the reader with PWS to share their feelings. (2005).

Michael and Marie, Children with Prader-Willi Syndrome, by Valerie Rush Sexton and Debbie Erbe Fortin, illustrated by Bonnie Branson.

Written by two teachers, this storybook is designed to be read to elementary school age children to educate classmates of special needs children about the need to understand and help create a friendly and safe environment for all children. (2003).

Sometimes I'm Mad, Sometimes I'm Glad—A Sibling Booklet, written by Janalee Heinemann, M.S.W., in the voice of a sibling of someone with PWS. Recognizes the range of feelings that arise in having a brother or sister with the syndrome, based on the author's observations in raising her son with PWS and his siblings. (1982).

Supportive Living Care Plan for an Adult with PWS in Placement. This comprehensive book/CD will help families create a plan that is specifically designed to help staff and supportive personnel provide predictable, consistent, and accountable care and advocacy for the adult with PWS. This is available in both a notebook format and in a changeable CD that can be adapted to explicitly meet the needs of each individual. (2002).

Video Products

"PWS—The Early Years" (42 minutes). This video offers help and practical suggestions for those families with a young child newly diagnosed with PWS. Genetics, medical, early intervention, and family issues are presented, personalized with family interviews. Although focusing on young children, this video is a wonderful resource for schools and families with children of all ages. PAL European version available. (2002).

"Prader-Willi Syndrome-An Overview for Health Professionals" (35 minutes). This outstanding medical overview video is a must for all health care professionals who are not "experts" on Prader-Willi syndrome. It deals with all the major genetics and health care issues of the child with PWS. PAL European version available. (2002). New DVD version available (revised 2004).

"Understanding Prader-Willi Syndrome" (18 minutes). A professionally produced video with good practical advice for individuals who work with persons who have PWS, designed to train service provider staff on the needs of individuals with PWS.

Publications in Spanish/Literatura en Español

Mi Hija tiene el síndrome de Prader-Willi¿Y ahora qué? by Carlos Molinet Sepulveda. The experience from a Chilean father's perspective of searching for answers about his daughter, who was born in 1988 with Prader-Willi syndrome in a country with no knowledge or resources on PWS. He tells a moving story about the power of love and perseverance. (2003).

Guia Para Familias y Professionales El Syndrome de Prader-Willi. Comprehensive book on the management of PWS, each chapter written by a specialist on the particular topic. Excellent reference tool. Reprinted thanks to Asociacion Espanola Prader-Willi. Softcover, 400 pages. Also available in CD format.

Sindrome de Prader-Willi: Guia Para Los Padres, Familiares Y Profesionales, by Moris Angulo, M.D. An overview of the syndrome in booklet form for parents and professionals. 16 pages (revised 2003).

Note: This is not a comprehensive list of PWSA's publications and videos, and available titles may change over time. For a current order form that includes all available products contact the PWSA (USA) office or visit the Association's Web site: www.pwsausa.org.

Index